Polymer Photovoltaics
Materials, Physics, and Device Engineering

RSC Polymer Chemistry Series

Editor-in-Chief:
Professor Ben Zhong Tang, *The Hong Kong University of Science and Technology, Hong Kong, China*

Series Editors:
Professor Alaa S. Abd-El-Aziz, *University of Prince Edward Island, Canada*
Professor Stephen Craig, *Duke University, USA*
Professor Jianhua Dong, *National Natural Science Foundation of China, China*
Professor Toshio Masuda, *Shanghai University, China*
Professor Christoph Weder, *University of Fribourg, Switzerland*

Titles in the Series:
1: Renewable Resources for Functional Polymers and Biomaterials
2: Molecular Design and Applications of Photofunctional Polymers and Materials
3: Functional Polymers for Nanomedicine
4: Fundamentals of Controlled/Living Radical Polymerization
5: Healable Polymer Systems
6: Thiol-X Chemistries in Polymer and Materials Science
7: Natural Rubber Materials: Volume 1: Blends and IPNs
8: Natural Rubber Materials: Volume 2: Composites and Nanocomposites
9: Conjugated Polymers: A Practical Guide to Synthesis
10: Polymeric Materials with Antimicrobial Activity: From Synthesis to Applications
11: Phosphorus-Based Polymers: From Synthesis to Applications
12: Poly(lactic acid) Science and Technology: Processing, Properties, Additives and Applications
13: Cationic Polymers in Regenerative Medicine
14: Electrospinning: Principles, Practice and Possibilities
15: Glycopolymer Code: Synthesis of Glycopolymers and their Applications
16: Hyperbranched Polymers: Macromolecules in-between of Deterministic Linear Chains and Dendrimer Structures
17: Polymer Photovoltaics: Materials, Physics, and Device Engineering

How to obtain future titles on publication:
A standing order plan is available for this series. A standing order will bring delivery of each new volume immediately on publication.

For further information please contact:
Book Sales Department, Royal Society of Chemistry, Thomas Graham House, Science Park, Milton Road, Cambridge, CB4 0WF, UK
Telephone: +44 (0)1223 420066, Fax: +44 (0)1223 420247
Email: booksales@rsc.org
Visit our website at www.rsc.org/books

Polymer Photovoltaics
Materials, Physics, and Device Engineering

Edited by

Fei Huang
South China University of Technology, Guangzhou, PR China
Email: msfhuang@scut.edu.cn

Hin-Lap Yip
South China University of Technology, Guangzhou, PR China
Email: msangusyip@scut.edu.cn

Yong Cao
South China University of Technology, Guangzhou, PR China
Email: yongcao@scut.edu.cn

THE QUEEN'S AWARDS
FOR ENTERPRISE:
INTERNATIONAL TRADE
2013

RSC Polymer Chemistry Series No. 17

Print ISBN: 978-1-84973-987-0
PDF eISBN: 978-1-78262-230-7
ISSN: 2044-0790

A catalogue record for this book is available from the British Library

Published by The Royal Society of Chemistry,
Thomas Graham House, Science Park, Milton Road,
Cambridge CB4 0WF, UK

Registered Charity Number 207890

For further information see our web site at www.rsc.org

Preface

To achieve broader deployment of photovoltaic (PV) technology, new genera-tion photovoltaic devices need to be produced from low-cost, non-toxic, and earth-abundant materials using environmentally friendly and scalable pro-cesses. Organics/polymer-based photovoltaics (OPV) have the potential to fulfil all these requirements and therefore have attracted significant global attention. When compared with inorganic-based photovoltaic cells, the advantages of OPVs include their potential to be manufactured at extremely high throughput with ultra low production cost using a solution-based, con-tinuous roll-to-roll coating process. Their features of ultra light weight and high flexibility also distinguish OPVs from traditional photovoltaic technol-ogy. With these unique properties, OPVs offer good form factors for vari-ous new applications, and therefore have extremely high commercial value. Over the past few years, the performance of OPVs has improved significantly, from ~7% in 2009 to ~12% in 2015, and now reaches a level that can be con-sidered for some specific PV applications. As the interest in polymer solar cell research continues to grow, this book aims to provide a comprehensive, up-to-date review of the recent advancements in OPV technology.

This book consists of 13 chapters contributed by leading experts in the OPV field covering the areas of new chemistry, design and synthesis of OPV mate-rials, interface engineering, photophysics, morphology control and charac-terizations, new device concepts and advanced manufacturing process. As the development of new materials is key to the continuous improvement of the performance of OPVs, a major portion of the book will be dedicated to discussing the design concept and synthesis of novel organic semiconduc-tors for OPV applications.

In **Chapter 1**, Li and Bo provide an overview of the key chemistry, includ-ing Stille coupling, Suzuki coupling and direct C–H arylation, for the syn-thesis of conjugated polymers used for OPVs. The reaction mechanism and

RSC Polymer Chemistry Series No. 17
Polymer Photovoltaics: Materials, Physics, and Device Engineering
Edited by Fei Huang, Hin-Lap Yip, and Yong Cao
© The Royal Society of Chemistry 2016
Published by the Royal Society of Chemistry, www.rsc.org

the pros and cons of each type of coupling reaction are discussed in detail. In **Chapter 2**, Hou and co-workers present the design criteria for developing state-of-the-art polymer donors for highly efficient OPVs. The polymer structure–device performance relationships for a few important classes of polymers are discussed. In **Chapter 3**, Matsuo provides a comprehensive discussion on the design concepts and synthesis of new fullerene acceptors for OPVs. The effects of chemical functionalization, multi-adduct modification and the crystal structure of fullerenes on the OPV device performance are also discussed. In **Chapter 4**, Yan and co-workers review the recent development of efficient n-type semiconducting polymers and their applications as electron acceptor materials for all polymer-based OPVs. The effects of molecular weight, side chains, crystallinity of polymers and processing conditions on the bulk heterojunction morphology, as well as the device performance, are thoroughly discussed. In **Chapter 5**, Welch and co-workers highlight the synthesis and structure of several important classes of small molecule donors that enable the fabrication of highly efficient OPVs, and their structure–property–function relationships are discussed. In **Chapter 6**, Huang and co-workers give an overview of interface engineering for OPVs. The functions and the design criteria for efficient interfacial materials and their effects on OPV performance are discussed. In **Chapter 7**, Choy explains the functions and effects of solution processed metal oxide-based interfacial materials for OPVs. Different preparation methods for the metal oxide interlayers and their performance in both conventional and inverted OPVs are compared. In addition, the concepts of using chemical and metal nanoparticle doping of the oxide interlayer to introduce plasmonic-electrical effects to improve the performance of OPVs are presented. In **Chapter 8**, Heeger and co-workers describe how they discovered the ultrafast electron transfer process in nanostructured bulk heterojunctions and also discuss the ultrafast experimental results for both the polymer:fullerene and small molecule:fullerene films. They also present a general strategy to produce highly ordered and aligned semiconducting polymer films with exceptional high charge carrier mobilities. In **Chapter 9**, Russell and co-workers review some of the significant advances that have been made in understanding and manipulating the morphology of the OPV bulk heterojunction (BHJ) layers. Several key characterization techniques for morphological study are discussed. In addition, the major factors that affect the BHJ morphologies are discussed from both thermodynamic and kinetic aspects. In **Chapter 10**, Zhong discusses the current understanding of the photocurrent generation mechanism in OPVs, covering the electronic processes including charge generation, charge recombination, and charge transport. In **Chapter 11**, Furlan and Janssen discuss the design, working principle, modeling and characterizations of multi-junction OPVs. An overview of the high performance multi-junction OPVs with respect to the material choice, interconnection layer and device configuration is also provided. In **Chapter 12**, Yip and Jen provide an overview of the state-of-the-art semitransparent OPVs for power generating window applications. Several important factors that are required to achieve high performance

semitransparent OPVs, including the development of new transparent electrodes, efficient low-bandgap materials, new device structures, and optical
engineering, are discussed. Finally, in **Chapter 13**, Youn and Guo discuss
the scalability issues of OPVs, covering cost evaluation, material choice, and
large-scale manufacturing of OPVs based on roll-to-roll coating processes.

With the comprehensive, up-to-date reviews on the recent advancements
in OPV technology from the leading experts around the world, we hope that
this book will provide a useful source of information and a practical guide for
scientists, engineers and graduate students that are interested in this field.

Fei Huang
Hin-Lap Yip
Yong Cao

Contents

RSC Polymer Chemistry Series No. 17
Polymer Photovoltaics: Materials, Physics, and Device Engineering
Edited by Fei Huang, Hin-Lap Yip, and Yong Cao
© The Royal Society of Chemistry 2016
Published by the Royal Society of Chemistry, www.rsc.org

CHAPTER 1

New Chemistry for Organic Photovoltaic Materials

CUIHONG LI[a] AND ZHISHAN BO[*a]

[a]College of Chemistry, Beijing Normal University, 100875 Beijing, People's Republic of China
*E-mail: zsbo@bnu.edu.cn

1.1 Introduction

Organic semiconducting materials have become the cornerstone of organic electronics, including photovoltaic cells, light-emitting diodes, field effect transistors, and electrochromic devices. The synthesis of new organic semiconducting materials and the development of new synthetic methods for preparing semiconducting organic materials are two important issues that have attracted great attention. In this chapter, we mainly focus on the new chemistry for the synthesis of conjugated polymers used for organic photovoltaics.

Palladium-mediated cross-coupling reactions such as Suzuki–Miyaura, Sonogashira, Heck, and Stille reactions have been widely used in the synthesis of π-conjugated semiconducting materials. Recently, some new π-conjugated donor–acceptor type copolymers have shown great prospects for photovoltaic cell applications. Power conversion efficiencies (PCEs) above 9% have been achieved for bulk heterojunction (BHJ) polymer solar cells (PSCs).[1] Conjugated polymers synthesized by Heck and Sonogashira coupling reactions are seldom used for organic photovoltaic (OPV) applications. Most high efficiency conjugated polymers for OPV are synthesized by Stille

RSC Polymer Chemistry Series No. 17
Polymer Photovoltaics: Materials, Physics, and Device Engineering
Edited by Fei Huang, Hin-Lap Yip, and Yong Cao
© The Royal Society of Chemistry 2016
Published by the Royal Society of Chemistry, www.rsc.org

and Suzuki cross-couplings. The classical synthetic routes toward donor–acceptor (D–A) type copolymers are palladium catalyzed AA/BB-type (hetero) aryl–(hetero)aryl cross-couplings of dihaloarylene monomers and suitably functionalized bifunctional aromatic counterparts, mostly arylene diboronic acids/diboronic esters (Suzuki-type coupling) or distannyl arylenes (Stille-type coupling). Recently, the C–H arylation cross-coupling reaction, the so-called direct arylation for the syntheses of π-conjugated polymers, has been reported.[2] Direct C–H arylation has been expected as an alternative route to replace the widely used Stille and Suzuki reactions. Only limited examples of D–A type conjugated polymers synthesized by the direct C–H arylation have been reported as donor materials to achieve high PCE in PSCs. In this chapter, we summarize the synthetic methods for D–A conjugated polymers, namely, Stille, Suzuki–Miyaura, and direct C–H arylation polycondensation. In addition, D–A conjugated polymers synthesized by these methods and used for photovoltaic applications will be described.

1.2   Stille Polycondensation

1.2.1   History and Mechanism of the Stille Coupling Reaction

Eaborn and Kosugi developed the first examples of cross-coupling reactions between organostannanes and electrophilic partners between 1976 and 1977.[3] Soon the body of work was well known, when it became established as the title of the Stille coupling in 1978.[4] Stille and co-workers reported the preparation of ketones from acyl chlorides and organostannanes by the use of palladium-catalyzed cross-coupling.[5] Following this, the Stille reaction quickly took its place as one of the most useful protocols for forming sp^2 carbon–carbon bonds in organometallic chemistry. Yu and co-workers further developed this methodology in 1993 for use in polycondensation reactions for heteroaromatic diblock copolymers.[6] They optimized reaction conditions and prepared high molecular weight copolymers.

  The reaction mechanism itself is known to be far more complex and has been the subject of extensive work.[7] The generally accepted process involves an oxidative addition step, a transmetalation step, and a reductive elimination step, as shown in Scheme 1.1. The Pd(0) species is the active catalyst. Thus, the whole reaction includes Pd(0)-mediated cross-coupling of organohalides, triflates, and acyl chlorides with organostannanes. The Pd(II) catalysts used in Stille reactions are reduced to Pd(0) by the organostannane monomers, enabling entry into the catalytic cycle. The detailed catalytic cycle steps are as follows: (1) oxidative addition: the organohalide or triflate oxidatively adds to the Pd(0) active catalyst and forms a Pd(II) intermediate [PdL$_2$R^{1}X] (L = ligand; R^1 = alkenyl, alkynyl, aryl; X = Br, I, Cl, or OTf); (2) transmetalation, which is generally regarded as the rate-determining step and is the most complex and thus has been the subject of much debate. It is generally accepted as a process of cleavage of the Sn–C bond by an electrophilic Pd(II) complex and ligand substitution on a Pd(II) complex; (3) reductive

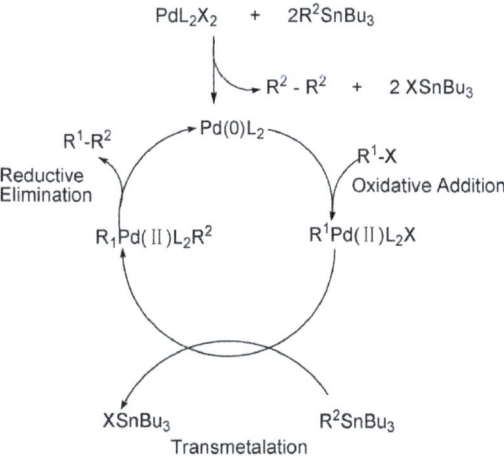

Scheme 1.1 General mechanism of the Stille reaction.

elimination, the final step in the process, which generates the desired product and allows the palladium catalyst to reenter the catalytic cycle.

1.2.2 The Reaction Catalyst, Ligand and Solvent

The effects of catalytic systems on cross-coupling reactions have been extensively studied. $Pd(PPh_3)_4$ is the most commonly employed catalyst in the Stille reaction. For $Pd(PPh_3)_4$, ligand PPh_3 is easily oxidized by traces of oxygen in reaction system to its oxide, Ph_3PO.[8] Excess PPh_3 can inhibit the Stille reaction process. Researchers have developed a more air-stable source of palladium-(0), $[Pd_2(dba)_3]$, which has been widely used in Stille cross-couplings. Additionally, the other catalyst systems have also been widely used in the Stille reactions, such as benzyl(chloro)bis(triphenylphosphine) palladium-(II), bis(acetonitrile)palladium(II) dichloride, 1,1′-bis(diphenylphosphino)ferrocene palladium(II) dichloride and allylpalladium(II) chloride dimer. Meanwhile, the ligand in catalyst systems has been widely studied because it plays a critical role in the kinetics of the Stille reaction. Ligands such as $(4\text{-MeO-}C_6H_4)_3P$, PPh_3, tri(2-furyl)phosphine, and $AsPh_3$ were employed with $Pd_2(dba)_3$ as the catalyst precursors to optimize the reaction conditions.

Benzene, toluene, xylene, mesitylene, tetrahydrofuran (THF), N,N-dimethylformamide (DMF), N-methyl pyrrolidone (NMP), dimethyl sulfoxide (DMSO), dioxane, and chloroform are widely used solvents for Stille cross-coupling reactions. Toluene is generally a good choice for Stille polymerization carried out at temperatures above 120 °C. The higher boiling chlorobenzene can be used in some microwave-assisted Stille polycondensations owing to the need for higher temperatures, in excess of 200 °C. Highly polar solvents such as DMF and NMP can help solubilize the resulting polymer, which can function as a catalyst-stabilizing effect for the palladium center. Therefore, they can be used in mixed solvent systems with toluene or another cosolvent, for example toluene/DMF or toluene/NMP. For the mixed solvent systems, a high yield of high molecular weight polymers can be obtained.[6b] Solvents such as THF and dioxane can work as catalyst stabilizers and solvents for the resulting polymers.

1.2.3   Monomers

Difunctional monomers, such as diorganohalides/ditriflates and distannanes, are generally used in Stille couplings. According to the reaction mechanism, diorganohalides or ditriflates carrying electron-withdrawing groups can facilitate the first oxidative addition step. For the second transmetalation step, the process may also be facilitated by organotin compounds with electron-rich properties. Thus, high molecular weight polymers can be prepared by electron-rich organotin compounds and an electron-deficient halide or triflate. Different halides exhibit discrepant reactivity. Generally, diiodo monomers show higher reactivity than dibromo monomers and dichloro monomers.

1.2.4   Advantages of the Stille Polycondensation

The major advantages of Stille polycondensation are the tolerance of many functional groups and the mild reaction conditions. These features are especially important for the synthesis of conjugated polymers bearing functional groups. The organotin monomers can be conveniently prepared, and they are far less sensitive to oxygen and moisture than many other organometallic compounds, *e.g.*, Grignard reagents, organozinc and organolithium reagents.[7a] Stille polycondensation is broadly applicable to the synthesis of a wide variety of donor–acceptor conjugated polymers for OPVs, and provides a facile route to prepare high molecular weight, narrowly dispersed polymers under mild conditions.

1.2.5   Disadvantages of the Stille Polycondensation

First, the purification of many tin compounds is difficult because of their instability under silica gel column chromatography. Difficulties in the purification of monomers can pose a substantial problem in the synthesis of high molecular weight polymers, because the purity of monomers is crucial to achieving a precise stoichiometry between two monomers. Second, Stille couplings require the use of highly toxic organostannyls, which are not environmentally friendly. Generally, the organotin compounds are prepared using chlorotrimethylstannane or chlorotributylstannane; chlorotrimethylstannane is more reactive than chlorotributylstannane.[9] Trimethyltin compounds are often more easily purified by recrystallization than their tributyltin counterparts, but the toxicity of chlorotrimethylstannane is 100 times greater than that of chlorotributylstannane. Moreover, the trimethyltin derivatives tend to be more volatile when they are exposed to air. The difficulty in purification and the high toxicity of organostannyls make the use of Stille cross-couplings an unwelcome choice.

1.2.6 Examples of Synthesis of D–A Conjugated Polymers by Stille Coupling

The Stille polycondensation mainly involves the coupling reactions of ditin compounds with dihalide compounds and can be performed using conventional heating. Significantly, microwave irradiation was found to improve the number average molecular weight (M_n) and yield, and decrease the polydispersity index (PDI). This section will describe several examples in which D–A low bandgap conjugated polymers synthesized by Stille cross-couplings were used as donor materials in single junction bulk heterojunction polymer solar cells to achieve high PCE (>7%).

Liang *et al.* reported the synthesis of polymer P_1 by Stille polycondensation between a dibromo compound and ditin compound.[10] Pd(PPh$_3$)$_4$ was used as the catalyst, toluene and DMF (4 : 1) were used as the solvent, and the polymerization was carried out at 120 °C for 12 h under N$_2$ protection. PTB7 was obtained with average molecular weight (M_w) of 97.5 kDa and a PDI of 2.1. A PCE over 7% was obtained for BHJ PSCs based on PTB7. Later, the PCE was increased to 9.2% after device optimization (Scheme 1.2).[11]

P_2 was prepared by Wei *et al.* with Stille polymerization of a dibromo compound and ditin compound using [Pd$_2$dba$_3$/P(o-tolyl)$_3$] as the catalyst precursor.[12] The M_n of 9.7 kg mol^{-1} with a PDI of 1.4 was determined by gel permeation chromatography (GPC) using chloroform as the eluent. P_2 is readily soluble in hot chlorinated solvents such as chloroform, chlorobenzene, and dichlorobenzene. With the blends of P_2:PC$_{71}$BM (1:1.5, w/w) as the active layer, a PCE of 7.3% was achieved using 1,6-diiodohexane (DIH) as the processing additive (Scheme 1.3).[12]

Chu *et al.* reported a new alternating copolymer of dithienosilole and thienopyrrole-4,6-dione (P_3), which was synthesized by Stille coupling of a ditin compound and dibromo compound in refluxing toluene/dimethylformamide (10:1) with Pd(PPh$_3$)$_4$ as the catalyst.[13] The purified polymer has a M_n of 28 kDa and a PDI of 1.6, as determined by GPC using chlorobenzene (CB) as the eluent. P_3

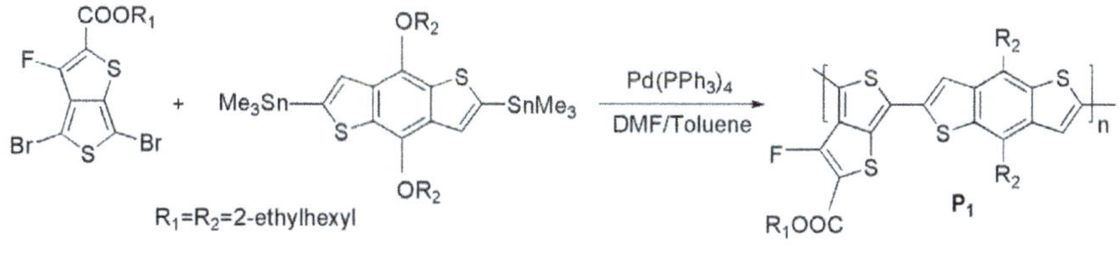

Scheme 1.2 Synthesis of polymer P_1 by Stille polycondensation.

could be readily dissolved in chlorinated solvents even at room temperature. When blended with PC$_{71}$BM, **P$_3$** exhibited a PCE of 7.3% (Scheme 1.4).

You *et al.* reported the first fluorinated D–A conjugated polymers applied in PSCs with an exceptional performance. **P$_4$** was synthesized by Stille coupling of a ditin compound and dibromo compound.[14] The polymerization was carried out at 120 °C for 20 min under microwave irradiation with Pd$_2$dba$_3$ and P(*o*-tolyl)$_3$ as the catalyst precursors and *o*-xylene as the solvent. **P$_4$** was obtained in a yield of 89% with M_n of 33.8 kDa. A PCE of 7.2% was obtained for **P$_4$** T:PC$_{61}$BM BHJ based PSCs (Scheme 1.5).

Fluorinated polymer **P$_5$** was synthesized in a similar method by You *et al.*[15] PSCs based on **P$_5$** showed a PCE above 7% when blended with PC$_{61}$BM as the

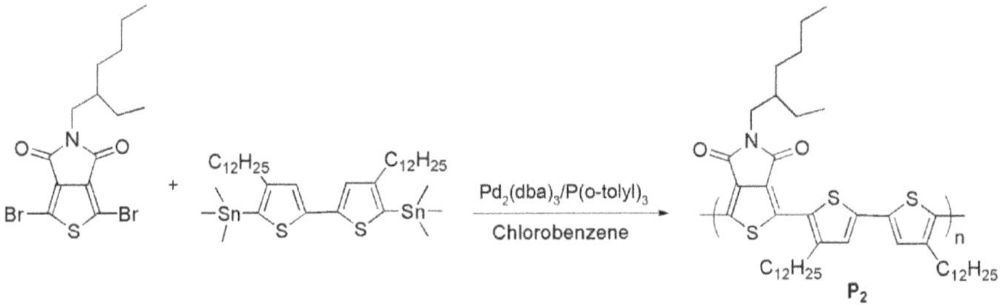

Scheme 1.3   Synthesis of polymer **P$_2$**.

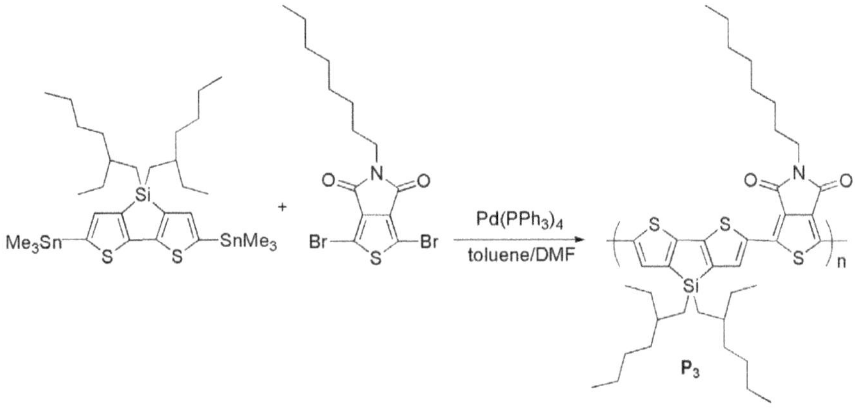

Scheme 1.4   Synthesis of polymer **P$_3$**.

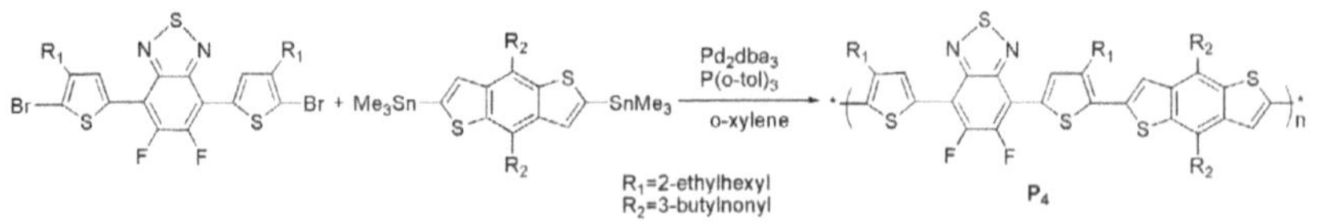

Scheme 1.5   Synthesis of polymer **P$_4$**.

active layer, and a PCE above 6% is still maintained at an active layer thickness of 1 μm (Scheme 1.6).

The copolymer **P₆** was obtained through Stille coupling polymerization of two monomers with a yield of 86%.[16] **P₆** has a M_n of 40.5 kDa with a PDI of 3.20. PSCs based on **P₆** showed a PCE of 6.00% when blended with $PC_{71}BM$ as the active layer. A PCE of 8.4% was achieved from the inverted PSC by using a PFN–Br interfacial layer to modify the ZnO electron extraction layer (Scheme 1.7).[17]

Amb *et al.* reported the synthesis of the first dithienogermole (DTG)-containing conjugated polymers by Stille polycondensation and their photovoltaic performance.[18] When **P₇**:$PC_{70}BM$ blends are utilized in inverted bulk heterojunction solar cells, the cells display a PCE of 7.3%. In inverted PSCs, when surface-modified ZnO–polymer nanocomposites were used as the electron-transporting layer, a PCE of 7.4% was achieved (Scheme 1.8).[19]

Huo *et al.* developed new poly[benzo(1,2-b:4,5-*b'*)dithiophene-co-thieno-(3,4-*b*)thiophene] (PBDTTT) derivatives having the thienyl substituted [benzo(1,2-b:4,5-*b'*)dithiophene] (BDT) and the alkylcarbonyl-substituted

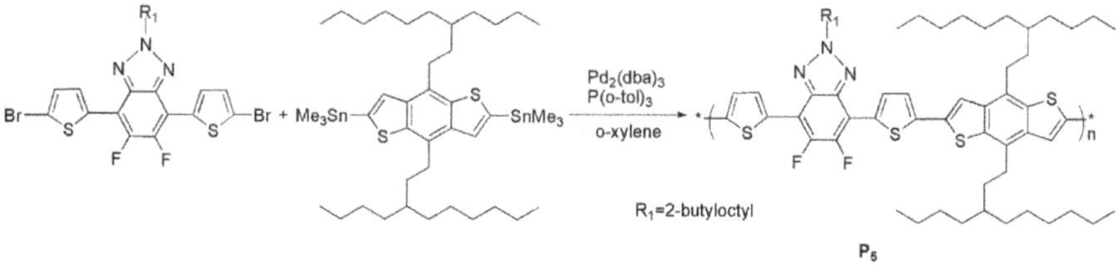

Scheme 1.6   Synthesis of polymer **P₅**.

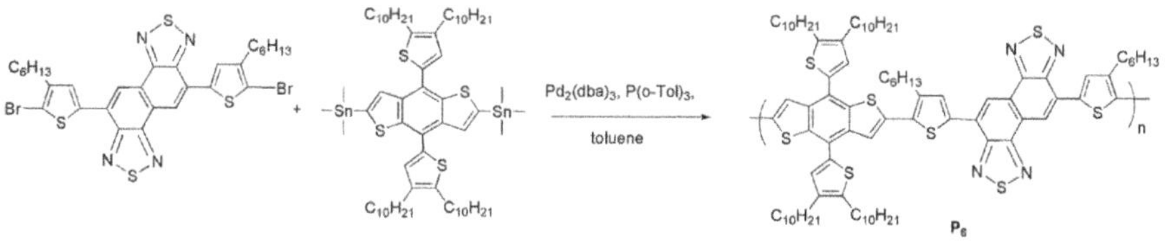

Scheme 1.7   Synthesis of polymer **P₆**.

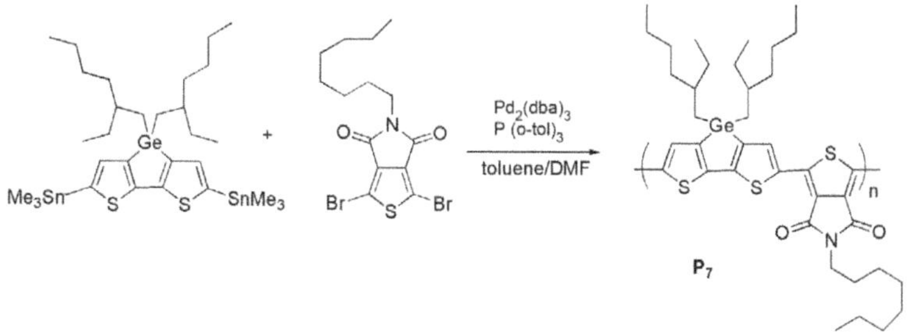

Scheme 1.8   Synthesis of polymer **P₇**.

thieno-[3,4-*b*]thiophene (TT-C).[20] The polymer **P₈** was prepared by Stille coupling of the bis(trimethyltin) BDT monomers and the bromides (TTC) in a solvent mixture of toluene and DMF (5:1) with Pd(PPh₃)₄ as the catalyst. PSCs based on **P₈** and PC₇₀BM reached a PCE of 7.59%. A PCE of 8.79% for a single-junction BHJ PSC was obtained with metallic nanoparticles (NPs) embedded in the active layer.[21] The PCE was further increased to 9.13% in inverted PSCs by using 1,8-diiodooctane (DIO) as the processing additive (Scheme 1.9).

Xu *et al.* reported the synthesis of **P₉** by Stille coupling of IDTT-di-Tin and diiodo-DFBT with Pd₂dba₃ and P(*o*-tol)₃ as catalyst precursors.[22] PSC devices based on **P₉** showed an improved PCE of 7.03% without the use of any additives or post-solvent/thermal annealing processes (Scheme 1.10).

Wu *et al.* reported the synthesis of polymer **P₁₀** through Stille coupling of the bis(trimethyltin) monomers and the dibromo monomers.[23] **P₁₀**:PC₇₁BM based PSCs exhibit a PCE above 7.79% without any further treatment such as the use of additives or annealing in device fabrication (Scheme 1.11).

Dong *et al.* prepared a copolymer **P₁₁** by Stille coupling. The **P₁₁**-based device showed an impressive PCE of 7.11% in inverted PSCs (Scheme 1.12).[24]

P₁₂ was synthesized by microwave-assisted Stille coupling using Pd₂dba₃ and P(*o*-tolyl)₃ as the catalyst precursor, and chlorobenzene (CB) as the solvent. PSCs based on **P₁₂**:PC₇₁BM blends afforded PCE up to 7.2% without thermal annealing or the use of processing additives (Scheme 1.13).[25]

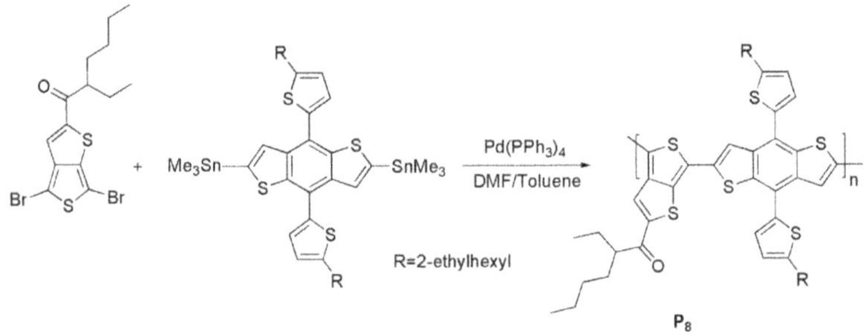

Scheme 1.9   Synthesis of polymer **P₈**.

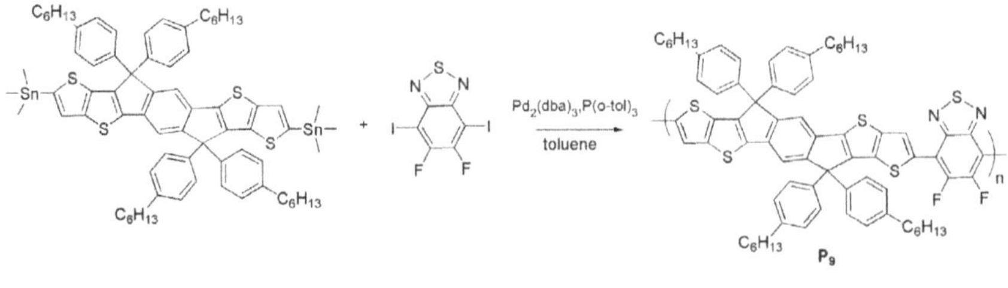

Scheme 1.10   Synthesis of polymer **P₉**.

Son *et al.* reported the synthesis of polymer **P**$_{13}$ by Stille coupling using Pd(PPh$_3$)$_4$ as the catalyst and DMF and anhydrous toluene (1:4) as the reaction cosolvent. After optimization of the polymer's solubility and morphological compatibility with PC$_{71}$BM, PSCs achieved a PCE of 7.6% (Scheme 1.14).[26]

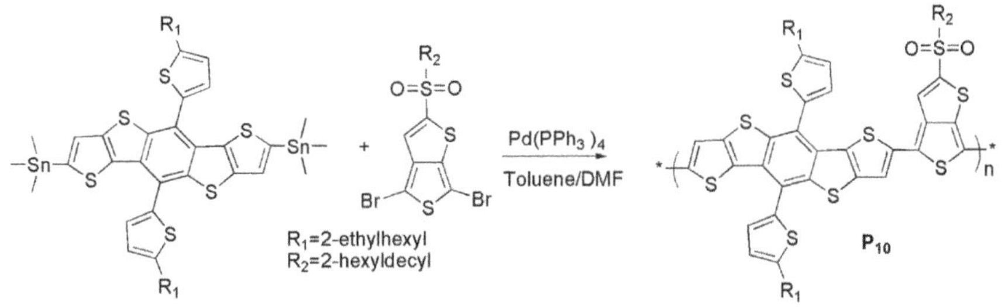

Scheme 1.11 Synthesis of polymer **P**$_{10}$.

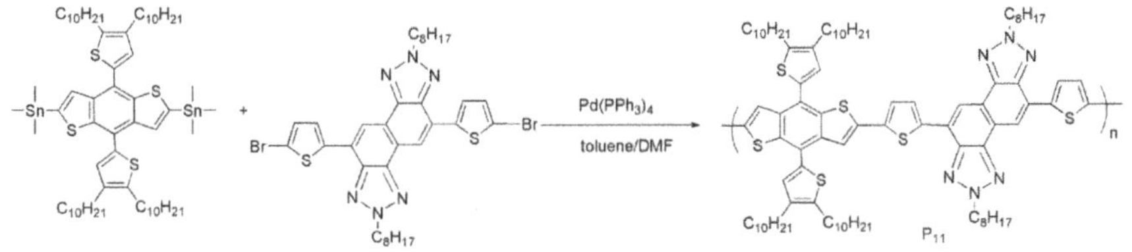

Scheme 1.12 Synthesis of polymer **P**$_{11}$.

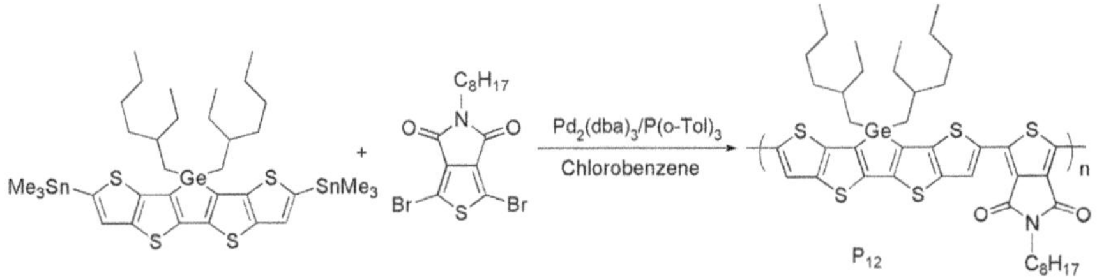

Scheme 1.13 Synthesis of polymer **P**$_{12}$.

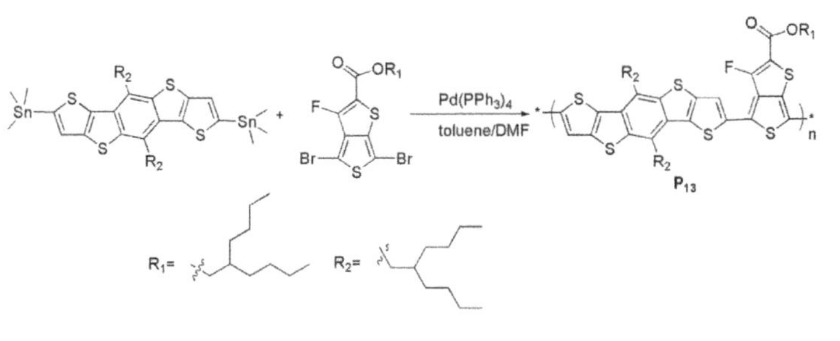

Scheme 1.14 Synthesis of polymer **P**$_{13}$.

Liao *et al.* reported the synthesis of **P$_{14}$** by Stille coupling with Pd(PPh$_3$)$_4$ as the catalyst in a solvent mixture of toluene and DMF (5:1). Fullerene derivative (PCBE–OH)-doped ZnO nanometer-thick film (40 nm) was used as the cathode interfacial layer for the effective collection of electrons. PSCs with the **P$_{14}$**:PC$_{71}$BM active layer and the ZnO–C$_{60}$ modified cathode give a PCE of 9.35% (Scheme 1.15).[1]

Wang *et al.* reported the synthesis of **P$_{15}$** by Stille coupling with Pd(PPh$_3$)$_4$ as the catalyst. **P$_{15}$** can dissolve in chloroform, toluene and 1,2,4-trichlorobenzene (TCB) at room temperature. The best device performance, with a relatively high PCE of 8.30%, was obtained for **P$_{15}$**:PC$_{71}$BM (1:1.5) (Scheme 1.16).[27]

Li *et al.* prepared polymer **P$_{16}$** using Stille coupling. A toluene/DMF (10:1, v/v) solvent mixture was used to obtain high molecular weight materials. PSCs based on **P$_{16}$**:PC$_{71}$BM furnished a PCE of 7.1% (Scheme 1.17).[28]

Hou *et al.* prepared a new copolymer, **P$_{17}$**, with Stille coupling. PSCs based on **P$_{17}$**:PC$_{71}$BM (1:1.5, w/w) gave the best PCE of 8.07% (Scheme 1.18).[29]

Chen *et al.* reported the synthesis of **P$_{18}$** by Stille coupling.[30] With **P$_{18}$** as the donor and PC$_{71}$BM as the acceptor in inverted PSCs, the highest PCE of 7.64% was achieved with an active layer 230 nm thick (Scheme 1.19).

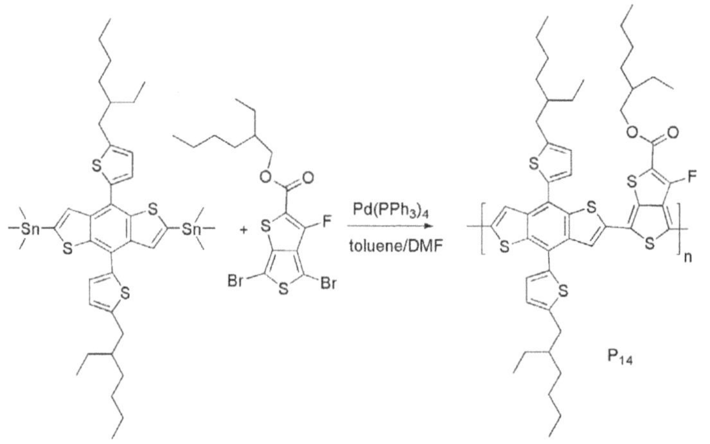

Scheme 1.15   Synthesis of polymer **P$_{14}$**.

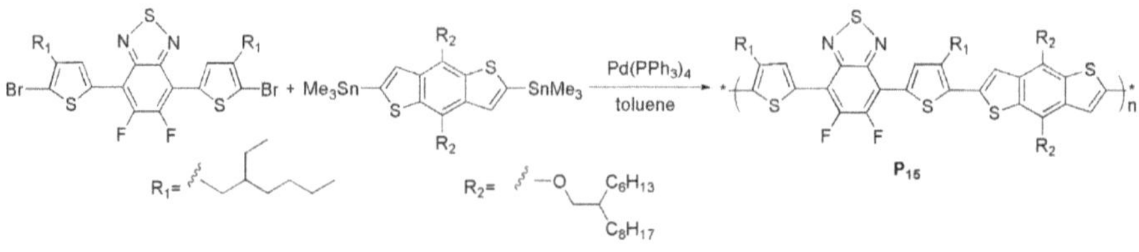

Scheme 1.16   Synthesis of polymer **P$_{15}$**.

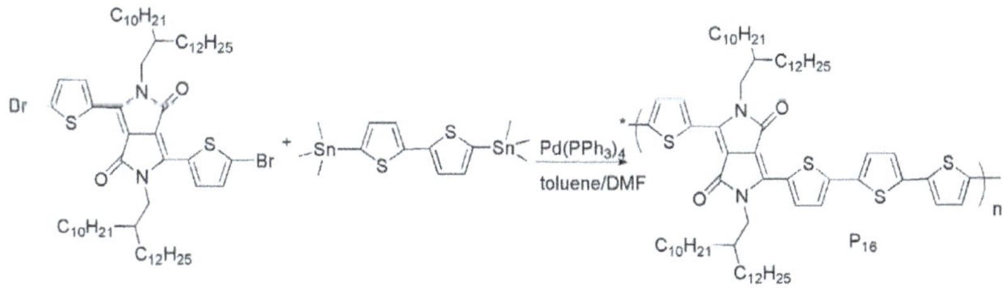

Scheme 1.17 Synthesis of polymer P$_{16}$.

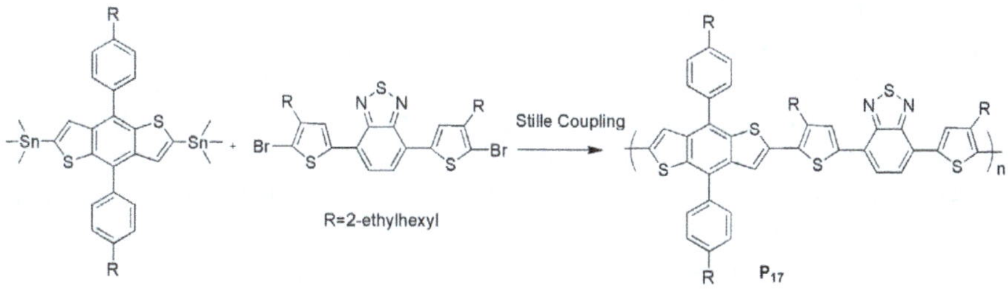

Scheme 1.18 Synthesis of polymer P$_{17}$.

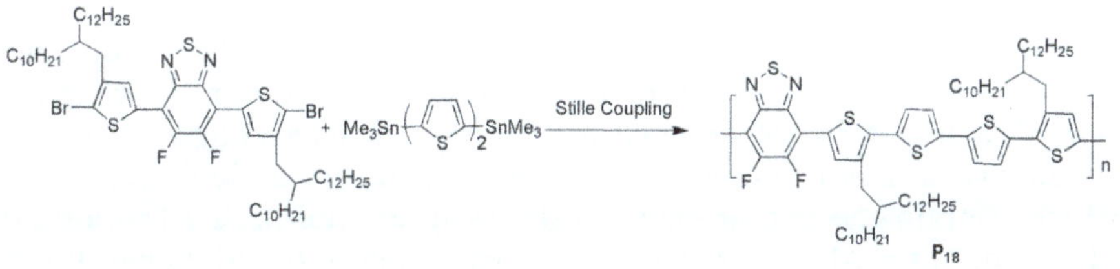

Scheme 1.19 Synthesis of polymer P$_{18}$.

1.3 Suzuki Polycondensation

1.3.1 History and Mechanism of the Suzuki Coupling Reaction

Over the past two decades, Suzuki polycondensation has become one of the most efficient methods for the synthesis of conjugated polymers. As another important cross-coupling protocol, the Suzuki–Miyaura cross-coupling reaction was invented by Suzuki and co-workers in 1979.[31] The scope of the Suzuki reaction for synthetic applications has been surveyed in several excellent reviews by Kotha, Lahiri, Kashinath, Miyaura and Fu.[32] The Suzuki–Miyaura cross-coupling reaction provided deeper insights into how to connect two specific sp^2-hybridized C-atoms more efficiently and under milder conditions. The Suzuki–Miyaura cross-coupling reaction was first used by Schlueter *et al.* to prepare poly(*para*-phenylene)s.[33]

Since then, Suzuki polycondensation (SPC) has become one of the most powerful and widely used methodologies for the synthesis of conjugated polymers.

1.3.2 Mechanism of the Suzuki Coupling Reaction

The catalytic cycle of Suzuki coupling is thought to follow a sequence involving the oxidative addition of an aryl halide to a Pd(0) complex to form an arylpalladium(ii) halide intermediate, the transmetalation with a boronic acid, and reductive elimination of the resulting diarylpalladium complex to afford the corresponding biaryl and to regenerate the Pd(0) complex. Oxidative addition is often the rate-limiting step, and it is not surprising that the relative reactivity of aryl halides decreases in the order I > Br > Cl. The role of the base in these reactions is to facilitate the transmetalation of the boronic acid by forming a more reactive boronate species that can interact with the Pd center and transmetalate in an intramolecular fashion (Scheme 1.20, path A).[34] Alternatively, it has also been proposed that the base replaces the halide in the coordination sphere of the palladium complex and facilitates an intramolecular transmetalation (path B).[35] In fact, the exact nature of the actual catalyst remains ambiguous.[36]

Suzuki polycondensation is generally believed to be of the step-growth type involving so-called AA/BB and AB approaches. In the AA/BB case, two different monomers are required, each of which carries either two boronic acids (or esters) or two leaving groups such as halogen or triflate. When two aromatic monomers are combined, polyarylene backbones which contain the two aromatic residues in an alternating fashion are obtained. In the AB case, the monomer carries both functional groups at the same time.

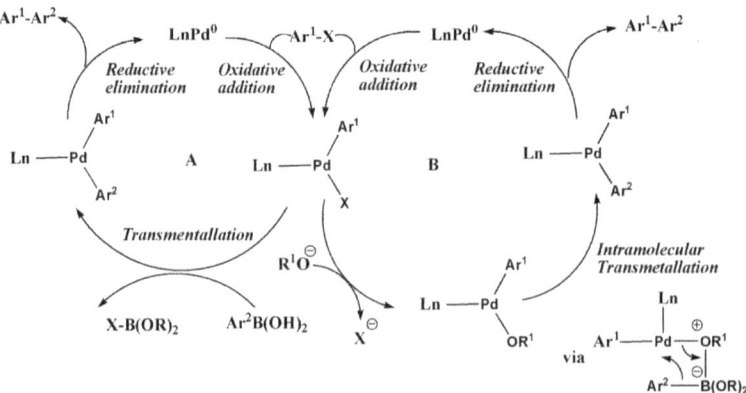

Scheme 1.20   General mechanism of the Suzuki reaction.

1.3.3 Catalyst, Ligand and Solvents

Almost all Suzuki polycondensations published to date in the literature use 1–3 mol% of catalyst, mostly Pd[P(*p*-tolyl)$_3$]$_3$, Pd(PPh$_3$)$_4$ or *in situ* prepared Pd[P(*o*-tolyl)$_3$]$_2$. Most catalysts for the Suzuki polycondensation employ tri-arylphosphine ligands. New ligands, which include Buchwald's biaryl-based phosphines,[37] Beller's diadamantyl phosphines,[38] Fu's tri(*tert*-butyl)phosphine,[39] and Hartwig's pentaphenylated ferrocenyl phosphines,[40] have been developed for Suzuki–Miyaura cross-coupling reactions. Buchwald-type ligand has been applied to polymerize dichloro monomers using Suzuki polycondensation.

Because of the hydrolytic deboronation of 2,5-thiophenebis(boronic acid pinacol ester) under standard Suzuki–Miyaura cross-coupling conditions, most attempts to synthesize thiophene-containing conjugated polymers from electron-rich 2,5-thiophene bis(boronic acid pinacol ester) and aryl dibromides by Suzuki polycondensation have failed to afford high molecular weight polymers. Bo designed and synthesized a new thiophene-containing phosphorous compound, **L1**, which was used as the ligand for a zerovalent palladium catalyst for Suzuki polycondensation of 2,5-thiophenebis(boronic acid pinacol ester) and aryl dihalides.[41] High molecular weight thiophene-containing conjugated polymers were successfully synthesized for the first time from thiophene based diboronic acid ester monomers. The new catalytic system can enhance the SPC reaction rate and cope with steric hindrance imparted by the monomers. With the new ligand SPC proceeded very rapidly, and high molecular weight fourth generation dendronized polymers could be obtained in a short time. This method should be also of great interest for the synthesis of pharmaceutical and agrochemical compounds and natural products (Scheme 1.21).

The solvent systems will affect the progress of the polycondensation. Most Suzuki polycondensation are carried out in biphasic mixtures of organic solvents such as toluene, xylene, THF, or dioxane and an aqueous medium

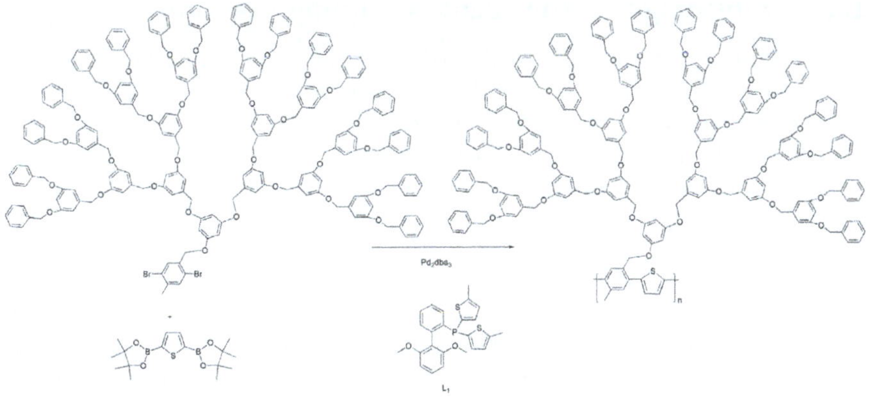

Scheme 1.21 Synthesis of polymers by the Suzuki coupling reaction.

containing the base. The commonly used bases include K_3PO_4, K_2CO_3, NaHCO$_3$, KOH, KF, and sodium *tert*-butoxide. However, the choice of base is still empirical, and no general rule for their selection has been established at present. The other solvent systems, in particular homogeneous ones, have been less explored. Phase transfer catalysts (PTCs) such as tetraalkylammonium salts (tetraethylammonium hydroxide) have also been tried.

1.3.4   Monomers

Aryl halides (bromides or iodides) and triflates substituted with electron-withdrawing groups are suitable substrates for the cross-coupling reaction. Aryl triflates and sulphonates are regarded as the synthetic equivalents of aryl halides. Triflates are, however, thermally labile, prone to hydrolysis and expensive to prepare. Aryl sulphonates are an attractive option because they are easily prepared from phenols, are more stable than triflates and are cheap and easily available starting materials. Suzuki polycondensations are carried out mainly between aryl (heteroaryl) halides and aryl (heteroaryl) boronic acids (esters). Similar to other step-growth polymerization, monomer purity is a key issue for Suzuki polycondensation, especially when the AA/BB approach is used. Boronic acids easily form partially and fully dehydrated products, which makes it difficult to reach the correct 1:1 stoichiometry. Alternatively, the corresponding cyclic boronic esters are widely used, because the commonly used boronic pinacol esters can be easily purified by silica gel column chromatography. Free boronic acids tend to be more reactive than their ester analogs. The solubility of boronic acids *vs.* esters in solvents also influences relative reactivity. The higher reactivity of the acids can be counteracted by their lower solubility.

1.3.5   Advantages of the Suzuki Coupling Reaction

The palladium catalyzed Suzuki–Miyaura coupling reaction is one of the most efficient methods for the construction of C–C bonds. The coupling can be carried out under mild reaction conditions, which will not be affected by the presence of water and heat. The reaction can tolerate a broad range of functionality and yield non-toxic byproducts. The boron-containing byproducts are easily separated from the reaction mixture and handled when compared with other organometallic reagents, especially in large-scale production. Additionally, the diverse boronic acids are commercially available and environmentally friendly. Consequently, the cross-coupling reaction has been realized in diverse applications, not only in academic laboratories but also in industry. These desirable features make the Suzuki–Miyaura reaction an important tool in medicinal chemistry as well as in the large-scale synthesis of pharmaceuticals and fine chemicals.

1.3.6 Drawbacks of the Suzuki Coupling Reaction

The Suzuki–Miyaura reaction suffers from a few key drawbacks, the first of which is the requirement for basic conditions. A number of monomers may be unstable in basic conditions, thus rendering this methodology impractical for these applications, or require more complex protection–deprotection strategies. Also, the Suzuki–Miyaura reaction requires a two-phase system; thus, polymers that rapidly decrease in solubility as molecular weight increases may form precipitate in poor yields or display very low molecular weights and high polydispersities under Suzuki–Miyaura conditions, which is disadvantageous for photovoltaic application.

1.3.7 Examples of the Suzuki Coupling Reaction

We will provide a class of low bandgap polymers applied in BHJ PSCs that are synthesized by Suzuki polycondensation. This section cannot possibly cover all these different polymers. Emphasis will be placed upon important classes of conjugated polymers based on bridged phenylenes, for example poly(2,7-fluorene), poly(2,7-carbazole), and poly(2,7-dibenzosilole) based D–A conjugated polymers.

In 2003, Svensson *et al.* reported a low bandgap polymer, P_{19}, and its application in PSCs. P_{19} was synthesized by Suzuki polycondensation of 2,7-fluorenediboronic acid pinacol ester and benzothiadiazole based dibromide using $Pd(PPh_3)_4$ as the catalyst and toluene and 20% aqueous tetraethylammonium hydroxide as the reaction media (Scheme 1.22).[42]

Similarly, Blouin *et al.* reported the synthesis of P_{20} by Suzuki coupling of 2,7-carbazolediboronic acid pinacol ester and benzothiadiazole based dibromide using Pd_2dba_3 and $P(o\text{-tol})_3$ as the catalyst precursors. PSCs based on P_{20} showed a PCE of 3.6%.[43] After device optimization by Heeger *et al.*, the PCE was enhanced to 6.1% (Scheme 1.23).[44]

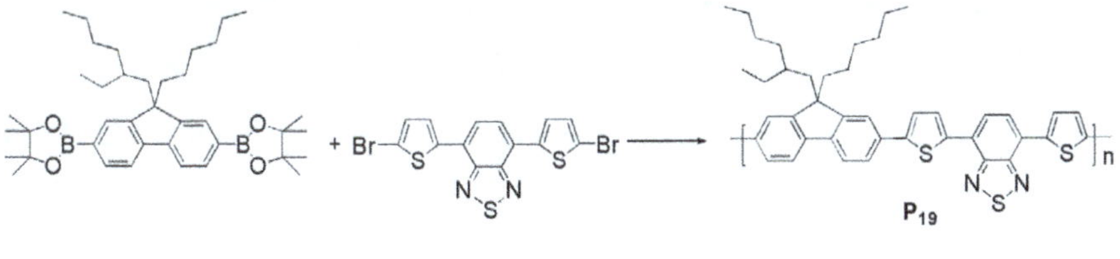

Scheme 1.22 Synthesis of polymer P_{19}.

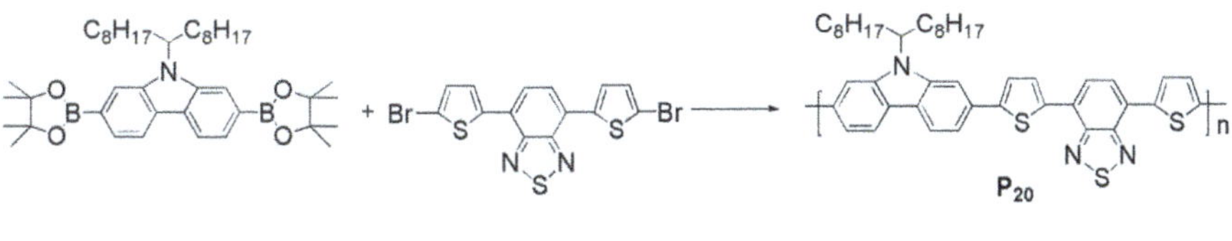

Scheme 1.23 Synthesis of polymer P_{20}.

Bo *et al.* reported the synthesis of **P$_{21}$** by Suzuki polycondensation of 2,7-carbazolediboronic acid pinacol ester and 5,6-bis(octyloxy)benzothiadiazole based dibromide using Pd(PPh$_3$)$_4$ as the catalyst precursor and a biphasic mixture of THF–toluene (5:1)/aqueous NaHCO$_3$ as the reaction medium. **P$_{21}$** based PSCs exhibit a PCE of 5.4% (Scheme 1.24).[45]

He *et al.* reported a copolymer, **P$_{22}$**, which was synthesized by Suzuki polycondensation. With **P$_{22}$**:PC$_{71}$BM (1:4) as the active layer, solar cells with a PFN/Al bilayer cathode displayed PCEs up to 6.07% (Scheme 1.25).[46]

P$_{23}$ was first reported as donor material for PSCs by Leclerc *et al.* in 2007.[47] **P$_{23}$** was prepared with a M_n of 15 kDa by Suzuki polycondensation of 2,7-silafluorenediboronic acid pinacol ester and benzothiadiazole based dibromide. Preliminary device experiments with this copolymer gave a PCE of 1.6%. In parallel, Cao *et al.* reported the synthesis of the same polymer with a high molecular weight (M_n = 79 kDa). The PCE of **P$_{23}$** based PSCs was increased to 5.4% with a polymer:fullerene ratio of 1:2 (Scheme 1.26).[48]

Bo *et al.* reported the synthesis of a series of **D–A** alternating conjugated polymers **P$_{24-31}$** with 5,6-bis(octyloxy)benzothiadiazole as the acceptor unit by Suzuki polycondensation of dibromo monomers and 5,6-bis(octyloxy)benzothiadiazole based diboronic acid pinacol ester. **P$_{24}$**, with a M_n of 102 kg mol^{-1} and a PDI of 1.66, gave a PCE of 5.08% in devices with **P$_{24}$**:PC$_{71}$BM as the active

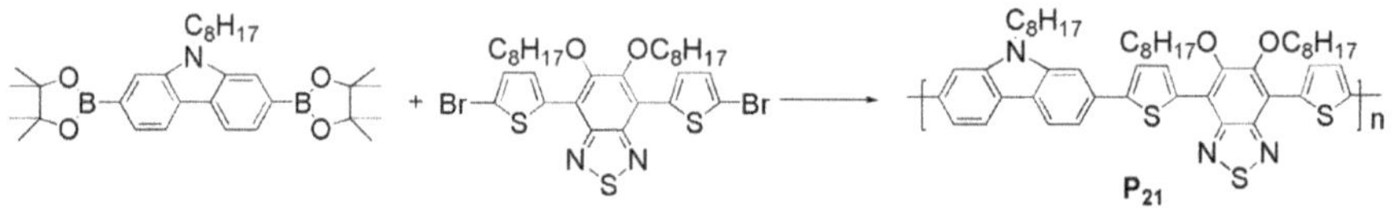

Scheme 1.24   Synthesis of polymer **P$_{21}$**.

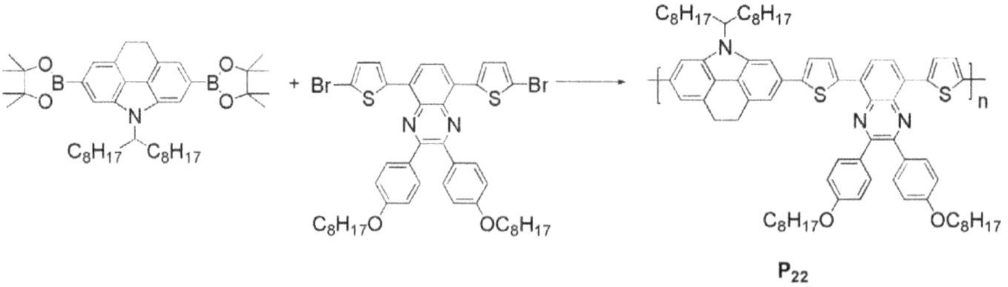

Scheme 1.25   Synthesis of polymer **P$_{22}$**.

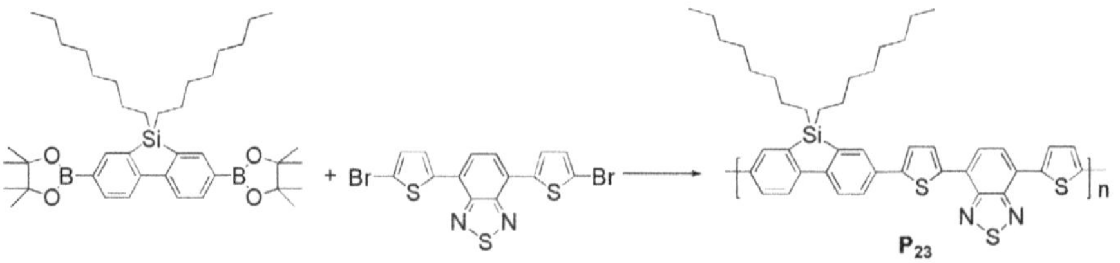

Scheme 1.26   Synthesis of polymer **P$_{23}$**.

layer. When TiO$_x$ was used as an electron-blocking layer, the PCE was further increased to 6.05%.[49] The 9-arylidene-9*H*-fluorene based D–A conjugated polymers **P$_{25}$** and **P$_{26}$** were synthesized to investigate the influence of alkyloxy position on the performance of PSCs. High molecular weight **P$_{25}$** (**HMW-P$_{25}$**) and low molecular weight **P$_{25}$** (**LMW-P$_{25}$**) were used to investigate the influence of molecular weight on the performance of PSCs. **HMW-P$_{25}$:PC$_{71}$BM**-based PSCs showed a PCE of 6.52%; **LMW-P$_{25}$:PC$_{71}$BM** based PSCs showed poor photovoltaic performance, with a PCE of only 2.75%; and **P$_{26}$:PC$_{71}$BM** based PSCs gave a PCE of only 2.51%.[50] Polymer solar cells with **P$_{27}$:PC$_{71}$BM** as the active layer demonstrate a PCE of 2.23% with a high open circuit voltage (V_{OC}) of 0.96 V (Schemes 1.27–1.30).[51]

P$_{28}$, with a M_n of 27.7 kg mol^{-1} and a PDI of 3.1, gave a PCE of 4.48% when a **P$_{28}$:PC$_{71}$BM** blend (1:3, by weight) was used as the active layer.[52] **P$_{29}$**, with 3,6-difluorocarbazole as the donor unit, has a M_n of 9.1 kg mol^{-1} and a PDI of 2.63. Polymer solar cells based on **P$_{29}$** and PC$_{71}$BM demonstrate a PCE of 4.8%.[53] **P$_{30}$**, with 9-alkylidene-9*H*-fluorene as the donor unit and 5,6-bis(octyloxy) benzothiadiazole as the acceptor unit, is of planar structure. PSCs with a blend of **P$_{30}$** and PC$_{71}$BM as the active layer demonstrate a PCE of 6.2%.[54] **P$_{31}$**,

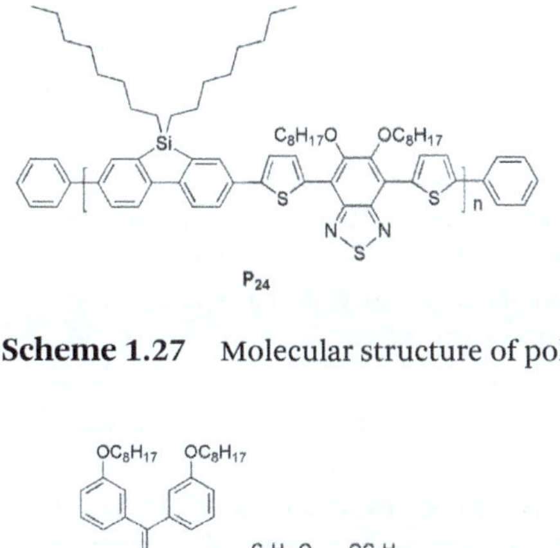

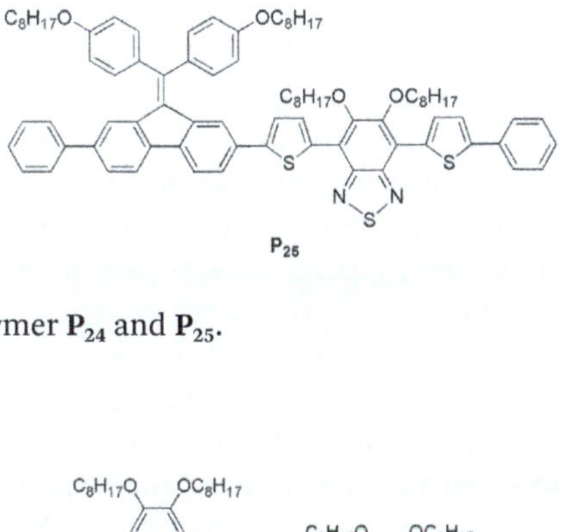

Scheme 1.27   Molecular structure of polymer **P$_{24}$** and **P$_{25}$**.

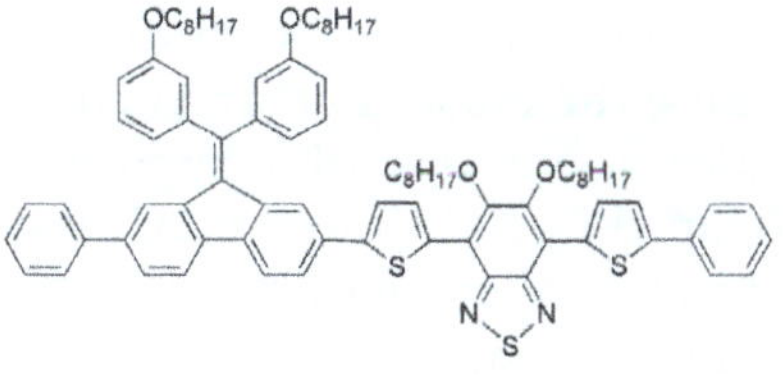

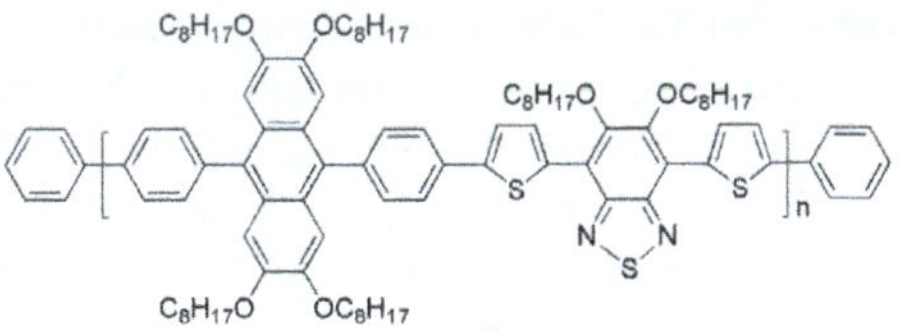

Scheme 1.28   Molecular structure of polymer **P$_{26}$** and **P$_{27}$**.

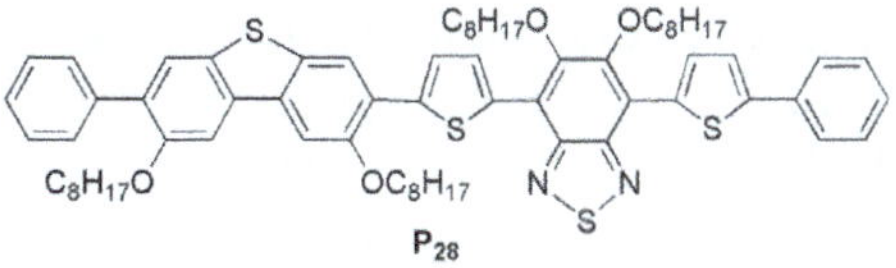

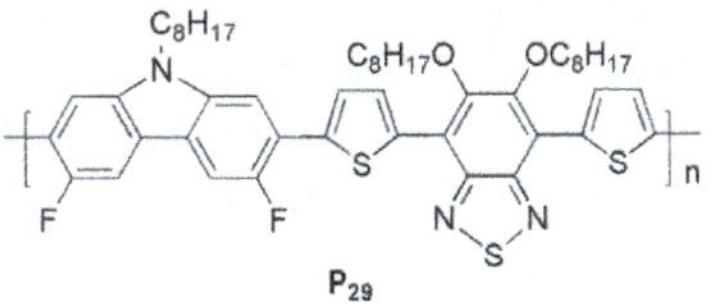

Scheme 1.29   Molecular structure of polymer **P$_{28}$** and **P$_{29}$**.

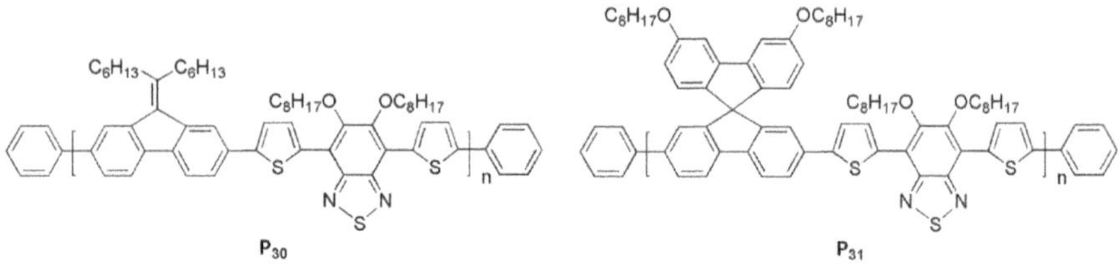

Scheme 1.30 Molecular structure of polymer **P₃₀** and **P₃₁**.

with spirobifluorene as the donor unit and 5,6-bis(octyloxy)benzothiadiazole as the acceptor unit, was synthesized and applied in PSCs. PSCs based on the blend films of **P₃₁** and PC$_{71}$BM show a high open-circuit voltage of 0.94 V and a PCE of 4.6% without any post-treatment.[55]

1.4   C–H Activation/Direct Arylation Polycondensation

Conventional synthesis of the π-conjugated polymers mainly relies on transition metal-catalyzed cross-couplings, such as Stille and Suzuki couplings (*vide supra*). The preparation of organotin or organoboron monomers used for Stille or Suzuki coupling requires multistep reactions and tedious purification. Additionally, Stille coupling requires the use of very toxic reagents as well as generating toxic byproducts, which is not environmentally friendly. In particular, some important classes of heteroaryl organotin or organoboron reagents are not readily accessible and may even be too unstable to undergo the coupling processes, which may confine the variety of polymer libraries to some extent. In recent years, transition metal-catalyzed direct C–H arylation of non-preactivated arenes with aryl halides or pseudohalides, called "direct arylation", has attracted much attention and worldwide interest. The direct arylations, with the advantages of synthetic simplicity and atom economy, without the use of troublesome and toxic (hetero)aryl organometallic intermediates, would be one of the most ideal routes toward π-conjugated polymers. These reactions are mostly developed for the synthesis of small molecules. Indeed, up until now, only a few publications have reported the use of conjugated polymers synthesized by direct arylation in PSCs.

1.4.1   History and Mechanism of the C–H Activation Polycondensation

In the last decade, Fagnou *et al.* detailed efficient coupling reactions between electron-deficient aromatic rings and arylhalides catalyzed by palladium complex in the presence of phosphine ligand and base, a synthetic reaction now termed "direct arylation".[56] Recently, this reaction has been applied to the synthesis of conjugated polymers, for example regioregular poly(3-hexylthiophene) (P3HT), although it required rigorous heating in a THF solution at 120 °C and under high pressure.

1.4.2 Mechanistic Insight

The mechanism of C–H activation has been studied experimentally and computationally, and possible pathways include electrophilic aromatic substitution, Heck-type coupling and concerted metalation–deprotonation (CMD).[57] Most heterocycles, such as thiophenes and indoles, are believed to follow a base-assisted CMD pathway. Two catalytic cycles for a CMD coupling of bromobenzene and thiophene using a palladium/phosphine catalytic system and cesium carbonate are shown in Schemes 1.31 and 1.32. Scheme 1.31 depicts a carboxylate-mediated process, while Scheme 1.32 represents the reaction process without a carboxylate additive. In the presence of carboxylate as the additive (Scheme 1.31), first the carbon–halogen bond is formed by oxidative addition, and then complex **1** is formed by exchange of the halogen ligand for the carboxylate anion. Complex **1** then deprotonates the thiophene substrate while simultaneously forming a metal–carbon bond by going through transition state **1-TS**. The phosphine ligands, or the solvent, can recoordinate to the metal center, following Pathway 1, or the carboxylate group can remain coordinated throughout the entire process (Pathway 2). Finally, reductive elimination renders the aryl coupled product. In the case of no carboxylate additive, after oxidative addition, the reaction follows one of the two pathways shown in Scheme 1.32. If a bidentate phosphine is

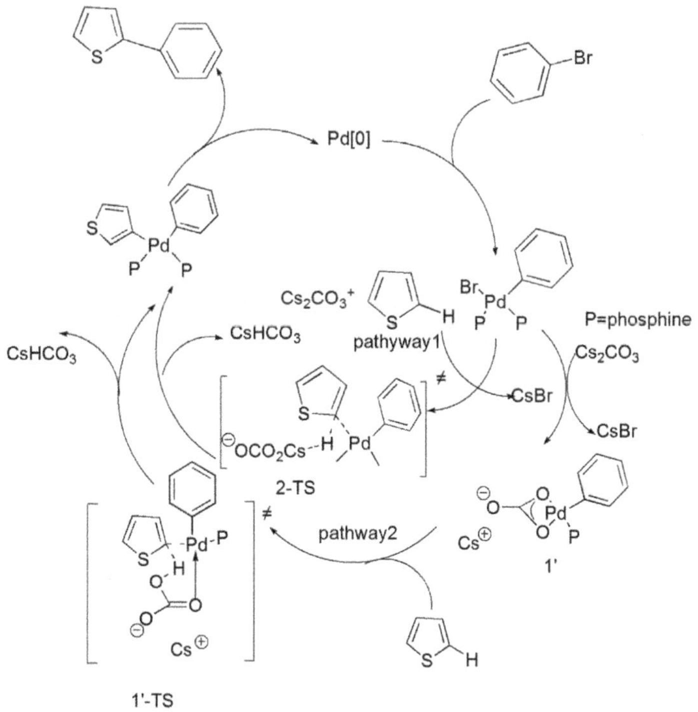

Scheme 1.31 General mechanism of the C–H activation reaction with a carboxylate-mediated process.

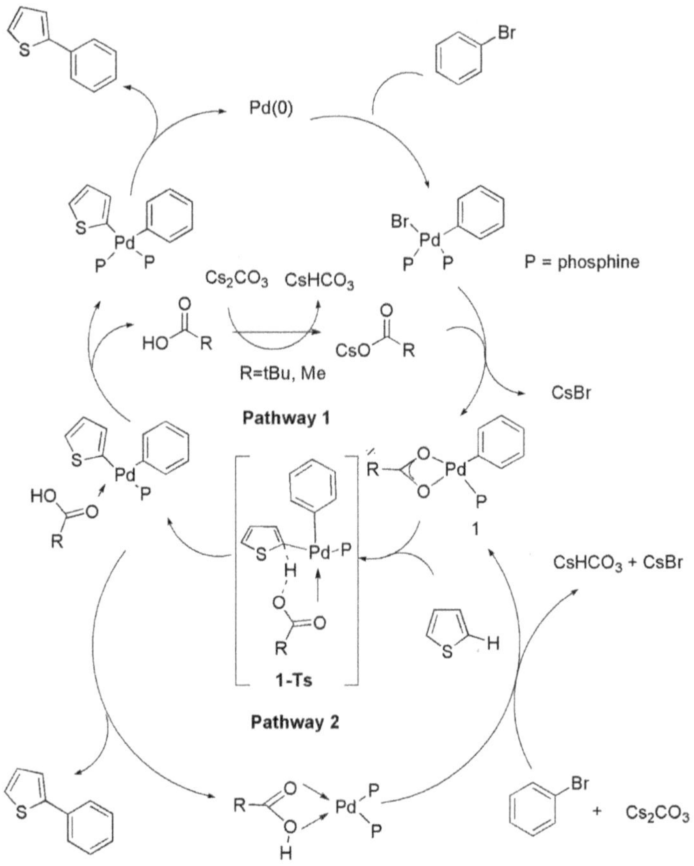

Scheme 1.32   General mechanism of the C–H activation reaction without a carboxylate additive.

employed, C–H activation of thiophene can follow Pathway 1, where deprotonation is assisted intermolecularly (2-TS). When a monodentate phosphine
is used, the reaction may follow Pathway 1 or Pathway 2. The latter mechanism most closely resembles Pathway 2 in Scheme 1.31 where the carbonate
coordinates to the metal center to give the zwitterionic species. Here, deprotonation occurs intramolecularly through transition state 1'-TS. Reductive
elimination then renders 2-phenylthiophene.

1.4.3   Catalysts, Additive and Solvents

In most polymerization examples, $Pd(OAc)_2$ was used as the palladium catalyst. Some research groups have exploited the highly stable Herrmann–Beller
catalyst. Catalytic addition of pivalic acid can aid the C–H activation. Some
groups have found that polymerization in the absence of phosphine ligand
can give high molecular weight materials.

Polar (dimethylacetamide (DMAC), DMF, NMP, THF) and nonpolar (toluene) solvents are suitable for these reactions and should be selected according to the polymer solubility. DMAc is not a suitable solvent for polymerization and always gives brown soluble material. Addition of carboxylic acid may be beneficial in nonpolar solvents owing to the high polarity of the C–H bond transition states. However, C–H bond activations have been accomplished in toluene and xylenes without carboxylic acids.[58]

1.4.4 Monomers

Several electron-rich monomers and electron-deficient monomers can be used for direct arylation polymerization. Direct arylation of 2-halo-3-alkylthiophenes for the preparation of polymers (P3RT) was first published in the late 1990s.[59] Monomers and dimers can be used to synthesize poly(3,4-ethylenedioxythiophene) (PEDOT) and poly(3,4-propylenedioxythiophene) (PProDOT) by direct arylation polymerization with aryl bromides.[58a] Interestingly, functional groups were tolerant to direct arylation polymerization conditions and polymers were obtained in relatively low molecular weights (M_n = 3–4 kDa) when EDOT monomers were substituted with different functional groups.[60] In addition, 2,2′-bithiophene or fused bithiophene based monomers can undergo facile C–H bond activation.

Electron-deficient thiophene-based monomers including thieno[3,4-*c*] pyrrole-4,6-dione (TPD), furo[3,4-*c*]pyrrole-4,6-dione (FPD) monomers, diketopyrrolopyrrole and isoindigo can be copolymerized with brominated thiophene-based electron-rich monomers such as benzodithiophene, dithienosilole, and dithienogermole by direct arylation polymerization. Additionally, electron deficient 4,4′-dinonyl-2,2′-bithiazole and 1,2,4,5-tetrafluorobenzene are highly reactive toward direct arylation polymerization.[61]

1.4.5 Advantages of the Direct Arylation Polycondensation

Direct arylation should bring significant benefits in the synthesis of donor–acceptor polymers by decreasing the number of monomer preparation steps, producing only acid as a byproduct, and freeing of undesired waste from the metal-containing reagents. This is particularly significant because the traditional Suzuki and Stille couplings inherently possess a number of problems: synthesis of monomers is time-consuming, purifying the organotin monomers is difficult, and organotin compounds are highly toxic. Therefore, it is highly desirable to develop new direct arylation methodology that is environmentally friendly, scalable, and high yielding for mass production, to meet the requirement for synthesis of donor–acceptor polymers used for PSCs.

1.4.6 Drawbacks of the Direct Arylation Polycondensation

For it to be a reliable synthetic methodology with a wide application range, there are still several problems related to direct arylation polycondensation reactions that need to be solved. First, the low molecular weight of the polymer obtained is disadvantageous for the photovoltaic application. Second, the poor selectivity of the C–H bond in direct arylation will resulted in branched and cross-linked polymer structures that affect the solubility and optoelectronic properties of polymers.

1.4.7 Examples of the Direct Arylation Polycondensation

The first example of synthesis of poly(3-alkylthiophene)s by direct arylation was reported in 1999.[59] Recently, the syntheses of high molecular weight poly(fluorene-*alt*-tetrafluorobenzene), poly(fluorene-*alt*-dithiophene), and poly(thienopyrroledione-*alt*-dithiophene) by direct arylation were published.[61b,62] Here, we provide a few examples to show the use of direct arylation to synthesize low bandgap conjugated polymers that are promising donor materials for PSCs.

A series of **DTDPP**-based copolymers with an alternating D–A sequence and homopolymer **P₃₂** were synthesized by the direct arylation polycondensation. The polymer was endowed with the common features of excellent π-conjugation and ideal planarity, which lead to remarkably low band gaps (1.22 eV). Photovoltaic properties of these polymers were not reported (Scheme 1.33).[63]

P₃₃, with a higher molecular weight, up to 70 kDa, was synthesized *via* palladium catalyzed direct arylation between dithiophene derivative and dibromo compound with K_2CO_3, $Pd(OAc)_2$ (10 mol%) and pivalic acid (30 mol%) in a small volume of *N*-methylpyrrolidone (NMP) in a 10 mL Schlenk tube. For the first time, OPV characteristics of polymers synthesized *via* direct arylation were compared to those synthesized *via* Suzuki coupling. A blend ratio **P₃₃**:PC₆₁BM of 1:3 was used to fabricate OPV devices and give the highest PCE of 2.24%, which represents a moderate enhancement in performance obtained *via* Suzuki-coupled polymer (2.01%) (Scheme 1.34).[64]

Direct arylation polycondensation using the phosphine-free catalytic system can be adapted to the synthesis of bithiazole-based alternating copolymers (**P₃₄**). In comparison with conventional polycondensation *via* other cross-coupling reactions, the polycondensation proceeded with a reduced amount of Pd catalyst (2 mol%) in a short reaction time (10 min to 3 h). Owing to the difference in reactivity of the C–H bond, controlling the reaction time was effective for suppressing the side reaction at the unexpected C–H bond (Scheme 1.35).[65]

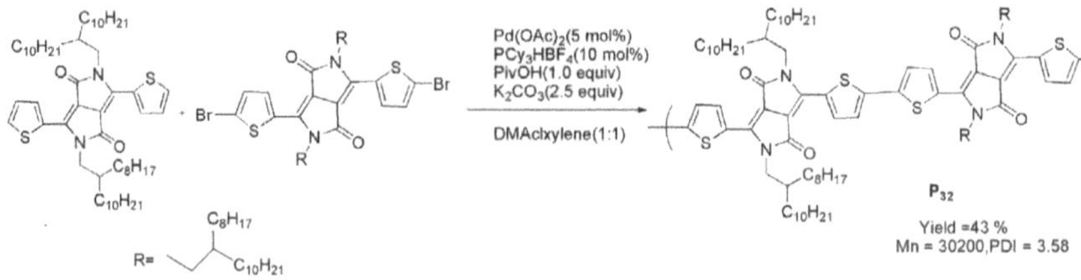

Scheme 1.33   Synthesis of polymer **P₃₂**.

High molecular weight **P₃₅**, using nonactivated dithiophene derivative and dibromo derivative as coupling monomers, was obtained in good yields (up to 80%) with a M_n of up to 40 kDa (Scheme 1.36).[66]

Wang *et al.* reported that direct arylation polycondensation of 2-bromo-3-hexylthiophene using Herrmann's catalyst and a triarylphoshine ligand yielded regioregular poly(3-hexylthiophene) with a high molecular weight.[67] This catalytic system also can be used to direct arylation polycondensation of a thieno[3,4-*c*]pyrrole-4,6-dione derivative with a dibromobithiophene derivative, to give D–A polymer **P₃₆**. Photovoltaic characterization was not tried (Scheme 1.37).[62b]

A 2,2′-bithiophene-based monomer has more than one type of reactive C–H bond, although reactions typically occur in the 5- and 5′-positions first. Attempts to copolymerize 2,2-bithiophene with dibromo compound rendered materials with low solubility due to cross-linking in the 3,3′- and 4,4′-positions of the bithiophene monomer. To evade these side reactions, the other reactive C–H bonds on the thiophene monomers were blocked with methyl groups. **P₃₇** and **P₃₈** were synthesized by direct heteroarylation polymerization. However, the film absorption onsets were hypsochromically shifted compared to the "unprotected" bithiophene analogs. The methyl groups indeed cause twisting in the polymer backbone and disrupt conjugation.

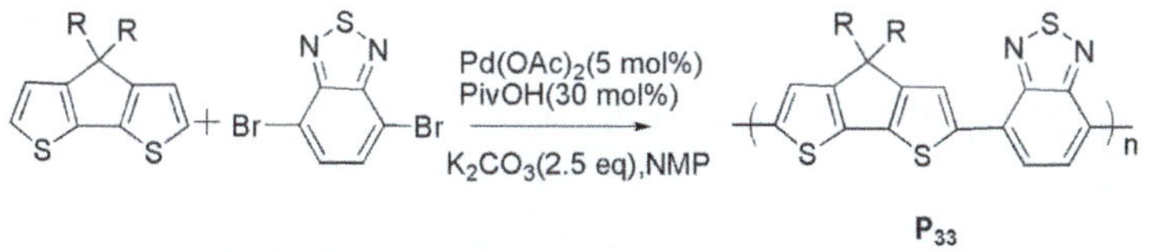

Scheme 1.34 Synthesis of polymer **P₃₃**.

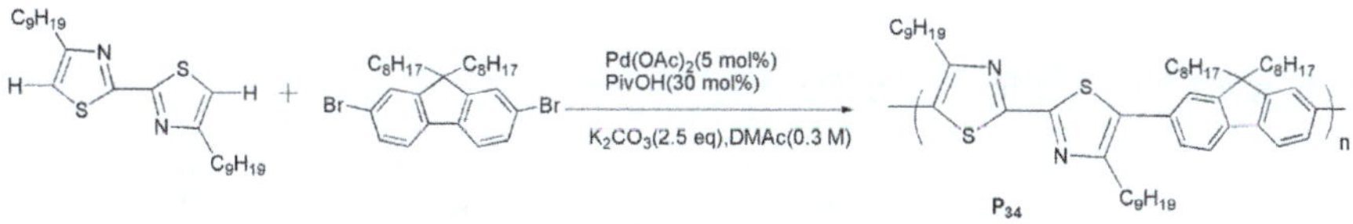

Scheme 1.35 Synthesis of polymer **P₃₄**.

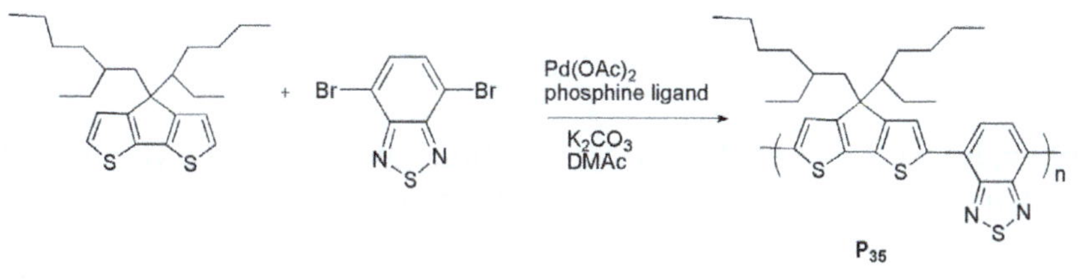

Scheme 1.36 Synthesis of polymer **P₃₅**.

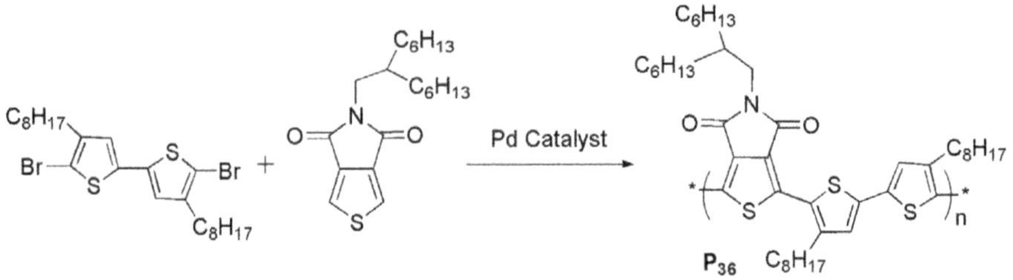

Scheme 1.37 Synthesis of polymer P$_{36}$.

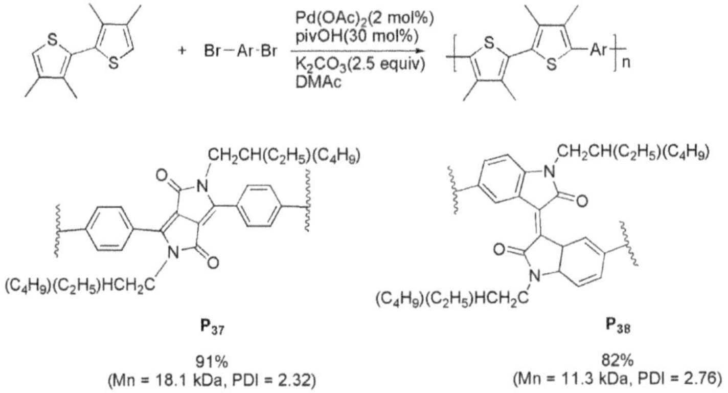

Scheme 1.38 Synthesis of polymer P$_{37}$ and P$_{38}$.

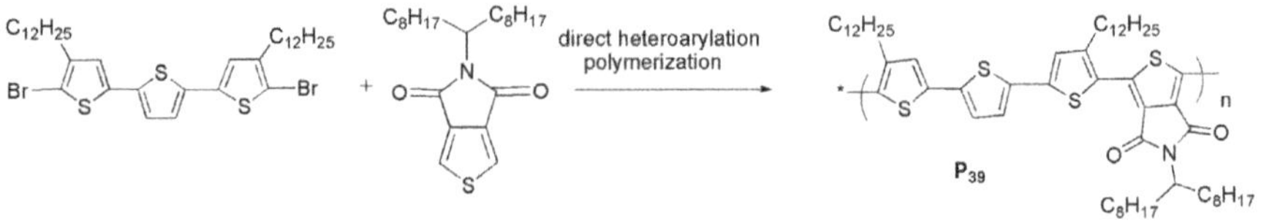

P$_{37}$

91%
(Mn = 18.1 kDa, PDI = 2.32)

P$_{38}$

82%
(Mn = 11.3 kDa, PDI = 2.76)

Scheme 1.39 Synthesis of polymer P$_{39}$.

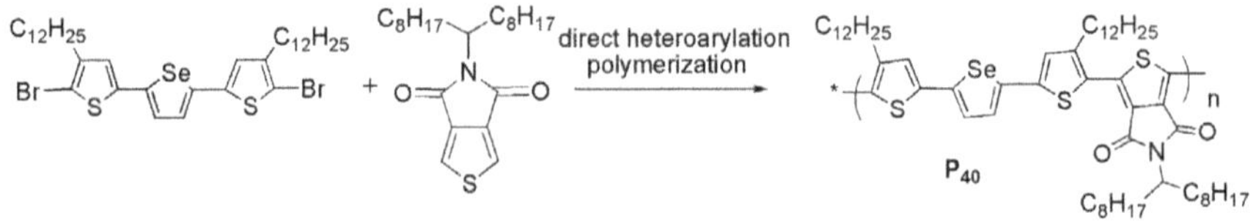

Scheme 1.40 Synthesis of polymer P$_{40}$.

Although protection of the 3,3',4,4'-positions of bithiophene circumvented cross-linking on this monomer, the planarity of the copolymers was disrupted and optimal packing properties diminished (Scheme 1.38).[68]

A D–A conjugated polymer, P$_{39}$, with a M_n of 41 kDa was synthesized by Heeger *et al.* using a direct heteroarylation procedure. A relatively high PCE of over 6% was obtained from BHJ solar cells based on a photoactive film comprising a composite of **P₃₉** and PC₇₁BM. This is the best result for polymers prepared *via* the direct heteroarylation method and used for BHJ solar cells (Scheme 1.39).[69]

A selenophene–TPD copolymer, **P₄₀**, with a M_n of 36 kDa and a PDI of 1.97 was synthesized using the direct heteroarylation polymerization in a 94% yield. PSCs with the blend **P₄₀** and PC₇₁BM as the active layer gave a PCE of about 5.8% (Scheme 1.40).[70]

References

1. S. H. Liao, H. J. Jhuo, Y. S. Cheng and S. A. Chen, Fullerene Derivative-Doped Zinc Oxide Nanofilm as the Cathode of Inverted Polymer Solar Cells with Low-Bandgap Polymer (PTB7-Th) for High Performance, *Adv. Mater.*, 2013, **25**(34), 4766–4771.

2. L. G. Mercier and M. Leclerc, Direct (Hetero)Arylation: A New Tool for Polymer Chemists, *Acc. Chem. Res.*, 2013, **46**(7), 1597–1605.

3. (a) D. Azarian, S. S. Dua, C. Eaborn and D. R. M. Walton, Reactions of Organic Halides with R₃mmr₃ Compounds (M = Si, Ge, Sn) in Presence of Tetrakis(Triarylphosphine) Palladium, *J. Organomet. Chem.*, 1976, **117**(3), C55–C57; (b) M. Kosugi, K. Sasazawa, Y. Shimizu and T. Migita, Reactions of Allyltin Compounds. 3. Allylation of Aromatic Halides with Allyltributyltin in Presence of Tetrakis(Triphenylphosphine)Palladium(0), *Chem. Lett.*, 1977, (3), 301–302; (c) M. Kosugi, Y. Shimizu and T. Migita, Alkylation, Arylation, and Vinylation of Acyl Chlorides by Means of Organotin Compounds in Presence of Catalytic Amounts of Tetrakis(Triphenylphosphine)Palladium(0), *Chem. Lett.*, 1977, **12**, 1423–1424.

4. D. Milstein and J. K. Stille, General, Selective, and Facile Method for Ketone Synthesis from Acid-Chlorides and Organotin Compounds Catalyzed by Palladium, *J. Am. Chem. Soc.*, 1978, **100**(11), 3636–3638.

5. V. Farina, S. R. Baker, D. A. Benigni and C. Sapino, Palladium-Catalyzed Coupling between Cephalosporin Derivatives and Unsaturated Stannanes—A New Ligand for Palladium Chemistry, *Tetrahedron Lett.*, 1988, **29**(45), 5739–5742.

6. (a) Z. N. Bao, W. K. Chan and L. P. Yu, Synthesis of Conjugated Polymer by the Stille Coupling Reaction, *Chem. Mater.*, 1993, **5**(1), 2–3; (b) Z. N. Bao, W. K. Chan and L. P. Yu, Exploration of the Stille coupling reaction for the syntheses of functional polymers, *J. Am. Chem. Soc.*, 1995, **117**(50), 12426–12435.

7. (a) J. K. Stille, The Palladium-Catalyzed Cross-Coupling Reactions of Organotin Reagents with Organic Electrophiles, *Angew. Chem., Int. Ed.*, 1986, **25**(6), 508–523; (b) P. Espinet and A. M. Echavarren, The mechanisms of the Stille reaction, *Angew. Chem., Int. Ed.*, 2004, **43**(36), 4704–4734.

8. V. Farina and B. Krishnan, Large Rate Accelerations in the Stille Reaction with Tri-2-Furylphosphine and Triphenylarsine as Palladium Ligands—Mechanistic and Synthetic Implications, *J. Am. Chem. Soc.*, 1991, **113**(25), 9585–9595.

9. B. A. Pearlman, S. R. Putt and J. A. Fleming, Olefin Synthesis by Reaction of Stabilized Carbanions with Carbene Equivalents. 1. Use of (Iodomethyl)Tributylstannane for Methylenation of Sulfones, *J. Org. Chem.*, 1985, **50**(19), 3622–3624.

10. (a) Y. Y. Liang, Z. Xu, J. B. Xia, S. T. Tsai, Y. Wu, G. Li, C. Ray and L. P. Yu, For the Bright Future-Bulk Heterojunction Polymer Solar Cells with Power Conversion Efficiency of 7.4%, *Adv. Mater.*, 2010, **22**(20), E135–E138; (b) Y. Y. Liang, D. Q. Feng, Y. Wu, S. T. Tsai, G. Li, C. Ray and L. P. Yu, Highly Efficient Solar Cell Polymers Developed *via* Fine-Tuning of Structural and Electronic Properties, *J. Am. Chem. Soc.*, 2009, **131**(22), 7792–7799.

11. (a) Z. C. He, C. M. Zhong, X. Huang, W. Y. Wong, H. B. Wu, L. W. Chen, S. J. Su and Y. Cao, Simultaneous Enhancement of Open-Circuit Voltage, Short-Circuit Current Density, and Fill Factor in Polymer Solar Cells, *Adv. Mater.*, 2011, **23**(40), 4636–4643; (b) Z. C. He, C. M. Zhong, S. J. Su, M. Xu, H. B. Wu and Y. Cao, Enhanced power-conversion efficiency in polymer solar cells using an inverted device structure, *Nat. Photonics*, 2012, **6**(9), 591–595; (c) C. Liu, K. Wang, X. W. Hu, Y. L. Yang, C. H. Hsu, W. Zhang, S. Xiao, X. Gong and Y. Cao, Molecular Weight Effect on the Efficiency of Polymer Solar Cells, *ACS Appl. Mater. Interfaces*, 2013, **5**(22), 12163–12167; (d) S. J. Liu, K. Zhang, J. M. Lu, J. Zhang, H. L. Yip, F. Huang and Y. Cao, High-Efficiency Polymer Solar Cells *via* the Incorporation of an Amino-Functionalized Conjugated Metallopolymer as a Cathode Interlayer, *J. Am. Chem. Soc.*, 2013, **135**(41), 15326–15329; (e) L. Y. Lu, T. Xu, W. Chen, J. M. Lee, Z. Q. Luo, I. H. Jung, H. I. Park, S. O. Kim and L. P. Yu, The Role of N-Doped Multiwall Carbon Nanotubes in Achieving Highly Efficient Polymer Bulk Heterojunction Solar Cells, *Nano Lett.*, 2013, **13**(6), 2365–2369; (f) H. Q. Zhou, Y. Zhang, J. Seifter, S. D. Collins, C. Luo, G. C. Bazan, T. Q. Nguyen and A. J. Heeger, High-Efficiency Polymer Solar Cells Enhanced by Solvent Treatment, *Adv. Mater.*, 2013, **25**(11), 1646–1652.

12. M. S. Su, C. Y. Kuo, M. C. Yuan, U. S. Jeng, C. J. Su and K. H. Wei, Improving Device Efficiency of Polymer/Fullerene Bulk Heterojunction Solar Cells Through Enhanced Crystallinity and Reduced Grain Boundaries Induced by Solvent Additives, *Adv. Mater.*, 2011, **23**(29), 3315–3319.

13. T. Y. Chu, J. P. Lu, S. Beaupre, Y. G. Zhang, J. R. Pouliot, S. Wakim, J. Y. Zhou, M. Leclerc, Z. Li, J. F. Ding and Y. Tao, Bulk Heterojunction Solar Cells Using Thieno[3,4-*c*]pyrrole-4,6-dione and Dithieno[3,2-*b*:2′,3′-*d*] silole Copolymer with a Power Conversion Efficiency of 7.3%, *J. Am. Chem. Soc.*, 2011, **133**(12), 4250–4253.

14. H. X. Zhou, L. Q. Yang, A. C. Stuart, S. C. Price, S. B. Liu and W. You, Development of Fluorinated Benzothiadiazole as a Structural Unit for a Polymer Solar Cell of 7% Efficiency, *Angew. Chem., Int. Ed.*, 2011, **50**(13), 2995–2998.

15. S. C. Price, A. C. Stuart, L. Q. Yang, H. X. Zhou and W. You, Fluorine Substituted Conjugated Polymer of Medium Bandgap Yields 7% Efficiency in Polymer–Fullerene Solar Cells, *J. Am. Chem. Soc.*, 2011, **133**(20), 4625.

16. M. Wang, X. W. Hu, P. Liu, W. Li, X. Gong, F. Huang and Y. Cao, Donor–Acceptor Conjugated Polymer Based on Naphtho[1,2-*c*:5,6-*c*]bis[1,2,5]

thiadiazole for High-Performance Polymer Solar Cells, *J. Am. Chem. Soc.*, 2011, **133**(25), 9638–9641.

17. T. B. Yang, M. Wang, C. H. Duan, X. W. Hu, L. Huang, J. B. Peng, F. Huang and X. Gong, Inverted polymer solar cells with 8.4% efficiency by conjugated polyelectrolyte, *Energy Environ. Sci.*, 2012, **5**(8), 8208–8214.

18. C. M. Amb, S. Chen, K. R. Graham, J. Subbiah, C. E. Small, F. So and J. R. Reynolds, Dithienogermole As a Fused Electron Donor in Bulk Heterojunction Solar Cells, *J. Am. Chem. Soc.*, 2011, **133**(26), 10062–10065.

19. C. E. Small, S. Chen, J. Subbiah, C. M. Amb, S. W. Tsang, T. H. Lai, J. R. Reynolds and F. So, High-efficiency inverted dithienogermole-thienopyrrolodione-based polymer solar cells, *Nat. Photonics*, 2012, **6**(2), 115–120.

20. L. J. Huo, S. Q. Zhang, X. Guo, F. Xu, Y. F. Li and J. H. Hou, Replacing Alkoxy Groups with Alkylthienyl Groups: A Feasible Approach To Improve the Properties of Photovoltaic Polymers, *Angew. Chem., Int. Ed.*, 2011, **50**(41), 9697–9702.

21. X. H. Li, W. C. H. Choy, L. J. Huo, F. X. Xie, W. E. I. Sha, B. F. Ding, X. Guo, Y. F. Li, J. H. Hou, J. B. You and Y. Yang, Dual Plasmonic Nanostructures for High Performance Inverted Organic Solar Cells, *Adv. Mater.*, 2012, **24**(22), 3046–3052.

22. Y. X. Xu, C. C. Chueh, H. L. Yip, F. Z. Ding, Y. X. Li, C. Z. Li, X. S. Li, W. C. Chen and A. K. Y. Jen, Improved Charge Transport and Absorption Coefficient in Indacenodithieno[3,2-*b*]thiophene-based Ladder-Type Polymer Leading to Highly Efficient Polymer Solar Cells, *Adv. Mater.*, 2012, **24**(47), 6356–6361.

23. Y. Wu, Z. J. Li, W. Ma, Y. Huang, L. J. Huo, X. Guo, M. J. Zhang, H. Ade and J. H. Hou, PDT-S-T: A New Polymer with Optimized Molecular Conformation for Controlled Aggregation and pi-pi Stacking and Its Application in Efficient Photovoltaic Devices, *Adv. Mater.*, 2013, **25**(25), 3449–3455.

24. Y. Dong, X. W. Hu, C. H. Duan, P. Liu, S. J. Liu, L. Y. Lan, D. C. Chen, L. Ying, S. J. Su, X. Gong, F. Huang and Y. Cao, A Series of New Medium-Bandgap Conjugated Polymers Based on Naphtho[1,2-*c*:5,6-*c*]bis(2-octyl-[1,2,3] triazole) for High-Performance Polymer Solar Cells, *Adv. Mater.*, 2013, **25**(27), 3683–3688.

25. H. L. Zhong, Z. Li, F. Deledalle, E. C. Fregoso, M. Shahid, Z. P. Fei, C. B. Nielsen, N. Yaacobi-Gross, S. Rossbauer, T. D. Anthopoulos, J. R. Durrant and M. Heeney, Fused Dithienogermolodithiophene Low Bandgap Polymers for High-Performance Organic Solar Cells without Processing Additives, *J. Am. Chem. Soc.*, 2013, **135**(6), 2040–2043.

26. H. J. Son, L. Y. Lu, W. Chen, T. Xu, T. Y. Zheng, B. Carsten, J. Strzalka, S. B. Darling, L. X. Chen and L. P. Yu, Synthesis and Photovoltaic Effect in Dithieno[2,3-*d*:2′,3′-*d*′]Benzo[1,2-*b*:4,5-*b*′]dithiophene-Based Conjugated Polymers, *Adv. Mater.*, 2013, **25**(6), 838–843.

27. N. Wang, Z. Chen, W. Wei and Z. H. Jiang, Fluorinated Benzothiadiazole-Based Conjugated Polymers for High-Performance Polymer Solar Cells without Any Processing Additives or Post-treatments, *J. Am. Chem. Soc.*, 2013, **135**(45), 17060–17068.

28. W. W. Li, K. H. Hendriks, A. Furlan, W. S. C. Roelofs, M. M. Wienk and R. A. J. Janssen, Universal Correlation between Fibril Width and Quantum Efficiency in Diketopyrrolopyrrole-Based Polymer Solar Cells, *J. Am. Chem. Soc.*, 2013, **135**(50), 18942–18948.

29. M. J. Zhang, Y. Gu, X. Guo, F. Liu, S. Q. Zhang, L. J. Huo, T. P. Russell and J. H. Hou, Efficient Polymer Solar Cells Based on Benzothiadiazole and Alkylphenyl Substituted Benzodithiophene with a Power Conversion Efficiency over 8%, *Adv. Mater.*, 2013, **25**(35), 4944–4949.

30. Z. H. Chen, P. Cai, J. W. Chen, X. C. Liu, L. J. Zhang, L. F. Lan, J. B. Peng, Y. G. Ma and Y. Cao, Low Band-Gap Conjugated Polymers with Strong Interchain Aggregation and Very High Hole Mobility Towards Highly Efficient Thick-Film Polymer Solar Cells, *Adv. Mater.*, 2014, **26**(16), 2586–2591.

31. (a) N. Miyaura and A. Suzuki, Stereoselective Synthesis of Arylated (E)-Alkenes by the Reaction of Alk-1-Enylboranes with Aryl Halides in the Presence of Palladium Catalyst, *J. Chem. Soc., Chem. Commun.*, 1979, **19**, 866–867; (b) N. Miyaura, T. Yanagi and A. Suzuki, The Palladium-Catalyzed Cross-Coupling Reaction of Phenylboronic Acid with Haloarenes in the Presence of Bases, *Synth. Commun.*, 1981, **11**(7), 513–519; (c) N. Miyaura, K. Yamada and A. Suzuki, New Stereospecific Cross-Coupling by the Palladium-Catalyzed Reaction of 1-Alkenylboranes with 1-Alkenyl or 1-Alkynyl Halides, *Tetrahedron Lett.*, 1979, **20**(36), 3437–3440.

32. (a) S. Kotha, K. Lahiri and D. Kashinath, Recent applications of the Suzuki–Miyaura cross-coupling reaction in organic synthesis, *Tetrahedron*, 2002, **58**(48), 9633–9695; (b) P. Lloyd-Williams and E. Giralt, Atropisomerism, biphenyls and the Suzuki coupling: peptide antibiotics, *Chem. Soc. Rev.*, 2001, **30**(3), 145–157; (c) A. F. Littke and G. C. Fu, Palladium-catalyzed coupling reactions of aryl chlorides, *Angew. Chem., Int. Ed.*, 2002, **41**(22), 4176–4211.

33. M. Rehahn, A. D. Schluter, G. Wegner and W. J. Feast, Soluble Poly(*para*-Phenylene)s. 2. Improved Synthesis of Poly(*para*-2,5-Di-Normal-Hexylphenylene) *via* Pd-Catalyzed Coupling of 4-Bromo-2,5-Di-Normal-Hexylbenzeneboronic Acid, *Polymer*, 1989, **30**(6), 1060–1062.

34. A. A. C. Braga, N. H. Morgon, G. Ujaque and F. Maseras, Computational characterization of the role of the base in the Suzuki-Miyaura cross-coupling reaction, *J. Am. Chem. Soc.*, 2005, **127**(25), 9298–9307.

35. N. Miyaura, Cross-coupling reaction of organoboron compounds *via* base-assisted transmetalation to palladium(II) complexes, *J. Organomet. Chem.*, 2002, **653**(1–2), 54–57.

36. N. T. S. Phan, M. Van Der Sluys and C. W. Jones, On the nature of the active species in palladium catalyzed Mizoroki-Heck and Suzuki-Miyaura couplings—Homogeneous or heterogeneous catalysis, a critical review, *Adv. Synth. Catal.*, 2006, **348**(6), 609–679.

37. (a) S. D. Walker, T. E. Barder, J. R. Martinelli and S. L. Buchwald, A rationally designed universal catalyst for Suzuki-Miyaura coupling processes, *Angew. Chem., Int. Ed.*, 2004, **43**(14), 1871–1876; (b) T. E. Barder, S. D.

Walker, J. R. Martinelli and S. L. Buchwald, Catalysts for Suzuki-Miyaura coupling processes: Scope and studies of the effect of ligand structure, *J. Am. Chem. Soc.*, 2005, **127**(13), 4685–4696.

38. A. Zapf, R. Jackstell, F. Rataboul, T. Riermeier, A. Monsees, C. Fuhrmann, N. Shaikh, U. Dingerdissen and M. Beller, Practical synthesis of new and highly efficient ligands for the Suzuki reaction of aryl chlorides, *Chem. Commun.*, 2004, **1**, 38–39.

39. A. F. Littke and G. C. Fu, A convenient and general method for Pd-catalyzed Suzuki cross-couplings of aryl chlorides and arylboronic acids, *Angew. Chem., Int. Ed.*, 1998, **37**(24), 3387–3388.

40. J. P. Stambuli, R. Kuwano and J. F. Hartwig, Unparalleled rates for the activation of aryl chlorides and bromides: Coupling with amines and boronic acids in minutes at room temperature, *Angew. Chem., Int. Ed.*, 2002, **41**(24), 4746–4748.

41. M. F. Liu, Y. L. Chen, C. Zhang, C. H. Li, W. W. Li and Z. S. Bo, Synthesis of thiophene-containing conjugated polymers from 2,5-thiophenebis(boronic ester)s by Suzuki polycondensation, *Polym. Chem.*, 2013, **4**(4), 895–899.

42. M. Svensson, F. L. Zhang, S. C. Veenstra, W. J. H. Verhees, J. C. Hummelen, J. M. Kroon, O. Inganas and M. R. Andersson, High-performance polymer solar cells of an alternating polyfluorene copolymer and a fullerene derivative, *Adv. Mater.*, 2003, **15**(12), 988–991.

43. N. Blouin, A. Michaud and M. Leclerc, A low-bandgap poly(2,7-carbazole) derivative for use in high-performance solar cells, *Adv. Mater.*, 2007, **19**(17), 2295–2300.

44. S. H. Park, A. Roy, S. Beaupre, S. Cho, N. Coates, J. S. Moon, D. Moses, M. Leclerc, K. Lee and A. J. Heeger, Bulk heterojunction solar cells with internal quantum efficiency approaching 100%, *Nat. Photonics*, 2009, **3**(5), 297–302.

45. R. P. Qin, W. W. Li, C. H. Li, C. Du, C. Veit, H. F. Schleiermacher, M. Andersson, Z. S. Bo, Z. P. Liu, O. Inganas, U. Wuerfel and F. L. Zhang, A Planar Copolymer for High Efficiency Polymer Solar Cells, *J. Am. Chem. Soc.*, 2009, **131**(41), 14612–14613.

46. Z. C. He, C. Zhang, X. F. Xu, L. J. Zhang, L. Huang, J. W. Chen, H. B. Wu and Y. Cao, Largely Enhanced Efficiency with a PFN/Al Bilayer Cathode in High Efficiency Bulk Heterojunction Photovoltaic Cells with a Low Bandgap Polycarbazole Donor, *Adv. Mater.*, 2011, **23**(27), 3086–3089.

47. P. L. T. Boudreault, A. Michaud and M. Leclerc, A new poly(2,7-dibenzosilole) derivative in polymer solar cells, *Macromol. Rapid Commun.*, 2007, **28**(22), 2176–2179.

48. E. G. Wang, L. Wang, L. F. Lan, C. Luo, W. L. Zhuang, J. B. Peng and Y. Cao, High-performance polymer heterojunction solar cells of a polysilafluorene derivative, *Appl. Phys. Lett.*, 2008, **92**(3), 033307.

49. J. S. Song, C. Du, C. H. Li and Z. S. Bo, Silole-Containing Polymers for High-Efficiency Polymer Solar Cells, *J. Polym. Sci., Polym. Chem.*, 2011, **49**(19), 4267–4274.

50. Q. Liu, C. H. Li, E. Q. Jin, Z. Lu, Y. C. Chen, F. H. Li and Z. S. Bo, 9-Arylidene-9*H*-Fluorene-Containing Polymers for High Efficiency Polymer Solar Cells, *ACS Appl. Mater. Interfaces*, 2014, **6**(3), 1601–1607.

51. X. Gong, C. H. Li, Z. Lu, G. W. Li, Q. Mei, T. Fang and Z. S. Bo, Anthracene-Containing Wide-Band-Gap Conjugated Polymers for High-Open-Circuit-Voltage Polymer Solar Cells, *Macromol. Rapid Commun.*, 2013, **34**(14), 1163–1168.

52. E. Q. Jin, C. Du, M. Wang, W. W. Li, C. H. Li, H. D. Wei and Z. S. Bo, Dibenzothiophene-Based Planar Conjugated Polymers for High Efficiency Polymer Solar Cells, *Macromolecules*, 2012, **45**(19), 7843–7854.

53. C. Du, W. W. Li, Y. Duan, C. H. Li, H. L. Dong, J. Zhu, W. P. Hu and Z. S. Bo, Conjugated polymers with 2,7-linked 3,6-difluorocarbazole as donor unit for high efficiency polymer solar cells, *Polym. Chem.*, 2013, **4**(9), 2773–2782.

54. C. Du, C. H. Li, W. W. Li, X. Chen, Z. S. Bo, C. Veit, Z. F. Ma, U. Wuerfel, H. F. Zhu, W. P. Hu and F. L. Zhang, 9-Alkylidene-9*H*-Fluorene-Containing Polymer for High-Efficiency Polymer Solar Cells, *Macromolecules*, 2011, **44**(19), 7617–7624.

55. M. Wang, C. H. Li, A. F. Lv, Z. H. Wang and Z. S. Bo, Spirobifluorene-Based Conjugated Polymers for Polymer Solar Cells with High Open-Circuit Voltage, *Macromolecules*, 2012, **45**(7), 3017–3022.

56. (a) L. C. Campeau, D. J. Schipper and K. Fagnou, Site-selective Sp(2) and benzylic Sp(3) palladium-catalyzed direct arylation, *J. Am. Chem. Soc.*, 2008, **130**(11), 3266–3267; (b) M. Lafrance and K. Fagnou, Palladium-catalyzed benzene arylation: Incorporation of catalytic pivalic acid as a proton shuttle and a key element in catalyst design, *J. Am. Chem. Soc.*, 2006, **128**(51), 16496–16497; (c) D. R. Stuart and K. Fagnou, The catalytic cross-coupling of unactivated arenes, *Science*, 2007, **316**(5828), 1172–1175.

57. S. I. Gorelsky, D. Lapointe and K. Fagnou, Analysis of the Palladium-Catalyzed (Aromatic)C–H Bond Metalation-Deprotonation Mechanism Spanning the Entire Spectrum of Arenes, *J. Org. Chem.*, 2012, **77**(1), 658–668.

58. (a) C. Y. Liu, H. Zhao and H. Yu, Efficient Synthesis of 3,4-Ethylenedioxythiophene (EDOT)-Based Functional pi-Conjugated Molecules through Direct C–H Bond Arylations, *Org. Lett.*, 2011, **13**(15), 4068–4071; (b) K. Ueda, S. Yanagisawa, J. Yamaguchi and K. Itami, A General Catalyst for the beta-Selective C–H Bond Arylation of Thiophenes with Iodoarenes, *Angew. Chem., Int. Ed.*, 2010, **49**(47), 8946–8949.

59. M. Sevignon, J. Papillon, E. Schulz and M. Lemaire, New synthetic method for the polymerization of alkylthiophenes, *Tetrahedron Lett.*, 1999, **40**(32), 5873–5876.

60. H. C. Zhao, C. Y. Liu, S. C. Luo, B. Zhu, T. H. Wang, H. F. Hsu and H. H. Yu, Facile Syntheses of Dioxythiophene-Based Conjugated Polymers by Direct C–H Arylation, *Macromolecules*, 2012, **45**(19), 7783–7790.

61. (a) W. Lu, J. Kuwabara, T. Iijima, H. Higashimura, H. Hayashi and T. Kanbara, Synthesis of pi-Conjugated Polymers Containing Fluorinated Arylene Units *via* Direct Arylation: Efficient Synthetic Method of

Materials for OLEDs, *Macromolecules*, 2012, **45**(10), 4128–4133; (b) W. Lu, J. Kuwabara and T. Kanbara, Polycondensation of Dibromofluorene Analogues with Tetrafluorobenzene *via* Direct Arylation, *Macromolecules*, 2011, **44**(6), 1252–1255.

62. (a) Y. Fujinami, J. Kuwabara, W. Lu, H. Hayashi and T. Kanbara, Synthesis of Thiophene- and Bithiophene-Based Alternating Copolymers *via* Pd-Catalyzed Direct C–H Arylation, *ACS Macro Lett.*, 2012, **1**(1), 67–70; (b) P. Berrouard, A. Najari, A. Pron, D. Gendron, P. O. Morin, J. R. Pouliot, J. Veilleux and M. Leclerc, Synthesis of 5-Alkyl[3,4-*c*]thienopyrrole-4,6-dione-Based Polymers by Direct Heteroarylation, *Angew. Chem., Int. Ed.*, 2012, **51**(9), 2068–2071.

63. Q. Guo, J. X. Dong, D. Y. Wan, D. Wu and J. S. You, Modular Establishment of a Diketopyrrolopyrrole-Based Polymer Library *via* Pd-Catalyzed Direct C–H (Hetero)arylation: A Highly Efficient Approach to Discover Low-Bandgap Polymers, *Macromol. Rapid Commun.*, 2013, **34**(6), 522–527.

64. S. W. Chang, H. Waters, J. Kettle, Z. R. Kuo, C. H. Li, C. Y. Yu and M. Horie, Pd-Catalysed Direct Arylation Polymerisation for Synthesis of Low-Bandgap Conjugated Polymers and Photovoltaic Performance, *Macromol. Rapid Commun.*, 2012, **33**(22), 1927–1932.

65. W. Lu, J. Kuwabara and T. Kanbara, Synthesis of 4,4′-dinonyl-2,2′-bithiazole-based copolymers *via* Pd-catalyzed direct C–H arylation, *Polym. Chem.*, 2012, **3**(12), 3217–3219.

66. S. Kowalski, S. Allard and U. Scherf, Synthesis of Poly(4,4-dialkyl-cyclopenta[2,1-*b*:3,4-*b′*]dithiophene-*alt*-2,1,3-benzothiadiazole) (PCPDTBT) in a Direct Arylation Scheme, *ACS Macro Lett.*, 2012, **1**(4), 465–468.

67. (a) Q. F. Wang, R. Takita, Y. Kikuzaki and F. Ozawa, Palladium-Catalyzed Dehydrohalogenative Polycondensation of 2-Bromo-3-hexylthiophene: An Efficient Approach to Head-to-Tail Poly(3-hexylthiophene), *J. Am. Chem. Soc.*, 2010, **132**(33), 11420–11421; (b) Q. F. Wang, M. Wakioka and F. Ozawa, Synthesis of End-capped Regioregular Poly(3-hexylthiophene)s *via* Direct Arylation, *Macromol. Rapid Commun.*, 2012, **33**(14), 1203–1207.

68. J. Kuwabara, Y. Nohara, S. J. Choi, Y. Fujinami, W. Lu, K. Yoshimura, J. Oguma, K. Suenobu and T. Kanbara, Direct arylation polycondensation for the synthesis of bithiophene-based alternating copolymers, *Polym. Chem.*, 2013, **4**(4), 947–953.

69. J. Jo, A. Pron, P. Berrouard, W. L. Leong, J. D. Yuen, J. S. Moon, M. Leclerc and A. J. Heeger, A New Terthiophene-Thienopyrrolodione Copolymer-Based Bulk Heterojunction Solar Cell with High Open-Circuit Voltage, *Adv. Energy Mater.*, 2012, **2**(11), 1397–1403.

70. D. H. Wang, A. Pron, M. Leclerc and A. J. Heeger, Additive-Free Bulk-Heterojuction Solar Cells with Enhanced Power Conversion Efficiency, Comprising a Newly Designed Selenophene-Thienopyrrolodione Copolymer, *Adv. Funct. Mater.*, 2013, **23**(10), 1297–1304.

CHAPTER 2

New Polymer Donors for Polymer Solar Cells

LONG YE[a,b], SUNSUN LI[a,b], AND JIANHUI HOU*[a]

[a]State Key Laboratory of Polymer Physics and Chemistry, Beijing National Laboratory for Molecular Sciences, Institute of Chemistry, Chinese Academy of Sciences, Beijing 100190, P. R. China; [b]University of Chinese Academy of Sciences, Beijing 100049, P. R. China
*E-mail: hjhzlz@iccas.ac.cn

2.1   Introduction

Polymer solar cells (PSCs) have been the subject of extensive study owing to their promising potentials of easy fabrication, high flexibility and light weight when compared with other photovoltaic technologies.[1–4] In the past decade, great achievements have been obtained in developing new active layer materials with broad absorption bands, appropriate molecular energy levels and high mobilities. In order to realize more efficient PSCs, many strategies including interface engineering, morphology control and innovative device architectures have been devoted to optimizing conjugated photovoltaic polymers, thus promoting power conversion efficiencies (PCEs) to new heights.[5–12] Clearly, the application of novel photovoltaic polymers with superior photovoltaic properties is one of the main driving forces in improving the photovoltaic performance of PSCs.[5]

In this chapter, we describe and discuss the recent advances in the polymeric photovoltaic donor materials in PSC devices. First, we summarize

RSC Polymer Chemistry Series No. 17
Polymer Photovoltaics: Materials, Physics, and Device Engineering
Edited by Fei Huang, Hin-Lap Yip, and Yong Cao
© The Royal Society of Chemistry 2016
Published by the Royal Society of Chemistry, www.rsc.org

and discuss the requirements and design strategies of donor polymers in highly efficient PSCs. Design strategies such as backbone tuning, side chain optimization and post-production are introduced along with several excellent examples. Second, the design concepts and representative donor–acceptor (D–A) copolymers will be discussed. We will provide insights into the correlation between device performance and detailed material properties, and rational guidance for molecular designing and fine-tuning of novel photovoltaic polymers. Notably, the blooming developments of novel two-dimensional conjugated polymers are emphasized. Third, novel terpolymers have been introduced, which indicates that conjugated terpolymers exhibit great potential for OPV application. Finally, we summarize the future directions and approaches to developing higher performance donor polymers for photovoltaic applications.

In 1995, Heeger and co-workers introduced the bulk-heterojunction (BHJ) structure by simply blending polymer and fullerene as active materials, which is considered the best PSC device architecture.[13,14] The extensive studies of donor polymer photovoltaic materials have been well discussed and reviewed.[15-25] Although thousands of donor polymers with different backbones and side groups have been developed, synthesized and applied in PSCs, donor polymers can be roughly classified into two types according to their structures, namely the classical and novel donor polymers.

Three types of the first used homopolymer donor materials, *i.e.* the derivatives of poly(1,4-phenylene vinylene) (PPVs), polythiophene (PTs), poly(thienylene vinylene) (PTVs) *etc.*,[26-31] and are regarded as the classical polymer donors (see Scheme 2.1). Among the classical donor polymers, regioregular poly(3-hexyl) thiophene (P3HT) is the most widely studied polymer in the

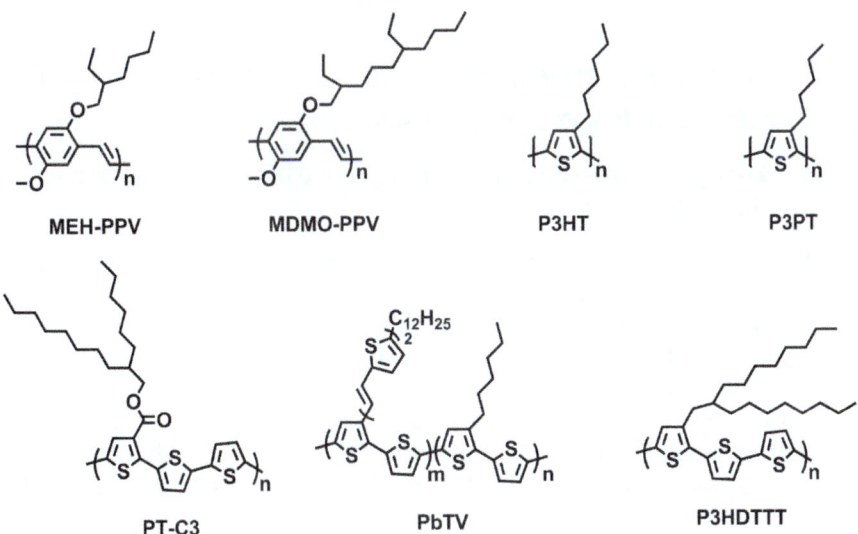

Scheme 2.1   The molecular structures of classical polymer donors.

field of PSC.[30,31] The photovoltaic performance of P3HT/PCBM has been widely studied by numerous groups. It should be noted that the photovoltaic properties of P3HT/PCBM based PSC devices are strongly dependent on the molecular weight and regioregularity of P3HT as well as the device fabrication method.[31] In this regard, various strategies have been developed to optimize the performance of P3HT based PSCs. The breakthrough in the photovoltaic performance of P3HT/PCBM was reported by Yang *et al.*[32] and Heeger *et al.*[33] in 2005, respectively. Yang and co-workers[32] introduced the slow-growth method to optimize the morphology of P3HT/PCBM, and a high PCE of 4.4% was realized, which was the highest value reported in the past decade. As an alternative, Heeger *et al.*[33] utilized a process of post-annealing, and P3HT/PCBM based PSC devices with PCE surpassing 5% were achieved. In 2009, Li *et al.* introduced a novel fullerene acceptor, namely Indene-C_{60} bisadduct (ICBA), with high-lying LUMO level in the P3HT-based PSCs, which promoted the performance of P3HT to a new height.[34] After that, solvent additives and device engineering methods were utilized and promoted the PCE of P3HT/ICBA to over 6%.[35–37] Other P3HT-analogue polymers, for example PbTV,[38] poly(3-pentylthiophene) (P3PT),[39] P3HDTTT,[40] PT-C3,[41] *etc.* were also explored in the device fabrications owing to their optimized molecular energy or absorption characteristics. However, PCEs of ~7% represent the bottleneck of PSCs based on the classical donor polymers owing to the limited absorption ranges and molecular energy levels. In recent years, the novel donor polymers have played important roles in promoting the PCEs of the PSC field. In the following section, we will provide representative examples for donor–acceptor copolymers based on two or more conjugated building blocks. For instance, silole-containing polymers, indacenodithiophene-based polymers, benzodithiophene-based polymers, *etc.* will be introduced and discussed.

2.2 Design Requirements and Strategies for Highly Efficient Polymer Donors

2.2.1 Design Requirements for Highly Efficient Polymer Donors

The basic requirements in the molecular design of high efficiency photovoltaic polymers have been discussed and summarized in several reviews. It is known that the PCE of a PSC is defined as PCE = $(J_{SC} \times V_{OC} \times FF)/P_{in}$, in which J_{SC} is the short-circuit current density, V_{OC} is the open-circuit voltage, FF is the fill factor, and P_{in} is the input power in the form of solar radiation. Consequently, simultaneous optimization of the above-mentioned parameters is necessary to maximize the PCE. The V_{OC} of the PSC can be roughly approximated by the difference between the HOMO level of the donor and the LUMO of the acceptor, even though other factors are also found to influence the V_{OC}. The value of FF can be mainly attributed to the active layer morphology and charge transport through the bulk. The J_{SC} of the solar cell is related to the product of the spectral absorption breadth and the absorption intensity of the active layer. Basically, the photoelectrical conversion of PSCs can be classified into five physical processes: (i) light absorption and exciton generation; (ii) exciton diffusion to the D–A interface; (iii) charge separation; (iv) charge transport; (v) charge collection. From the view of the device physical process, important factors including solubility, absorption, energy level, and mobility as well as morphology should be considered when developing high performance donor polymers for photovoltaic applications.

2.2.1.1 Solubility

The main factor relating to the ease of material processibility and photovoltaic performance is the polymer solubility in common organic solvents such as chloroform (CF), chlorobenzene (CB), and dichlorobenzene (DCB). Good solution processibility can result in good film-forming properties and a desirable penetrating network, which is helpful for achieving high J_{SC} and excellent FF. The degree of solubility of a given polymer is governed by several structural factors, including the degree of polymerization, the chain length of the aromatic groups, the polarity of the attached substituents, backbone rigidity, polymer regioregularity, and intermolecular interactions. Therefore, from the viewpoint of molecular design, appropriate solubility is the first requirement for novel conjugated polymers.

2.2.1.2 Absorption Spectrum

The factors limiting the PCE of the PSCs include the low exploitation of sunlight due to the narrower absorption band of the absorption spectra of conjugated polymers in comparison with the solar irradiation spectrum. To attain high J_{SC}, the absorption spectrum of a novel polymer is needed to match well with the solar radiation spectrum. From a photo-harvesting point of view, the donor polymer should have a broad and intense absorption with a high extinction coefficient (ε), on the order of 10^{-5} cm^{-1}, in order to maximize the incident photon harvest. Often the film or solution extinction coefficient is the preliminary parameter used to evaluate the performance of novel photovoltaic polymers. Narrowing the bandgaps of donor polymers is helpful for absorbing more solar light, which provides the possibility of high J_{SC}. Up to now, great efforts have been devoted to developing low bandgap (LBG) polymers.

2.2.1.3 Energy Level

The value of V_{OC} for a PSC can be expressed by the empirical equation $V_{OC} = [E_{HOMO}(D) - E_{LUMO}(A) - 0.3 \text{ eV}]/e$, where e is the elementary charge, $E_{HOMO}(D)$ is the highest occupied molecular orbital (HOMO) of the polymer and 0.3 eV is an empirical value for efficient charge separation. The lowest unoccupied molecular orbital (LUMO) of [6,6]-phenyl-C_{61}(or C_{71})-butyric acid methyl ester (PCBM) is generally considered to be −3.9 eV. To overcome the exciton binding energy (~0.3 eV), the LUMO level of polymers should be −3.6 eV. In 2006, Scharber *et al.*[42] proposed a simple relation in correlating the HOMO level and band gap of the polymer with the device performance (V_{OC}, PCE) of photovoltaic polymers, which can be used as a guideline for the selection and design of new polymer donors for highly efficient PSCs, towards 10% PCE. Appropriately lowering the HOMO level and reducing the voltage loss, such as ΔLUMO, might be the key to simultaneous achievement of high V_{OC} and J_{SC}.

2.2.1.4   Mobility

In fact, narrowing the band gap alone is not necessarily enough to achieve the expected J_{SC}. Other parameters, such as charge carrier mobility, intermolecular interaction and molecular packing, have influence on J_{SC} as well. Hole mobility (μ_h) is a significant parameter in evaluating the photovoltaic performance of novel donor polymers. In addition, balanced mobility should also be considered because FF is related to the ratio of hole-to-electron mobility of the D–A blends to some extent. Since the typical electron mobility of PCBM is ~10^{-4} cm^2 V^{-1} s^{-1}, the μ_h should be at least 10^{-4} cm^2 V^{-1} s^{-1} for the high performance donor polymers.

2.2.1.5   Morphology

Besides the intrinsic properties of photovoltaic polymers, such as bandgaps and molecular energy levels, morphological properties of the polymer/PCBM blends, including crystallinity, domain size and domain purity, are also of great importance for the photovoltaic performance of the PSC devices.[9–12] The charge transfer, transport and collect are strongly dependent on the bulk and vertical nano-scale morphology of the active layer.[10,11] An optimal morphology is crucial for photovoltaic polymers in achieving high efficiency in BHJ PSC devices.

   In order to achieve ideal morphology in photoactive layers, the blend films can be treated with thermal annealing and solvent-annealing processes. Alternatively, for high-performance polymer:fullerene blends, the film morphology is effectively controlled by using mixed solvent systems. A typical example is the low bandgap polymer poly(diketopyrrolopyrrole-terthiophene) (PDPP3T; Figure 2.1a); Ye *et al.* incorporated three functional solvents in a novel ternary solvent system (Figure 2.1b), which dramatically promoted the PCE up to 6.71%.[12]

   Clearly, to obtain an ideal donor polymer, broader and stronger absorption, high hole mobility, and favorable morphology as well as suitable energy levels should be realized.

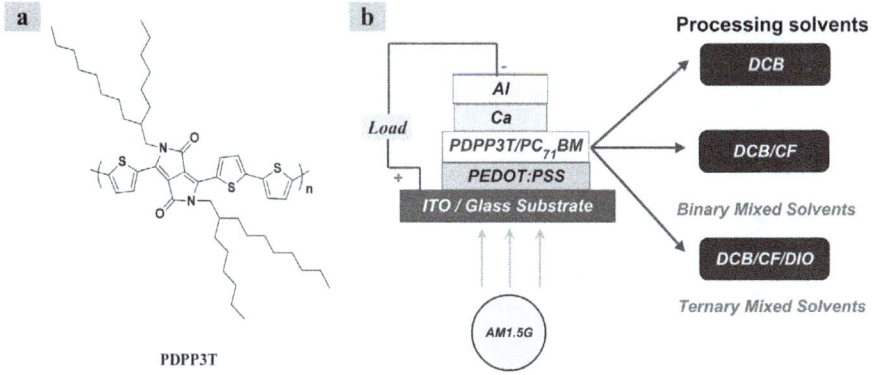

Figure 2.1 (a) The molecular structure of PDPP3T; (b) Ternary solvent processing for PDPP3T-based PSC.

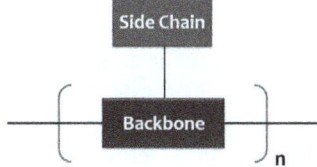

Figure 2.2 The structural composition of a typical photovoltaic polymer.

2.2.2 Design Strategies for Highly Efficient Polymer Donors

To meet the above requirements of highly efficient polymer donors, research-ers have explored various methods to optimize the existing and design novel molecular structures. Typically, the variables of a conjugated polymer can be classified into the backbone and side chain, as depicted in Figure 2.2. Besides the degree of polymerization (n) and the molecular weight, two major design strategies involve optimizing either backbones or side chains.

2.2.2.1 Tuning the Backbone

Usually, planar and rigid backbones are preferable because these types of backbone structure have small reorganization energies and tend to pack closely in solid films through strong intermolecular interactions, and high charge-carrier mobility. Accordingly, low bandgap D–A conjugated polymers which have planar and rigid backbones should be promising for fabricating high performance PSCs.

In this section, several examples, such as introducing novel building blocks, functional substituent or pi-spacers, will be introduced to overview the strategies for backbone design. Benzo[1,2-*b*:4,5-*b*′]dithiophene (BDT) units, first introduced into photovoltaic polymers by Hou and co-workers in 2008,[43] have been used extensively as building blocks as well as electron donor

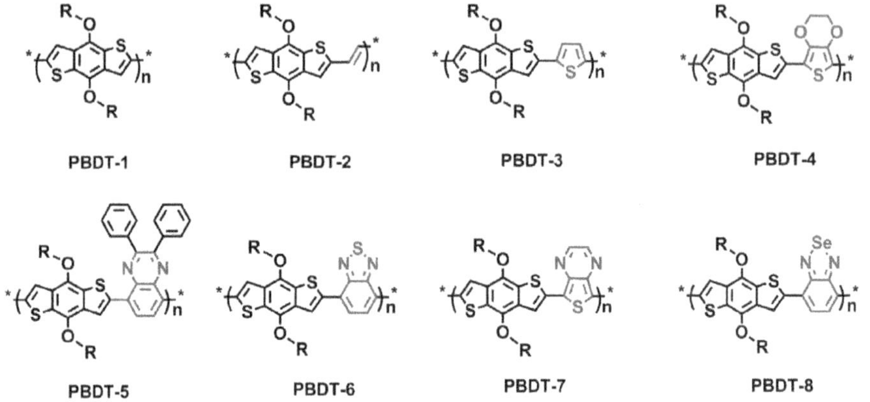

PBDT-1   PBDT-2   PBDT-3   PBDT-4

PBDT-5   PBDT-6   PBDT-7   PBDT-8

Scheme 2.2   Tuning the backbone of BDT-based polymers with tunable acceptor units.

units for conjugated copolymers over the past five years.[44] The optimizations of BDT-based polymers provide excellent examples for the development of backbones. To alter the bandgap and energy level in BDT-based polymers, different electron acceptor units such as thiophene, benzothiadiazole (BT), thieno[3,4-b]pyrazine (TPZ), etc. have been explored (see Scheme 2.2).[43] The bandgaps of these BDT-based polymers are located in the range of 1.1–2.0 eV and their HOMO (−4.65 to −5.16 eV) and LUMO (−2.66 to −3.46 eV) energy levels could also be tuned effectively. The absorption edges were also tuned from 600 nm to 1100 nm. This work provided valuable insights into bandgap and molecular energy level modulation *via* changing backbone structure in conjugated polymers.

To tune the planarity of BDT, a novel building moiety of BDT derivate, dithieno[2,3-d:2′,3′-d′]benzo[1,2-b:4,5-b′]dithiophene (DTBDT), was utilized as a linearly fused conjugated unit.[45] Owing to the linear conformation of the backbone, the copolymer of DTBDT and TT, PDT-S-T, exhibited the best performance, with PCE up to 7.79%, due to the favorable and ordered molecular packing. Another simple strategy is to introduce π bridges such as thiophene, thieno[3,2-b]thiophene, etc. into the backbone of conjugated polymers (see Scheme 2.3). PBDTTT-S-T has zigzagged linear backbones due to the conformational nature of BDT and TT-S. By introducing thiophene as π spacers, the planarity of the PBDTTT-S backbone is improved. As a result, the hole mobility and PCE are significantly improved.[46] Notably, the effect of T-BDT-T is quite similar to that of DTBDT, and the PCEs of PBDTDTT-S and PDT-S-T were both improved to ~7.8%. Guo *et al.* extended the conjugation of π bridges in PBDTBT-T by replacing thiophene with fused thiophene, which exhibited little influence on the molecular energy levels and can be seen as a feasible strategy to improve the photovoltaic properties of D–pi–A conjugated polymers by enhancing intermolecular interactions.[47] A similar trend was also observed in PBDPP-T and PBDTDTBT-T systems.[48,49] These results demonstrated that extending pi bridges might be an effective way to tune the chain curvature and thus promote the photovoltaic performance of D–A conjugated polymers.

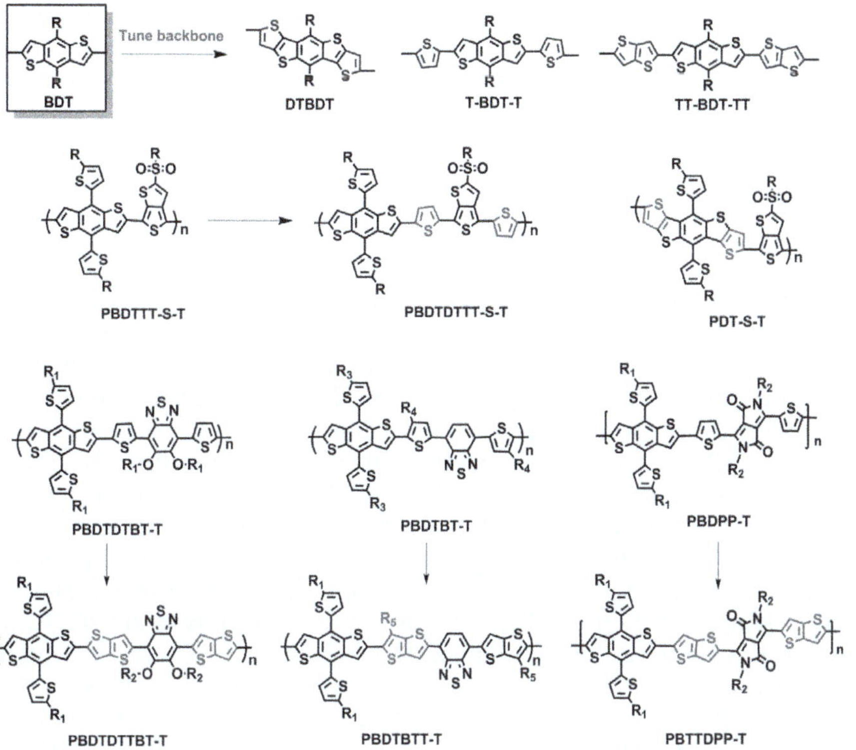

Scheme 2.3 Tuning the backbone of BDT-based polymers with extended pi-conjugation.

Apart from the above methods, changing end-groups is a useful method to tune the backbone, which significantly tunes the surface energy of conjugated polymers. Cho *et al.*[50] tuned the end-groups of poly(3-hexylthiophene) (P3HT), and a variety of end-group functionalized P3HT were developed. From the optimized P3HT–CF$_3$–PCBM blend morphology, they obtained enhanced PCE of 4.5% with a high FF of 69%. Therefore, altering the end functional group of novel donor polymers may be an efficient strategy to promote the overall efficiency.

2.2.2.2 Tuning the Side Chain

The solution-processability of conjugated polymers in organic solvents has classically been achieved by modulating the side chains attached to the well-defined backbone. Side chains have played important roles in the molecular design and fine optimization of photovoltaic polymers, as recently summarized by Pei *et al.* and Bao *et al.*[51,52] The positive roles of an ideal side chain should include optimizing solubility, extending the pi conjugation and thereby tuning molecular packing. In order to obtain good solubility, it is important to have long side chains on the backbone. However, bulky side chains have a negative effect on the carrier mobility, while a side chain with insufficient length limits the solubility and the processability. Therefore, from the viewpoint of molecular design, choosing an appropriate size/shape/substitute/branch position for the side chain is a critical issue for novel conjugated polymers.

The bis-octyl-substituted PFDTBT (BisO-PFDTBT) was previously shown to have poor solubility, and as a result, only a moderate PCE (~2%) was obtained.[53] In 2008, Hou *et al.* tuned the alkyl chains in the classical PFDTBT backbone. In their work, two PFDTBT-based polymers with the same polymer backbone as PFDTBT but different side chains, namely, BisEH-PFDTBT and BisDMO-PFDTBT, were studied to investigate the side-chain effects (see Scheme 2.4).[54] Notably, the saturated alkyl chains have little influence on the molecular energy levels of PFDTBT. From quantum-chemical calculations, both the ethyl groups are in the proximity of the conjugated backbone, and thus decrease the probability of pi–pi stacking in BisEH-PFDTBT originating from the steric effect. On the contrary, there are two small methyl groups located on the third and seventh carbons of BisDMO-PFDTBT, which not only decreased the steric effect, but also increased the solubility and hole mobility.

The substitute position also plays important roles in molecular design. Huang *et al.* reported a pair of D–A type low bandgap polymers based on thieno[3,2-*b*]thiophene and quinoxaline, namely PTTQx-*m* and PTTQx-*p*.[55] Interestingly, only a slight difference in the chemical structure, *i.e.*, whether the alkoxy group is attached to the *para-* or *meta-*position on the peripheral phenyl rings, resulted in a significant difference in their V_{OC}. Relative to the parameters (HOMO, V_{OC}, PCE) of PTTQx-*p*, PTTQx-*m* delivered lower HOMO level, higher V_{OC} and thereby higher PCE, correspondingly. The results provide insights into rationally selecting the optimal substitute position of the alkoxy group and molecular energy level modulation by changing the position of electron-donating side groups. Therefore, the length, branch point, and position are three issues for soluble alkyl chain engineering.

Besides the length, type, branch position and substitute positions of side chains, the terminal groups of alkyl chains could be also tuned. Cao *et al.* introduced polar groups into the conjugated backbone and developed a class of water/alcohol soluble conjugated polymers.[56,57] Qian *et al.* synthesized bromine-functionalized low bandgap polymers, PBDTTT-Br25 and PBDTTT-Br50, for the purpose of photocross-linking and stabilizing the film morphology. The ultraviolet (UV)-sensitive bromine (Br) group attached to the alkyl chain does not disturb the pi–pi stacking of the backbone. As a result, the stability and solvent resistance of the device are enhanced when compared with cells based on noncross-linkable PBDTTT. The PCEs of the PSCs based on PBDTTT-Br25 and PBDTTT-Br50 reached 5.17% and 4.48%, respectively, which were obviously improved in comparison with that (4.26%) of the PSCs based on the control polymer PBDTTT.[58] This approach could offer possibilities for constituting multilayer devices with various functionalities.

The introduction of conjugated side chains is demonstrated to be a useful synthetic strategy for tuning the electronic properties of donor polymers, with wide applicability. In 2011, conjugated side chains such as alkylthienyl

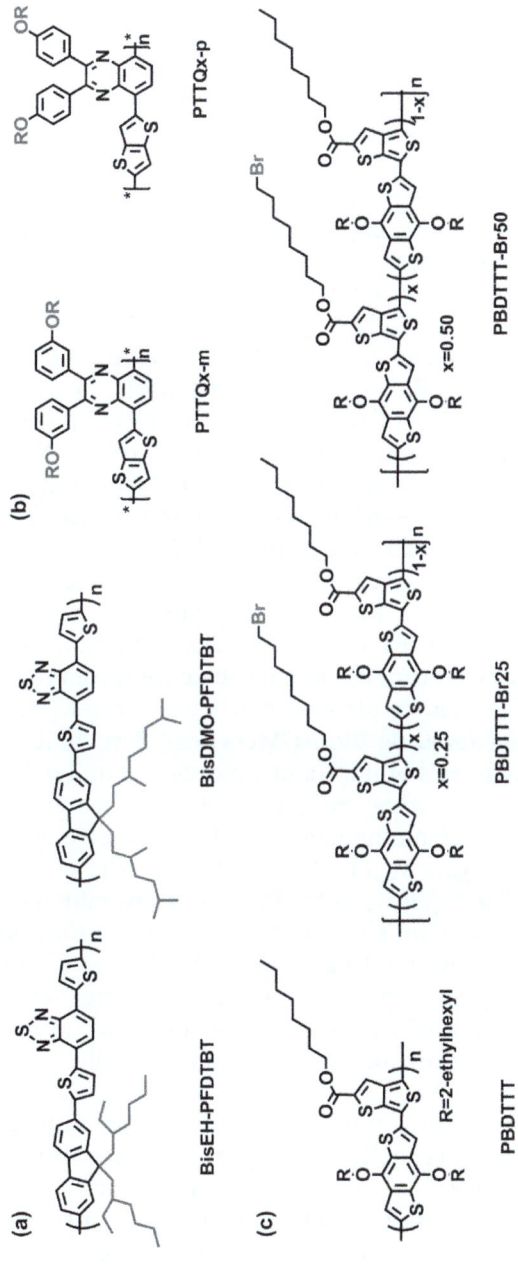

Scheme 2.4 Typical example of tuning the side chains of (a) PFDTBT-based polymers; (b) PTTQ-based polymers; (c) PBDTTT-based polymers.

were successfully introduced into the side chains of BDT units by Hou and co-workers.[59] The novel polymers based on these alkylthienyl-substituted BDT, including PBDTTT-C-T[59] and PBDTDTTT-S-T,[46] delivered improved thermal stabilities, lower HOMO and LUMO energy levels, higher hole mobilities, and significantly improved photovoltaic performance in comparison with those of their corresponding alkoxyl-substituted BDT counterparts.

Introducing functional substituents such as fluorine atoms, or a cyano, carbonyl or sulfonyl group, is also an important approach to tuning the backbone of photovoltaic polymers.[60–65] Take the thieno[3,4-*b*]thiophene (TT) moiety as an example, as shown in Scheme 2.5. The PSC based on the BDT and carboxyl group substituted TT copolymer, *i.e.*, PBDTTT-E, as donor exhibited very promising photovoltaic properties, including high J_{SC} and FF. However, the V_{OC} of PBDTTT-E was only 0.6 V and was the limiting factor of the PCE. Several functional substituents have been incorporated in the PBDTTT based backbones to increase V_{OC}. For instance, Hou *et al.* replaced the ester with a carbonyl group in TT, and the resulting polymer, PBDTTT-C, reached a high PCE of 6.58% with a improved V_{OC} of 0.70 V.[60] Liang *et al.* realized a higher value of V_{OC} (0.74 V) when fluorine was introduced to the thieno[3,4-*b*]thiophene unit, and a PCE of 6.1% was recorded in a PTB4-based polymer.[61] Hou and co-workers optimized PBDTTT-C by incorporating the fluorine atom in the TT unit to lower its HOMO level, thus a high V_{OC} of 0.76 V and an impressive PCE of 7.73% were achieved.[62] Similarly, Yu *et al.* optimized PBDTTT-E with fluorine substitution in the TT moiety and obtained a high PCE of 7.40%.[63] The stepwise molecular evolution from PBDTTT-E to PBDTTT-CF is an excellent example of tuning the molecular energy levels of conjugated polymers. Clearly, the addition of more than one electron-withdrawing group is effective in further lowering the HOMO energy level and increasing the V_{OC}. However, the synthesis of fluorinated monomers and polymers is quite tedious and costly. To avoid the complexity of introducing fluorine, Huang *et al.* developed a novel method by introducing a strong electron-withdrawing group, sulfonyl, into PBDTTT backbone. The resulting polymer, PBDTTT-S, exhibited a higher V_{OC} of 0.76 V and a desirable PCE of 6.22%.[64] Therefore, sulfonyl is a promising candidate to replace the combination of ester (or ketone) and fluorine. In a recent work, Hou *et al.* introduced the cyano group into conjugated polymers to tune the energy levels.[65] It can be concluded that the photovoltaic properties, particularly the energy levels, of conjugated polymers are tunable by introducing suitable functional groups as substituents.

Interestingly, the introduction of fluorine atoms with high electron affinity to donor polymers such as an alkylthienyl substituted BDT moiety is also effective in further lowering both the HOMO and the LUMO energy levels of PBDTTT polymers to attain higher V_{OC}. As listed in Table 2.1, the synergistic effect of fluorination is an excellent example of modulating the molecular energy level and improving the device performance of PBDTTT-based polymers, and realized an excellent PCE of over 8.6% (see Scheme 2.6).[66]

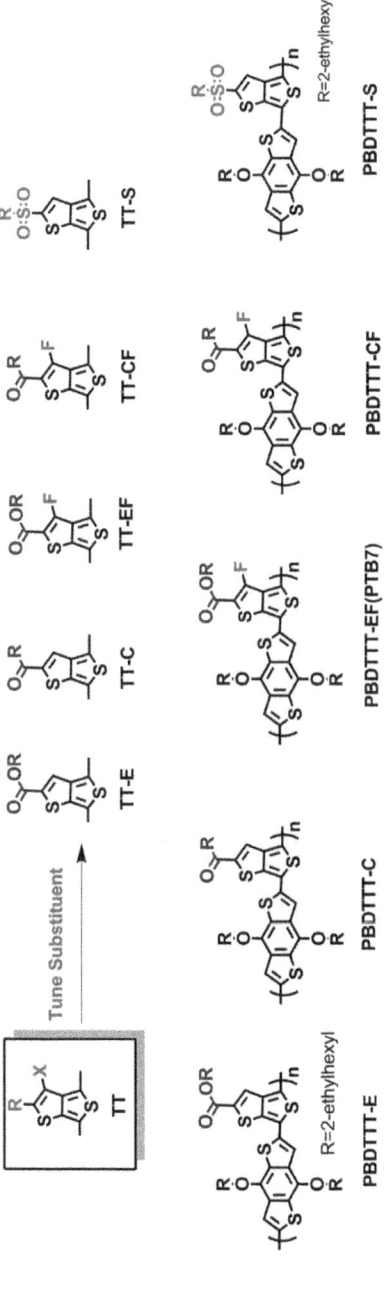

Scheme 2.5 Typical examples of tuning the substituents of PBDTTT-based polymers.

Table 2.1 The photovoltaic results of polymers resulting from side chain modulation.

Material	V_{OC} [V]	J_{SC} [mA cm^{-2}]	FF [%]	PCE [%]	Ref.
BisEH-PCDTBT	0.95	8.4	44	3.5	54
BisDMO-PCDTBT	0.97	9.1	51	4.5	54
PTTQx-*p*	0.60	8.22	59.4	2.93	55
PTTQx-*m*	0.73	9.87	55.1	3.98	55
PBDTTT	0.59	12.97	55.7	4.26	58
PBDTTT-Br25	0.61	14.88	57.0	5.17	58
PBDTTT-Br50	0.60	17.97	50.0	4.48	58
PBDTTT-E	0.62	13.2	63.0	5.15	60
PBDTTT-C	0.70	14.7	64.1	6.58	60
PTB7	0.74	14.50	69.0	7.4	63
PBDTTT-CF	0.76	15.2	66.9	7.73	62
PBDTTT-S	0.76	14.1	58.0	6.3	64
PBT-0F	0.56	12.2	66.7	4.5	66
PBT-1F	0.60	14.3	65.7	5.6	66
PBT-2F	0.74	14.4	67.7	7.2	66
PBT-3F	0.78	15.2	72.4	8.6	66

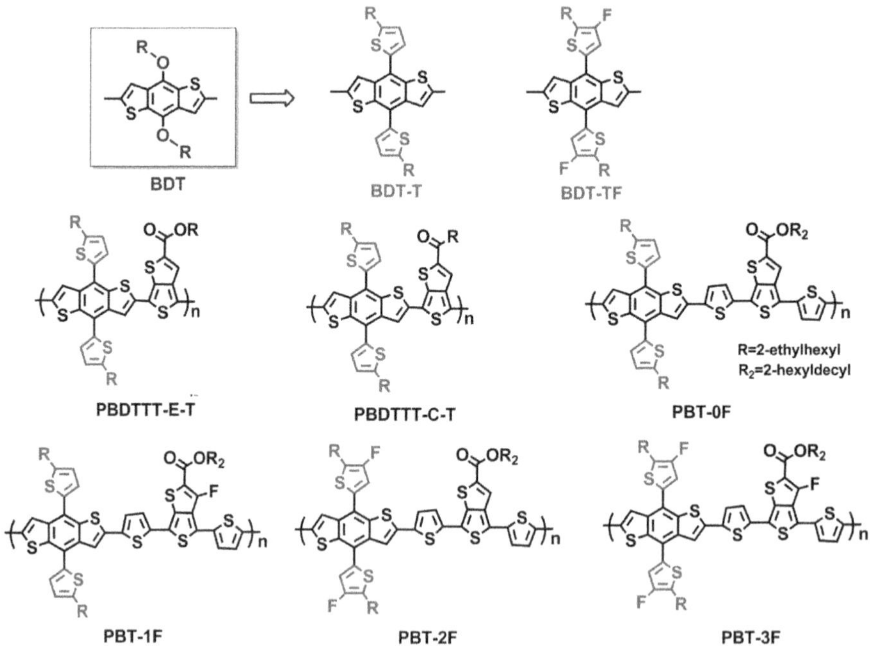

Scheme 2.6 A typical example of tuning the side chains of PBDTTT-based polymers.

2.2.2.3 Post-Production

Apart from tuning the backbone and side chain, post-production is another feasible strategy for achieving high performance.[67–69] Post-polymerization modification is an attractive approach for preparation and optimization of conjugated polymers, allowing incorporation of functionality incompatible

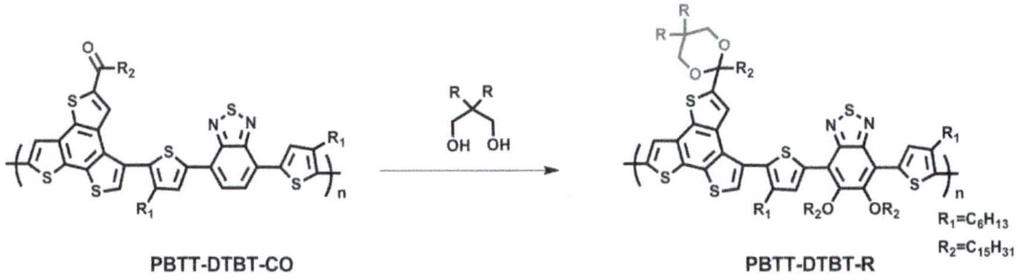

Scheme 2.7 A typical example of post-production of conjugated polymers.

with the polymerization. Several chemical reactions, such as Michael-type addition, radical thiol addition, *etc.*, have been successfully utilized to prepare functional polymers *via* post-production techniques. Moreover, new synthetic concepts such as click chemistry also exhibit potential for postpolymerization modification. Recently, Nielsen and collaborators reported a novel post-polymerization ketalization approach (see Scheme 2.7) and obtained superior processability, leading to significantly enhanced PCEs.[67] This work demonstrated a new and very versatile approach to structural modification of the soluble side chains of donor polymers, therefore allowing quick and efficient assessment of the photovoltaic properties of a polymeric donor candidate.

By tuning the side chains or/and backbone of conjugated polymers, a variety of high-performance photovoltaic polymers have been developed and applied in PSCs in recent years.

2.3 Novel D–A Copolymers for Polymer Solar Cells

2.3.1 Design Considerations for D–A Polymer Donors

Donor–acceptor (D–A) conjugated polymers have become the most successful class of polymer donor materials for PSCs because they usually have a low bandgap and their HOMO and LUMO energy levels can been effectively tuned by choosing donor (D) and acceptor (A) building blocks with appropriate electron donating or accepting strength. Since the pioneer work of Andersson *et al.*,[53] various novel D–A copolymers based on numerous donor units and acceptor units have been synthesized and utilized in PSCs. D–A copolymerization is the most important strategy to broaden the absorption and tune the energy levels of the conjugated polymers. As shown in Figure 2.3, the D–A copolymers possess an intramolecular charge transfer absorption band in the longer wavelength direction, so that the absorption of the copolymers is broadened. Usually, a π-bridge (such as thiophene) is inserted between the donor and acceptor units to reduce the steric hindrance between the donor and acceptor units and to improve the planarity of the copolymers. Moreover, the HOMO level of the copolymers mainly depends on the donor unit, and their LUMO levels are mainly related to the acceptor unit, so the electronic energy levels of the copolymers can be easily tuned by selecting suitable donor and acceptor units. For instance, incorporating electron-withdrawing

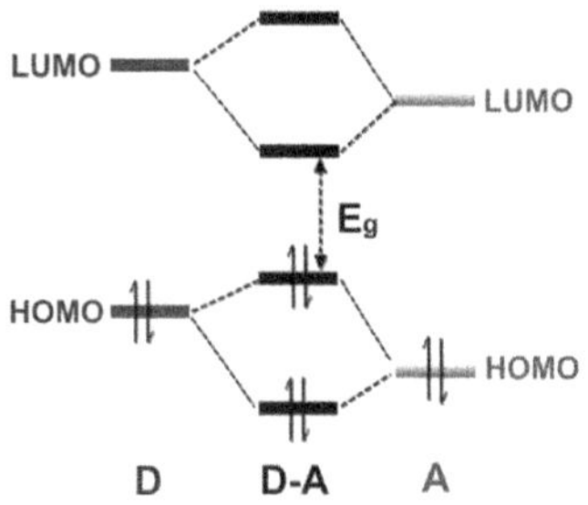

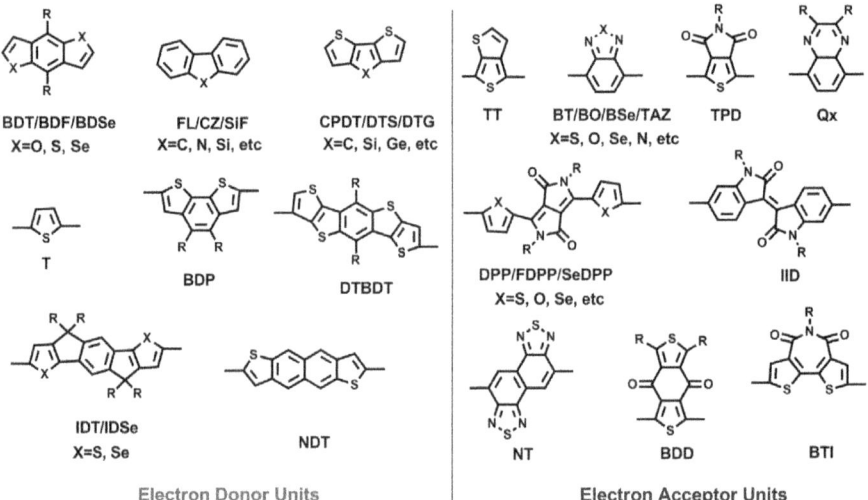

Figure 2.3   The push–pull effect of a D–A copolymer.

Scheme 2.8   The famous building blocks of D–A copolymers.

atoms in the donor unit would lead to a lower HOMO energy level. You *et al.* proposed an empirical rule for the design of high performance D–A copolymers.[70] The employment of weak donors and strong acceptors *via* innovative structural modification achieves a deeper HOMO and a lower bandgap, thereby enhancing both V_{OC} and J_{SC}.

It is known that conjugated copolymers are utilized by different functional building blocks. The common electron donor units and accepter units are drawn in Scheme 2.8. Owing to their flexibility and large number, we can roughly divide the novel D–A copolymers into five types on the basis of the donor units. Herein, the five types of donor unit have been exemplified by a range of thiophene, bridged and fused rings based on thiophene and/ or phenyl. Moreover, the corresponding polymer donors will be reviewed and discussed. In the following section, we will mainly focus on the rapid progress of high efficiency D–A copolymers during the past five years. For every type of donor unit, a series of representative D–A copolymers will be introduced and discussed.

2.3.2 D–A Copolymers Based on Thiophene Units

Although thiophenes or fused thiophenes are the simplest donor units, they play important roles in constituting highly efficient photovoltaic polymers with interesting and superior properties, as depicted in Scheme 2.9. As listed in Table 2.2, most of these polythiophene derivatives exhibit high FF of over 70%.

Diketopyrrolopyrrole (DPP) and its derivatives are strong electron acceptor units, and when copolymerized with thiophene units, deep low bandgap (1.3–1.4 eV) polymers such as PDPP3T, PDPP2FT, and PDPP3MT can be obtained.[71–74] PDPP3T was first synthesized and applied in organic photovoltaic devices by Janssen's group in 2009.[71] After optimization with regard to processing solvent and molecular weight, the moderate PCE of 4.7% was increased to ~7.0% by Hou's group[12] in 2012 and Janssen's group in 2013,[74] respectively. Noticeably, these red absorber materials exhibited potential applications in tandem and triple junction PSCs.[74,75] Thieno[3,4-*c*]pyrrole-4,6-dione (TPD) and TPD analogues such as BDD (benzo[1,2-*c*:4,5-*c'*] dithiophene-4,8-dione) and BTI (bithiopheneimide) also copolymerized with thiophene analogues to actively construct highly efficient polymers such as PBTTPD, PBT1, PTPD3T and PBTI3T. In 2011, Wei *et al.* designed a crystalline polymer PDTTPD, which was constructed by bithiophene (2T) and TPD.[76] A preliminary PCE of 5.0% and V_{OC} up to 0.94 V was achieved for the PBTTPD:PC$_{71}$BM system. Incorporating 1,6-diiodohexane as solvent additive in the blend resulted in the formation of substantially enhanced polymer crystallinity and smaller as well as better dispersed PC$_{71}$BM domains. As anticipated, an improved J_{SC} as high as 12.1 mA cm^{-2} and a PCE of 7.3% were recorded. Marks and co-workers developed two high performance polymers, PTDP3T and PBTI3T.[77] In particular, the terthiophene (3T) and BTI copolymer, PBTI3T, yielded an excellent PCE of over 8.6% and an exceptionally high FF, approaching 80%. The high FF of PBTI3T is comparable to that of their inorganic counterparts. The extremely high FFs originated from the highly ordered, closely packed and properly oriented active-layer microstructures with optimal horizontal phase separation and vertical phase gradation, which is beneficial for efficient charge collection, and eliminated bulk as well as interfacial bimolecular recombination. This work depicted an excellent example of the enhancement of high FF in PSC by integrating complementary materials design, synthesis, processing and device engineering strategies. The π electrons of quaterthiophene (4T) units can be delocalized effectively through the alternative electron push–pull effect, allowing a low bandgap to be realized. PBT1 is a successful example of the use of molecular structure as a tool to realize optimal photovoltaic performance with high polymer content, thus enabling efficient absorption in very thin films.[78] As a result, a high efficiency of 6.88% was recorded in the PBT1-based PSC with a 75 nm active layer. Similar to DPP, isoindigo (IID) is also a strong electron-deficient unit with a rigid and planar structure, and has been used to construct high mobility and low bandgap conjugated polymers for PSCs. In 2011, Wang *et al.* developed a high performance IID-based photovoltaic polymer, P3TI, by

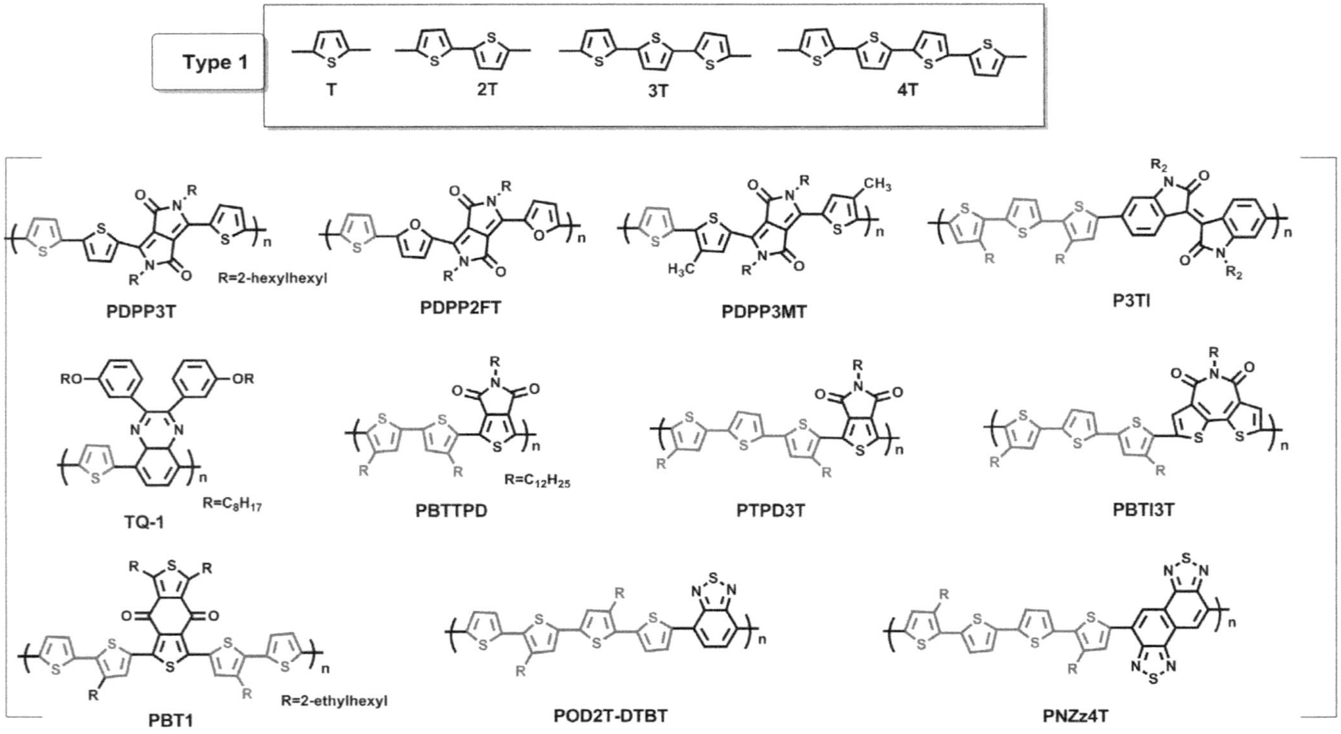

Scheme 2.9   Representative D–A copolymers based on thiophene units.

Table 2.2   Representative D–A copolymers based on thiophene units.

Material	V_{OC} [V]	J_{SC} [mA cm^{-2}]	FF [%]	PCE [%]	Ref.
PDPP3T	0.66	15.41	65.92	6.71	12
PDPP2FT	0.65	14.8	64	6.5	72
PDPP3MT	0.60	17.8	66	7.0	73
PBTTPD	0.92	13.1	61	7.3	76
PTPD3T	0.80	12.5	79.6	7.90	77
PBTI3T	0.86	12.9	77.8	8.66	77
PBT1	0.83	11.57	71	6.88	78
P3TI	0.70	13.1	69	6.3	79
TQ1	0.89	10.5	64	6.0	80
POD2D2T-DTBT	0.72	12.3	70.5	6.26	81
PNTz4T	0.76	12	69	6.3	82

copolymerizing IID with 3T.[79] The PSC-based P3TI/PC$_{71}$BM realized a high internal quantum efficiency (IQE) of 87% and a high PCE of 6.3%. Wang and co-workers also designed a thiophene-based blue polymer, PTQ-1, with PCE up to 6% and V_{OC} of 0.9 V, indicating the potential in tandem PSCs as blue absorbers.[80]

Fused thiophenes also performed well with other acceptor units such as benzothiadiazole (BT) and naphthobisthiadiazole (NT). Given that the strong electron affinity of BT may offer deep HOMO energy levels and broad absorption ranges, in 2011, Chen and co-workers developed a novel low-bandgap polymer, POD2T-DTBT, which was synthesized *via* the Stille method from benzothiadiazole (BT) and bithiophene (2T).[81] They found that POD2T-DTBT give excellent performance in polymer field-effect transistor (PFET) and PSC applications. PFETs based on POD2T-DTBT achieved excellent hole mobilities of 0.2 cm^2 V^{-1} s^{-1} and PSCs based on POD2T-DTBT:PC$_{71}$BM demonstrated promising PCE of 6.26%. In 2012, Osaka *et al.*[82] introduced a doubly BT-fused ring, namely, a naphthobisthiadiazole (NT) unit, into polythiophenes, and also investigated their versatile applications in PFETs and PSCs. Owing to the highly π-extended structure and strong electron affinity of NT, the NT-based polymer (PNTz4T) affords a smaller bandgap and a deeper HOMO level than the BT-based polymer. Interestingly, PNTz4T exhibited not only high field-effect mobilities of ~0.56 cm^2 V^{-1} s^{-1} but also high photovoltaic properties, with PCE up to 6.3%, which originates mostly from a more linear backbone and better molecular ordering. These impressive results demonstrated great promise for the use of thiophene and fused thiophenes as building units for high-performance conjugated polymers for both PFETs and PSCs.

2.3.3   D–A Copolymers Based on Bridged Biphenyl Derivatives

The fluorene (FL) based D–A copolymers may be the first and most classical D–A copolymers based on bridged biphenyl derivatives (see Scheme 2.10). In 2003, Andersson *et al.* reported the photovoltaic properties of a D–A copolymer containing a fluorine (FL) donor unit and BT with a thiophene π-bridge (DTBT); the PCE of the PSC based on PFDTBT:PCBM reached 2.2%.[53]

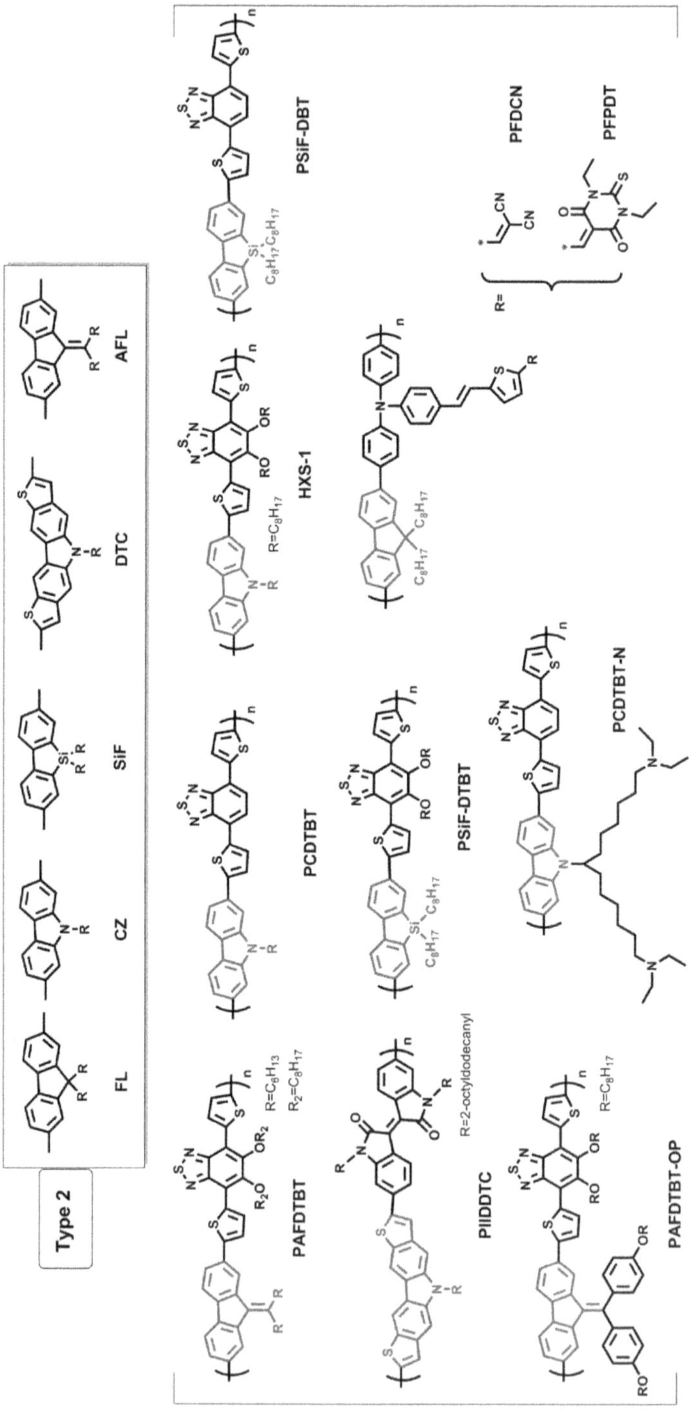

Scheme 2.10 High performance D–A copolymers based on bridged biphenyl analogues.

Although the V_{OC} of this polymer is as high as 1 V, the overall performance is relatively poor. In 2009, Hou *et al.* carefully optimized the side chain of PFDTBT-based polymers, and the result polymer, BisDMO-PFDTBT, achieved a promoted PCE of 4.5%.[54] In 2008, Wang, Cao and co-workers introduced the silicon atom into the FL unit and designed a novel polysilafluorene derivative. By copolymerizing the novel silafluorene (SiF) unit and DTBT, a PCE up to 5.4% was achieved.[83] Bo and co-workers further optimized the photovoltaic performance by attaching two flexible alkoxyl chains to the BT units and realized a PCE over 6%.[84] These results indicated that the silole-containing polymers are promising candidates for high-performance photovoltaic polymers, as listed in Table 2.3. In addition, Bo *et al.* also altered the side groups of FL; the resulting polymer, PAFDTBT, achieved a high PCE of 6.2% and excellent FF of over 70%.[85] Very recently, 9-arylidene-9*H*-fluorene containing PAFDTBT polymers have been synthesized and utilized by Bo's group to study the influence of molecular weight and the position of alkoxy chains on the photovoltaic performance. The best PCE of 6.52% was achieved by PSC based on PAFDTBT-OP with high molecular weight.[86]

Another representative polymer is PCDTBT, which was copolymerized by carbazole (CZ) and DTBT in Leclerc's group in 2007.[87] PCDTBT had a relatively low HOMO level (−5.5 eV) and a moderate bandgap of 1.88 eV. The initial PCE of PCDTBT-based PSC reached 3.6% in a typical bulk-heterojunction PSC and has been subsequently improved by Heeger and co-workers, to above 6%, by thickness optimization and control of the morphology of the BHJ layer.[88] Bo *et al.* further optimized the molecular structure of PCDTBT by attaching two flexible alkoxyl chains to the BT unit and designed a novel planar polymer, HXS-1, with high PCE of over 5%.[89] Recently, Huang *et al.* synthesized an alcohol soluble PCDTBT-N for environmentally friendly solvent processed PSCs by introducing pendant tertiary amino groups in PCDTBT.[90] Although D–A blends containing amino groups exhibited no photovoltaic properties, PCDTBT-N performed well as a buffer layer to improve electron collection in PSCs. Experimental and calculated results revealed that the amino groups act as hole traps and disable hole transport in the active layer and thereby reduce the photovoltaic performance.

Table 2.3   Representative D–A copolymers based on FL units.

Material	V_{OC} [V]	J_{SC} [mA cm^{-2}]	FF [%]	PCE [%]	Ref.
BisDMO-PFDTBT	0.97	9.1	51	4.5	54
PSiFDBT	0.90	9.5	50.7	5.4	83
PSiF-DTBT	0.91	11.5	58	6.05	84
PAFDTBT	0.89	9.9	70	6.2	85
PAFDTBT-OP	0.90	12.72	57	6.52	86
PCDTBT	0.89	6.9	56	3.6	87
HXS-1	0.81	9.6	69	5.4	89
PFDCN	0.99	9.62	50	4.74	92
PFPDT	0.99	9.61	46	4.37	92
PIIDDTC	0.78	15.2	69	8.2	93

Various electron-deficient fused heteroarenes were systematically copolymerized with 2,7-carbazole toward rational design of photovoltaic conjugated polymers by Leclerc *et al.* in 2008.[91] In 2009, Huang, Jen and co-workers designed two new carzole-based polymers. Both exhibit excellent photovoltaic properties with a PCE as high as 4.74%.[92] In contrast to the linear D–A conjugated polymers, these newly designed polymers have a two-dimensional conjugated structure with their tunable acceptors located at the end of D–A side chains and connected to the donors on the main chain through an efficient π-bridge. This design strategy provided great flexibility in fine-tuning the absorption spectra and energy levels of the photovoltaic polymers. A breakthrough in bridged biphenyl units was made by Geng and collaborators.[93] Recently, they designed a five-ring-fused aromatic unit, namely dithieno[3,2-*b*;6,7-*b*]carbazole (DTC), which is structurally related to FL. Optimized by inverted device structures, PIIDDTC achieved an extremely high PCE of 8.2%. Notably, PIIDDTC is the first IID-based polymer with a high PCE beyond 7%, and PIIDDTC is also the first amorphous polymer with a PCE over 8%.

2.3.4   D–A Copolymers Based on Bridged Bithiophene Derivatives

Cyclopenta[2,1-*b*;3,4-*b'*]dithiophene (CPDT) is also a classical D building block, developed by Brabec and coworkers in 2006.[94] In 2007, Bazan *et al.* introduced a solvent additive to optimize the morphology of the CPDT-based polymer PCPDTBT; a two-fold enhancement was observed in PCPDTBT-based PSCs. A high PCE of 5.5% was recorded, and regarded as a milestone in the PSC field.[95] Driven by the high performance of PCPDTBT and PSiF-DBT, Hou *et al.* designed and synthesized a low bandgap silole-containing polymer, PSBTBT, in 2008.[96] As an alternative, Neher *et al.* recently introduced fluorine into PCPDTBT.[97] As a consequence of reduced non-geminate recombination, a high J_{SC} of 14 mA cm^{-2}, an increased V_{OC} of 0.74 V, and a FF of 58% are achieved, affording an impressive PCE of 6.16%. The excellent device performance and the low bandgap render this new polymer highly promising for the construction of efficient polymer-based tandem solar cells. By utilizing 1,3,5-trichlorobenzene as host solvent to manipulate F-PCPDTBT:PC$_{71}$BM morphology, the single-junction PSC based on F-PCPDTBT exhibited both excellent air stability and a high PCE of 6.6%, which represents one of the highest PCEs for CPDT-based PSCs.[98] Benefits from the broad absorption of PSBTBT and F-PCPDTBT, PCE up to 7%, were achieved in Yang's group and Jen's group by utilizing these polymers as efficient red absorbers for tandem PSCs.[98,99]

Afterwards, DTS-based polymers attract much attention in the PSC field (see Scheme 2.11 and Table 2.4). In 2011, Tao *et al.* synthesized a PDTS-TPD and realized a high PCE of 7.3%, which was the highest value among silole-containing polymers.[100] Zhang *et al.*[101] copolymerized DTS with the thiazolothiazole (TTz) acceptor unit and the PSC comprising PDTS-TTz:PC$_{71}$BM as active layers reached a good PCE of 5.6%. Cui *et al.*[102] also designed a strong electron-withdrawing unit, naphtho[2,3-*c*]thiophene-4,9-dione (NTDO),

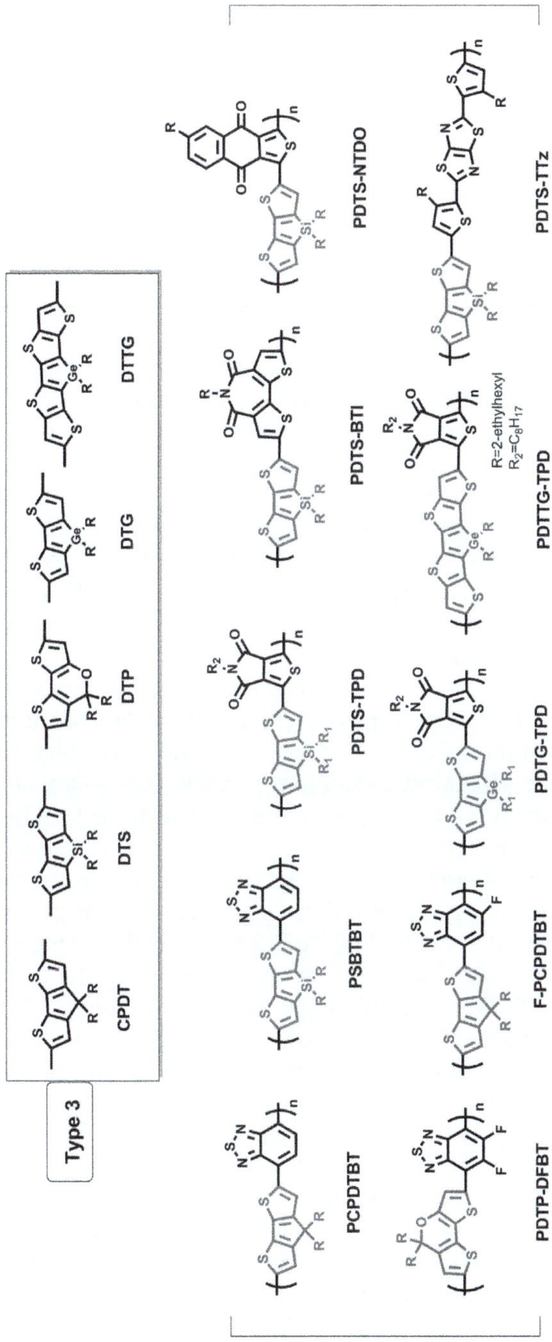

Scheme 2.11 High performance D–A copolymers based on bridged dithiophene analogues.

Table 2.4 Representative D–A copolymers based on DTS analogues.

Material	V_{OC} [V]	J_{SC} [mA cm^{-2}]	FF [%]	PCE [%]	Ref.
PCPDTBT	0.62	16.2	55.0	5.5	95
PSBTBT	0.68	12.7	55	5.1	96
F-PCPDTBT	0.74	14.08	58	6.16	97
PDTS-TPD	0.88	12.2	68	7.3	100
PDTS-TTz	0.77	11.9	61	5.58	101
PDTS-NTDO	0.88	9.24	64	5.21	102
PDTS-BTI	0.80	12.81	62.3	6.41	103
PDTG-TPD	0.85	12.6	68	7.3	104
PDTTG-TPD	0.81	13.85	64	7.2	106
PDTP-DFBT	0.70	18.0	63	8.0	107

which was copolymerized with DTS to construct a low bandgap polymer, PDTS-NTDO. The PCE of the PDTS-NTDO-based PSC reached 5.2% and a high V_{OC} of 0.88 V was realized, due to the low-lying HOMO level. Marks *et al.*[103] synthesized a new series of bithiopheneimide (BTI)-based D–A copolymers for efficient PSCs. Among these, PSCs featuring BTI and DTS copolymer as donor and PC$_{71}$BM as acceptor exhibited promising device performance with PCE up to 6.41% and high V_{OC} of over 0.80 V. The BTI analogue, TPD-based PSC exhibited 0.08 V higher V_{OC} with an enhanced PCE of 6.83%, which was mainly attributed to the lower-lying HOMO induced by the higher imide group density in the backbone.

Dithienogermole (DTG) was introduced by Reynolds and co-workers in order to further improve the intermolecular interactions of silole-containing polymers. The dithienogermole–thienopyrrolodione copolymer PDTG-TPD displayed an absorption shift to 735 nm, and a higher HOMO level than the analogous copolymer containing the commonly utilized DTS heterocycle.[104] When PDTG-TPD was utilized in inverted PSCs, the cells displayed an average PCE of 7.3%, compared with 6.6% for the DTS-containing PSCs prepared under identical conditions. Inverted PSCs based on DTG and TPD copolymer (PDTG-TPD) delivered a certified PCE of 7.4%, which is the highest efficiency reported to date for DTG-based polymers.[105] Following the great success of PDTS-TPD and PDTG-TPD, dithienogermolodithiophene (DTTG) was also incorporated as building blocks in TPD-based polymers owing to the potential for extended conjugation length and improved coplanarity. Very recently, Heeney *et al.*[106] reported the first synthesis of a novel DTG analogue, DTTG, in which two thieno[3,2-*b*]thiophene units are bridge-linked by germanium (Ge). The DTTG and TPD copolymer PDTTG-TPD has a low bandgap of 1.75 eV combined with a low-lying HOMO level. PSC devices comprising PDTTG-TP-D:PC$_{71}$BM as active layers afforded a PCE up to 7.2% without the need for any optimization process. It should be also noted that the study of bridged bithiophene provides considerable synthetic scope for further improvement of performance by altering the bridging atoms, such as Si and Ge.

Recently, a dithieno[3,2-*b*:2′,3′-*d*]pyran (DTP)-containing polymer, PDTP-DFBT, was utilized in photovoltaic polymers and achieved new record PCEs in the PSC field.[107] The electron-donating property of the DTPy unit was found to be the strongest among the most frequently used donor units such as benzodithiophene (BDT) or cyclopentadithiophene (CPDT) units. When the DTPy unit was polymerized with the strongly electron-deficient difluoro-benzothiadiazole (DFBT) unit, a regiorandom polymer (PDTP-DFBT, E_g^{opt} = 1.38 eV) was obtained. In comparison with benzodithiophene (BDT) or cyclopentadithiophene (CPDT) units, the DTP based polymer PDTP-DFBT shows significantly improved solubility and processibility, as well as PCE, when compared with the BDT or CPDT based polymers. Excellent performance in single and double junction solar cells with excellent PCE, respectively reaching 8.0%[107] and 10.6%,[108] demonstrated that the DTP unit is a promising building block for high-performance photovoltaic materials.

2.3.5 D–A Copolymers Based on Benzodithiophene Analogues

Benzodithiophene derivatives have played vital roles in the PSC and related fields in recent years. Among them, benzo[1,2-*b*:4,5-*b'*]dithiophene (BDT) has proved to be the most successful building block for highly efficient photovoltaic polymers (PCE over 7%; see Scheme 2.12). In 2008, Hou *et al.* first introduced the BDT unit to the synthesis and application of low bandgap photovoltaic polymers.[43] Liang *et al.* copolymerized BDT with thieno[3,4-*b*] thiophene (TT) and designed a series of BDT and TT copolymers, namely the PTB series.[61,63,109] Among the PTB series, PTB7 is the best performing photovoltaic polymer and is widely used in device physics as well as in morphology studies.[63] Hou *et al.* also designed a series of BDT and TT copolymers that show high efficiency, such as PBDTTT-CF.[62] It should be mentioned that PTB7[63] and PBDTTT-CF[62] were the first two polymers with PCE exceeding 7%, which pushed PSC research to new heights. Notably, molecular weight is a key factor in achieving high performance in PTB7-based PSCs, especially high J_{SC}. Recently, Gong *et al.*[110] tested the photovoltaic performance of PTB7 with different molecular weights and PCE up to 8.5% was obtained when the M_w was 128 kg mol^{-1}. The PCE enhancement is attributed to the enhanced light absorption and increased charge carrier mobility with high M_w, and a proper phase separation in the BHJ composite. This work demonstrated that the molecular weight of the donor polymer also plays an important role in the photovoltaic performance of PSCs.

Following the pioneering work on PBDTTT-CF and PTB7, PCEs over 7% were frequently reported with either new materials or novel device optimization techniques, as summarized in Table 2.5. Besides TT, numerous building blocks, including thieno[3,4-*c*]pyrrole-4,6-dione (TPD),[111-115] FTAZ,[116] benzothiadiazole (BT),[117,118] TFQ,[119] benzoxadiazole (BO),[120] and diketopyrrolopyrrole (DPP)[121-123] have been successfully combined with the BDT unit to constitute highly efficient polymers. For instance, Leclerc *et al.* synthesized BDT and TPD copolymers with PCE up to 5.5%.[111] Almost simultaneously, the same polymer was published by the Jen group, Xie group, and Fréchet

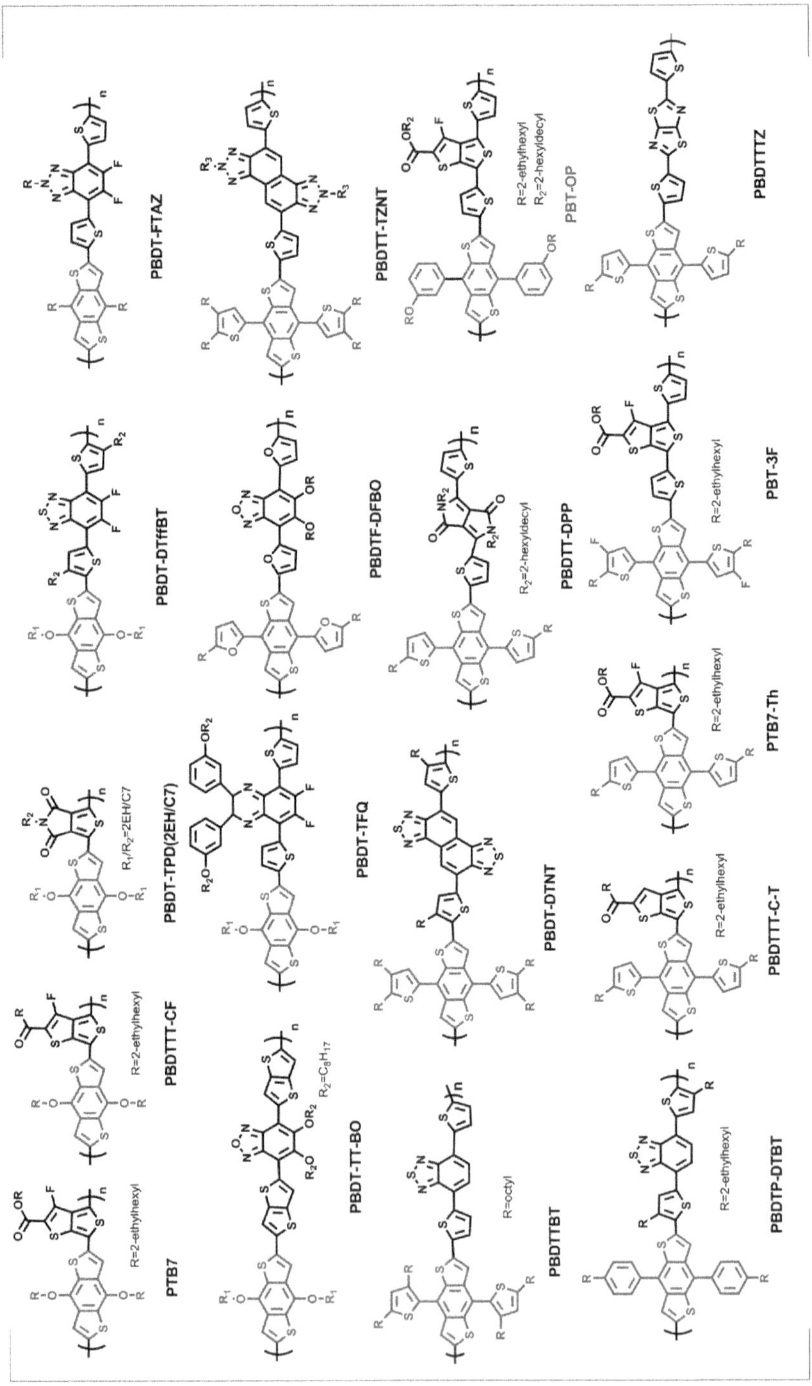

Scheme 2.12 High performance D–A copolymers based on BDT.

Table 2.5 Representative D–A copolymers based on BDT.

Material	V_{OC} [V]	J_{SC} [mA cm^{-2}]	FF [%]	PCE [%]	Ref.
PBDTTT-CF	0.76	15.2	66.9	7.73	62
PTB7	0.74	14.5	68.97	7.4	63
PBDTTPD	0.97	12.6	70	8.5	115
PBDT-FTAZ	0.79	12.45	72.2	7.1	116
PBnDTDTffBT	0.91	12.91	61.2	7.2	117
PBDTDTBTff	0.78	15.38	69.2	8.3	118
PBDT-TFQ	0.76	18.2	58.1	8.0	119
PBDT-TT-BO	0.76	13.87	66.6	7.05	120
PBDTTT-C-T	0.74	17.48	58.7	7.59	59
PBDTT-SeDPP	0.69	16.8	62.0	7.2	122
PBDTTBT	0.92	10.7	57.5	5.66	128
PBDTTTZ	0.85	10.4	59.0	5.22	129
PBDT-DTNT	0.80	11.71	61.0	6.0	131
PBDT-TZNT	0.90	10.79	62.0	6.10	132
PBDTP-DTBT	0.88	12.94	70.9	8.07	133
PBDTBDD	0.86	10.68	72.27	6.67	134
PTB7-Th	0.79	14.02	69.1	9.35	138
PBDTF-DFBO	0.83	12.7	62.0	7.0	139

group.[112–114] After optimization with side alkyl chains in both of the BDT and TPD units, a high PCE of over 8.5% was realized by Fréchet and co-workers.[115] You *et al.* designed several efficient BDT-containing polymers, namely, PBnDT-FTAZ[116] and PBDTDTffBT.[117] Interestingly, the broad bandgap polymer PBnDT-FTAZ exhibited PCE above 7% when blended with PC$_{61}$BM, and PCE above 6% are still obtained even at thicknesses up to 1 μm.[116] The superior performance originates from their high hole mobility and low HOMO levels. Very recently, alkyl chain engineering was successfully utilized in PBDT-DTffBT, which dramatically promoted the PCE of PBDT-DTffBT up to 8.3%.[118] A medium bandgap fluorinated quinoxaline-based polymer, PBDT-TFQ, was also designed and synthesized.[119] A high PCE of 8.0% was obtained, with a V_{OC} of 0.76 V, an extremely high J_{SC} of 18.2 mA cm^{-2}, and a FF of 58.1% in PBDT-TFQ:PC$_{71}$BM based PSCs. The resulting copolymer revealed an extremely high J_{SC}, arising from the higher hole mobility of PBDT-TFQ, together with better morphology for efficient exciton dissociation and charge transport. Recently, a TT-bridged polymer, P(BDT-TT-BO), featuring a BO acceptor unit was designed and synthesized by Li and co-workers.[120] The PCE of PSC with PBDT-TT-BO as donor reached 7.05%, which is the highest result in benzoxadiazole-containing conjugated polymers and comparable to that of their benzothiadiazole counterparts. The photovoltaic performance of BDT-based polymers was further optimized by several research groups from the device aspect.[124–127] For instance, by employing the polymer interlayer as cathodic buffer layer, PTB7-based PSC devices with PCE up to 9.2% were realized by Cao and co-workers.

Recently, Hou *et al.* proposed a novel approach to enhance the pi-conjugation of the BDT unit at two dimensions.[59,128] A novel and large class of copolymers based on the two-dimensional conjugated BDT units (collectively called BDT-Ar units) were developed.[129–135] By introducing the BDT-Ar units, the PCE of several novel polymers, including PBDTTT-C-T, PBDTTDTT-S-T, PDT-S-T, PBDTP-DTBT, and PBT-3F, was increased to 8–9% by the Hou group. The PBDTP-DTBT based PSCs exhibited a PCE up to 8.07% with a $V_{OC} = 0.88$ V, a $J_{SC} = 12.94$ mA cm^{-2} and a FF = 70.9% under irradiation with air mass 1.5G and 100 mW cm^{-2}. Interestingly, with only 0.5% of DIO by volume, the PBDTP-DTBT based D–A blends exhibited the best performance, due to the well-tuned morphology.[133] To ameliorate the relatively low voltage in alkylthienyl substituted BDT and thiophene bridged TT based copolymers, Hou *et al.* further demonstrated a synergistic effect of introducing fluorine atoms in lowering the molecular energy levels, and the HOMO and LUMO levels, of copolymers of alkylthienyl substituted BDT and thiophene bridged TT. When three fluorine atoms were introduced to both the donor and the acceptor unit, the PSC device based on the trifluorinated polymer (PBT-3F) demonstrated an extremely high PCE of 8.6% and significantly improved V_{OC} of 0.78 V, which held the efficiency record for PSC utilizing the conventional device configuration.[66] Choy *et al.* employed a dual plasmonic approach and increased the PCE of PBDTTT-C-T to 8.8%.[136] Huang *et al.* also developed two high performance polymers, PBDT-DTNT[131] and PBDT-TZNT,[132] which were copolymerized by BDT-Ar with DTNT and DTNZ, respectively. Afterwards, Gong *et al.* improved the PCE of PBDT-DTNT to 8.4% through inverted architecture.[137] Chen and co-workers recently developed an efficient inverted PSC based on a novel BDT-Ar polymer, PTB7-Th, which obtained a record PCE of 9.35% with interfacial optimization.[138] Ge *et al.* incorporated alkylfuranyl as conjugated side groups in BDT-based polymers and designed a high performance polymer, PBDTF-DFBO, which is the best-performing furan-containing polymer to date.[139] Improvements to the newly developed small bandgap polymers based on BDT-Ar in tandem PSC devices have raised PCE to ~9.5% in Yang's group.[121–123] Notably, when combined with BDT-Ar based polymers, PSCs employing bis-PCBM[140] and a non-fullerene acceptor[141] achieved 6.07% and 4.03%, respectively. This work demonstrates that BDT-Ar based polymers exhibit wide applicability in various photovoltaic devices.

Inspired by the great success of BDT, various BDT analogues, such as BDF, BDSe, NDT, and NNT, were designed and synthesized for constituting D–A copolymers (see Scheme 2.13).[142–147] Hou *et al.* designed and introduced a benzo[1,2-*b*:4,5-*b'*]difuran (BDF) unit into D–A copolymers. Given that the dihedral angle of BDF is lower than that of BDT, the steric hindrance is significantly reduced in BDF.[142] By introducing conjugated side chains into the BDF unit, a high PCE of up to 6.26% was achieved in PBDFTTCF-T based polymers (see Table 2.6).[143] Recently, Hou *et al.* designed a novel BDT analogue, namely DTBDT, and synthesized a series of DTBDT based copolymers. Among these, PDT-S-T exhibited a dramatic PCE of 7.79%.[144] Yu *et al.* copolymerized DTBDT and TT, and a highly efficient polymer, PTDBD2, was obtained with a high V_{OC} of ~0.9 V.[145] Similarly, naphthodithiophene (NNT) has an extended pi-system and larger planar heteroarene structure; this may form highly ordered pi–pi

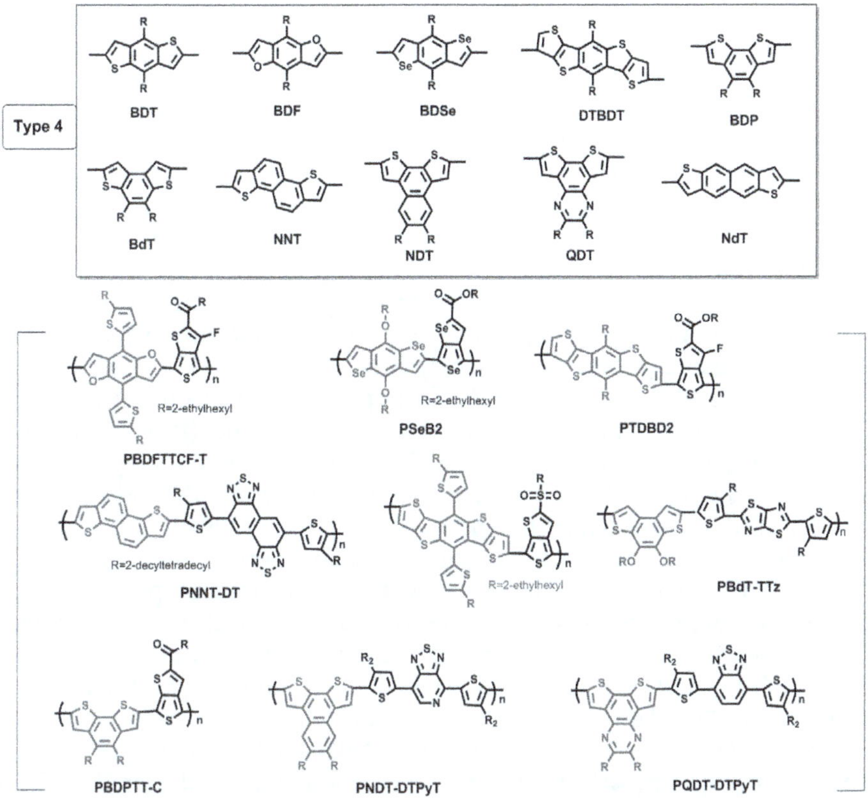

Scheme 2.13 High performance D–A copolymers based on BDT derivatives.

Table 2.6 Photovoltaic results of high performance D–A copolymers based on BDT derivatives.

Material	V_{OC} [V]	J_{SC} [mA cm^{-2}]	FF [%]	PCE [%]	Ref.
PBDFDTBT	0.78	11.77	54.6	5.01	142
PBDFTTCF-T	0.78	13.04	61.55	6.26	143
PDT-S-T	0.73	16.63	64.13	7.79	45
PSeB2	0.64	16.8	64	6.87	144
PTDBD2	0.89	13.0	65.3	7.6	145
PNDT-DPP	0.76	13.34	68	6.92	146
PNNT-BT	0.82	15.6	69	8.2	147
PBDPTT-C	0.82	11.76	54.1	5.21	148
PBdT-TTz	0.90	5.50	68.7	3.40	149
PNDT-DTBT	0.64	13.50	62	5.27	150
PNDT-DTPyT	0.71	14.16	61.7	6.20	150
PQDT-DTPyT	0.75	13.49	55.1	5.57	150

stacking between the polymer backbones and reduce steric repulsion, which is beneficial for charge separation and transport. Introducing linear alkyl chains improved the solubility as well as giving rise to a change in the orientational order with no alteration of the energy levels, which resulted in quite impressive PCEs of over 8% in a conventional single-junction PSC device.[147] Surprisingly, the introduction of linear alkyl chains led to a drastic change in polymer orientation to the face-on motif, which was beneficial for charge transport in solar cells and promoted the photovoltaic performance. In addition, PSCs based on NNT–BT yielded PCE as high as 8.2% with thickness up to 300 nm. These results indicated that this polymer platform is of particular interest in the understanding of molecular packing and carrier transport in high performance photoelectronic polymers.

As isomers of benzo[1,2-*b*:4,5-*b'*]-dithiophene (BDT), benzo[2,1-*b*:3,4-*b'*] dithiophene (BDP) and benzo[1,2-*b*:4,3-*b'*]-dithiophene (BdT) have been shown to be useful conjugated building blocks in photovoltaic polymer materials (see Scheme 2.13). For instance, by copolymerizing BDP and TT, a high PCE of 5.21% was obtained for the resulting polymer, PBDPTT-C.[148] Zhang *et al.* have synthesized a variety of BdT-based polymers.[149] Among these, the BdT and TTz copolymers achieved a PCE of 3.4%. You *et al.* proposed a design strategy of "weak donor–strong acceptor" to develop ideal polymers with both a low HOMO level and a small bandgap for PSCs, in order to achieve both high V_{OC} and high J_{SC}.[70] Two "weak donors", naphtho[2,1-*b*:3,4-*b'*]dithiophene (NDT) and dithieno[3,2-*f*:2',3'-*h*]quinoxaline (QDT), were copolymerized with DTPyT; the BHJ devices based on PNDT-DTPyT and PQDT-DTPyT as the donor polymer (with $PC_{61}BM$ as the acceptor) exhibited overall efficiencies of 6.2% and 5.6%, respectively.[150]

2.3.6   D–A Copolymers Based on Indacenodithiophene Analogues

Owing to the dramatic interest in BDT, a high level of fusion of the aromatic on the BDT unit, such as indacenodithiophene (IDT), has also been actively explored, as shown in Scheme 2.14. The coplanar and rigid geometries of IDT-like ladder units can suppress the rotational disorder and lower the reorganization energy, which leads to higher charge mobility.[151] Moreover, these units facilitated π-electron delocalization and thereby increased the effective conjugation length, offering an effective approach to reducing the bandgap as well as tuning molecular energy levels. As the most successful ladder-type unit, IDT has been widely used in organic field-effect transistors and organic solar cells in recent years. Ting and collaborators synthesized random and alternating copolymers based on IDT and BT, and the corresponding PCE reached 6.4%.[152,153] Jen *et al.* did systematic work on IDT-based PSCs and developed several high performance IDT based D–A copolymers, including PIDT-diphQ, PIDT-phanQ and PIDT-DFBT (see Table 2.7).[154,155] Obviously, the PIDT-phanQ polymer showed greater hole mobility than PIDT-diphQ owing to the enhanced planarity of the phenanthrenequninoxaline unit. Huang *et al.*

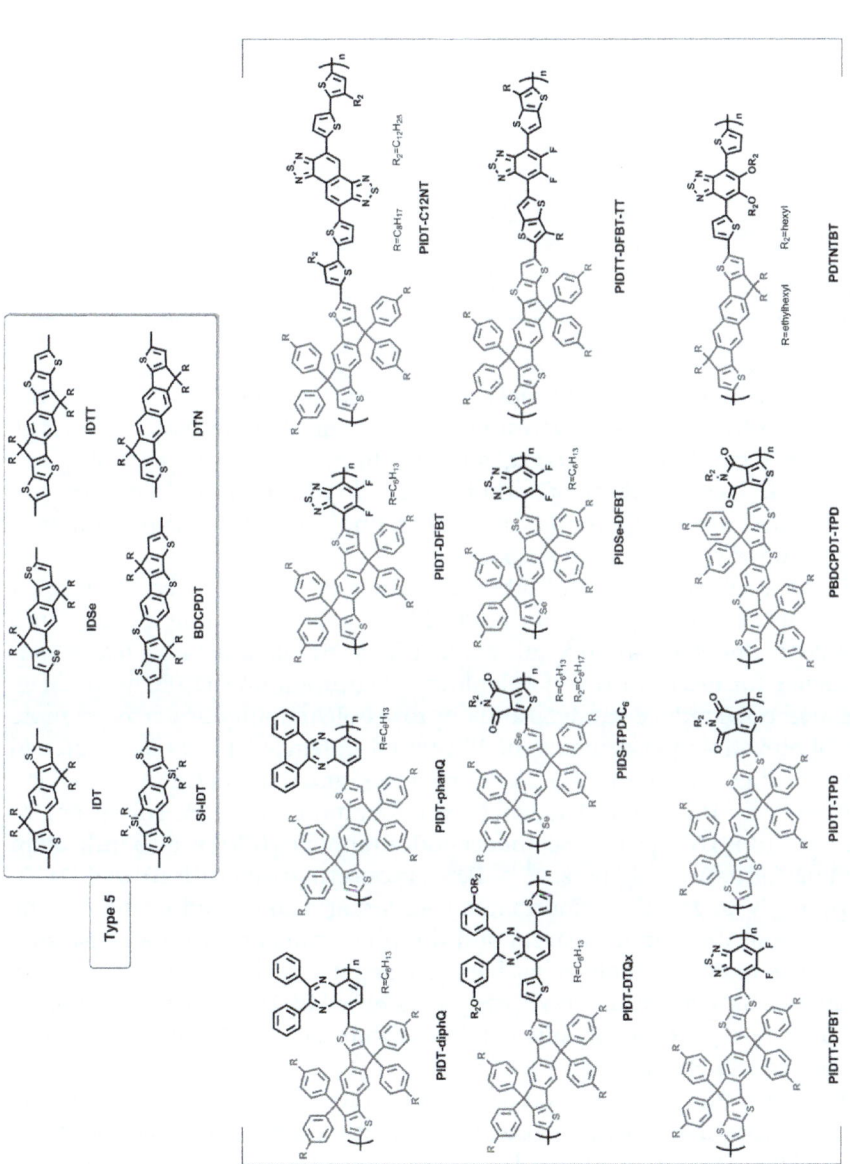

Scheme 2.14 High performance D–A copolymers based on IDT and its derivatives.

Table 2.7   Photovoltaic results of D–A copolymers based on IDT analogues.

Material	V_{OC} [V]	J_{SC} [mA cm^{-2}]	FF [%]	PCE [%]	Ref.
PIDT-diphQ	0.87	10.9	60	5.69	154
PIDT-phanQ	0.87	11.2	64	6.2	154
PIDT-DFBT	0.97	11.2	55	5.97	155
PIDT-DTQx	0.87	12.34	70.23	7.51	156
PIDT-C12NT	0.90	10.21	55	5.05	157
PIDS-TPD-C$_6$	0.92	9.77	51	4.6	158
PIDSe-DFBT	0.89	13.7	56.3	6.8	159
PIDSi-BT	0.88	9.93	52	4.5	162
PIDTT-TPD	0.90	7.99	60	4.3	163
PIDTT-DFBT	0.95	12.21	61	7.03	164
PIDTT-DFBT-TT	0.96	11.9	63	7.2	166
PDTNTBT	0.84	9.96	60	5.0	168
PBDCPDT-TPD	0.87	12.21	62.2	6.6	169

synthesized IDT with naphtho[1,2-c:5,6-c]bis(1,2,5-thiadiazole) (NT); the resulting polymer, PIDT-C12NT, exhibited a desirable PCE of 5.05%.[156] Hou *et al.* copolymerized IDT with DTQx, and the highest PCE, up to 7.5%, was achieved by a combination of optimizing the donor:acceptor (D:A) ratio and the thermal annealing conditions. Interestingly, the optimal performance was obtained when the D:A ratio was 1:4.[157]

Carbon, selenophene, and other analogues of IDT have been developed in recent years. Cheng *et al.* introduced indacenodiselenophene (IDSe) into D–A copolymers for photovoltaic applications and obtained a moderate performance for PIDT-TPD-C$_6$.[158] Selenium substitution on the IDT is an effective method to reduce bandgap and improve hole mobility, and may promote the photovoltaic performance of IDT-based polymers. Jen *et al.* improved the molecular weight of IDSe-based polymers, and a high PCE of up to 6.8% was realized for PIDSe-DFBT, which exhibited more than 10% enhancement over that of PIDT-DFBT.[159] Silaindacenodithiophene (IDSi) is also utilized in high performance polymers.[160–162] IDSi is copolymerized with BT and DTBT, respectively, to give their fluorinated counterparts DFBT and DTffBT.[162] The influence of the thienyl spacers and fluorine atoms on molecular packing and active layer morphology has been investigated with regard to device performance. Bulk heterojunction (BHJ) PSCs based on IDSi polymers achieved a PCE of 4.5%, and the organic field-effect transistor (OFET) hole mobilities as high as 0.28 cm^2 V^{-1} s^{-1} are achieved.

Recently, a novel ladder-type donor (IDTT) was developed by substituting the two outward thiophenes of the IDT unit with two thieno[3,2-b]thiophenes. The polymer derived from IDTT possesses longer effective conjugation and better planarity, which may improve both electron delocalization along the polymer backbone and charge mobility. Cheng *et al.* copolymerized IDTT with different electron-deficient acceptors, including BT, FBT and TPD, respectively.[163] The highest PCE of 4.3% was achieved for the PSC device incorporating PIDTT-TPD:PC$_{71}$BM (1:4, wt%) as the active layer. Jen *et al.*

developed a high performance IDTT-based polymer, PIDTT-DFBT, which exhibited a high PCE of 7.03% with a large V_{OC} of 0.95 V without any device optimization.[164] Impressively, Jen and co-workers also demonstrated for the first time that PCE exceeding 7% can be processed from non-halogenated cosolvents in both conventional and inverted PSC devices of PIDTT-DFBT.[165] Further optimization of PIDTT-DFBT was realized through incorporation of an octylthieno[3,2-*b*]thiophene π-bridge between an IDTT donor unit and DFBT acceptor unit. With incorporation of a fullerene–self-assembled monolayer (SAM) modified ZnO as the n-type layer and a GO/MoO$_3$ bilayer as the p-type layer, inverted PSCs based on PIDTT-DFBT-TT showed a higher PCE of 7.3% as well as improved V_{OC} of 0.97 V.[166] Very recently, Jen *et al.* developed a flexible PSC based on PIDTT-DFBT in a microcavity configuration using a TeO$_2$/Ag semitransparent electrode to confine the optical field within the device, with significant performance improvements and reaching a impressive PCE of 8.56%, indicating the great potential of IDT-based photovoltaic polymers.[167]

Similarly, Ma *et al.* developed a novel ladder-type unit, dithienonaphthalene (DTN). Conventional PSC based on PDTNTBT:PC$_{71}$BM (1 : 3, wt/wt) exhibited a PCE of 4.8%.[168] By utilizing an inverted PSC device, the PCE was improved to 5.0%. Cheng, Hsu and co-workers integrated the CPDT and BDT moieties into a coplanar conjugated unit and developed a new ladder-type unit, namely BDCPDT.[169] Owing to the coplanar, symmetrical, and extended conjugated structure of BDCPDT, strong donor–acceptor interactions along the conjugated backbone can be preserved, leading to more red-shifted and broader absorption spectra and higher hole mobility. Utilizing DMSO as the processing additive and MoO$_3$ as the p-type buffer layer, the PSC device based on PBDCPDT-TPD achieved a high PCE of 6.6%, which is one of the highest performances among TPD-based copolymers.[169] Soon after, an inverted tandem PSC incorporating a BDCPDT-based polymer achieved a V_{OC} of 1.62 V and a PCE of 7.08%.[170]

2.4 Novel Terpolymer Donors for Polymer Solar Cells

2.4.1 Design Considerations for Terpolymer Donors

It is well known that D–A copolymers have been investigated intensively with the aim of matching a greater part of the solar spectrum, and thus harvesting more photon flux, to produce high PCEs. In most cases, low bandgap D–A polymers achieved by the donor–acceptor (D–A) approach do not possess broad absorption, but instead the absorption maxima are red-shifted, decreasing the number of absorbed photons in the visible region and ultimately limiting the achievable photocurrent. Three promising methods to extend the light absorption range of D–A copolymers have been reported: (i) fabrication of tandem PSCs,[171] (ii) use of a ternary (two donors and one acceptor) blend consisting of different photoactive components which have a complementary absorption range,[172,173] and (iii) synthesis of a random copolymer consisting of different monomers (at least three) showing different absorption ranges.[174] Although impressive advances in PCE of over 9% have been achieved using a tandem device structure, the fabrication technique is still challenging because it is not always possible to select an appropriate interconnection layer material. The second approach, the use of a multiple chromophore mixture, such as ternary blend in a single active layer, is also useful for achieving broad light absorption because the absorption range can be widened when two different components with a complementary absorption range are used as active layer materials. However, the successful increase of the current density has rarely been reported owing to the difficulty in morphology control of the ternary blend. From a molecular design perspective, developing a "multi-component" copolymerization synthetic approach could potentially afford macromolecular materials encompassing all of the aforementioned desired parameters. One promising and feasible approach to broaden the absorption of the solar spectrum is to develop random "three-component" copolymers, *i.e.*, terpolymers with complementary absorption spectra. In terms of the number of donor units, terpolymers can be classified into two major categories. The terpolymers in the first category are copolymerized by two different donor units and one acceptor unit, namely $D-A_1-A_2$ terpolymers. The second category (D_1-D_2-A terpolymers) is terpolymer based on one donor unit and two different acceptor units.

Rational selection of electron-deficient or electron-rich building blocks with different electron-withdrawing or -donating strengths is critical to achieving complementary and broad visible light absorption, thus maximizing photon harvesting. How to develop random terpolymers with broad light absorption and high efficiency is still a key challenge for molecular design in the PSC field. Moreover, designing and synthesizing terpolymers is critical to understanding the fundamental effects of donor and acceptor units on the polymer packing structure, the optical and electrical properties, and the electron distribution in the terpolymers.

2.4.2   Novel Terpolymers Based on One Donor Unit

In 2008, two low bandgap terpolymers based on thiophene–phenylene–thiophene (IDT) with suitable energy levels were designed and synthesized for application in bulk-heterojunction PSCs.[152] The absorption spectral, electrochemical, field effect hole mobility and photovoltaic properties of IDT derivatives were investigated and compared with poly(3-hexylthiophene) (P3HT). Both possessed broader absorption spectra, low-lying HOMO levels and high absorption coefficients. Furthermore, both terpolymers showed sufficient hole mobility [3.4×10^{-3} cm^2 V^{-1} s^{-1}], allowing efficient charge extraction and a good fill-factor for PSC application. High PCE of 4.4% was obtained under simulated solar light AM1.5G (100 mW cm^{-2}) from PIDT-TBTT-based PSC, which is superior to that of the analogous P3HT cell (3.9%) under the identical condition.

In recent work, the electron-rich units, alkylthienyl-substituted benzodithiophenes (BDTT), are actively utilized to construct $D-A_1-A_2$ terpolymers

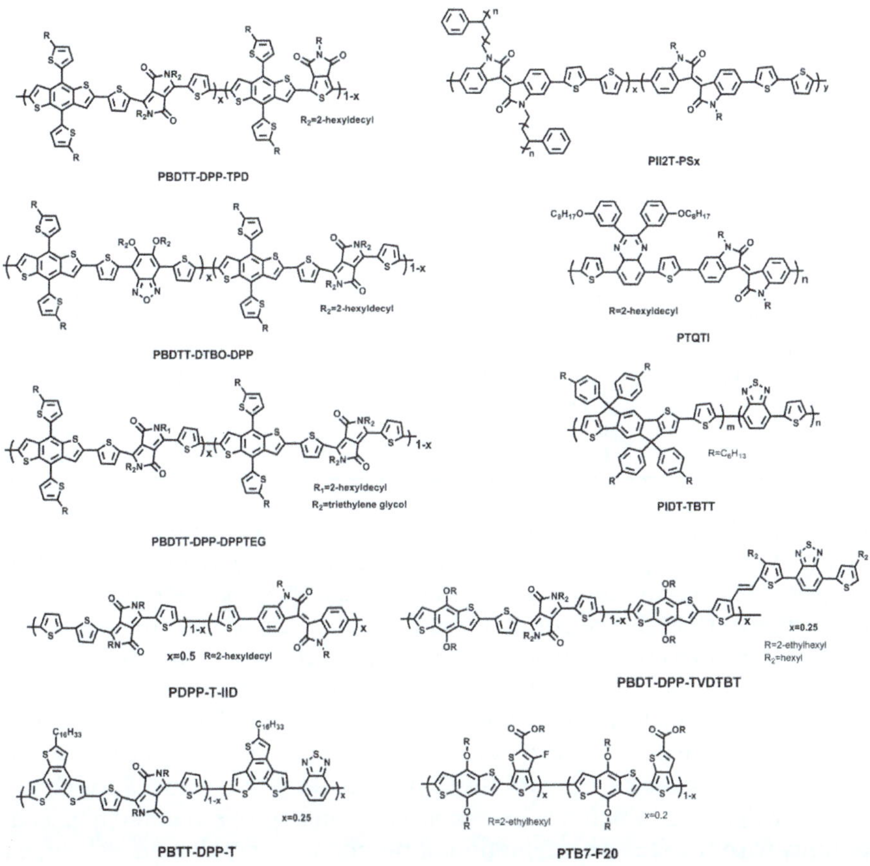

Scheme 2.15 Typical examples of D–A$_1$–A$_2$ terpolymers.

(see Scheme 2.15).[175–177] A series of new D–A conjugated random copolymers, PBDTT-BO-DPP, were designed and successfully synthesized by Jiang *et al.* They comprised electron-rich alkylthienyl-substituted benzodithiophene (BDTT) units in conjugation with electron-deficient 2,1,3-benzooxadiazole (BO) and diketopyrrolopyrrole (DPP) moieties, which have complementary light absorption behavior (300–900 nm). At the molar ratio of 1:0.5:0.5 (*i.e.*, $x = 0.5$), the random copolymers exhibited a value of V_{OC} of 0.73 V, a high J_{SC} of 17 mA cm^{-2}, a FF of 0.55, and a promising PCE of 6.8%.[175] In 2012, Nielson *et al.* designed a series of benzotrithiophene-containing random terpolymers for PSCs.[178] Through variations of the two other components in the terpolymers, the absorption profile and the molecular energy levels were optimized and the maximum PCE were nearly doubled relative to the parent D–A copolymer. Jo, Russell and collaborators synthesized random conjugated copolymers consisting of DPP and isoindigo (IID) as co-electron accepting units in D–A conjugated copolymers in 2013.[179] The representative DPP-based polymer (PDPP3T) shows quite broad absorption over 900 nm with a high-lying

Table 2.8 Photovoltaic results for novel terpolymers.

Materials	V_{OC} [V]	J_{SC} [mA cm^{-2}]	FF [%]	PCE [%]	Ref.
PIDT-TBTT	0.81	10.2	53	4.4	152
PBDTT-DPP-DTBT	0.73	17	55	6.8	175
PBDTT-DPP-DPPTEG	0.72	14.3	68	7.0	177
PBTT-DPP-BT	0.68	10.95	69	5.14	178
PDPP-T-IID	0.77	13.52	58	6.04	179
PTB7-F20	0.60	14.5	67.5	5.85	180
PTQTI	0.81	11.56	54	5.03	181
PII2T-PS10	0.96	11.6	63	7.0	183
PBDT-DPP-TVDTBT	0.78	14.13	48	5.29	184
P3HTT-DPP	0.57	13.87	63	4.94	185
P3HTT-TPD-TPD	0.50	16.37	61	4.92	186
TTV1	0.78	12.20	68	6.46	188
TTV2	0.42	22.65	52	4.99	188
PDPP3T*alt*TPT	0.75	15.9	67	8.0	189

HOMO level around 5.2 eV, whereas the representative isoindigo-based polymer (PIT) absorbs photons shorter than 750 nm in wavelength with a low-lying HOMO level of around 5.9 eV. Hence, it could be expected that random copolymerization of DPP and IID would tune the absorption range and the HOMO energy level of the resulting polymer. Furthermore, the novel terpolymer exhibited not only broad and intense light absorption from 500 nm to 900 nm but also a low-lying HOMO level. The X-ray diffraction (XRD) characterization revealed that the random copolymers predominantly take the face-on orientation in the thin film, which is likely to beneficial for charge transport. The combination of favorable absorption properties and high crystallinity with the face-on orientation of the random terpolymers delivered a promising PCE of 6.04%. Gong *et al.* systematically investigated the effect of fluorine units on the electronic properties and the thin film morphologies of PTB7-based terpolymers.[180] An optimal PCE of 5.85% was observed from PSCs made of terpolymers with 20% fluorine units. The detailed photovoltaic parameters are listed in Table 2.8. More fluorine units coupled with thieno[3,4-*b*]thiophene adversely affect the phase morphology to coarsen it, which in turn reduces the device performance.

In 2013, Wang and co-workers proposed an alternating D–A$_1$–A$_2$ terpolymer concept for the design of low bandgap polymers for high performance PSCs.[181,182] In this case, the alternating terpolymer exhibited an extended absorption spectrum and enhanced J_{SC} when compared with its original polymers. It also showed superiority over the random copolymer. Notably, this is the highest PCE (over 6%) achieved by an alternating D–A$_1$–A$_2$ terpolymer containing two electron-deficient moieties in the repeating unit.[182] The simple concept proposed in this work opened the door to the design of novel terpolymers with wide absorption spectra for efficient PSCs. Bao *et al.* presented a side chain engineering approach, *via* random copolymerization, to gain good processability while maintaining high charge transport and photovoltaic performance for isoindigo (IID)-based polymers.[183] The goal was achieved by the incorporation of tunable ratios of repeating units with short weight polystyrene (PS) side chains into the copolymer. In their work, a series of poly(IID-dithiophene)-based conjugated polymers with varying amounts of low molecular weight PS side chains were synthesized. PSCs fabricated with these PS-containing copolymers exhibited significantly improved performances (PCE > 7% and V_{OC} > 0.95 V), compared with the highest reported performance (PCE = 6.3% and V_{OC} = 0.70 V) based on similar IID-containing polymers. Clearly, the synthesis, processing, and device performance of PS-containing copolymers represented a novel approach in molecular engineering to achieving a balance between the optical/electronic properties and solubility/processibility of reproducible conjugated polymer donors.

Similarly, Tan *et al.* also fabricated three novel random terpolymers by copolymerizing a BDT donor with an electron-deficient diketopyrrolo[3,4-*c*] pyrrole (DPP) unit and a thiophene-vinylene-dithienyl-benzothiadiazole (TVDTBT) side group on a polymer backbone. By tuning the ratio of DPP and TVDTBT in the terpolymers, the optical properties and energy levels of these random terpolymers can be rationally controlled.[184] Accordingly, the terpolymers exhibited a complementary absorption range of 300–900 nm with a high absorption coefficient as well as a deep HOMO level. Bulk-heterojunction PSC fabricated from PBDT-DPP-TVDTBT/PC$_{61}$BM blends exhibited a desirable PCE of 5.29%.

2.4.3 Novel Terpolymers Based on Two Donor Units

With the goal of ameliorating the absorption and photovoltaic performance of P3HT, a successful attempt to develop random terpolymers with broad absorption as well as higher efficiency was conducted by Thompson and co-workers in 2011.[185] In their work, the absorption range of polythiophenes is greatly extended, and thereby significantly enhanced J_{SC} and PCE were recorded for the resulting terpolymers, as listed in Table 2.8.[186] These polymers also exhibit great potential for studies of device physics.[187]

In 2012, Zhou *et al.* reported a feasible approach to modifying two kinds of D$_1$–D$_2$–A type polymer based on a thiophene containing tris(thienylenevinylene) (TTV) conjugated side chain (see Scheme 2.16). The resulting copolymers, TTV1 and TTV2, showed significant improvement (25% and 43%) in PCE in comparison to the original polymers.[188] Impressively, the J_{SC} of TTV2 reached 22.6 mA cm^{-2}, which should be the highest record for PSCs to date. This design strategy opens up a new possibility to modify the existing D–A type photovoltaic polymers for better performance. A novel terpolymer design strategy was recently proposed by Janssen *et al.* to produce a photovoltaic polymer with tailored energy levels and optical bandgap.[189] A high molecular weight terpolymer, PDPP3T*alt*DPP, realized high PCE of up to 8.0% in PSC employing PC$_{71}$BM as acceptor. Relative to the PCE of parent copolymers PDPP3T (7.1%) and PDPPTPT (7.4%), the PCE of PDPP3T*alt*TPT exhibited more than 7% improvement, which demonstrated that the terpolymer could outperform the two parent copolymers when applied the alternating terpolymer strategy.

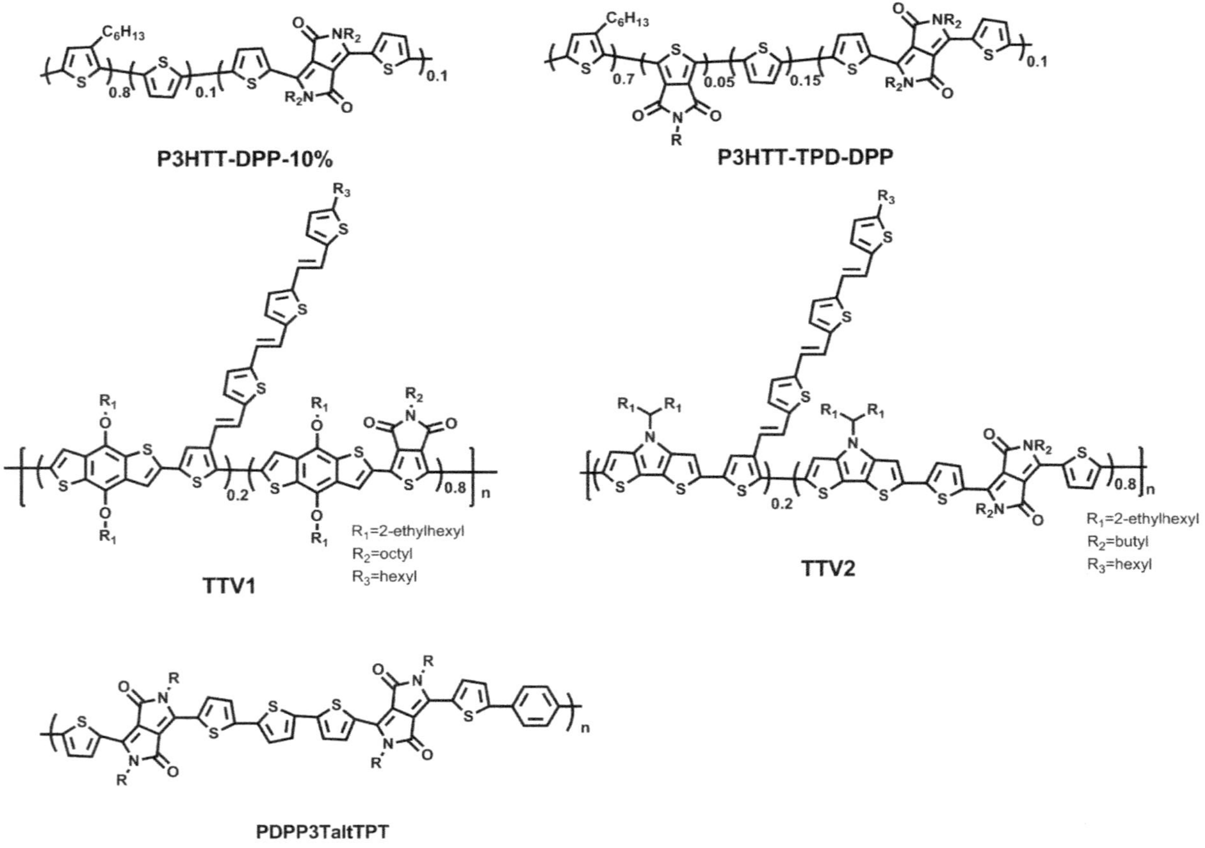

Scheme 2.16 Typical examples of D_1–D_2–A terpolymers.

Owing to the versatile nature of the random copolymerization-based approach involving terpolymers, it could be readily extended to other conjugated polymeric systems, either upon optimizing the side-chain ratio or the donor–acceptor type to achieve significantly improved performance and highly tunable morphology in PSCs.

2.5 Summary and Outlook

In summary, novel donor polymers for high performance PSCs should have at least three features: (i) a low bandgap and strong absorption spectrum for harvesting solar light efficiently; (ii) appropriate HOMO and LUMO levels for efficient charge separation in the blend films with fullerene derivatives which are usually used as electron acceptors; (iii) high mobility for fast exciton diffusion and hole transport as well as efficient charge collection. The correlation of the photovoltaic performance of donor polymers with their molecular properties, including solubility, absorption spectra, bandgap, energy levels, and mobility, has been discussed in detail. Moreover, we have summarized the recent progress in design, synthesis and optimization of novel donor polymers and tried to reveal the correlations of molecular structures and photovoltaic performance.

Recent reviews have also pointed out that intrinsic parameters, including dielectric constants, binding energy, and dipole moments, should also be considered in the molecular design of donor polymers.[190,191] Rational design of high performance polymer donors is still a challenge in the PSC field. Notably, the recent optimizations and device applications of novel donor polymer materials boost the PCEs of PSCs to above 9%, demonstrating their commercial potential as an energy source. Although recent advances look impressive in publications, until now the commercialization of PSCs has been hampered by the difficulty of converting laboratory cell figures into reliable industrial modules. The key challenge is to develop polymer donors exhibiting the required technical and economic characteristics that can be conveniently used in an industrial environment. The well established strategies for the design of materials for efficient laboratory PSCs are not sufficient when industrial panels are concerned. Besides, other requirements must be met in the near future; for example, the donor polymers must be easily accessible in few synthetic steps from low cost raw materials; they need to be stable and soluble enough to afford ink formulations processible with roll-to-roll compatible equipment; and the active layers should achieve a stable morphology under ambient conditions.[192–194]

References

1. M. Helgesen, R. Sondergaard and F. C. Krebs, *J. Mater. Chem.*, 2010, **20**, 36.
2. M. Kaltenbrunner, M. S. White, E. D. Glowacki, T. Sekitani, T. Someya, N. S. Sariciftci and S. Bauer, *Nat. Commun.*, 2012, **3**, 770.

3. Y. H. Zhou, C. Fuentes-Hernandez, T. M. Khan, J. C. Liu, J. Hsu, J. W. Shim, A. Dindar, J. P. Youngblood, R. J. Moon and B. Kippelen, *Sci. Rep.*, 2013, **3**, 1536.

4. R. Sondergaard, M. Hosel, D. Angmo, T. T. Larsen-Olsen and F. C. Krebs, *Mater. Today*, 2012, **15**, 36.

5. L. Dou, J. You, Z. Hong, Z. Xu, G. Li, R. A. Street and Y. Yang, *Adv. Mater.*, 2013, **25**, 6642.

6. Z. He, H. Wu and Y. Cao, *Adv. Mater.*, 2014, **26**, 1006.

7. H. L. Yip and A. K. Y. Jen, *Energy Environ. Sci.*, 2012, **5**, 5994.

8. G. Li, R. Zhu and Y. Yang, *Nat. Photonics*, 2012, **6**, 153.

9. F. Liu, Y. Gu, X. Shen, S. Ferdous, H.-W. Wang and T. P. Russell, *Prog. Polym. Sci.*, 2013, **38**, 1990.

10. L.-M. Chen, Z. Hong, G. Li and Y. Yang, *Adv. Mater.*, 2009, **21**, 1434.

11. L. Ye, Y. Jing, X. Guo, H. Sun, S. Zhang, M. Zhang, L. Huo and J. Hou, *J. Phys. Chem. C*, 2013, **117**, 14920.

12. L. Ye, S. Zhang, W. Ma, B. Fan, X. Guo, Y. Huang, H. Ade and J. Hou, *Adv. Mater.*, 2012, **24**, 6335.

13. G. Yu, J. Gao, J. C. Hummelen, F. Wudl and A. J. Heeger, *Science*, 1995, **270**, 1789.

14. A. J. Heeger, *Adv. Mater.*, 2014, **26**, 10.

15. Y. J. Cheng, S. H. Yang and C. S. Hsu, *Chem. Rev.*, 2009, **109**, 5868.

16. J. Chen and Y. Cao, *Acc. Chem. Res.*, 2009, **42**, 1709.

17. X. Zhan and D. Zhu, *Polym. Chem.*, 2010, **1**, 409.

18. P. M. Beaujuge and J. M. J. Fréchet, *J. Am. Chem. Soc.*, 2011, **133**, 20009.

19. F. He and L. Yu, *J. Phys. Chem. Lett.*, 2011, **2**, 3102.

20. P.-L. T. Boudreault, A. Najari and M. Leclerc, *Chem. Mater.*, 2010, **23**, 456.

21. Y. Li, *Acc. Chem. Res.*, 2012, **45**, 723.

22. C. Duan, F. Huang and Y. Cao, *J. Mater. Chem.*, 2012, **22**, 10416.

23. H. Zhou, L. Yang and W. You, *Macromolecules*, 2012, **45**, 607.

24. Z. B. Henson, K. Mullen and G. C. Bazan, *Nat. Chem.*, 2012, **4**, 699.

25. R. L. Uy, S. C. Price and W. You, *Macromol. Rapid Commun.*, 2012, **33**, 1162.

26. H. Spanggaard and F. C. Krebs, *Sol. Energy Mater. Sol. Cells*, 2004, **83**, 125.

27. Y. Li and Y. Zou, *Adv. Mater.*, 2008, **20**, 2952.

28. S. E. Shaheen, C. J. Brabec, N. S. Sariciftci, F. Padinger, T. Fromherz and J. C. Hummelen, *Appl. Phys. Lett.*, 2001, **78**, 841.

29. Q. Zhou, L. Zheng, D. Sun, X. Deng, G. Yu and Y. Cao, *Synth. Met.*, 2003, **135–136**, 825.

30. M. T. Dang, L. Hirsch and G. Wantz, *Adv. Mater.*, 2011, **23**, 3597.

31. M. T. Dang, L. Hirsch, G. Wantz and J. D. Wuest, *Chem. Rev.*, 2013, **113**, 3734.

32. G. Li, V. Shrotriya, J. Huang, Y. Yao, T. Moriarty, K. Emery and Y. Yang, *Nat. Mater.*, 2005, **4**, 864.

33. W. Ma, C. Yang, X. Gong, K. Lee and A. J. Heeger, *Adv. Funct. Mater.*, 2005, **15**, 1617.

34. Y. He, H.-Y. Chen, J. Hou and Y. Li, *J. Am. Chem. Soc.*, 2010, **132**, 1377.

35. C.-Y. Chang, C.-E. Wu, S.-Y. Chen, C. Cui, Y.-J. Cheng, C.-S. Hsu, Y.-L. Wang and Y. Li, *Angew. Chem., Int. Ed.*, 2011, **50**, 9386.
36. S.-H. Liao, Y.-L. Li, T.-H. Jen, Y.-S. Cheng and S.-A. Chen, *J. Am. Chem. Soc.*, 2012, **134**, 14271.
37. X. Guo, C. Cui, M. Zhang, L. Huo, Y. Huang, J. Hou and Y. Li, *Energy Environ. Sci.*, 2012, **5**, 7943.
38. J. Hou, Z. Tan, Y. Yan, Y. He, C. Yang and Y. Li, *J. Am. Chem. Soc.*, 2006, **128**, 4911.
39. P.-T. Wu, H. Xin, F. S. Kim, G. Ren and S. A. Jenekhe, *Macromolecules*, 2009, **42**, 8817.
40. J. Hou, T. L. Chen, S. Zhang, L. Huo, S. Sista and Y. Yang, *Macromolecules*, 2009, **42**, 9217.
41. M. Zhang, X. Guo, Y. Yang, J. Zhang, Z.-G. Zhang and Y. Li, *Polym. Chem.*, 2011, **2**, 2900.
42. M. C. Scharber, D. Mühlbacher, M. Koppe, P. Denk, C. Waldauf, A. J. Heeger and C. J. Brabec, *Adv. Mater.*, 2006, **18**, 789.
43. J. Hou, M.-H. Park, S. Zhang, Y. Yao, L.-M. Chen, J.-H. Li and Y. Yang, *Macromolecules*, 2008, **41**, 6012.
44. L. Huo and J. Hou, *Polym. Chem.*, 2011, **2**, 2453.
45. Y. Wu, Z. Li, W. Ma, Y. Huang, L. Huo, X. Guo, M. Zhang, H. Ade and J. Hou, *Adv. Mater.*, 2013, **25**, 3449.
46. Y. Huang, X. Guo, F. Liu, L. J. Huo, Y. N. Chen, T. P. Russell, C. C. Han, Y. F. Li and J. H. Hou, *Adv. Mater.*, 2012, **24**, 3383.
47. X. Guo, M. Zhang, L. Huo, F. Xu, Y. Wu and J. Hou, *J. Mater. Chem.*, 2012, **22**, 21024.
48. G. Zuo, Z. Li, M. Zhang, X. Guo, Y. Wu, S. Zhang, B. Peng, W. Wei and J. Hou, *Polym. Chem.*, 2014, **5**, 1976.
49. S. Zhang, L. Ye, Q. Wang, Z. Li, X. Guo, L. Huo, H. Fan and J. Hou, *J. Phys. Chem. C*, 2013, **117**, 9550.
50. J. S. Kim, Y. Lee, J. H. Lee, J. H. Park, J. K. Kim and K. Cho, *Adv. Mater.*, 2010, **22**, 1355.
51. T. Lei, J.-Y. Wang and J. Pei, *Chem. Mater.*, 2013, **26**, 594.
52. J. Mei and Z. Bao, *Chem. Mater.*, 2013, **26**, 604.
53. M. Svensson, F. Zhang, S. C. Veenstra, W. J. H. Verhees, J. C. Hummelen, J. M. Kroon, O. Inganäs and M. R. Andersson, *Adv. Mater.*, 2003, **15**, 988.
54. M.-H. Chen, J. Hou, Z. Hong, G. Yang, S. Sista, L.-M. Chen and Y. Yang, *Adv. Mater.*, 2009, **21**, 4238.
55. Y. Huang, M. Zhang, L. Ye, X. Guo, C. C. Han, Y. Li and J. Hou, *J. Mater. Chem.*, 2012, **22**, 5700.
56. Z. He, H. Wu and Y. Cao, *Adv. Mater.*, 2014, **26**, 1006.
57. C. Duan, K. Zhang, C. Zhong, F. Huang and Y. Cao, *Chem. Soc. Rev.*, 2013, **42**, 9071.
58. D. Qian, Q. Xu, X. Hou, F. Wang, J. Hou and Z. Tan, *J. Polym. Sci., Part A: Polym. Chem.*, 2013, **51**, 3123.
59. L. J. Huo, S. Q. Zhang, X. Guo, F. Xu, Y. F. Li and J. H. Hou, *Angew. Chem., Int. Ed.*, 2011, **50**, 9697.

60. J. Hou, H.-Y. Chen, S. Zhang, R. I. Chen, Y. Yang, Y. Wu and G. Li, *J. Am. Chem. Soc.*, 2009, **131**, 15586.
61. Y. Liang, D. Feng, Y. Wu, S.-T. Tsai, G. Li, C. Ray and L. Yu, *J. Am. Chem. Soc.*, 2009, **131**, 7792.
62. H.-Y. Chen, J. Hou, S. Zhang, Y. Liang, G. Yang, Y. Yang, L. Yu, Y. Wu and G. Li, *Nat. Photonics*, 2009, **3**, 649.
63. Y. Liang, Z. Xu, J. Xia, S.-T. Tsai, Y. Wu, G. Li, C. Ray and L. Yu, *Adv. Mater.*, 2010, **22**, 135.
64. Y. Huang, L. Huo, S. Zhang, X. Guo, C. C. Han, Y. Li and J. Hou, *Chem. Commun.*, 2011, **47**, 8904.
65. L. Huo, Z. Li, X. Guo, Y. Wu, M. Zhang, L. Ye, S. Zhang and J. Hou, *Polym. Chem.*, 2013, **4**, 3047.
66. M. Zhang, X. Guo, S. Zhang and J. Hou, *Adv. Mater.*, 2014, **26**, 1118.
67. C. B. Nielsen, R. S. Ashraf, S. Rossbauer, T. Anthopoulos and I. McCulloch, *Macromolecules*, 2013, **46**, 7727.
68. B. J. Campo, J. Duchateau, C. R. Ganivet, B. Ballesteros, J. Gilot, M. M. Wienk, W. D. Oosterbaan, L. Lutsen, T. J. Cleij, G. de la Torre, R. A. J. Janssen, D. Vanderzande and T. Torres, *Dalton Trans.*, 2011, **40**, 3979.
69. J. Zhang, L. Wang, C. Li, Y. Li, J. Liu, Y. Tu, W. Zhang, N. Zhou and X. Zhu, *J. Polym. Sci., Part A: Polym. Chem.*, 2014, **52**, 691.
70. H. Zhou, L. Yang, S. Stoneking and W. You, *ACS Appl. Mater. Interfaces*, 2010, **2**, 1377.
71. J. C. Bijleveld, A. P. Zoombelt, S. G. J. Mathijssen, M. M. Wienk, M. Turbiez, D. M. de Leeuw and R. A. J. Janssen, *J. Am. Chem. Soc.*, 2009, **131**, 16616.
72. A. T. Yiu, P. M. Beaujuge, O. P. Lee, C. H. Woo, M. F. Toney and J. M. J. Frechet, *J. Am. Chem. Soc.*, 2012, **134**, 2180.
73. W. W. Li, K. H. Hendriks, W. S. C. Roelofs, Y. Kim, M. M. Wienk and R. A. J. Janssen, *Adv. Mater.*, 2013, **25**, 3182.
74. W. W. Li, A. Furlan, K. H. Hendriks, M. M. Wienk and R. A. J. Janssen, *J. Am. Chem. Soc.*, 2013, **135**, 5529.
75. J. Jo, J. R. Pouliot, D. Wynands, S. D. Collins, J. Y. Kim, T. L. Nguyen, H. Y. Woo, Y. M. Sun, M. Leclerc and A. J. Heeger, *Adv. Mater.*, 2013, **25**, 4783.
76. M. S. Su, C. Y. Kuo, M. C. Yuan, U. S. Jeng, C. J. Su and K. H. Wei, *Adv. Mater.*, 2011, **23**, 3315.
77. X. G. Guo, N. J. Zhou, S. J. Lou, J. Smith, D. B. Tice, J. W. Hennek, R. P. Ortiz, J. T. L. Navarrete, S. Y. Li, J. Strzalka, L. X. Chen, R. P. H. Chang, A. Facchetti and T. J. Marks, *Nat. Photonics*, 2013, **7**, 825.
78. D. P. Qian, W. Ma, Z. J. Li, X. Guo, S. Q. Zhang, L. Ye, H. Ade, Z. A. Tan and J. H. Hou, *J. Am. Chem. Soc.*, 2013, **135**, 8464.
79. E. G. Wang, Z. F. Ma, Z. Zhang, K. Vandewal, P. Henriksson, O. Inganas, F. L. Zhang and M. R. Andersson, *J. Am. Chem. Soc.*, 2011, **133**, 14244.
80. E. G. Wang, L. T. Hou, Z. Q. Wang, S. Hellstrom, F. L. Zhang, O. Inganas and M. R. Andersson, *Adv. Mater.*, 2010, **22**, 5240.
81. K. H. Ong, S. L. Lim, H. S. Tan, H. K. Wong, J. Li, Z. Ma, L. C. H. Moh, S. H. Lim, J. C. De Mello and Z. K. Chen, *Adv. Mater.*, 2011, **23**, 1409.
82. I. Osaka, M. Shimawaki, H. Mori, I. Doi, E. Miyazaki, T. Koganezawa and K. Takimiya, *J. Am. Chem. Soc.*, 2012, **134**, 3498.

83. E. G. Wang, L. Wang, L. F. Lan, C. Luo, W. L. Zhuang, J. B. Peng and Y. Cao, *Appl. Phys. Lett.*, 2008, **92**, 033307.
84. J. S. Song, C. Du, C. H. Li and Z. S. Bo, *J. Polym. Sci., Part A: Polym. Chem.*, 2011, **49**, 4267.
85. C. Du, C. H. Li, W. W. Li, X. Chen, Z. S. Bo, C. Veit, Z. F. Ma, U. Wuerfel, H. F. Zhu, W. P. Hu and F. L. Zhang, *Macromolecules*, 2011, **44**, 7617.
86. Q. Liu, C. H. Li, E. Q. Jin, Z. Lu, Y. C. Chen, F. H. Li and Z. S. Bo, *ACS Appl. Mater. Interfaces*, 2014, **6**, 1601.
87. N. Blouin, A. Michaud and M. Leclerc, *Adv. Mater.*, 2007, **19**, 2295.
88. S. H. Park, A. Roy, S. Beaupre, S. Cho, N. Coates, J. S. Moon, D. Moses, M. Leclerc, K. Lee and A. J. Heeger, *Nat. Photonics*, 2009, **3**, 297.
89. R. P. Qin, W. W. Li, C. H. Li, C. Du, C. Veit, H. F. Schleiermacher, M. Andersson, Z. S. Bo, Z. P. Liu, O. Inganas, U. Wuerfel and F. L. Zhang, *J. Am. Chem. Soc.*, 2009, **131**, 14612.
90. C. H. Duan, W. Z. Cai, B. B. Y. Hsu, C. M. Zhong, K. Zhang, C. C. Liu, Z. C. Hu, F. Huang, G. C. Bazan, A. J. Heeger and Y. Cao, *Energy Environ. Sci.*, 2013, **6**, 3022.
91. N. Blouin, A. Michaud, D. Gendron, S. Wakim, E. Blair, R. Neagu-Plesu, M. Belletete, G. Durocher, Y. Tao and M. Leclerc, *J. Am. Chem. Soc.*, 2008, **130**, 732.
92. F. Huang, K. S. Chen, H. L. Yip, S. K. Hau, O. Acton, Y. Zhang, J. D. Luo and A. K. Y. Jen, *J. Am. Chem. Soc.*, 2009, **131**, 13886.
93. Y. Deng, J. Liu, J. Wang, L. Liu, W. Li, H. Tian, X. Zhang, Z. Xie, Y. Geng and F. Wang, *Adv. Mater.*, 2014, **26**, 471.
94. D. Muhlbacher, M. Scharber, M. Morana, Z. G. Zhu, D. Waller, R. Gaudiana and C. Brabec, *Adv. Mater.*, 2006, **18**, 2884.
95. J. Peet, J. Y. Kim, N. E. Coates, W. L. Ma, D. Moses, A. J. Heeger and G. C. Bazan, *Nat. Mater.*, 2007, **6**, 497.
96. J. H. Hou, H. Y. Chen, S. Q. Zhang, G. Li and Y. Yang, *J. Am. Chem. Soc.*, 2008, **130**, 16144.
97. S. Albrecht, S. Janietz, W. Schindler, J. Frisch, J. Kurpiers, J. Kniepert, S. Inal, P. Pingel, K. Fostiropoulos, N. Koch and D. Neher, *J. Am. Chem. Soc.*, 2012, **134**, 14932.
98. C. Y. Chang, L. J. Zuo, H. L. Yip, Y. X. Li, C. Z. Li, C. S. Hsu, Y. J. Cheng, H. Z. Chen and A. K. Y. Jen, *Funct. Mater.*, 2013, **23**, 5084.
99. J. Yang, R. Zhu, Z. R. Hong, Y. J. He, A. Kumar, Y. F. Li and Y. Yang, *Adv. Mater.*, 2011, **23**, 3465.
100. T. Y. Chu, J. P. Lu, S. Beaupre, Y. G. Zhang, J. R. Pouliot, S. Wakim, J. Y. Zhou, M. Leclerc, Z. Li, J. F. Ding and Y. Tao, *J. Am. Chem. Soc.*, 2011, **133**, 4250.
101. M. J. Zhang, X. Guo and Y. F. Li, *Adv. Energy Mater.*, 2011, **1**, 557.
102. C. H. Cui, X. Fan, M. J. Zhang, J. Zhang, J. Min and Y. F. Li, *Chem. Commun.*, 2011, **47**, 11345.
103. X. G. Guo, N. J. Zhou, S. J. Lou, J. W. Hennek, R. P. Ortiz, M. R. Butler, P. L. T. Boudreault, J. Strzalka, P. O. Morin, M. Leclerc, J. T. L. Navarrete, M. A. Ratner, L. X. Chen, R. P. H. Chang, A. Facchetti and T. J. Marks, *J. Am. Chem. Soc.*, 2012, **134**, 18427.

104. C. M. Amb, S. Chen, K. R. Graham, J. Subbiah, C. E. Small, F. So and J. R. Reynolds, *J. Am. Chem. Soc.*, 2011, **133**, 10062.

105. C. E. Small, S. Chen, J. Subbiah, C. M. Amb, S. W. Tsang, T. H. Lai, J. R. Reynolds and F. So, *Nat. Photonics*, 2012, **6**, 115.

106. H. L. Zhong, Z. Li, F. Deledalle, E. C. Fregoso, M. Shahid, Z. P. Fei, C. B. Nielsen, N. Yaacobi-Gross, S. Rossbauer, T. D. Anthopoulos, J. R. Durrant and M. Heeney, *J. Am. Chem. Soc.*, 2013, **135**, 2040.

107. L. Dou, C.-C. Chen, K. Yoshimura, K. Ohya, W.-H. Chang, J. Gao, Y. Liu, E. Richard and Y. Yang, *Macromolecules*, 2013, **46**, 3384.

108. J. B. You, L. T. Dou, K. Yoshimura, T. Kato, K. Ohya, T. Moriarty, K. Emery, C. C. Chen, J. Gao, G. Li and Y. Yang, *Nat. Commun.*, 2013, **4**, 1446.

109. Y. Y. Liang and L. P. Yu, *Acc. Chem. Res.*, 2010, **43**, 1227.

110. C. Liu, K. Wang, X. Hu, Y. Yang, C.-H. Hsu, W. Zhang, S. Xiao, X. Gong and Y. Cao, *ACS Appl. Mater. Interfaces*, 2013, **5**, 12163.

111. Y. Zou, A. Najari, P. Berrouard, S. Beaupré, B. Réda Aïch, Y. Tao and M. Leclerc, *J. Am. Chem. Soc.*, 2010, **132**, 5330.

112. C. Piliego, T. W. Holcombe, J. D. Douglas, C. H. Woo, P. M. Beaujuge and J. M. J. Fréchet, *J. Am. Chem. Soc.*, 2010, **132**, 7595.

113. Y. Zhang, S. K. Hau, H.-L. Yip, Y. Sun, O. Acton and A. K. Y. Jen, *Chem. Mater.*, 2010, **22**, 2696.

114. G. Zhang, Y. Fu, Q. Zhang and Z. Xie, *Chem. Commun.*, 2010, **46**, 4997–4999.

115. C. Cabanetos, A. El Labban, J. A. Bartelt, J. D. Douglas, W. R. Mateker, J. M. J. Fréchet, M. D. McGehee and P. M. Beaujuge, *J. Am. Chem. Soc.*, 2013, **135**, 4656.

116. S. C. Price, A. C. Stuart, L. Yang, H. Zhou and W. You, *J. Am. Chem. Soc.*, 2011, **133**, 4625.

117. H. Zhou, L. Yang, A. C. Stuart, S. C. Price, S. Liu and W. You, *Angew. Chem., Int. Ed.*, 2011, **50**, 2995.

118. N. Wang, Z. Chen, W. Wei and Z. Jiang, *J. Am. Chem. Soc.*, 2013, **135**, 17060.

119. H. C. Chen, Y. H. Chen, C. C. Liu, Y. C. Chien, S. W. Chou and P. T. Chou, *Chem. Mater.*, 2012, **24**, 4766.

120. X. Wang, P. Jiang, Y. Chen, H. Luo, Z. Zhang, H. Wang, X. Li, G. Yu and Y. Li, *Macromolecules*, 2013, **46**, 4805.

121. L. T. Dou, J. B. You, J. Yang, C. C. Chen, Y. J. He, S. Murase, T. Moriarty, K. Emery, G. Li and Y. Yang, *Nat. Photonics*, 2012, **6**, 180.

122. L. T. Dou, W. H. Chang, J. Gao, C. C. Chen, J. B. You and Y. Yang, *Adv. Mater.*, 2013, **25**, 825.

123. L. T. Dou, J. Gao, E. Richard, J. B. You, C. C. Chen, K. C. Cha, Y. J. He, G. Li and Y. Yang, *J. Am. Chem. Soc.*, 2012, **134**, 10071.

124. Z. C. He, C. M. Zhong, X. Huang, W. Y. Wong, H. B. Wu, L. W. Chen, S. J. Su and Y. Cao, *Adv. Mater.*, 2011, **23**, 4636.

125. Z. C. He, C. M. Zhong, S. J. Su, M. Xu, H. B. Wu and Y. Cao, *Nat. Photonics*, 2012, **6**, 591.

126. C. H. Duan, K. Zhang, X. Guan, C. M. Zhong, H. M. Xie, F. Huang, J. W. Chen, J. B. Peng and Y. Cao, *Chem. Sci.*, 2013, **4**, 1298.

127. S. Liu, K. Zhang, J. Lu, J. Zhang, H.-L. Yip, F. Huang and Y. Cao, *J. Am. Chem. Soc.*, 2013, **135**, 15326.

128. L. J. Huo, J. H. Hou, S. Q. Zhang, H. Y. Chen and Y. Yang, *Angew. Chem., Int. Ed.*, 2010, **49**, 1500.

129. L. J. Huo, X. Guo, S. Q. Zhang, Y. F. Li and J. H. Hou, *Macromolecules*, 2011, **44**, 4035.

130. R. M. Duan, L. Ye, X. Guo, Y. Huang, P. Wang, S. Q. Zhang, J. P. Zhang, L. J. Huo and J. H. Hou, *Macromolecules*, 2012, **45**, 3032.

131. M. Wang, X. W. Hu, P. Liu, W. Li, X. Gong, F. Huang and Y. Cao, *J. Am. Chem. Soc.*, 2011, **133**, 9638.

132. Y. Dong, X. Hu, C. Duan, P. Liu, S. Liu, L. Lan, D. Chen, L. Ying, S. Su, X. Gong, F. Huang and Y. Cao, *Adv. Mater.*, 2013, **25**, 3683.

133. M. Zhang, Y. Gu, X. Guo, F. Liu, S. Zhang, L. Huo, T. P. Russell and J. Hou, *Adv. Mater.*, 2013, **25**, 4944.

134. D. P. Qian, L. Ye, M. J. Zhang, Y. R. Liang, L. J. Li, Y. Huang, X. Guo, S. Q. Zhang, Z. A. Tan and J. H. Hou, *Macromolecules*, 2012, **45**, 9611.

135. M. Zhang, X. Guo, W. Ma, S. Zhang, L. Huo, H. Ade and J. Hou, *Adv. Mater.*, 2014, **26**, 2089–2095.

136. X. H. Li, W. C. H. Choy, L. J. Huo, F. X. Xie, W. E. I. Sha, B. F. Ding, X. Guo, Y. F. Li, J. H. Hou, J. B. You and Y. Yang, *Adv. Mater.*, 2012, **24**, 3046.

137. T. B. Yang, M. Wang, C. H. Duan, X. W. Hu, L. Huang, J. B. Peng, F. Huang and X. Gong, *Energy Environ. Sci.*, 2012, **5**, 8208.

138. S.-H. Liao, H.-J. Jhuo, Y.-S. Cheng and S.-A. Chen, *Adv. Mater.*, 2013, **25**, 4766.

139. Y. Wang, Y. Liu, S. Chen, R. Peng and Z. Ge, *Chem. Mater.*, 2013, **25**, 3196.

140. L. Ye, S. Zhang, D. Qian, Q. Wang and J. Hou, *J. Phys. Chem. C*, 2013, **117**, 25360.

141. X. Zhang, Z. Lu, L. Ye, C. Zhan, J. Hou, S. Zhang, B. Jiang, Y. Zhao, J. Huang, S. Zhang, Y. Liu, Q. Shi, Y. Liu and J. Yao, *Adv. Mater.*, 2013, **25**, 5791.

142. L. J. Huo, Y. Huang, B. H. Fan, X. Guo, Y. Jing, M. J. Zhang, Y. F. Li and J. H. Hou, *Chem. Commun.*, 2012, **48**, 3318.

143. L. J. Huo, L. Ye, Y. Wu, Z. J. Li, X. Guo, M. J. Zhang, S. Q. Zhang and J. H. Hou, *Macromolecules*, 2012, **45**, 6923.

144. H. A. Saadeh, L. Lu, F. He, J. E. Bullock, W. Wang, B. Carsten and L. Yu, *ACS Macro Lett.*, 2012, **1**, 361.

145. H. J. Son, L. Lu, W. Chen, T. Xu, T. Zheng, B. Carsten, J. Strzalka, S. B. Darling, L. X. Chen and L. Yu, *Adv. Mater.*, 2013, **25**, 838.

146. Q. Peng, Q. Huang, X. Hou, P. Chang, J. Xu and S. Deng, *Chem. Commun.*, 2012, **48**, 11452.

147. I. Osaka, T. Kakara, N. Takemura, T. Koganezawa and K. Takimiya, *J. Am. Chem. Soc.*, 2013, **135**, 8834.

148. L. Huo, X. Guo, Y. Li and J. Hou, *Chem. Commun.*, 2011, **47**, 8850.

149. M. Zhang, Y. Sun, X. Guo, C. Cui, Y. He and Y. Li, *Macromolecules*, 2011, **44**, 7625.

150. H. Zhou, L. Yang, S. C. Price, K. J. Knight and W. You, *Angew. Chem., Int. Ed.*, 2010, **49**, 7992.

151. I. McCulloch, R. S. Ashraf, L. Biniek, H. Bronstein, C. Combe, J. E. Donaghey, D. I. James, C. B. Nielsen, B. C. Schroeder and W. Zhang, *Acc. Chem. Res.*, 2012, **45**, 714.

152. C. P. Chen, S. H. Chan, T. C. Chao, C. Ting and B. T. Ko, *J. Am. Chem. Soc.*, 2008, **130**, 12828.

153. Y.-C. Chen, C.-Y. Yu, Y.-L. Fan, L.-I. Hung, C.-P. Chen and C. Ting, *Chem. Commun.*, 2010, **46**, 6503.

154. Y. Zhang, J. Zou, H.-L. Yip, K.-S. Chen, D. F. Zeigler, Y. Sun and A. K.-Y. Jen, *Chem. Mater.*, 2011, **23**, 2289.

155. Y. X. Xu, C. C. Chueh, H. L. Yip, F. Z. Ding, Y. X. Li, C. Z. Li, X. Li, W. C. Chen and A. K. Y. Jen, *Adv. Mater.*, 2012, **24**, 6356.

156. M. Wang, X. Hu, L. Liu, C. Duan, P. Liu, L. Ying, F. Huang and Y. Cao, *Macromolecules*, 2013, **46**, 3950.

157. X. Guo, M. Zhang, J. Tan, S. Zhang, L. Huo, W. Hu, Y. Li and J. Hou, *Adv. Mater.*, 2012, **24**, 6536.

158. H.-H. Chang, C.-E. Tsai, Y.-Y. Lai, W.-W. Liang, S.-L. Hsu, C.-S. Hsu and Y.-J. Cheng, *Macromolecules*, 2013, **46**, 7715.

159. J. J. Intemann, K. Yao, H.-L. Yip, Y.-X. Xu, Y.-X. Li, P.-W. Liang, F.-Z. Ding, X. Li and A. K.-Y. Jen, *Chem. Mater.*, 2013, **25**, 3188.

160. J.-Y. Wang, S. K. Hau, H.-L. Yip, J. A. Davies, K.-S. Chen, Y. Zhang, Y. Sun and A. K. Y. Jen, *Chem. Mater.*, 2010, **23**, 765.

161. R. S. Ashraf, Z. Chen, D. S. Leem, H. Bronstein, W. Zhang, B. Schroeder, Y. Geerts, J. Smith, S. Watkins, T. D. Anthopoulos, H. Sirringhaus, J. C. de Mello, M. Heeney and I. McCulloch, *Chem. Mater.*, 2010, **23**, 768.

162. B. C. Schroeder, Z. Huang, R. S. Ashraf, J. Smith, P. D'Angelo, S. E. Watkins, T. D. Anthopoulos, J. R. Durrant and I. McCulloch, *Adv. Funct. Mater.*, 2012, **22**, 1663–1670.

163. H.-H. Chang, C.-E. Tsai, Y.-Y. Lai, D.-Y. Chiou, S.-L. Hsu, C.-S. Hsu and Y.-J. Cheng, *Macromolecules*, 2012, **45**, 9282.

164. Y. X. Xu, C. C. Chueh, H. L. Yip, F. Z. Ding, Y. X. Li, C. Z. Li, X. S. Li, W. C. Chen and A. K. Y. Jen, *Adv. Mater.*, 2012, **24**, 6356.

165. C.-C. Chueh, K. Yao, H.-L. Yip, C.-Y. Chang, Y.-X. Xu, K.-S. Chen, C.-Z. Li, P. Liu, F. Huang, Y. Chen, W.-C. Chen and A. K. Y. Jen, *Energy Environ. Sci.*, 2013, **6**, 3241.

166. J. J. Intemann, K. Yao, Y.-X. Li, H.-L. Yip, Y.-X. Xu, P.-W. Liang, C.-C. Chueh, F.-Z. Ding, X. Yang, X. Li, Y. Chen and A. K. Y. Jen, *Adv. Funct. Mater.*, 2014, **24**, 1465.

167. K.-S. Chen, H.-L. Yip, J.-F. Salinas, Y.-X. Xu, C.-C. Chueh and A. K. Y. Jen, *Adv. Mater.*, 2014, **26**, 3349–3354.

168. Y. Ma, Q. Zheng, Z. Yin, D. Cai, S.-C. Chen and C. Tang, *Macromolecules*, 2013, **46**, 4813.

169. H.-H. Chang, C.-E. Tsai, Y.-Y. Lai, W.-W. Liang, S.-L. Hsu, C.-S. Hsu and Y.-J. Cheng, *Macromolecules*, 2013, **46**, 7715.

170. Y.-L. Chen, W.-S. Kao, C.-E. Tsai, Y.-Y. Lai, Y.-J. Cheng and C.-S. Hsu, *Chem. Commun.*, 2013, **49**, 7702.

171. S. Sista, Z. R. Hong, L. M. Chen and Y. Yang, *Energy Environ. Sci.*, 2011, **4**, 1606.
172. L. Yang, L. Yan and W. You, *J. Phys. Chem. Lett.*, 2013, 1802.
173. T. Ameri, P. Khoram, J. Min and C. J. Brabec, *Adv. Mater.*, 2013, **25**, 4245.
174. P. P. Khlyabich, B. Burkhart, A. E. Rudenko and B. C. Thompson, *Polymer*, 2013, **54**, 5267.
175. J.-M. Jiang, H.-C. Chen, H.-K. Lin, C.-M. Yu, S.-C. Lan, C.-M. Liu and K.-H. Wei, *Polym. Chem.*, 2013, **4**, 5321.
176. T. E. Kang, H.-H. Cho, H. J. Kim, W. Lee, H. Kang and B. J. Kim, *Macromolecules*, 2013, **46**, 6806.
177. W.-H. Chang, J. Gao, L. Dou, C.-C. Chen, Y. Liu and Y. Yang, *Adv. Energy. Mater.*, 2014, **4**, 1300864.
178. C. B. Nielsen, R. S. Ashraf, B. C. Schroeder, P. D'Angelo, S. E. Watkins, K. Song, T. D. Anthopoulos and I. McCulloch, *Chem. Commun.*, 2012, **48**, 5832.
179. J. W. Jung, F. Liu, T. P. Russell and W. H. Jo, *Energy Environ. Sci.*, 2013, **6**, 3301.
180. H. Wang, X. Yu, C. Yi, H. Ren, C. Liu, Y. Yang, S. Xiao, J. Zheng, A. Karim, S. Z. D. Cheng and X. Gong, *J. Phys. Chem. C*, 2013, **117**, 4358.
181. W. Sun, Z. Ma, D. Dang, W. Zhu, M. R. Andersson, F. Zhang and E. Wang, *J. Mater. Chem. A*, 2013, **1**, 11141.
182. D. Dang, W. Chen, R. Yang, W. Zhu, W. Mammo and E. Wang, *Chem. Commun.*, 2013, **49**, 9335.
183. L. Fang, Y. Zhou, Y.-X. Yao, Y. Diao, W.-Y. Lee, A. L. Appleton, R. Allen, J. Reinspach, S. C. Mannsfeld and Z. Bao, *Chem. Mater.*, 2013, **25**, 4874.
184. Y. Huang, M. Zhang, H. Chen, F. Wu, L. Zhang and S. Tan, *J. Mater. Chem. A*, 2014, **2**, 5218–5223.
185. P. P. Khlyabich, B. Burkhart, C. F. Ng and B. C. Thompson, *Macromolecules*, 2011, **44**, 5079.
186. B. Burkhart, P. P. Khlyabich and B. C. Thompson, *ACS Macro. Lett.*, 2012, **1**, 660.
187. K. Li, P. P. Khlyabich, L. Li, B. Burkhart, B. C. Thompson and J. C. Campbell, *J. Phys. Chem. C*, 2013, **117**, 6940.
188. E. Zhou, J. Cong, K. Hashimoto and K. Tajima, *Energy Environ. Sci.*, 2012, **5**, 9756.
189. K. H. Hendriks, G. H. L. Heintges, V. S. Gevaerts, M. M. Wienk and R. A. J. Janssen, *Angew. Chem., Int. Ed.*, 2013, **52**, 8341.
190. L. J. A. Koster, S. E. Shaheen and J. C. Hummelen, *Adv. Energy Mater.*, 2012, **2**, 1246.
191. R. C. Chiechi and J. C. Hummelen, *ACS Macro Lett.*, 2012, **1**, 1180.
192. D. J. Burke and D. J. Lipomi, *Energy Environ. Sci.*, 2013, **6**, 2053.
193. R. Po, A. Bernardi, A. Calabrese, C. Carbonera, G. Corso and A. Pellegrino, *Energy Environ. Sci.*, 2014, **7**, 925.
194. M. C. Scharber and N. S. Sariciftci, *Prog. Polym. Sci.*, 2013, **38**, 1929.

CHAPTER 3

Fullerene Derivatives as Electron Acceptors in Polymer Solar Cells

YUTAKA MATSUO[*a]

[a]Department of Chemistry, School of Science, The University of Tokyo, 7-3-1 Hongo, Bunkyo-ku, Tokyo 113-0033, Japan
*E-mail: matsuo@chem.s.u-tokyo.ac.jp

3.1 Design Concepts of Fullerene Acceptors

Fullerene derivatives are commonly used as electron acceptors in organic solar cells (OSCs).[1-3] Fullerenes offer many advantages over other electron-accepting molecules because of their high electron affinity,[4] fast electron transfer from donor materials,[5] small reorganization energy,[6] and slow back electron transfer.[7] In addition, three-dimensional π-conjugation networks and the spherical shape of the fullerene cores are favorable for electron transport and nano-sized phase separation, respectively. Derivatization of fullerenes enables us to tune their electronic properties, solubility, and miscibility with a given donor in order to obtain high performance OSCs. For instance, in OSCs the open-circuit voltage (V_{OC}) is proportional to the difference in energy levels between the lowest unoccupied molecular orbital (LUMO) levels of the acceptor and the highest occupied molecular orbital (HOMO) of the donor.[8-10] By raising the LUMO level of fullerene derivatives we can increase the potential difference between the HOMO of the donor and the LUMO of the fullerene, and thus increase the V_{OC}.

RSC Polymer Chemistry Series No. 17
Polymer Photovoltaics: Materials, Physics, and Device Engineering
Edited by Fei Huang, Hin-Lap Yip, and Yong Cao
© The Royal Society of Chemistry 2016
Published by the Royal Society of Chemistry, www.rsc.org

Two main approaches are routinely used to raise the LUMO levels (*i.e.* decrease the energy level) of fullerene derivatives. One is to introduce electron-donating addends onto the fullerene core, while the alternative method is to reduce the π-conjugation of the fullerene through the addition of organic addends onto fullerenes. Inserting additional addends causes the sp^2 carbon atoms of fullerenes to change to sp^3 carbon atoms, thus losing conjugated π-electrons at the derivatized positions. The latter approach of introducing multiple adducts generally leads to stronger shifts in the LUMO level than in the case of introducing electron-donating addends. This chapter focuses on syntheses, design concepts, and photovoltaic performance of fullerene-based electron acceptors to obtain high power conversion efficiency (PCE) in OSCs. In particular, because the performance of fullerene acceptors has been summarized in previous review papers,[8-12] this chapter is written from a chemist's viewpoint.

3.2 PCBM

PCBM ([6,6]-phenyl-C_{61}-butyric acid methyl ester) was originally synthesized for the purpose of derivatization of fullerenes by Hummelen, Wudl, and co-workers.[13] A functional group, such as the ester group, was conceived to have a large variation of derivatives for wide application in both biological and materials science fields. The C_3H_6 chain between the "methano" carbon atom and the ester group was considered to be of sufficient length to possess solubility in various organic solvents. One phenyl group was employed for synthetic convenience to stabilize the intermediate hydrazone and diazo compounds. Phenyl ketones are easily prepared by Friedel–Crafts acylation of arenes with acid chlorides or nucleophilic addition of aryl Grignard reagents to acid anhydrides (Scheme 3.1). Hydrazones derived from phenyl ketones with hydrazine are more stable than those from alkyl ketones and aldehydes. Mono-aryl-substituted diazoalkanes have acceptable stability compared with non-substituted or alkyl-substituted diazoalkanes. In addition, the products, aryl-alkyl-methanofullerenes such as PCBMs have higher solubility than diphenylmethanofullerenes[14,15] and dihydromethanofullerenes.[16] Heeger and co-workers used PCBMs as soluble fullerene derivatives to construct blended thin-films with a poly(*p*-phenylene-vinylene) donor polymer.[17] In this important work, PCBM played a crucial role in creating an appropriate bulk heterojunction (BHJ) structure for efficient charge separation. $PC_{61}BM$ and its C_{70} analogue $PC_{71}BM$ are still standard materials that are most widely used in BHJ OSCs, because of their good photovoltaic performance, stability, solubility, and availability.

3.2.1 Synthesis of PCBM

PCBMs are synthesized as follows (Scheme 3.1). First, tosylhydrazone is prepared by the reaction of ketone and tosylhydrazine. Tosylhydrazone is a white microcrystalline powder, which can be purified by recrystallization.

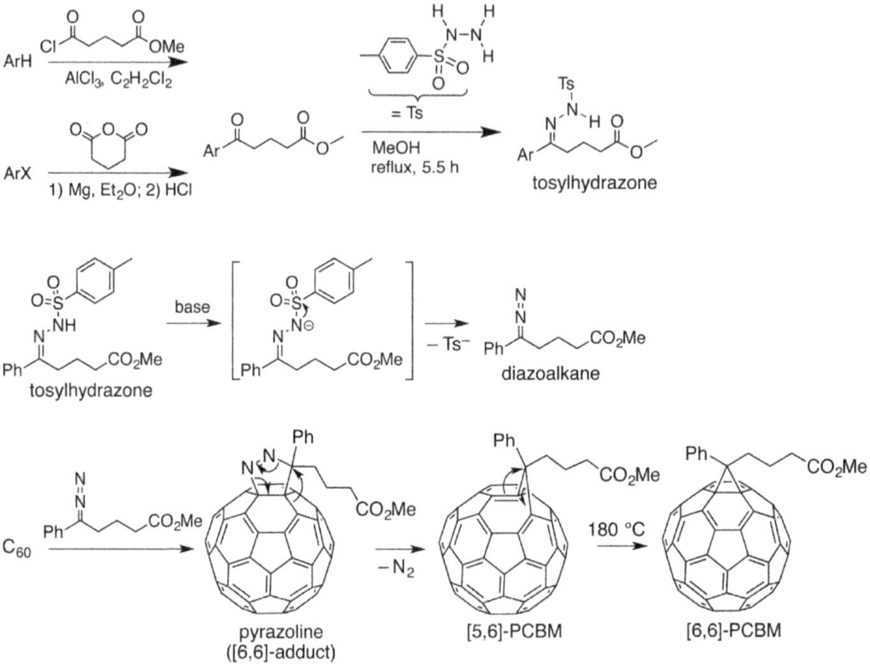

Scheme 3.1   Synthesis of PCBM.

Tosylhydrazone is subjected to deprotonation of the hydrogen atom on the nitrogen atom with base, giving diazoalkane through elimination of the tosyl anion. Diazoalkane reacts with C_{60} to give pyrazoline, which is a five-membered 6,6-adduct of fullerene. Upon heating, elimination of dinitrogen, N_2, produces a fulleroid-type 5,6-open PCBM. Heating or light irradiation to 5,6-PCBM gives thermodynamically stable 6,6-PCBM. Overall chemical yield from C_{60} is typically 50–60%. The product 6,6-PCBM can be easily purified by silica gel column chromatography owing to the existence of a polar ester group.

3.2.2   Fundamental Properties of PCBMs

Reported LUMO levels of PCBM vary between −3.7 eV and −4.3 eV, because of different measurement methods, either in solution or solid, and different equations used to estimate the values.[9] One of the most convenient techniques to determine the LUMO level of fullerenes is cyclic voltammetry.[18] The LUMO levels of fullerene derivatives can be estimated from the first reduction potential. Because reduction potentials depend on the choice of the reference electrode, and vary slightly as a result of different electrochemical cell settings and choice of solvent, the use of an internal reference compound is required for accurate determination of the redox potentials of fullerenes, and any other molecule for that matter. Ferrocene is the most widely used reference compound because of its stable redox behavior. After the cyclic voltammetry, ferrocene is added to the electrochemical cell, and scanned to have an oxidation potential of ferrocene in this experimental setting for determination of the half-wave reduction potentials $(E_{1/2}^{red})$ of the samples against a ferrocene–ferrocenium couple (Fc–Fc$^+$). To estimate the LUMO levels of fullerene derivatives, we use the following equation: LUMO level = $-(4.8 + E_{1/2}^{red}$ vs. Fc/Fc$^+$) eV. It is to be noted that the reduction potential is also varied by the choice of solvent. The use of the same solvent is necessary for a comparison of the LUMO levels. For instance, reduction potentials in benzonitrile are usually *ca.* 0.1 V shifted to the negative direction compared with those in tetrahydrofuran (THF). PCBM shows $E_{1/2}^{red} = -1.00$ V vs. Fc/Fc$^+$ in THF, giving the LUMO level of −3.80 eV according to the above-mentioned equation. A similar LUMO level (*ca.* −3.8 eV) of PCBM has been obtained by near-ultraviolet inverse photoemission spectroscopy using ultra-low energy electrons.[19] To avoid confusion, LUMO levels determined by cyclic voltammetry are discussed in this chapter.

PCBM is usually regarded as a partly crystalline fullerene derivative. Differential scanning calorimetry analysis of PCBM shows an exo-thermic peak at 195 °C due to thermally driven crystallization.[20] X-ray crystallographic structures of PCBM containing chlorobenzene or 1,2-dichlorobenzene in their unit cells have been reported by Hummelen *et al.*[21] The moderately crystalline nature of PCBM in the thin-films provides respectably high electron mobility, whereas over-crystallization or over-aggregation decrease the photovoltaic performance and device stability. This undesired aggregation behavior is, in some cases, inhibited by the use of hard donor materials, such as poly(3-hexylthiophene) (P3HT) and small molecule donors,[22] or modification of PCBM to make it less crystalline (see below).

3.2.3 PCBM Derivatives in Photovoltaic Applications

To achieve high V_{OC} in OSCs, raising the LUMO level of PCBM was attempted by installation of electron-donating groups to PCBM (Scheme 3.2).[23] One to three methoxy groups, methylthio groups, or ethylene dioxy groups were attached to the phenyl group of PCBM. Electron-donating groups were installed in the ketone starting materials, and modified PCBMs were synthesized. Except for compounds 7 and 9 that have low solubility, higher V_{OC} was obtained by the use of compounds with higher LUMO levels. However, the maximum increase in V_{OC} observed was only 30 mV, when compound 6 was used with MDMO-PPV (poly[2-methoxy-5-(3,7-dimethyloctyloxy)-*p*-phenylenevinylene]). Further increases in V_{OC} have been achieved by changing the size or shape of π-conjugate systems of fullerene derivatives (*vide infra*).

To achieve highly stable BHJ thin-film morphology, amorphous PCBM derivatives have been developed by Jen and co-workers (Scheme 3.3).[24] They synthesized PCBM derivatives having bulky aromatic groups such as triphenylamine (TPA) and 9,9-dimethylfluorene (MF). The TPA-PCBM (10) and MF-PCBM (11) obtained have high glass transition temperatures, T_g, of

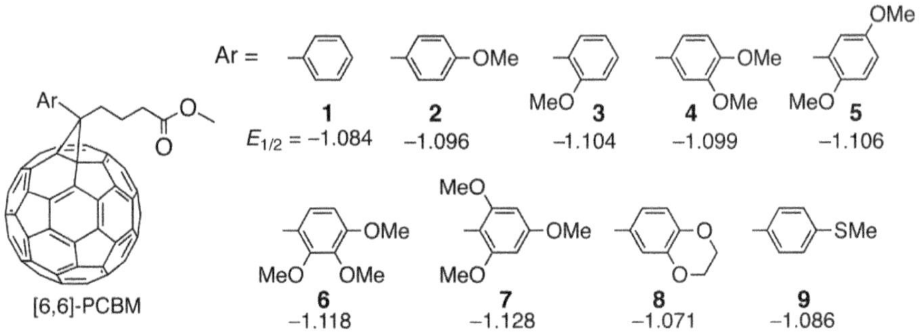

Scheme 3.2 PCBM derivatives. Values under compound numbers are the first half-wave reduction potentials against a standard ferrocene–ferrocenium couple, determined by cyclic voltammogram in 1,2-dichlorobenzene : acetonitrile (4 : 1) solution.

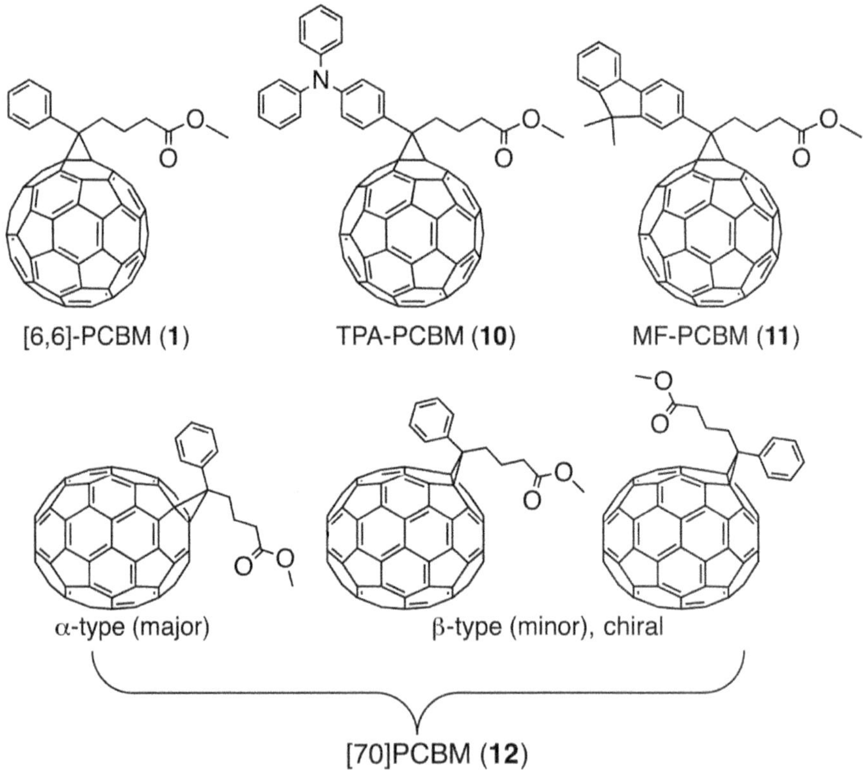

Scheme 3.3 Amorphous PCBMs.

170 °C and 180 °C, respectively, which are superior to that of PCBM (T_g = 122 °C). Compounds **10** and **11** reduce the amount of fullerene aggregation. These three compounds showed similar PCE in P3HT-based BHJ OSCs after thermal annealing at 150 °C for several tens of minutes, while amorphous PCBMs **10** and **11** showed much improved thermal stability even after thermal

annealing at the same temperature for 10 hours. This is contrary to the fact that long thermal annealing of the P3HT:PCBM devices causes a significant drop in PCE.

3.2.4 [70]PCBM

[70]PCBM (PC$_{71}$BM, **12**) is a C$_{70}$ acceptor frequently used for OSCs, especially with low bandgap polymer-based solar cells, to give high PCE.[25] [70]PCBM is synthesized as a mixture of isomers containing a major α-type isomer (*ca.* 85%) and minor β-type isomers (Scheme 3.3). Because separation of a single isomer from a mixture is difficult, [70]PCBM is used as a mixture in photovoltaic applications. Although C$_{70}$ derivatives are much more expensive than their C$_{60}$ analogues, C$_{70}$ acceptors have a favorable light absorption property. The measured extinction coefficient of [70]PCBM at 470 nm is 20 000 mol^{-1} cm^{-1}, which is nearly 20 times larger than that of [60]PCBM.[26] At the longer wavelength region, [70]PCBM has a larger extinction coefficient than [60]PCBM (at 650 nm, 2000 mol^{-1} cm^{-1} for [70]PCBM, <1000 mol^{-1} cm^{-1} for [60]PCBM). A regioisomeric mixture of [70]PCBM has similar or slightly lower-lying LUMO level (−3.82 eV) than [60]PCBM (−3.80 eV) on the basis of the above-mentioned cyclic voltammetric technique ($E^{red}_{1/2}$ = −0.98 V *vs.* Fc–Fc$^+$).

3.2.5 Mix-PCBM

The production of fullerenes produces not only C$_{60}$ but also C$_{70}$, as well as higher fullerenes. The separation of this mixture of fullerenes is time consuming and expensive. Mix-PCBM, the mixture of [60]PCBM (*ca.* 85%) and [70]PCBM (*ca.* 15%), has been reported as a promising electron acceptor for OSCs (Figure 3.1 and Table 3.1).[27] The use of mix-PCBM is cost effective because it is made from a cheap mixture of C$_{60}$ and C$_{70}$. Applying mix-PCBM in a standard P3HT:PCBM device showed PCE = 3.34% with V_{OC} = 0.63 V, short-circuit current density (J_{SC}) = 9.70 mA cm^{-2}, and fill factor (FF) = 0.55. In comparison, the P3HT:[60] PCBM device showed very similar performance (PCE = 3.27, V_{OC} = 0.63 V, J_{SC} = 8.28, FF = 0.63) under similar experimental conditions, with the mix-PCBM device giving higher J_{SC} but lower FF.

The inclusion of [70]PCBM prevents [60]PCBM aggregation and affords more stable devices. Mix-PCBM forms smaller domains than [60]PCBM during thermal annealing, because of lower crystallinity resulting from the nature of a mixture. This brings about a larger interface area between donor and acceptor materials, giving higher J_{SC}. In addition, the higher light absorption property of [70]PCBM may contribute to the higher J_{SC}. On the other hand, the lower charge transport ability of mix-PCBM likely comes from lower crystallinity due to the nature of the mixture giving lower FF. More importantly, the mix-PCBM device was more stable to heating at 150 °C than the [60]PCBM device, due to little morphological change, which was characterized by atomic force microscope and ultraviolet-visible (UV-vis) absorption measurements.

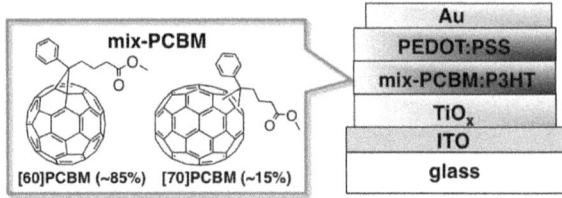

Figure 3.1   Device using mix-PCBM (ref. 27) ©AIP Publishing.

Table 3.1   Optimal device performance of P3HT:mix-PCBM devices under AM1.5G illumination (100 mW cm^{-2}) (ref. 27).

Entry	Acceptor (donor:acceptor)	Device type	Annealing time (min)	V_{OC} (V)	J_{SC} (mA cm^{-2})	FF	PCE (%)
1	Mix-PCBM (5:4)	Inverted	5	0.63	8.55	0.60	3.23
2	[60]PCBM (5:4)	Inverted	10	0.63	8.28	0.63	3.27
3	Mix-PCBM (5:3)	Inverted	5	0.63	9.70	0.55	3.34
4	Mix-PCBM (5:4)	Normal	5	0.64	8.64	0.63	3.42
5	[60]PCBM (5:4)	Normal	10	0.62	8.24	0.67	3.45
6	Mix-PCBM (5:3)	Normal	5	0.64	8.72	0.64	3.59

3.3   1,4-Di(organo)fullerene

We have proposed that changing the shape of fullerene π-conjugated systems from a 1,2-addition pattern to 1,4-addition can raise the LUMO levels of fullerene derivatives to obtain high V_{OC} (Figure 3.2).[8,28] Fullerene derivatives with a 1,4-addition pattern have smaller π-conjugated systems than those with a 1,2-addition pattern, and thus the LUMO levels of such 1,4-diadducts are higher owing to their lower electron affinity. Thus, 1,4-di(organo)[60]fullerenes are an interesting class of fullerene acceptors. A wide variety of the same or two different organic addends are installed on the fullerene core to produce symmetric or asymmetric fullerene derivatives with the same or two different organic addends.

3.3.1   Silylmethylfullerene (SIMEF)

1,4-Bis(silylmethyl)[60]fullerenes (SIMEFs)[18,20,28–34] can be synthesized through a two-step reaction (Scheme 3.4). The first step is efficient dimethylformamide (DMF)-assisted monoaddition of silylmethyl Grignard reagent to C$_{60}$. In this reaction, 3 equiv. of Grignard reagent and 30 equiv. of DMF are generally used to produce the monoadducts, hydro(silylmethyl)[60]fullerene C$_{60}$(CH$_2$SiMe$_2$R^1)H. The products are obtained in 5–10 min at room temperature in 70–93% yield. DMF accelerates nucleophilic attack through coordination to magnesium in the Grignard reagents and stabilizes anionic reaction intermediates before acid treatment. The second step is nucleophilic substitution of silylmethyl halides with monoadduct anions, which are generated

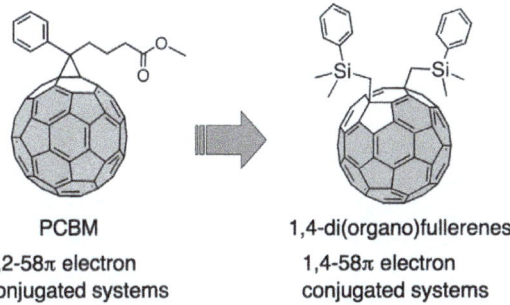

PCBM

1,2-58π electron
conjugated systems

1,4-di(organo)fullerenes

1,4-58π electron
conjugated systems

Figure 3.2 Shrinking π-electron conjugated systems by use of 1,4-addition (ref. 28).

1) R¹Me₂SiCH₂MgCl (3 eq)
DMF (30 eq)
1,2-Cl₂C₆H₄
25 °C, 5 min
2) H⁺

C₆₀

up to 93%

R¹ = alkyl and aryl groups

1) KO'Bu
2) R²Me₂SiCH₂Cl (20 eq)
KI (20 eq)
PhCN
110 °C, 8 h

up to 93%

R² = alkyl and aryl groups

SIMEF (13): R¹ = R² = Ph
SIMEF2 (14):
R¹ = Ph; R² = 2-MeOC₆H₄

Scheme 3.4 Synthesis of SIMEF.

by deprotonation of the monoadduct with base. This reaction proceeds in a regioselective manner to afford 1,4-diadducts $C_{60}(CH_2SiMe_2R^1)(CH_2SiMe_2R^2)$ in overall yields of 50–80% from C_{60}. As the carbon–silicon coupling reaction is much easier than the carbon–carbon one, a variety of substituted-silyl-methyl halides can be easily prepared from a commercially available chloro-methylsilyl chloride starting material. Various SIMEF derivatives have been synthesized with various R^1 and R^2 groups, which enables the electronic properties, solubility, crystal packing structure, and thermal properties of SIMEFs to be selectively tuned.

SIMEF (**13**), having two phenyl groups on the silicon atoms, has a higher LUMO level (−3.74 eV) than PCBM (−3.80 eV), because of the smaller size of the fullerene π-electron conjugated system. SIMEF bearing one phenyl group and one 2-methoxyphenyl group (SIMEF2, **14**) has a slightly higher LUMO level (−3.72 eV) due to the electron-donating nature of the methoxy group on the phenyl ring. The thermal crystallization behavior of SIMEF is especially important, with SIMEF being transformed from an amorphous solid to a crystalline solid at 150 °C. In this crystalline solid, the C_{60} cores are aligned in straight columnar arrays (Figure 3.3). Close inspection of the columns through single crystal X-ray diffraction (XRD) analysis shows that the organic addends on one fullerene tend to pack together with the addends on the adjacent fullerene, while the fullerene cores also align well with each other.

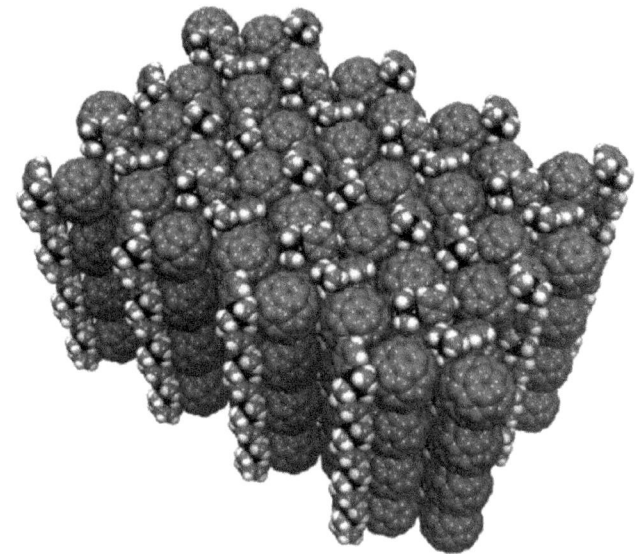

Figure 3.3   Crystal packing structure of SIMEF (ref. 20).

SIMEF exhibits relatively high mobility as an n-type organic semiconductor: 5.8×10^{-2} cm^2 V^{-1} s^{-1} (field-effect transistor) and 8.0×10^{-3} cm^2 V^{-1} s^{-1} (space-charge limited current, SCLC). SIMEFs give high V_{OC} without any decrease in J_{SC} in P3HT-based BHJ polymer solar cells, because of their high LUMO levels and high electron mobility.[30] In addition, SIMEF plays an important role in the formation of thermally driven columnar crystal formation of tetrabenzopor-phyrin (BP) donor in BP:SIMEF three-layered p–i–n small-molecule OSCs.[20]

3.3.2   1,4-Di(aryl)fullerene

1,4-Diaryl[60]fullerenes generally have higher stability but lower solubility than 1,4-dialkyl[60]fullerenes.[35] Various methods to obtain 1,4-dialkyl[60] fullerenes, 1,4-$C_{60}Ar_2$, have been reported (Scheme 3.5). (1) Oxidation of C_{60} using *m*-chloroperoxybenzoic acid produces fullerene epoxide $C_{60}O$.[36] Puri-fied $C_{60}O$ is treated with trifluoroborane diethyl etherate in arenes to gener-ate $C_{60}(OBF_3)^+$, which reacts with electron-rich arenes to produce 1,4-$C_{60}Ar_2$.[37] $C_{60}Ar(OH)$ is also obtained as an intermediate. (2) Reaction of arylhydrazine with sodium nitrite $NaNO_2$ gives 1,4-arylfullerenol, 1,4-$C_{60}Ar(OH)$,[38] which is reacted with arenes in the presence of *p*-toluenesulfonic acid to obtain 1,4-$C_{60}Ar_2$.[38,39] (3) A Friedel–Crafts type reaction of C_{60} with arenes in the pres-ence of $AlCl_3$ and a small amount of H_2O produces a hydroarylated product, $C_{60}Ar_2H_2$,[40] which is subjected to deprotonation with base and oxidation with copper salt to obtain 1,4-$C_{60}Ar_2$.[35] (4) Rhodium-catalyzed monoarylation of C_{60} using arylboronic acids gives a monoadduct, $C_{60}ArH$,[41] and then palladi-um-catalyzed cross-coupling reaction of the monoadduct with aryl halides

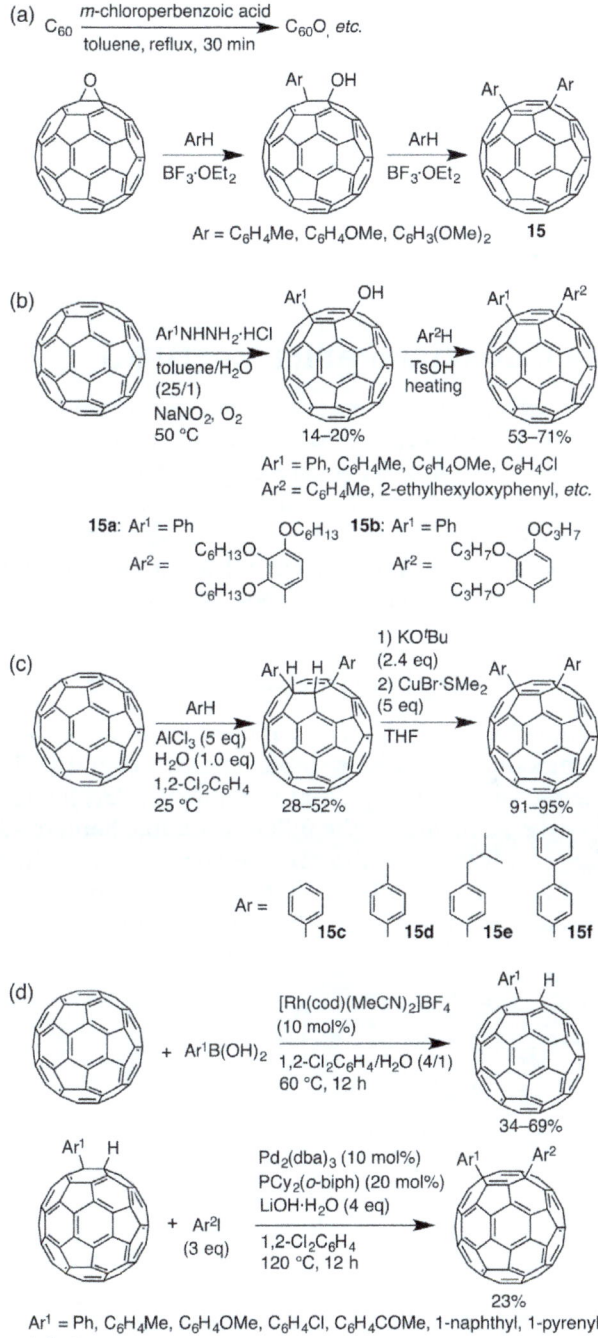

Scheme 3.5   Synthesis of 1,4-di(aryl)[60]fullerenes. (a) Starting from fullerene oxide. (b) Reaction with arylhydrazine. (c) Friedel–Crafts type reaction. (d) Rhodium-catalyzed monoarylation.

affords 1,4-$C_{60}Ar_2$.[42] In general, 1,4-aryl[60]fullerenes have lower LUMO levels than 1,4-alkyl[60]fullerenes, because of the electron-withdrawing nature of the aryl groups. The LUMO levels of 1,4-alkyl[60]fullerenes can be raised by introduction of electron-donating alkoxy groups on the phenyl group. Compound **15a**, with three alkoxy groups, has a 90 meV-higher LUMO level than PCBM. However, the reported performance of 1,4-diaryl[60]fullerene-based OSCs is moderate. The device using **15b** and P3HT also exhibited a PCE of 2.3%,[39] while a small-molecule OSC using $C_{60}(4\text{-Ph-}C_6H_4)_2$ (**15f**) as an acceptor and BP as a donor also showed 2.3% PCE.[35]

3.4 Diphenylmethanofullerene (DPM)

Martín and co-workers, in collaboration with several groups active in OSC research, have synthesized a series of diphenylmethanofullerene (DPM, **16**) derivatives, which have provided interesting insights into the relationship between fullerene structure and device performance.[43-46] DPMs are easily synthesized, show high stability, and have almost the same LUMO level as PCBM. The addition of an extra phenyl ring however leads to significant differences in the *J–V* characteristics observed for both polymer and small molecule OSCs.[22]

3.4.1 Synthesis of Diphenylmethanofullerene

Diphenyl(diazo)methane, Ph_2CN_2, is a typical 1,3-dipolar species (Scheme 3.6).[14] This compound is relatively stable, and can be purified by recrystallization.[47] The terminal nitrogen atom in diphenyl(diazo)methane partially has negative charge, and attacks the fullerene double bond to afford cyclization to produce an intermediate five-membered pyrazoline, which undergoes elimination of N_2 to give a [5,6]-adduct, fulleroid. Thermal treatment of the [5,6]-adduct produces a [6,6]-adduct, diphenylmethano[60]fullerene.

pyrazoline   [5,6]-adducts   [6,6]-adducts
                (fulleroides)   (methanofullerenes)

16: Ar = 4-$C_{12}H_{25}OC_6H_4$
(DPM-12)

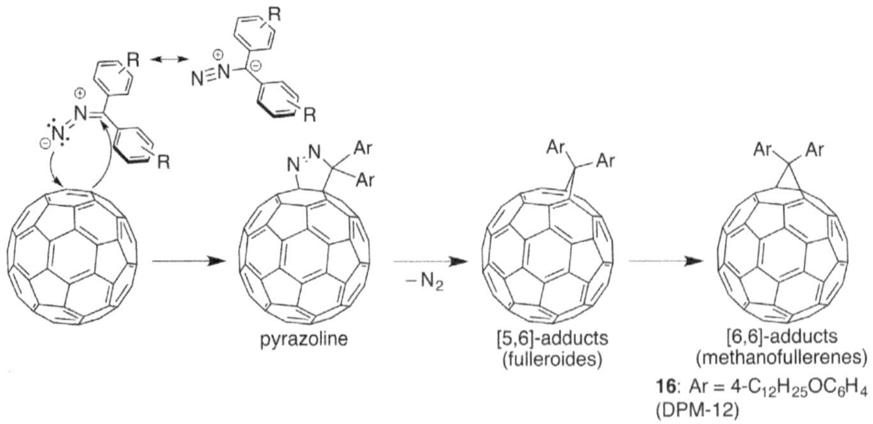

Scheme 3.6 Synthesis of diphenylmethano[60]fullerenes.

3.4.2 Photovoltaic Application

The first DPM molecule applied to OSCs was DPM-12, where 12 denotes the number of carbons in the alkyl chain attached to the phenyl rings, which was first studied in 2004 and showed a higher V_{OC} than PCBM with either a P3HT or OC_1C_{10}-PPV donor, even though both fullerene acceptors have the same LUMO level.[43] The increase in V_{OC} for P3HT:DPM-12 devices compared to P3HT:PCBM is approximately 100 mV.[43,44] Reducing the alkyl chain length from C_{12} to C_6 improved the electron mobility of P3HT:DPM-6 devices, which in turn improved the J_{SC}, and PCE improved from 2.3% to 2.6%.[48] The V_{OC} was also higher for P3HT:DPM-6 than the reference P3HT:PCBM devices, in this case only by ~50 mV, but J_{SC} was still lower for P3HT:DPM-6 devices.

Independent studies by Palomares[45] on DPM-12 and Bisquert[46] on DPM-6 have investigated the origin of the higher V_{OC} by transient optoelectronic techniques and impedance spectroscopy, respectively. Both studies observed a shift in the density of states (DOS) to higher potential for P3HT:DPM devices as well as slower non-geminate recombination dynamics. Palomares and co-workers assigned the positive shift in the DOS to the HOMO of P3HT, which may be due to DPM-12 impeding the crystallization of P3HT, as was observed by differential scanning calorimetry studies. Cyclic voltammetry of the active layer films also showed an increase in the oxidation potential for P3HT when blended with DPM-12, consistent with the positive shift in the DOS. Bisquert and co-workers, on the other hand, ascribe the positive shift in the DOS for P3HT:DPM-6 devices to the occupancy of the DOS in DPM-6 differing from that of PCBM. In the case of DPM-6 they find that the electronic band is fully occupied, while for PCBM charges reside in the tail of the DOS; these are lower in energy and thus lead to a lower potential difference between the quasi-Fermi levels that, together with the rate of non-geminate recombination, determines the V_{OC}.

Recently, C_{70} DPMs have been synthesized with C_4, C_6 and C_{12} alkyl chains and have been applied to small molecule BHJ OSCs where the donor molecule was 3,6-bis(5-(benzofuran-2-yl)thiophen-2-yl)-2,5-bis(2-ethylhexyl)pyrrolo[3,4-c]pyrrole-1,4-dione (DPP(TBFu)$_2$).[49] Similar to DPM:P3HT devices, [70]DPM showed higher V_{OC} than [70]PCBM due to longer charge carrier lifetimes, and the J_{SC} was seriously affected by the influence of DPM molecules on the molecular packing of DPP(TBFu)$_2$ following annealing. [70]DPM-6 devices were found to have the highest J_{SC}, with [70]DPM-4 devices having the lowest J_{SC}.

3.5 Fulleropyrrolidine

Fulleropyrrolidines (**17**) are obtained through a facile synthetic one-pot reaction, and can be easily purified with silica gel column chromatography. Fulleropyrrolidines offer a wide diversity of properties by changing substituents in the starting materials. Fulleropyrrolidines have 58π-electron conjugated systems with a 1,2-addition pattern, giving similar LUMO levels to PCBM.

Nevertheless, fulleropyrrolidines may have advantages over PCBM because of their potential cost-effectiveness and facile synthesis.

3.5.1 Synthesis of Fulleropyrrolidine

Fulleropyrrolidines are synthesized through the addition of azomethine ylides as 1,3-dipolar compounds to fullerene (Scheme 3.7).[50] This reaction is known as the Prato reaction. There are several methods of generating azomethine ylides. In the Prato reaction, a "decarboxylation route"[51] is employed with the use of easily available aldehydes and α-amino acid derivatives. Both starting materials undergo dehydrative condensation to give 5-oxazolidinones, which generate azomethine ylides with elimination of CO_2. Azomethine ylides react with C_{60} and produce fulleropyrrolidines in moderate yields.

3.5.2 Photovoltaic Applications

It has been reported that some fulleropyrrolidines show a similar or slightly higher photovoltaic performance compared with PCBM in P3HT-based BHJ devices. Itoh *et al.* have examined various fulleropyrrolidines, and found that the compound with an *ortho*-methoxy-substituted phenyl group gave the best performance among their tests.[52] Similar results have been obtained with *ortho*-methoxy-substituted PCBM,[23] SIMEF (SIMEF2, **14**),[32] and 1-alkyl-2-aryl-[60]fullerenes.[53] The higher electron-donating effect of the *ortho*-methoxy-substituted phenyl group compared with *para*-methoxy-substituted one is a rather unconventional observation in the context of standard organic chemistry, whereas through-space electron donation from the intimate electron-rich oxygen atom to the fullerene core is likely to contribute to raising LUMO levels.[53] The reported best PCE value of the P3HT:fulleropyrrolidine device was 3.44%, which was higher than that of the reference P3HT:PCBM device (2.53%) in their examination.[52]

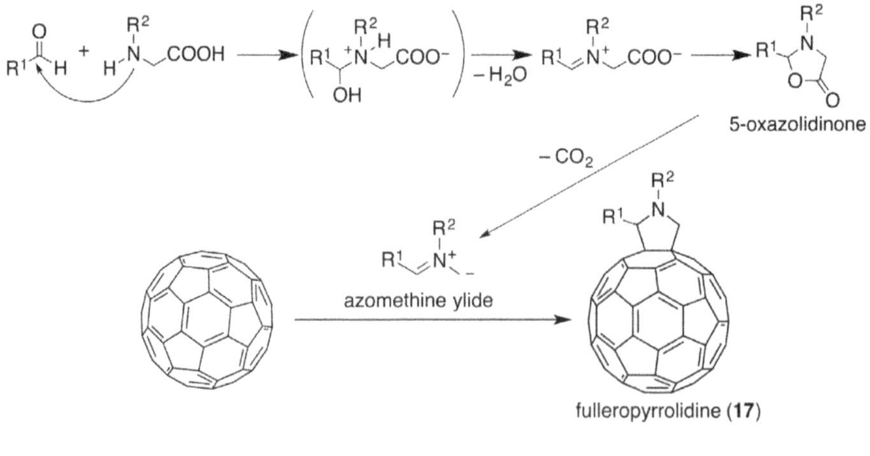

Scheme 3.7 Synthesis of fulleropyrrolidines.

3.6   56π-Electron Conjugated Fullerene Derivatives

Further shrinking of the π-electron conjugated systems of fullerene derivatives to have 56π-systems can effectively raise the LUMO levels of fullerene derivatives. The first 56π-electron fullerene acceptor tested in OSCs was bis-PCBM (**18**),[54] which was isolated from the reaction mixture in the synthesis of PCBM. The bis-PCBM showed high V_{OC} (0.72 V) because of its high-lying LUMO level. The PCE value of 4.5% was about a factor of 1.2 larger than that of the reference P3HT:PCBM device (3.8%) because of the increase in V_{OC}. Current OSC research more often uses 56π-fullerenes obtained by [4+2] Diels–Alder cycloadditions because of their high performance and easiness of synthesis.

3.6.1   Diels–Alder Reactions

Fullerene can act as a dienophile, and react with various dienes (Scheme 3.8a). Diels–Alder reactions of C_{60} with dienes produce a variety of [4 + 2] cycloadducts, while retro Diels–Alder reactions also can take place to give back the starting material, C_{60}. There are mainly two methods used to prevent the retro Diels–Alder reactions. One is the use of electron-withdrawing groups in dienes (Scheme 3.8a, right). The other is aromatization of the Diels–Alder products (Scheme 3.8b and c). A typical example is addition of *ortho*-quinodimethane, also known as *ortho*-xylylene.[55] 1,2-Bis(bromomethyl)benzene is reacted with potassium iodide for substitution of the bromo group by the iodo group in the presence of 18-crown-6 as a phase-transfer catalyst, followed by elimination of I_2, giving *ortho*-quinodimethane. This reactive diene proceeds to react with C_{60} to give an aromatized product, C_{60}(*ortho*-quinodimethane). Another example is the addition of isoindene.[56] Heating of indene causes isomerization to isoindene, which reacts with C_{60} to produce an aromatized compound, C_{60}(indene). In this chapter, C_{60}(*ortho*-quinodimethane) and C_{60}(indene) are abbreviated to C_{60}(QM) (**19**) and C_{60}(Ind) (**20**), respectively.

3.6.2   Indene-C_{60} Bis-Adducts (ICBA) and Related Compounds

Indene-C_{60} bis-adducts (ICBA) (also written as C_{60}(Ind)$_2$ (**21**) in this chapter; Scheme 3.9), was first invented by Plextronics and NanoC in 2006–2007,[57] and separately reported by Li *et al.* in 2010.[10,58] ICBA is synthesized in a one-pot reaction using fullerene and indene. ICBA is separated and purified as a mixture of regioisomers by means of chromatographic techniques. ICBA gives high V_{OC} and PCE especially in P3HT-based BHJ OSCs. With an optimized device structure performing solvent annealing and pre-thermal annealing at 150 °C for 10 min, the P3HT:ICBA device exhibited PCE of 6.48% with V_{OC} of 0.84 V, J_{SC} of 10.6 mA cm^{-2}, and FF of 72.7%.[59] ICBA was also used in an inverted structure OSC incorporating a zinc oxide electron collecting layer with a cross-linked fullerene interlayer, and showed PCE of 6.2%.[60] To improve absorption of visible light, a C_{70} congener, IC$_{70}$BA, has been developed. The PCE value of the P3HT:IC$_{70}$BA device was first reported as 5.64% (V_{OC} = 0.84 V).[61] With optimization of device

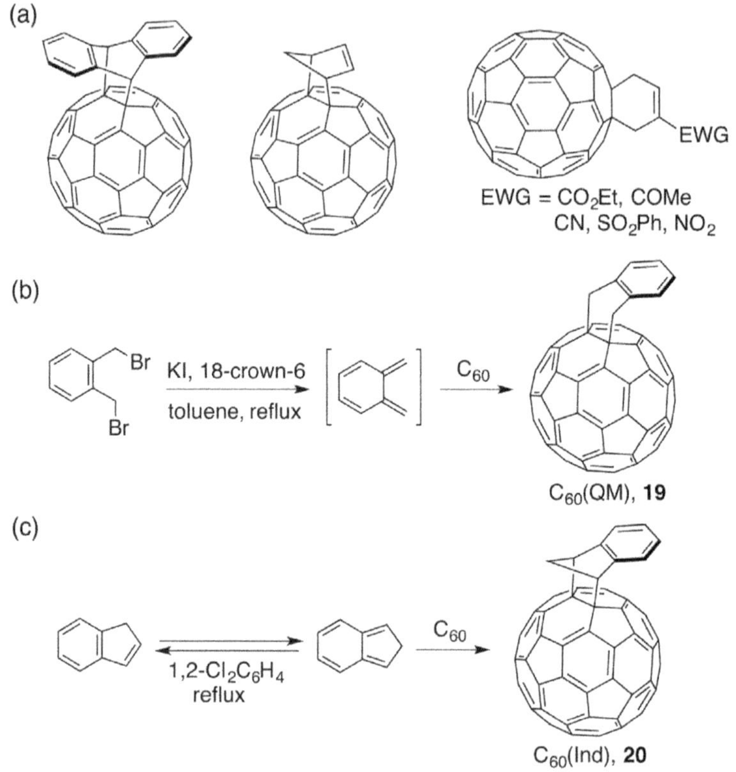

(a)

EWG = CO$_2$Et, COMe
CN, SO$_2$Ph, NO$_2$

(b)

C$_{60}$(QM), **19**

(c)

C$_{60}$(Ind), **20**

Scheme 3.8 Diels–Alder reactions used in the synthesis of fullerene acceptors. (a) Relatively unstable Diels–Alder adducts of anthracene and cyclopentadiene, as well as stable Diels–Alder adducts containing electron-withdrawing groups. (b) Diels–Alder reaction to obtain a quinodimethane adduct. (c) Diels–Alder reaction to obtain an indene adduct.

fabrication, the P3HT:IC$_{70}$BA (1:1, w/w) device with 3 vol% 1-chloronaphthalene and pre-thermal annealing at 150 °C for 10 min exhibited PCE of 7.40% with V_{OC} of 0.87 V, J_{SC} of 11.35 mA cm^{-2} and FF of 75.0%.[62]

On the basis of the success of ICBA, related 56π-electron conjugated fullerene acceptors have been developed. First, a C$_{60}$ bis-adduct of *ortho*-quinodimethane (*ortho*-xylylene) was reported by several research groups.[63–65] Although each research group gave this compound different names, such as bis-oQDMC$_{60}$,[63] OXCBA,[64] and NC$_{60}$BA (dihydronaphthyl C$_{60}$ bis-adduct),[65] this fullerene acceptor is written as C$_{60}$(QM)$_2$ (**22**) in the Scheme in this chapter to avoid confusion. The device using C$_{60}$(QM)$_2$ with P3HT showed PCE of 5.2–5.4%.[63–65] Other 56π-[60]fullerenes such as a DPM-type bis-adduct (PCE = 5.2%),[66] and compounds using substituted quinodimethanes (PCE = 4.58%)[67] and the thiophene ring instead of the benzene ring (PCE = 5.1%)[68] have been reported for use in P3HT-based devices. A C$_{70}$ bis-adduct of *ortho*-quinodimethane has also been reported by Wang *et al.*[69] A PCE value of 5.95% was obtained with a BHJ P3HT:C$_{70}$(QM)$_2$ device.

56π and 66π bis-adducts

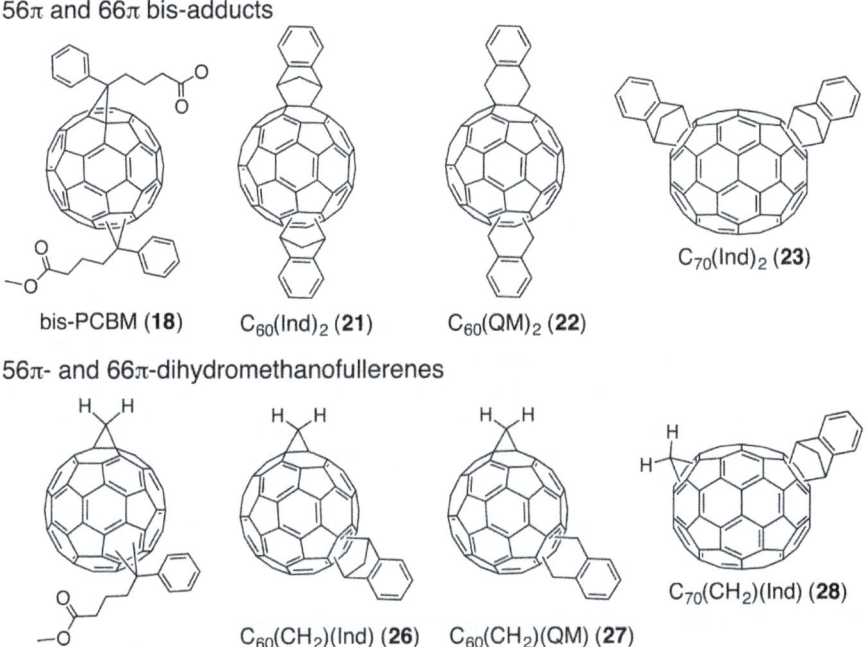

bis-PCBM (**18**)   C$_{60}$(Ind)$_2$ (**21**)   C$_{60}$(QM)$_2$ (**22**)   C$_{70}$(Ind)$_2$ (**23**)

56π- and 66π-dihydromethanofullerenes

methano-PCBM (**25**)   C$_{60}$(CH$_2$)(Ind) (**26**)   C$_{60}$(CH$_2$)(QM) (**27**)   C$_{70}$(CH$_2$)(Ind) (**28**)

Scheme 3.9   56π-[60]fullerene and 66π-[70]fullerene acceptors: bis-adducts and dihydromethanofullerenes.

In general, ICBA acts as a good electron acceptor alongside P3HT. The use of ICBA in the heterojunction devices using low-bandgap donor–acceptor copolymers is, however, a challenge because low-bandgap polymers tend to have low-lying LUMO levels, which can lead to poor exciton dissociation at the donor–acceptor interface. The offset in energy between donor and acceptor LUMO levels must exceed the Coulomb energy binding the exciton, which empirically has been observed to be 0.3 eV for many donor–acceptor systems. Several groups recently reported efficient PCE (5.35–5.4%) in the ICBA-based devices with donor–acceptor copolymers.[70,71] In these devices, high V_{OC} values of 0.92–1.03 V were obtained because of lowered HOMO levels of polymers due to the existence of acceptor units in the copolymers.

Another recent challenge is investigation of the effect of an isolated single isomer. The first attempt has been made by Imahori *et al.* using C$_{60}$(QM)$_2$ derivatives bearing polar alkoxycarbonyl groups for separation and purification.[72] The PCE values obtained were up to 1.44%, due to the use of bulky substituents. Successful data were also obtained by Wang *et al.*[73] They performed laborious chromatographic separation of each isomer of C$_{60}$(QM)$_2$ without using polar substituents, and found that the four isolated isomers had different photophysical and electrochemical properties as well as electron mobility. They obtained varying PCE values of 5.8, 6.3, 5.6, and 5.5%, which were higher than that for the C$_{60}$(QM)$_2$ mixture (5.3%). A single isomer of C$_{70}$(QM)$_2$ has been also successfully isolated and characterized crystallographically by

Wong *et al.*, and showed PCE of 5.9% on the P3HT-based BHJ OSCs.[74] An alternative method to obtain a single regioisomer has been proposed by us, with a regioselective nucleophilic addition reaction to C_{60} and C_{70}.[75–77] In this method, single isomers of 56π-[60]fullerene[77] and 66π-[70]fullerene[76] acceptors have been synthesized, and showed PCE of 3.4% and 3.3%, respectively.

3.7   Dihydromethanofullerene

In 2011, we, together with Nakamura, proposed a concept using a dihydromethano group as the smallest carbon addend to reduce steric hindrance to give high electron mobility in fullerene derivative solids to obtain high J_{SC}, FF, and PCE.[77,78] The dihydromethano group has higher stability than comparably small epoxy (–O–) and aziridino (–NH–) groups. A key to constructing the dihydromethano group on the fullerene is a DMF-assisted efficient monoaddition reaction of a silylmethyl group bearing the isopropoxy group on the silicon atom, as well as oxidation of the fullerene derivative's anion with copper oxidant to give an intermediate fullerene derivative cation, followed by spontaneous cyclization (Scheme 3.10).[77–82]

3.7.1   Synthesis of Dihydromethanofullerene

The DMF-assisted monoaddition reaction of a silylmethyl Grignard reagent is commonly the first step in the synthesis of SIMEF (see Schemes 3.4 and 3.10). A different point is the use of an electron-donating alkoxy group to accelerate the latter cyclization step. The product, monoadduct $C_{60}\{CH_2SiMe_2(O^iPr)\}H$, is usually obtained in high yield (*ca.* 80%). The monoadduct is subjected to deprotonation to generate a monoadduct anion $C_{60}\{CH_2SiMe_2(O^iPr)\}^-$, which is oxidized with copper(II) chloride to

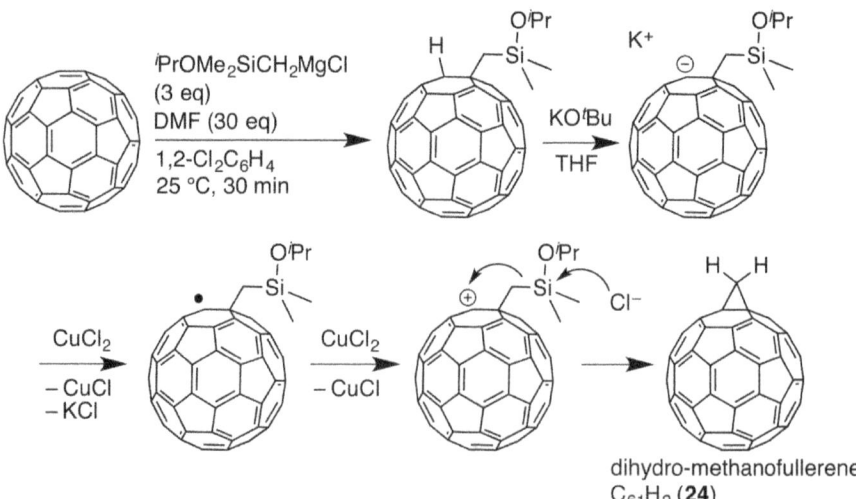

Scheme 3.10   Efficient synthesis of dihydromethanofullerene.

generate a key intermediate, the monoadduct cation $C_{60}\{CH_2SiMe_2(O^iPr)\}^+$. Through heating the cation, chloride attacks the silicon atom, and electron flows from the silylmethyl moiety to the fullerene cation to form a carbon–carbon bond in this cyclopropanation reaction to give dihydromethano[60] fullerene, $C_{61}H_2$ (**24**), in high yield (*ca.* 80% from the monoadduct).

3.7.2 56π-Dihydromethanofullerene

Methano-indene-fullerene, $C_{60}(CH_2)(Ind)$ (**26**), was synthesized in 47% yield by the Diels–Alder reaction of indene with **24**. The product **26** is a mixture of regioisomers; two isomers were successfully isolated by high-performance liquid chromatography (HPLC) separation and one (C_S-isomer) of two was crystallographically characterized (Figure 3.4).

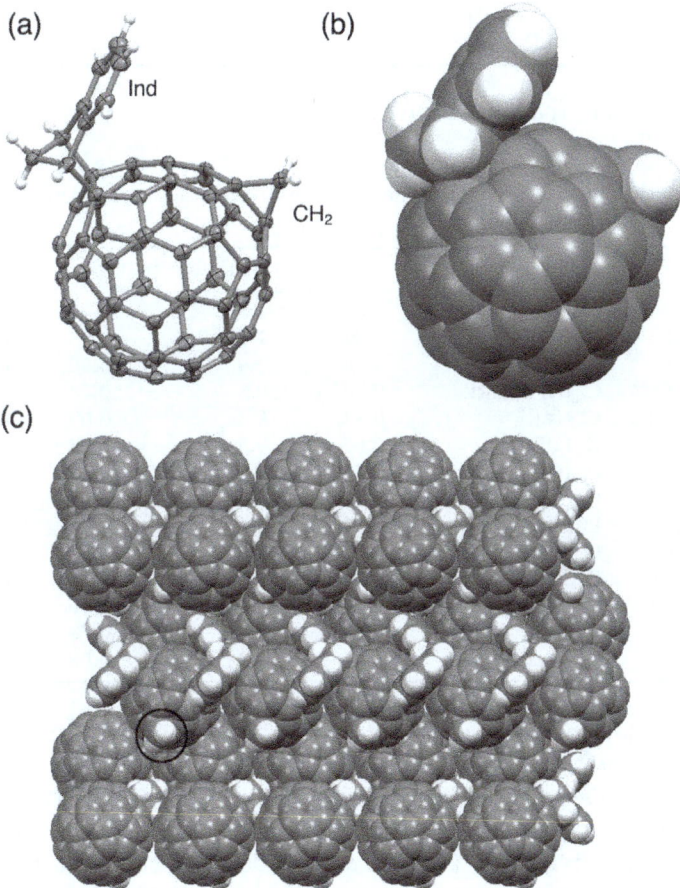

Figure 3.4 Molecular (a and b) and crystal packing (c) structures of methano-indene-fullerene, $C_{60}(CH_2)(Ind)$. A black circle denotes the dihydromethano group (ref. 78) © 2013 WILEY-VCH Verlag GmbH & Co. KGaA, Weinheim.

Table 3.2 LUMO levels, electron mobility, and solubility of 56π-dihydrometha-
nofullerenes (ref. 78).

Entry	π	Fullerene	LUMO level^a (eV)	Electron mobility^b (cm^2 V^{-1} s^{-1})	Solubility (wt%)
1^c	58π	$C_{60}(CH_2Ar)_2$	−3.60	—	—
2^c	56π	$C_{60}(CH_2)(CH_2Ar)_2$	−3.47	—	—
3	58π	$C_{60}(Ind)$	−3.76	9.7×10^{-3}	0.5
4	56π	$C_{60}(CH_2)(Ind)$	−3.66	8.0×10^{-4}	4.0
5	56π	$C_{60}(Ind)_2$	−3.63	6.3×10^{-4}	7.5
6	58π	$C_{60}(QM)$	−3.78	3.9×10^{-3}	0.3
7	56π	$C_{60}(CH_2)(QM)$	−3.66	4.8×10^{-3}	4.0
8	56π	$C_{60}(QM)_2$	−3.63	5.5×10^{-4}	5.0
9	58π	PCBM	−3.80	3.0×10^{-3}	1.5
10	56π	$PCBM(CH_2)$	−3.70	6.4×10^{-3}	7.0
11	56π	Bis-PCBM	−3.71	1.2×10^{-3}	>10
12	68π	$C_{70}(Ind)$	−3.76	1.2×10^{-2}	0.1
13	66π	$C_{70}(CH_2)(Ind)$	−3.65	4.4×10^{-3}	1.5
14	66π	$C_{70}(Ind)_2$	−3.62	1.4×10^{-3}	3.0

^{a}LUMO level = $-(4.8 + E_{1/2}{}^{red})$.
^{b}SCLC mobility for P3HT-containing BHJ films.
^{c}Ar = 4-t-BuC$_6$H$_4$.

Table 3.3 Photovoltaic performance of 56π-dihydromethanofullerene acceptors
in BHJ OSCs using P3HT (ref. 78).

Entry	π	Fullerene	V_{OC} (V)	J_{SC} (mA cm^{-2})	FF	PCE (%)
1	58π	$C_{60}(CH_2Ar)_2$	0.69	4.8	0.49	1.6
2	56π	$C_{60}(CH_2)(CH_2Ar)_2$	0.82	7.1	0.58	3.4
3	58π	$C_{60}(Ind)$	0.66	7.9	0.69	3.6
4	56π	$C_{60}(CH_2)(Ind)$	0.78	10.3	0.73	5.9
5	56π	$C_{60}(Ind)_2$	0.84	8.3	0.69	4.8
6	58π	$C_{60}(QM)$	0.65	9.5	0.68	4.2
7	56π	$C_{60}(CH_2)(QM)$	0.77	9.6	0.67	5.0
8	56π	$C_{60}(QM)_2$	0.85	8.3	0.67	4.7
9	58π	PCBM	0.61	9.1	0.72	4.0
10	56π	$PCBM(CH_2)$	0.73	9.0	0.71	4.6
11	56π	Bis-PCBM	0.76	8.2	0.71	4.4
12	68π	$C_{70}(Ind)$	0.70	8.6	0.70	4.2
13	66π	$C_{70}(CH_2)(Ind)$	0.79	11.1	0.73	6.4
14	66π	$C_{70}(Ind)_2$	0.85	9.3	0.68	5.4

The LUMO levels and electron mobility in BHJ films are summarized in
Table 3.2. 56π-Dihydromethanofullerenes have high-lying LUMO levels
as well as high electron mobility. $C_{60}(CH_2)(Ind)$ (**26**) and its C_{70} congener,
$C_{70}(CH_2)(Ind)$ (**28**) showed PCE of 5.9% and 6.4%, respectively, in BHJ
OSCs using a P3HT donor (Table 3.3). The data indicate that the addition
of one dihydromethano group raised the LUMO level by *ca.* 0.10 eV, and
thus increased V_{OC} of the P3HT-based BHJ device by *ca.* 0.11 V. In addition,
the sterically small dihydromethano group realized high electron mobility, and thus high J_{SC} and FF, rather than the trade-off typically found for

the 56π bis-adduct compounds. With this concept,[77-79,83] methano-PCBM (**25**),[80] $C_{60}(CH_2)(QM)$ (**27**),[84] and $C_{70}(CH_2)(Ind)$[85] have been synthesized, and used for investigation in BHJ OSCs, showing PCE of 3.81%, 5.74%, and 6.88%, respectively. The dihydromethano group was also utilized in the synthesis of 54π-[60]fullerene acceptors, showing PCE of 4.56–6.43% with high V_{OC} (0.94–1.00 V).[86,87]

3.8 Summary

Utilizing advantages of fast electron transfer from donor materials to fullerene and slow back electron transfer, various fullerene-based electron acceptors have been developed for use in solution processed BHJ OSCs. In the molecular design of fullerene acceptors, frontier orbital energy levels (*e.g.*, LUMO levels) and volume of organic addends are considered to have high V_{OC}, J_{SC}, and FF values. In addition, miscibility, crystallinity, and thermal properties in the BHJ films are considered to give efficient charge separation and transport. For the commercialization of OCSs, cost effectiveness and facile synthesis should be also considered.

Acknowledgements

The author thanks Dr James W. Ryan for fruitful discussions.

References

1. F. G. Brunetti, R. Kumar and F. Wudl, *J. Mater. Chem.*, 2010, **20**, 2934.
2. G. Li, R. Zhu and Y. Yang, *Nat. Photonics*, 2012, **6**, 153.
3. H. J. Son, B. Carsten, I. H. Jung and L. Yu, *Energy Environ. Sci.*, 2012, **5**, 8158.
4. Q. Xie, E. Perez-Cordero and L. Echegoyen, *J. Am. Chem. Soc.*, 1992, **114**, 3977.
5. N. S. Sariciftci, L. Smilowitz, A. J. Heeger and F. Wudl, *Science*, 1992, **258**, 1474.
6. H. Imahori, K. Hagiwara, T. Akiyama, M. Aoki, S. Taniguchi, T. Okada, M. Shirakawa and Y. Sakata, *Chem. Phys. Lett.*, 1996, **263**, 545.
7. H. Imahori, M. E. El-Khouly, M. Fujitsuka, O. Ito, Y. Sakata and S. Fukuzumi, *J. Phys. Chem. A*, 2001, **105**, 325.
8. Y. Matsuo, *Chem. Lett.*, 2012, **41**, 754.
9. C. Z. Li, H. L. Yip and A. K. Y. Jen, *J. Mater. Chem.*, 2012, **22**, 4161.
10. Y. Li, *Chem.–Asian J.*, 2013, **8**, 2316.
11. B. C. Thompson and J. M. J. Fréchet, *Angew. Chem., Int. Ed.*, 2008, **47**, 58.
12. Y. He and Y. Li, *Phys. Chem. Chem. Phys.*, 2011, **13**, 1970.
13. J. C. Hummelen, B. W. Knight, F. LePeq, F. Wudl, J. Yao and C. L. Wilkins, *J. Org. Chem.*, 1995, **60**, 532.
14. A. B. Smith III, R. M. Strongin, L. Brard, G. T. Furst, W. J. Romanow, K. G. Owens, R. J. Goldschmidt and R. C. King, *J. Am. Chem. Soc.*, 1995, **117**, 5492.

15. T. Suzuki, Q. Li, K. C. Khemani and F. Wudl, *J. Am. Chem. Soc.*, 1992, **114**, 7301.

16. A. B. Smith III, R. M. Strongin, L. Brard, G. T. Furst, W. J. Romanow, K. G. Owens and R. C. King, *J. Am. Chem. Soc.*, 1993, **115**, 5829.

17. G. Yu, J. Gao, J. C. Hummelen, F. Wudl and A. J. Heeger, *Science*, 1995, **270**, 1789.

18. Y. Matsuo, A. Iwashita, Y. Abe, C. Z. Li, K. Matsuo, M. Hashiguchi and E. Nakamura, *J. Am. Chem. Soc.*, 2008, **130**, 15429.

19. H. Yoshida, *MRS Symp. Proc.*, 2012, **1493**, 295.

20. Y. Matsuo, Y. Sato, T. Niinomi, I. Soga, H. Tanaka and E. Nakamura, *J. Am. Chem. Soc.*, 2009, **131**, 16048.

21. M. T. Rispens, A. Meetsma, R. Rittberger, C. J. Brabec, N. S. Sariciftci and J. C. Hummelen, *Chem. Commun.*, 2003, 2116.

22. B. Walker, C. Kim and T.-Q. Nguyen, *Chem. Mater.*, 2011, **23**, 470.

23. F. B. Kooistra, J. Knol, F. Kastenberg, L. M. Popescu, W. J. H. Verhees, J. M. Kroon and J. C. Hummelen, *Org. Lett.*, 2007, **9**, 551.

24. Y. Zhang, H. L. Yip, O. Acton, S. K. Hau, F. Huang and A. K. Y. Jen, *Chem. Mater.*, 2009, **21**, 2598.

25. M. M. Wienk, J. M. Kroon, W. J. H. Verhees, J. Knol, J. C. Hummelen, P. A. Hal and R. A. J. Janssen, *Angew. Chem., Int. Ed.*, 2003, **42**, 3371.

26. F. B. Kooistra, V. D. Mihailetchi, L. M. Popescu, D. Kronholm, P. W. M. Blom and J. C. Hummelen, *Chem. Mater.*, 2006, **18**, 3068.

27. Y. Santo, I. Jeon, K. S. Yeo, T. Nakagawa and Y. Matsuo, *Appl. Phys. Lett.*, 2013, **103**, 073306.

28. Y. Matsuo, *Pure Appl. Chem.*, 2012, **84**, 945.

29. N. Obata, Y. Sato, E. Nakamura and Y. Matsuo, *Jpn. J. Appl. Phys.*, 2011, **50**, 121603.

30. Y. Matsuo, J. Hatano, T. Kuwabara and K. Takahashi, *Appl. Phys. Lett.*, 2012, **100**, 063303.

31. Y. Matsuo, A. Ozu, N. Obata, N. Fukuda, H. Tanaka and E. Nakamura, *Chem. Commun.*, 2012, **48**, 3878.

32. H. Tanaka, Y. Abe, Y. Matsuo, J. Kawai, I. Soga, Y. Sato and E. Nakamura, *Adv. Mater.*, 2012, **24**, 3521.

33. Y. Matsuo, H. Oyama, I. Soga, T. Okamoto, H. Tanaka, A. Saeki, S. Seki and E. Nakamura, *Chem.–Asian J.*, 2013, **8**, 121.

34. H. Tamura and Y. Matsuo, *Chem. Phys. Lett.*, 2014, **598**, 81.

35. Y. Matsuo, Y. Zhang, I. Soga, Y. Sato and E. Nakamura, *Tetrahedron Lett.*, 2011, **52**, 2240.

36. Y. Tajima and K. Takeuchi, *J. Org. Chem.*, 2002, **67**, 1696.

37. Y. Tajima, T. Hara, T. Honma, S. Matsumoto and K. Takeuchi, *Org. Lett.*, 2006, **8**, 3203.

38. G.-W. Wang, Y.-M. Lu and Z.-X. Chen, *Org. Lett.*, 2009, **11**, 1507.

39. A. Varotto, N. D. Treat, J. Jo, C. G. Shuttle, N. A. Batara, F. G. Brunetti, J. H. Seo, M. L. Chabinyc, C. J. Hawker, A. J. Heeger and F. Wudl, *Angew. Chem., Int. Ed.*, 2011, **50**, 5166.

40. A. Iwashita, Y. Matsuo and E. Nakamura, *Angew. Chem., Int. Ed.*, 2007, **46**, 3513.

41. M. Nambo, R. Noyori and K. Itami, *J. Am. Chem. Soc.*, 2007, **129**, 8080.
42. M. Nambo and K. Itami, *Chem.–Eur. J.*, 2009, **15**, 4760.
43. I. Riedel, N. Martín, F. Giacalone, J. L. Segura, D. Chirvase, J. Parisi and V. Dyakonov, *Thin Solid Films*, 2004, **43**, 451.
44. I. Riedel, E. von Hauff, J. Parisi, N. Martín, F. Giacalone and V. Dyakonov, *Adv. Funct. Mater.*, 2005, **15**, 1979.
45. A. Sánchez-Díaz, M. Izquierdo, S. Filippone, N. Martín and E. Palomares, *Adv. Funct. Mater.*, 2010, **20**, 2695.
46. G. Garcia-Belmonte, P. P. Boix, J. Bisquert, M. Lenes, H. J. Bolink, A. La Rosa, S. Filippone and N. Martín, *J. Phys. Chem. Lett.*, 2010, **1**, 2566.
47. T. Oshima, H. Kitamura, T. Higashi, K. Kokubo and N. Seike, *J. Org. Chem.*, 2006, **71**, 2995.
48. H. J. Bolink, E. Coronado, A. Forment-Aliaga, M. Lenes, A. La Rosa, S. Filippone and N. Martín, *J. Mater. Chem.*, 2011, **21**, 1382.
49. E. J. Palomares, D. Fernández, A. Viterisi, James W. Ryan, F. Guispert-Guirado, S. Vidal, S. Filippone and N. Martín, *Nanoscale*, 2014, **6**, 5871.
50. M. Maggini, G. Scorrano and M. Prato, *J. Am. Chem. Soc.*, 1993, **115**, 9798.
51. O. Tsuge and S. Kanemasa, *Adv. Heterocycl. Chem.*, 1989, **45**, 231.
52. K. Matsumoto, K. Hashimoto, M. Kamo, Y. Uetani, S. Hayase, M. Kawatsura and T. Itoh, *J. Mater. Chem.*, 2010, **20**, 9226.
53. Y. Zhang, Y. Matsuo and E. Nakamura, *Org. Lett.*, 2011, **13**, 6058.
54. M. Lenes, G. J. A. H. Wetzelaer, F. B. Kooistra, S. C. Veenstra, J. C. Hummelen and P. W. M. Blom, *Adv. Mater.*, 2008, **20**, 2116.
55. P. Belik, A. Gügel, J. Spickermann and K. Müllen, *Angew. Chem., Int. Ed. Engl.*, 1993, **32**, 78.
56. A. Puplovskis, J. Kacens and O. Neilands, *Tetrahedron Lett.*, 1997, **38**, 285.
57. D. W. Laird, R. Stegamat, H. Richter, V. Vejins, L. Scott and T. A. Lada, *US Pat.* 8217260, 2007.
58. Y. He, H. Y. Chen, J. Hou and Y. Li, *J. Am. Chem. Soc.*, 2010, **132**, 1377.
59. G. Zhao, Y. He and Y. Li, *Adv. Mater.*, 2010, **22**, 4355.
60. Y.-J. Cheng, C.-H. Hsieh, Y. He, C.-S. Hsu and Y. Li, *J. Am. Chem. Soc.*, 2010, **132**, 17381.
61. Y. He, G. Zhao, B. Peng and Y. Li, *Adv. Funct. Mater.*, 2010, **20**, 3383.
62. X. Guo, C. Cui, M. Zhang, L. Huo, Y. Huang, J. Hou and Y. Li, *Energy Environ. Sci.*, 2012, **5**, 7943.
63. E. Voroshazi, K. Vasseur, T. Aernouts, P. Heremans, A. Baumann, C. Deibel, X. Xue, A. J. Herring, A. J. Athans, T. A. Lada, H. Richter and B. P. Rand, *J. Mater. Chem.*, 2011, **21**, 17345.
64. K. H. Kim, H. Kang, S. Y. Nam, J. Jung, P. S. Kim, C. H. Cho, C. Lee, S. C. Yoon and B. J. Kim, *Chem. Mater.*, 2011, **23**, 5090.
65. X. Meng, W. Zhang, Z. Tan, C. Du, C. Li, Z. Bo, Y. Li, X. Yang, M. Zhen, F. Jiang, J. Zheng, T. Wang, L. Jiang, C. Shu and C. Wang, *Chem. Commun.*, 2012, **48**, 425.
66. Y. J. Cheng, M. H. Liao, C. Y. Chang, W. S. Kao, C. E. Wu and C. S. Hsu, *Chem. Mater.*, 2011, **23**, 4056.

67. L. L. Deng, J. Feng, L. C. Sun, S. Wang, S. L. Xie, S. Y. Xie, R. B. Huang and L. S. Zheng, *Sol. Energy Mater. Sol. Cells*, 2012, **104**, 113.
68. C. Zhang, S. Chen, Z. Xiao, Q. Zuo and L. Ding, *Org. Lett.*, 2012, **14**, 1508.
69. X. Meng, W. Zhang, Z. Tan, Y. Li, Y. Ma, T. Wang, L. Jiang, C. Shu and C. Wang, *Adv. Funct. Mater.*, 2012, **22**, 2187.
70. H. Xin, S. Subramaniyan, T. W. Kwon, S. Shoaee, J. R. Durrant and S. A. Jenekhe, *Chem. Mater.*, 2012, **24**, 1995.
71. X. Guo, M. Zhang, L. Huo, C. Cui, Y. Wu, J. Houand and Y. Li, *Macromolecules*, 2012, **45**, 6930.
72. S. Kitaura, K. Kurotobi, M. Sato, Y. Takano, T. Umeyama and H. Imahori, *Chem. Commun.*, 2012, **48**, 8550.
73. X. Meng, G. Zhao, Q. Xu, Z. Tan, Z. Zhang, L. Jiang, C. Shu, C. Wang and Y. Li, *Adv. Funct. Mater.*, 2014, **24**, 158.
74. W. W. H. Wong, J. Subbiah, J. M. White, H. Seyler, B. Zhang, D. J. Jones and A. B. Holmes, *Chem. Mater.*, 2014, **26**, 1686.
75. Y. Matsuo and E. Nakamura, *Chem. Rev.*, 2008, **108**, 3016.
76. Z. Xiao, Y. Matsuo, I. Soga and E. Nakamura, *Chem. Mater.*, 2012, **24**, 2572.
77. Y. Zhang, Y. Matsuo, C. Li, H. Tanaka and E. Nakamura, *J. Am. Chem. Soc.*, 2011, **133**, 8086.
78. Y. Matsuo, J. Kawai, H. Inada, T. Nakagawa, H. Ota, S. Otsubo and E. Nakamura, *Adv. Mater.*, 2013, **25**, 6266.
79. C.-Z. Li, Y. Matsuo and E. Nakamura, *Tetrahedron*, 2011, **67**, 9944.
80. C.-Z. Li, S. Chien, H. Yip, C. Chueh, F. Chen, Y. Matsuo, E. Nakamura and A. K.-Y. Jen, *Chem. Commun.*, 2011, **47**, 10082.
81. Y. Abe, R. Hata and Y. Matsuo, *Chem. Lett.*, 2013, **42**, 1525.
82. Y. Abe, T. Yokoyama and Y. Matsuo, *Org. Electron.*, 2013, **14**, 3306.
83. Mitsubishi Chemical corporation and The university of Tokyo, Jpn. patent application number: 2011-192064, publication number: 2012-094829, 2011.
84. G. Ye, S. Chen, Z. Xiao, Q. Zuo, Q. Wei and L. Ding, *J. Mater. Chem.*, 2012, **22**, 22374.
85. D. He, C. Zuo, S. Chen, Z. Xiao and L. Ding, *Phys. Chem. Chem. Phys.*, 2014, **16**, 7205.
86. S. Chen, G. Ye, Z. Xiao and L. Ding, *J. Mater. Chem. A*, 2013, **1**, 5562.
87. D. He, X. Du, Z. Xiao and L. Ding, *Org. Lett.*, 2014, **16**, 612.

CHAPTER 4

Polymer Acceptors for All-Polymer Solar Cells

HE YAN*[a], CHRISTOPHER R. McNEILL*[b], AND CHENG MU[a]

[a]Department of Chemistry, The Hong Kong University of Science and Technology, Clear Water Bay, KowLoon, Hong Kong; [b]Department of Materials Engineering, Monash University, Wellington Road, Clayton, Victoria, 3800, Australia
*E-mail: hyan@ust.hk, christopher.mcneill@monash.edu

4.1 Introduction

Conventional inorganic solar cells can achieve high efficiencies, but are complicated and costly to produce. The desirability of a lower cost is driving the development of several third-generation solar cell technologies. Of these, the polymer solar cell (PSC) is particularly cheap to produce, because polymer solar panels can be fabricated using extremely high throughput roll-to-roll printing methods similar to those used to print newspapers.[1-6] State-of-the-art PSCs consist of a blend film of a polymer and a fullerene derivative, which function as an electron-donor and an electron-acceptor, respectively (Figure 4.1a). Although this type of polymer:fullerene solar cell achieves an impressive efficiency (9.2%[7] for single-junction cells), the fullerene acceptors have several disadvantages.

Fullerene materials, especially C_{60}-based derivatives, have relatively poor absorption properties. Given that fullerenes often account for at least 50% of the volume in a polymer:fullerene blend, the use of fullerenes significantly

RSC Polymer Chemistry Series No. 17
Polymer Photovoltaics: Materials, Physics, and Device Engineering
Edited by Fei Huang, Hin-Lap Yip, and Yong Cao
© The Royal Society of Chemistry 2016
Published by the Royal Society of Chemistry, www.rsc.org

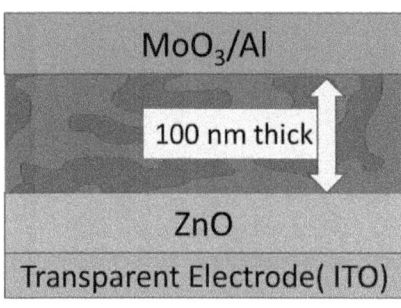
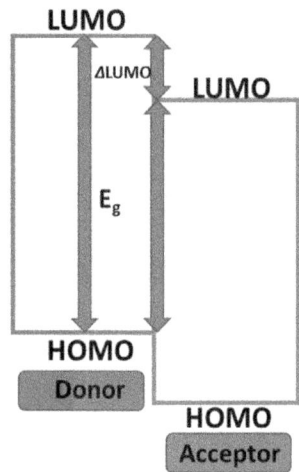

Figure 4.1   (a) Device architecture of inverted bulk heterojunction polymer solar cells and (b) energy level diagram for donor and acceptor.

undercuts the absorption strength of PSCs. By replacing fullerenes with a strongly absorbing polymer or small molecular dye, the absorption strength of a PSC can be dramatically enhanced. The poor absorption property of C_{60}-based fullerene derivatives is the reason why C_{70} derivatives are preferred in high-efficiency polymer:fullerene PSCs. Owing to the lower symmetry of its structure, C_{70} exhibits significantly stronger absorption than C_{60} at about 500 nm.[8] However, the absorption coefficient of C_{70} is still significantly lower than those of the best polymer materials and it is difficult to extend the absorption of C_{70} or any other fullerenes to near infrared spectral range. Importantly, the somewhat mediocre absorption property of C_{70} comes at a high cost. C_{70}-based derivatives are extremely expensive (~$2000 per gram, at least 10 times more expensive than gold) and are not acceptable for commercial applications.

In addition to the absorption problem and cost issue, another major drawback of fullerene materials is the limited tunability of their energy levels. Most monosubstituted fullerene derivatives have a lowest unoccupied molecular orbital (LUMO) level of about 4.0 eV. Changing the substitution groups on fullerene typically does not have a dramatic impact on its energy levels. Fullerene bisadducts (such as 1′,1″,4′,4″-tetrahydro-di[1,4]methanonaphthaleno[5,6]fullerene-C_{60} (ICBA))[9–11] and metafullerenes[12–14] have been developed that increase the LUMO of fullerenes by more than 0.2 eV. However, neither of these choices is a good solution. Metafullerenes are even more difficult and more expensive to synthesize than C_{70}. Fullerene biadducts contain multiple regioisomers that are practically impossible to separate. As a result, the widely used ICBA or bis(1-[3-(methoxycarbonyl)propyl]-1-phenyl)-[6,6]C_{62} (BisPCBM) are always a mixture of several isomers, which seriously harms the crystallinity and charge transport ability of the fullerene material, not to mention the high cost of ICBA.[15,16] Given the limited tunability of the energy levels of fullerenes, all donor polymer materials developed for PSCs have to be precisely matched with the energy levels of [6,6]-phenyl-C_{71}-butyric acid methyl ester ($PC_{71}BM$) to achieve the best performance. As predicted by Brabec's model,[17–19] the LUMO offset between the donor and acceptor materials is a critical parameter that determines the potential performance of a PSC (Figure 4.1b). The nearly fixed energy level of $PC_{71}BM$ is a constraint to the development of polymer materials.

In the hope of addressing these issues, PSCs based on non-fullerene acceptors have been developed. In general, there are two alternatives for the replacement of fullerenes in PSCs. One choice is a polymer acceptor and the other is a small molecular acceptor. Both the polymer and small molecule alternatives have easily tunable energy levels and potentially much greater light absorption properties than fullerenes. The challenge is to improve the PSC performance to a level comparable to that of polymer:-fullerene PSCs. The best efficiency levels achieved for PSCs based on polymers or small molecular acceptors are about 3–6%,[20–26] which are far behind those for polymer:fullerene PSCs. In this chapter, we will mainly review the work on polymer:polymer PSCs (or all-polymer solar cells, all-PSCs). Historically, polymer:polymer PSCs have attracted the most attention, and this area has seen a relatively rapid increase in cell efficiency in the past two years. The historical development of polymer:polymer PSCs will be briefly reviewed, but the main focus of this chapter is on recent advances in all-PSCs that have led to significantly enhanced PSC efficiency. This chapter will have two main parts, covering the material and morphological aspects of all-PSCs.

4.2 Materials Aspects for All-Polymer Solar Cells

There are two main themes along which the development all-PSCs can be logically followed and summarized. One theme follows the development of acceptor polymers and the other theme relates to donor polymers. One of the main challenges for all-PSCs is the lack of high-performance n-type polymeric semiconductors. In the polymeric semiconductor field, although there have been many reports of high-mobility p-type polymers, the development of high-performance n-type polymers has been relatively slow.[27] For this reason, the development of the all-PSC field has been partly driven by the development of n-type polymeric semiconductors, which is one of the main themes of development for all-PSCs. The second theme of development, which has given rise to most of the major advances in all-PSCs in the past two years, is the development of high-performance donor polymers for all-PSCs. Despite the impressive development in acceptor polymers for all-PSCs, donor polymers still play a critical role in their operation. As donor polymers make a significantly greater contribution to light absorption and thus the external quantum efficiencies (EQE) of the cells, the development of donor polymers has been the main reason for the achievement of 3–5% efficiency levels among all-PSCs in the past two years.[28–32] A main area of improvement for donor polymers is the reduction of optical bandgaps and thus the broadening of the absorption spectra of the polymers. Historically, the development of polymer:fullerene PSCs has been driven by the development of low bandgap donor polymers.[7,33–35] Several generations of donor polymers, such as poly[2-methoxy-5-(2′-ethylhexyloxy)-p-phenylene vinylene] (MEH-PPV), poly(3-hexylthiophene) (P3HT), poly[[4,8-bis[(2-ethylhexyl)oxy]benzo[1,2-b:4,5-b′]dithiophene-2,6-diyl][3-fluoro-2-[(2-ethylhexyl)carbonyl]thieno[3,4-b]thiophenediyl]] (PTB7), *etc.* (Figure 4.2), have been developed with decreasing bandgaps and dramatically improved efficiency (2.5,[36] 4.5,[37] and 9.2%,[7] respectively, for MEH-PPV, P3HT and PTB7). Similarly, the development of low bandgap donor polymers in the past two years has led to a rapidly increasing efficiency for all-PSCs. In this chapter, we summarise previous work on all-PSCs, mainly following the development of donor polymers with decreasing bandgaps, because this theme has led to the most important work on all-PSCs. The materials aspect of this chapter will be divided into three parts, for all-PSCs containing: (1) PPV or polyfluorene-based large bandgap (2.5 eV) donor polymers; (2) polythiophene-based large bandgap (1.9 eV) donor polymers; and (3) low bandgap donor polymers. For each of the three parts, we will also describe the acceptor polymers with various structural features including cyanated polyphenylenevinylenes (CN-PPV), benzothiadiazole (BT), and perylene- or naphthalenediimide (PDI or NDI). Owing to the extensive amount of work in all-PSCs, this book chapter will only cover the most important work in their development. Besides describing the main development of all-PSCs at various stages, we will also summarize the main advantages and disadvantages of each type of material at the end of their sections. These conclusion paragraphs are labelled as "mini-summary".

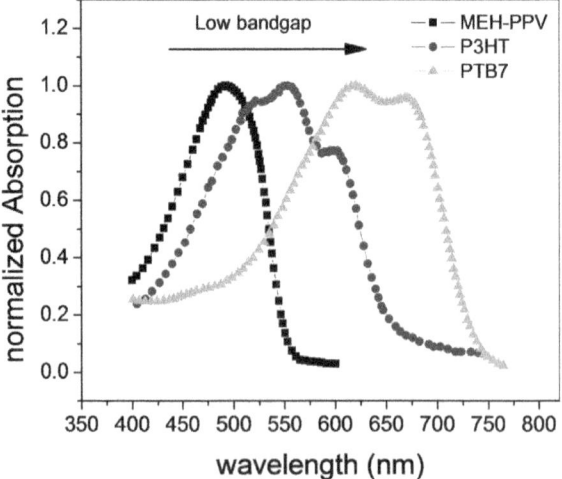

Figure 4.2   The ultraviolet-visible (UV-vis) absorption of MEH-PVV, P3HT and PTB7 in tetrahydrofuran (THF) solution.

4.2.1 All-PSCs Based on Large Bandgap (2–2.5 eV) Donor Polymers

Donor polymers based on MEH-PPV (Figure 4.3) enabled the first examples of all-PSCs, as reported by the groups of Friend[2] and Heeger[38] independently. In both studies, CN-PPV-based polymers were used as the acceptor, and efficient exciton dissociation was observed at the donor:acceptor interface. Note that CN-PPV (Figure 4.3) can be used as an acceptor polymer because the electron-withdrawing effect of the cyano group (CN) can effectively reduce the highest occupied molecular orbital (HOMO) and the lowest unoccupied molecular orbital (LUMO) levels of the polymer.

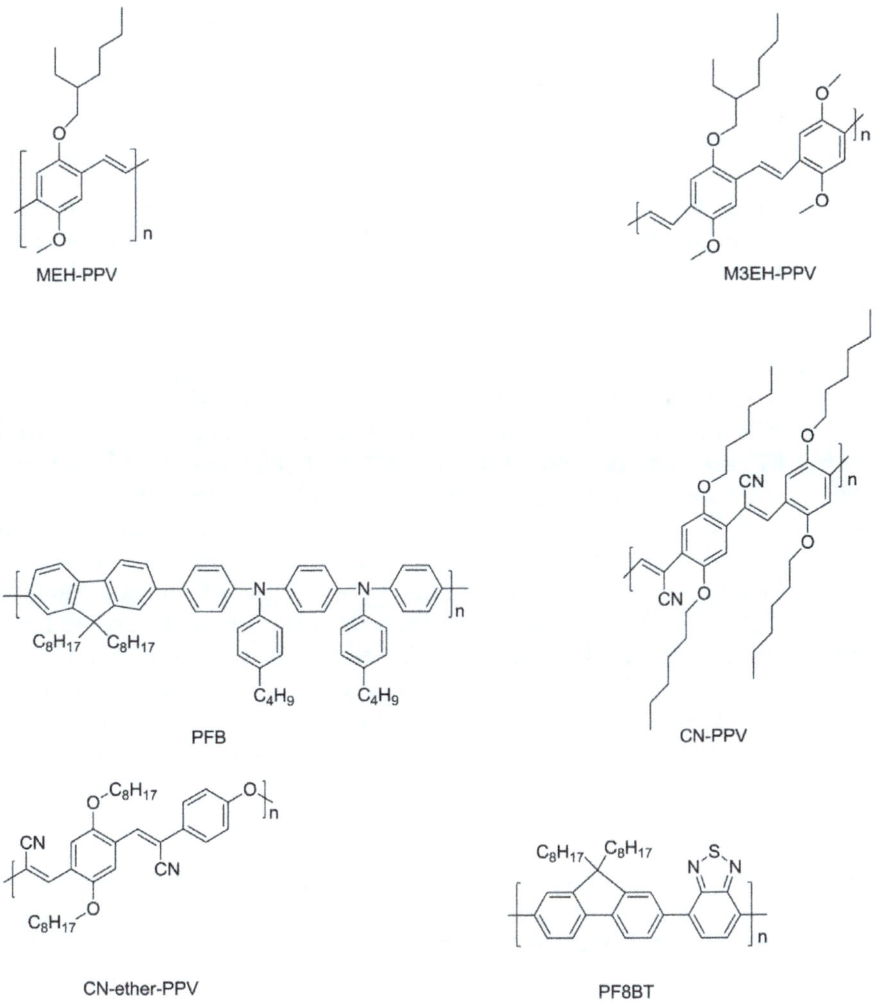

Figure 4.3 Chemical structures of donor and acceptor polymers used in all-polymer solar cells with large bandgap.

In the blend of two polymers, the one with lower energy levels tends to accept electrons and thus can function as an acceptor. Besides the observed exciton dissociation and photoluminescence quenching, both groups (Friend and Heeger) reported dramatically enhanced external quantum efficiency of the blend-based cells over the pure polymer-based cells. Although the power conversion efficiencies (PCE) of these early all-PSCs were low (<1%), this revolutionary work opened the interesting and important development of all-PSCs in the past 20 years.

Following the works of Friend and Heeger, Carter[39] and Kietzke[40] and their co-workers modified the structures of MEH-PPV and/or CN-PPV polymers and obtained all-PSCs with efficiencies between 1% and 2%. In Carter's report, an M3EH-PPV (Figure 4.3) donor polymer was combined with a CN-ether-PPV (Figure 4.3) acceptor polymer and yielded a PCE of about 1% using poly(3,4-ethylenedioxythiophene) polystyrene sulfonate (PEDOT:PSS) and Ca as anode and cathode respectively. In-depth investigation on the M3EH-PPV:CN-ether-PPV-based all-PSCs was later carried out by Kietzke and co-workers by comparing bilayer and blend devices. In Kietzke's work, chlorobenzene was used as solvent and led to interesting results. The authors believed that the spincoating of the polymer solution formed a vertically composition-graded layer due to the much lower solubility of M3EH-PPV in chlorobenzene than that of CN-ether-PPV. This type of "graded" blend structure offers better charge extractions than a uniformly distributed blend. As a result, a much higher efficiency (1.7%) was achieved.

Another important example of a large bandgap donor polymer is PFB (Figure 4.3), a copolymer that consists of fluorene and triarylamine building blocks. In the work by Arias,[41] the PFB donor polymer was mixed with a BT-based acceptor polymer, named PF8BT (Figure 4.3). Although the PFB:PF8BT-based all-PSCs cannot produce high efficiencies, the morphological study on this blend system provided interesting results that inspired morphological studies in the all-PSC field in the following years. In Arias's work, polymer blend films were obtained by spincoating from two different solvents (chloroform or xylene). It was found that the blend film coated from chloroform solution exhibits a fine and intimately mixed blend morphology, which led to effectively quenched photoluminescence. The quick drying time of chloroform solution prevents rearrangement of the polymer chains in the film-forming process and the formation of large domains. However, the xylene solution led to large polymer domains as evidenced by atomic force microscopy (AFM) images.

Mini-Summary (Section 4.2.1). Overall, although large bandgap donor polymers have produced all-PSCs with high open circuit voltage (VOC), the absorption range of the polymers is too narrow, which, combined with the low EQE of the cells, resulted in very low short circuit current density (JSC) based on today's standards. Therefore, it was clear that lower bandgap donor polymers are needed to produce higher efficiency all-PSCs. Another limitation with the reported PPV and PFB polymers is that they exhibit relatively poor hole transport ability, which is one of main factors that result in low fill factors (FF) for the all-PSCs. To enhance the hole transport ability of the donor polymer, it is desirable to adopt coplanar polymer structures that can form strong π–π stacking. With the emergence of thiophene-based conjugated polymers, the excellent charge transport ability of polythiophenes was soon recognized. Since then, there have been extensive research efforts in adopting polythiophenes for polymer:fullerene and polymer:polymer PSCs.

4.2.2 All-PSCs Based on Polythiophene Donor Polymers

4.2.2.1 POPT

One of the first examples of polythiophenes used in all-PSCs was a polythiophene substituted with an octylphenyl group, named POPT (Figure 4.4). There have been extensive studies on POPT:MEH-CN-PPV-based all-PSCs.[42–49] Note that some of these all-PSCs adopted a bilayer structure fabricated *via* a lamination approach.[50] Interestingly, the bilayer all-PSCs containing pure donor and pure acceptor layers exhibit poor efficiencies. However, when the pure donor and acceptor layers are doped with a low percentage of acceptor and donor polymers, respectively, the performance of the doped bilayer all-PSCs was dramatically enhanced. An impressive EQE of 29% and PCE of 1.9% were achieved, which were the best for all-PSCs at the time. Although the POPT may have reasonably good performance, it clearly has some drawbacks. For one, the octylphenyl group introduces significant twisting between benzene and thiophene; as a result, the π–π stacking of the polymer chains is not strong. To achieve high FF and high efficiency all-PSCs, it is important to use donor polymers with strong π–π stacking and high crystallinity.

4.2.2.2 P3HT

Following the success of P3HT:fullerene-based PSCs, P3HT (Figure 4.4) was extensively studied in all-PSCs. Earlier examples of P3HT-based all-PSCs used BT-based polymers as the acceptor. Later, after high-mobility NDI or PDI-based n-type polymers were developed, many examples of P3HT:NDI or PDI-based all-PSCs were reported.

4.2.2.2.1 P3HT:BT Polymer-Based All-PSCs. One of the first examples of P3HT:BT polymer-based all-PSCs was reported by Bradley and co-workers.[51] Among the various conditions explored using P3HT and PF8BT as donor and acceptor, the best EQE and PCE were obtained using xylene as solvent and LiF as the interlayer. The use of the LiF interlayer improved the device efficiency from 0.02% to 0.13%. As supported by time-of-flight mobility measurement, the low efficiency of the cell was mainly attributed to the poor electron mobility of PF8BT. Following this work, McNeill and co-workers[52] reported all-PSCs based on P3HT and a slightly modified BT-based polymer (named PF8TBT6). The structure of PF8TBT6 is mostly similar to PF8BT except that two 3-hexyl-thiophene rings are inserted on the two sides of BT (Figure 4.4).

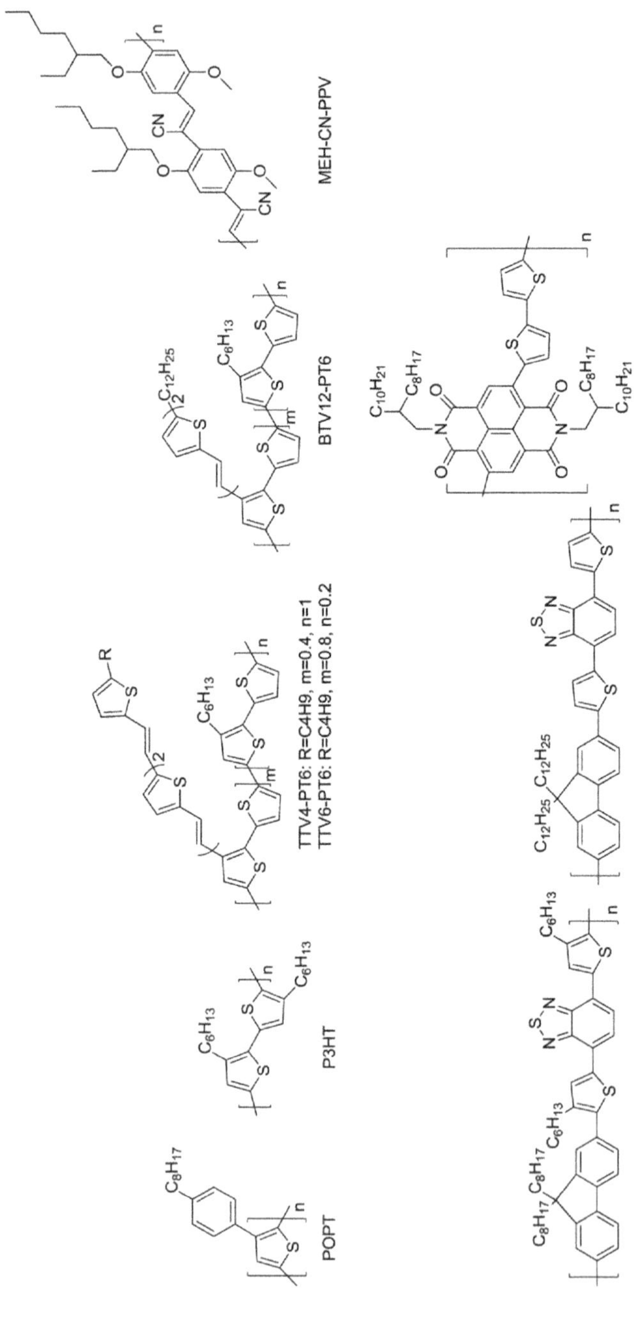

Figure 4.4   Chemical structures of polythiophene-based all-PSCs.

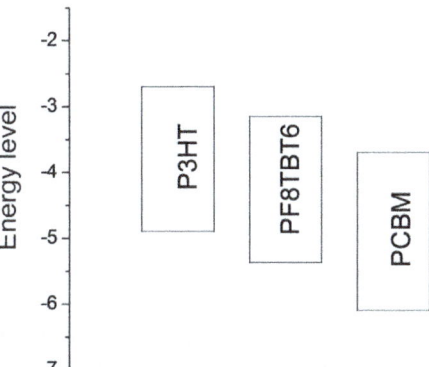

Figure 4.5   Energy level diagram showing the HOMO and LUMO levels of P3HT, PF8TBT6, and PCBM.

The insertion of the 3-hexyl-thiophene units changed the properties of the polymer dramatically and increased the efficiency of the all-PSCs to 1.7%. Interestingly, PF8TBT6 was demonstrated to function as an acceptor and a donor when it is mixed with P3HT and $PC_{61}BM$ respectively. This clearly demonstrates the general ambipolar nature of conjugated polymers and that the roles of donor and acceptor are determined on the basis of the relative position of the HOMO and LUMO levels of the two polymers in the blend (Figure 4.5). More important progress based on a P3HT:PF8TBT6 blend was made by Huck and co-workers,[53] who employed a double nano-imprint lithography (NIL) procedure to fabricate interdigitated P3HT:PF8TBT6 bilayer all-PSCs. The reported NIL procedure worked remarkably well, yielding a PCE of 1.85%, one of the highest efficiencies reported at the time. This approach was distinctively different from most other all-PSCs known and provided important inspiration in fabricating interdigitated PSC devices.

The best performance from P3HT:BT polymer-based all-PSCs was reported in 2011 by Ito and co-workers.[54] The authors used a slightly modified BT polymer named PF12TBT, which is identical to PF8TBT6 in terms of polymer backbone except that the alkyl chains on fluorene were changed from C_8 to C_{12} carbon chains and the C_6 chains on thiophene were removed. Removing the C_6 chains on thiophene should have a significant impact on the properties of the polymer, because the C_6 chain and the hydrogen on benzene are head-to-head and can create significant twisting between the two aromatic rings. For this reason, PF12TBT (Figure 4.4) likely exhibits better charge transport ability than PF8TBT6. In Ito's work, it was found that using chloroform as solvent led to a polymer blend with small domain sizes compared to the case of using chlorobenzene or dichlorobenzene as solvents. After a reasonably smooth P3HT:PF12TBT film is obtained, thermal annealing at 140 °C can increase the domain size slightly while still keeping the domains reasonably small. The authors believe that the thermal annealing is a "thermal purification" process that improves the purity of the polymer domains and thus enhancing the performance. Shortly after this report, Ito and co-workers further improved the efficiency of P3HT:P-F12TBT-based all-PSCs by using high molecular weight PF12TBT, and achieved 2.7% efficiency, the highest achieved at the time for all-PSCs.

Exceptional work on P3HT:PFTBT-based all-PSCs was reported by Verduzco and co-workers,[28] who synthesized a P3HT and PFTBT block-copolymer and achieved an impressive PCE of 3% using a pure film of the P3HT:PFTBT block-copolymer. This approach was distinctively different from most other all-PSCs that incorporate the blend of two polymers. Using the R-SoXS technique, the authors proved that the block copolymer can form microdomains with a domain spacing of about 18 nm, indicating that the average domain size is approximately 9 nm, which is quite comparable to the typical exciton diffusion length in organic semiconductors. This work offered a unique approach to achieving a polymer morphology with an optimal domain size for all-PSCs.

4.2.2.2.2   P3HT:NDI or PDI Polymer-Based All-PSCs. Since the initial reports on PDI and NDI-based organic n-type semiconductors by Marks and co-workers,[55] NDI and PDI have been regarded as two of the most promising candidates to construct high-performance n-type polymeric semiconductors. All-PSCs based on P3HT and PDI polymers were extensively studied by Hashimoto *et al.*[56] In their report, a series of PDI-based copolymers were synthesized and combined with P3HT to construct all-PSCs. However, the performance of P3HT:PDI polymer-based all-PSCs was very poor. Despite the fact that PDI is a superb building block for n-type semiconductors, PDI-based polymers did not appear to work as well as BT-based polymers when combined with P3HT. The authors showed evidence of large polymer domains found in P3HT:PDI polymer blend films.

An important breakthrough for the n-type polymeric semiconductor field was the development of an NDI-based polymer named PNDI2OD-T2 (Figure 4.4) or tradenamed as ActiveInk N2200 by Polyera Corporation.[27] N2200 can achieve an electron mobility up to 0.85 cm^2 V^{-1} s^{-1} in organic field-effect transistor (OFET) configuration, and demonstrated performance comparable to the best p-type polymeric semiconductors at the time. The report of N2200 inspired a wave of academic studies on the use of N2200 in all-PSCs. Loi and co-workers[57] reported P3HT:N2200-based all-PSCs with an efficiency of only 0.16%. In Loi's report, it was found that the blend P3HT:N2200 exhibits a balanced ability of hole and electron transports, as evidenced by the OFET mobility data. With balanced ambipolar mobilities, a good FF, up to 67%, was obtained for the all-PSCs. Despite these favorable charge transport conditions, the J_{SC} of the P3HT:N2200-based cells is extremely low. The authors attributed the low performance of P3HT:N2200-based cells to the poor morphology of P3HT:N2200 blend films and the large LUMO offset between P3HT and N2200.

Several other groups also attempted to fabricate P3HT:N2200-based all-PSCs, and all obtained poor results. Despite the success of N2200 as an OFET material, the performance of N2200 in all-PSCs turned out to be quite disappointing. Sirringhaus and co-workers[58] carried out systematic photophysics and morphological studies on N2200 to understand the fundamental reasons for the low performance of N2200 in all-PSCs. First, X-ray microscopy (scanning transmission X-ray microscopy, STXM) was used characterize the morphology of P3HT:N2200 blend films. STXM can reveal a great deal of morphological information that is unattainable using the conventional AFM technique. While AFM only reveals the surface topography of a polymer blend without being able to distinguish the chemical compositions of the two polymers, STXM can provide a material-sensitive image that reflects the domain size of the bulk of the blend film. In these studies, P3HT:N2200-based blend films were fabricated using three different solvents: chloroform, xylene, and dichlorobenzene. The blend films obtained from xylene and dichlorobenzene solutions clearly contain polymer domains as large as 2 μm, which is much larger than the optimal domain size for PSC operations. The blend film obtained from the chloroform solution has smaller domain sizes of about 200 nm, which is still not optimal. These results indicate that using a high boiling solvent will lead to larger domains because the polymers have more time to aggregate during the dying process of the film. Photophysics studies showed there was a rapid, initial geminate recombination of the charge population within 200 ps of the excitation. This was identified as another major reason for the low performance of the P3HT:N2200-based all-PSCs.

4.2.2.2.3 Problems with P3HT:N2200-based cells. These initial attempts in using N2200 as the acceptor in all-PSCs indicate that a polymer that exhibits great performance in OFET applications may not necessarily work well in all-PSCs. Besides the reasons revealed above, it is also important to note that OFETs and all-PSCs have completely different working mechanisms, which lead to different requirements for materials. For OFET application, the polymer should be highly crystalline and form large polymer aggregates or crystallites in the film. For PSCs, however, highly crystalline polymers may tend to form polymer domains that are too large for PSC operations. In addition, OFET and PSCs could also have different requirements on the LUMO levels of the polymer. To realize air-stable OFET devices, the LUMO value of the n-type polymer should be about 4.0 eV or larger. However, all-PSCs do not have such requirements on LUMO levels of the acceptor polymers. Depending on the LUMO level of the donor polymers, the LUMO levels of the acceptor polymers could be between 3.5 and 4.0 eV. If the acceptor LUMO value is too large, the V_{OC} of all-PSCs will be small. It is important to choose an acceptor polymer so that the LUMO offset between the donor and acceptor (ΔLUMO) is sufficiently large for exciton dissociation. It is commonly believed that exciton dissociation energy in organic materials is about 0.3 eV. Therefore, it would be ideal to use a set of donor and acceptor polymers with LUMO offset of about 0.3 eV, or slightly larger if necessary. From this perspective, P3HT and N2200 are not a good match, because their ΔLUMO is as large as 1.0 eV, resulting in a large V_{OC} loss. To realize the full potential of N2200 in all-PSCs, it is desirable to employ a donor polymer with lower-lying LUMO levels than P3HT.

To address the issue of large domain size for P3HT:N2200-based all-PSCs, Neher and co-workers[59] tried to optimize the processing conditions of P3HT:N2200 blend solutions to minimize the polymer domain size. As suggested above, N2200 has a strong aggregation tendency and forms large polymer domains. The authors confirmed this point from a different perspective by measuring the UV-vis absorption spectra of N2200 solution in different solvents. It was clearly demonstrated that a strong chlorinated solvent (*e.g.*, chloronaphthalene) can break the aggregation of the polymer chains in solution, as evidenced by the blue-shifted absorption of the solution. Inspired by this finding, the authors used a mixture of xylene and chloronaphthalene as the solvent and processed the polymer blend solution at elevated temperatures. When processing the solution at elevated temperature, the drying time of the solution is minimized; therefore, the polymer chains do not have sufficient time to aggregate or to form large domains. On the other hand, if the chloronaphthalene solution were to be processed at room temperature, it would take a long time for the solution to dry on a spincoater. During the long drying time, N2200 polymer chains will likely aggregate to form large domains. The xylene–chlorobenzene hot spincoating approach is quite successful, as it improved the efficiency of P3HT:N2200-based all-PSCs from 0.2% to 1.4%. An impressive fill factor of 65% was obtained, accompanied by a greatly enhanced J_{SC} (3.77 mA cm^{-2}), when compared with those in previous reports.

4.2.2.3   *Polythiophene with Conjugated Side Chains*

Since the initial report of P3HT in all-PSCs, another important direction of research was on polythiophene with conjugated side chains. These donor polymers, such as TTV4-PT6,[60] BTV12-PT6,[61] *etc.* (Figure 4.4) have two-dimensional conjugated structures, which are believed to enhance charge transport and facilitate the formation of well-connected conjugated networks by the donor polymer. A systematic comparison between P3HT and BTV12-PT6 (named PT1 in ref. 60) was carried out by Hashimoto and co-workers.[62] P3HT and PT1 are combined in six different PDI-based acceptor polymers and the PT1:PDI polymer-based all-PSCs were found to exhibit much better performances (best PCE 2.23%) than P3HT:PDI polymer-based all-PSCs (PCE < 0.5%). An interesting point was that PT1:PDI-based all-PSCs exhibit significantly higher V_{OC} than P3HT:PDI-based cells, owing to the lower-lying HOMO level of PT1. Although PT1 and P3HT have identical polymer backbones, the HOMO and LUMO levels of the two polymers appeared to be significantly different because PT1 contains randomly mixed comonomer units with C_6 carbon chains or thiophene-based conjugated side chains. Owing to the random nature of its structure, PT1 appears to be much less crystalline than P3HT, which is evidenced by the small polymer domains in PT1:PDI polymer blends. For nearly all six cases of PDI polymers, PT1:PDI polymer-based blend films exhibit smooth films and more uniform mixing than P3HT:PDI polymer blends. The authors believed that regioregular P3HT is too crystalline and resulted in large polymer domains that are undesirable for PSCs. This is another important reason to adopt random polymers with conjugated side chains, because the "randomness" can reduce the aggregation tendency of the polymer and facilitate the formation of a blend morphology with small polymer domains.

Mini-Summary (Section 4.2.2).   To summarize the overall research efforts in polythiophene-based all-PSCs, there are several key factors that need to be considered in order to improve the performance of all-PSCs beyond that achievable for polythiophene-based all-PSCs.

(i) The bandgap of polythiophenes is still too large. Ideally, low bandgap polymers should be adopted to extend the absorption range of the all-PSCs to at least 750 nm.

(ii) For the most commonly studied polythiophene (P3HT), although it has many advantages such as high crystallinity, easy processability and availability, its high crystallinity also causes many issues. In particular, when P3HT was blended with another highly crystalline acceptor polymer, N2200, it appeared to be highly challenging to obtain an optimal polymer:polymer blend morphology with reasonably small domain sizes.

(iii) Lastly, the LUMO level of polythiophenes (3.1 eV for P3HT) is quite high, which results in large LUMO offsets with many acceptor polymers such as state-of-the-art NDI or PDI-based acceptor polymers. Large LUMO offsets seriously limit the potential of P3HT-based all-PSCs.

With these arguments, it is important to develop all-PSCs based on lower bandgap donor polymers that have lower-lying LUMO levels than polythiophenes.

4.2.3   All-PSCs Based on Medium or Low Bandgap Polymers

The improvement of all-PSC efficiency has been relatively slow (from 1% to 2.5%) for nearly 15 years. However, there have been several important reports of higher efficiency all-PSCs in the past year, with the PCE approaching 5% within a year or so. In this section, several recent reports of high-efficiency all-PSCs will be summarized. The main difference of these high efficiency all-PSCs from previous work is that donor polymers with lower bandgaps and lower-lying LUMO levels are used. These new donor polymers not only increase the J_{SC} of the cells by broadening the absorption range, they also allow for the achievement of reasonably large V_{OC} by reducing the LUMO offset between the donor and acceptor polymers, and thus the V_{OC} loss. The performance of these high-efficiency all-PSCs, along with some previous examples, is summarized in Table 4.1.

Jenekhe and co-workers[29] reported a thiazolothiazole-based copolymer (PSEHTT,[27]) with an optical bandgap of 1.8 eV (Figure 4.6). The reduced bandgap extends the absorption onset of the cell by about 50–60 nm compared to P3HT-based cells. Compared to P3HT:N2200-based cells, the V_{OC} of the PSE-HTT-based cells is increased by about 0.15 V, and the bandgap of PSEHTT is decreased by 0.1 V. The reason PSEHTT-based cells can achieve higher V_{OC} and smaller bandgap at the same time is because PSEHTT has significantly lower-lying LUMO level than P3HT. In this work, an NDI–selenophene copolymer

Table 4.1   Photoactive blend composition and device performance for all-PSCs based on low bandgap polymers.

	PCE (%)	J_{SC} (mA cm^{-2})	V_{OC} (V)	E_g (eV)	V_{OC} loss (eV)	ΔLUMO (eV)	Ref.
PNDIS-HD:PSEHTT	3.26	7.78	0.76	1.65	0.89	0.7	29
PTQ1:N2200	4.1	8.85	0.84	1.77	0.93	0.57	31
TTV7:PC-NDI	3.68	7.71	0.88	1.73	0.85	0.39	63
Pil-2T-PSS:P(TP)	4.4	9.0	1.04	1.65	0.61	0.04	32

was used as the acceptor polymer and achieved the best results among the various conditions reported. It was shown that the NDI–selenophene-based copolymer has higher electron mobility than a NDI–thiophene-based copolymer, which is believed to be the main reason for the enhanced performance for NDI–selenophene-based cells. At about the same time, Tajima and co-workers reported a random copolymer containing benzodithiophene, thiophene, and thienothiophene (TT) units, named TTV7 (Figure 4.6), which achieved a high PCE of 3.68%.[63] The addition of TT comonomer reduces the bandgap of the polymer and shifts the absorption edge of the polymer to about 700 nm. Furthermore, the cell has an impressive V_{OC} of 0.88 V, which is in agreement with the lower-lying HOMO level of the polymer. Although 80% of the building blocks of TTV7 are identical to those of PTB7, the properties of TTV7 appear to be distinctly different because it is a random copolymer containing 20% of thiophene comonomer with conjugated side chains. The TTV7-based cells clearly outperformed the PTB7-based cells. As revealed by AFM studies, the TTV7-based polymer blends exhibit significantly smaller domains than PTB7-based blends, with the domain size in PTB7 films clearly larger than the optimal size for PSC operation. It is likely that the random nature of TTV7 slightly reduces its crystallinity compared to the PTB7 polymer and allows TTV7 to form smaller polymer domains. In Tajima's report, it was also demonstrated that the DIO additive has a dramatic impact on all-PSC performance, with the PCE increased from 1.6% to 3.68%. Another case of all-PSCs with 4.1% PCE was reported by Ito and co-workers based on a polymer, named PTQ1 (Figure 4.6), with similar optical properties to the previous two cases in this paragraph.[31] When PTQ1 was combined with N2200, it was found that the hole mobility of PTQ1 is significantly lower than the electron mobility of N2200, which resulted in unbalanced charge transport. The authors therefore reduced the weight percentage of the N2200 polymer and achieved significantly higher fill factor and efficiency for the all-PSCs. Besides this, another important reason why PTQ1–N2200-based cells could outperform all-PSCs in previous reports is that PTQ1 and N2200 form a blend film with very small feature size by AFM. Although AFM does not reveal complete information on the bulk morphology of a polymer:polymer blend, it is likely that the PTQ1:N2200 blend has one of best morphologies among all-PSCs reported to date.

In all three cases of the all-PSCs above, the LUMO offsets between the donor and acceptor polymers were smaller than in P3HT:N2200-based all-PSCs. The value of LUMO offset is one of the most important parameters

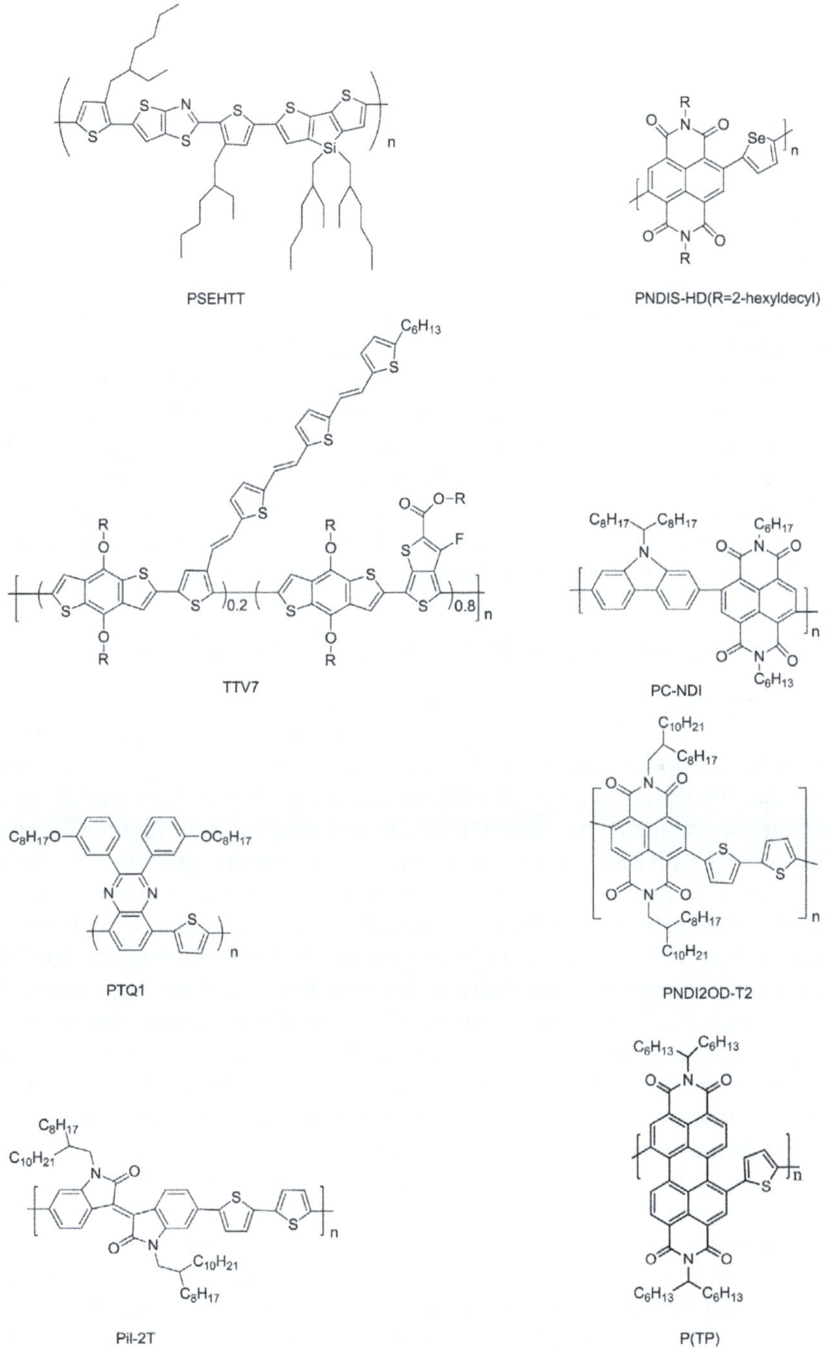

Figure 4.6 Chemical structures of donor and acceptor polymers used in all-polymer solar cells with low bandgap.

that determine the potential performance of a PSC. With a small value of LUMO offset, the V_{OC} loss of a PSC will be small, and it is possible for a PSC to achieve high V_{OC} and low bandgap at the same time. For this reason, the three cases of high-efficiency all-PSCs all exhibit higher V_{OC} and lower bandgap than P3HT:N2200-based all-PSCs, which can be partly attributed to the small LUMO offset values in these cases.

Mostly recently, Bao[32] and co-workers demonstrated an all-PSC with 4.4% efficiency, by combining two donor and acceptor polymers, (Pil-2T) and P(TP) (Figure 4.6), with a nominal LUMO offset value of only 0.1 eV. This is exceptional because it is commonly believed that a LUMO value of 0.3 eV is needed for exciton dissociation in PSCs. For state-of-the-art polymer:fullerene PSCs, the LUMO offset value is typically around 0.4 eV. Demonstrating reasonably high efficiency in all-PSCs with such a small LUMO offset value is a major advance for the PSC field. As the benefit of the small LUMO offset value, a high V_{OC} of 1.04 V was observed for the cell with an optical bandgap of 1.65 V. The V_{OC} loss is only about 0.6 V, which is the best known value for PSCs today. This important result raised many important fundamental questions, such as whether exciton dissociation in a polymer:polymer blend could require much less energy than in polymer:fullerene-based systems.

4.3   Morphology of Polymer:Polymer Blends

As for polymer:fullerene blends, and as highlighted above, optimizing the performance of polymer solar cells with non-fullerene acceptors critically depends on optimization of the active layer morphology. Conceptually, blends using non-fullerene small molecule acceptors will have similar phase separation behavior to polymer:fullerene blends. For polymer:fullerene blends, structure formation is often driven by polymer crystallization and/or fullerene aggregation,[64,65] with the high polymer:fullerene miscibility resulting in mixed phases even in equilibrium structures.[66] All-polymer blends, on the other hand, are distinct from polymer:fullerene blends in that both donor and acceptor are long chain molecules. Because of the high molecular weights of both donor and acceptor, polymer:polymer donor:acceptor pairs are generally expected to be immiscible. This immiscibility results in a low entropy of mixing, with relatively pure domains much larger than the exciton diffusion length the favored equilibrium structure. Films prepared *via* rapid solvent evaporation (such as during spincoating) lead to non-equilibrium structures, with the possibility of achieving nanostructured blends of the right dimensions *via* arresting of phase separation. However, the tendency of all-polymer blends to macro-phase segregate on a length scale larger than desired has been problematic in a number of systems.[67] Control of morphology will be improved by a deeper understanding of the morphology of all-polymer blends (as distinct from polymer:fullerene blends) and what factors influence structure formation on the sub-100 nm length scale.

In this section, parameters and challenges facing morphology optimization in all-polymer blends are discussed. The discussion is restricted to the morphology of blends prepared by dissolving donor and acceptor polymers in a common solution and solution processing. However, it should be noted that there unique opportunities afforded by the use of polymer acceptors for morphology optimization including block-copolymers[28] and nano-imprint lithography,[53] alluded to above. While these novel approaches allow for better control of morphology and undoubtedly assist in understanding morphology–performance relationships, any commercial all-polymer technology will likely utilize a blend of donor and acceptor deposited from solution.

4.3.1 Solution Deposition

The active layer in all-polymer blends is generally prepared by dissolving both donor and acceptor polymers in a common solution and spincoating. In solution, the solvent molecules dilute the polymer:polymer interactions, and the system (if sufficiently dilute) can be regarded as having one phase. During spincoating (or other solution deposition processes) the solvent evaporates, leading to an increase in polymer:polymer interactions. The driving force for phase separation derives from the fact that the enthalpic gain of pure (or rich) phases compensates for the reduction in entropy in going from a mixed to a de-mixed system. While phase separation proceeds as the solvent evaporates, eventually enough solvent will have evaporated such that the polymer chains become immobile and the morphology is frozen in.

Numerous studies have noted that the morphology and performance of all-polymer solar cells is sensitive to the choice of solvent used.[68–72] As an example, Figure 4.7 presents electron microscopy images of PTB7:N2200

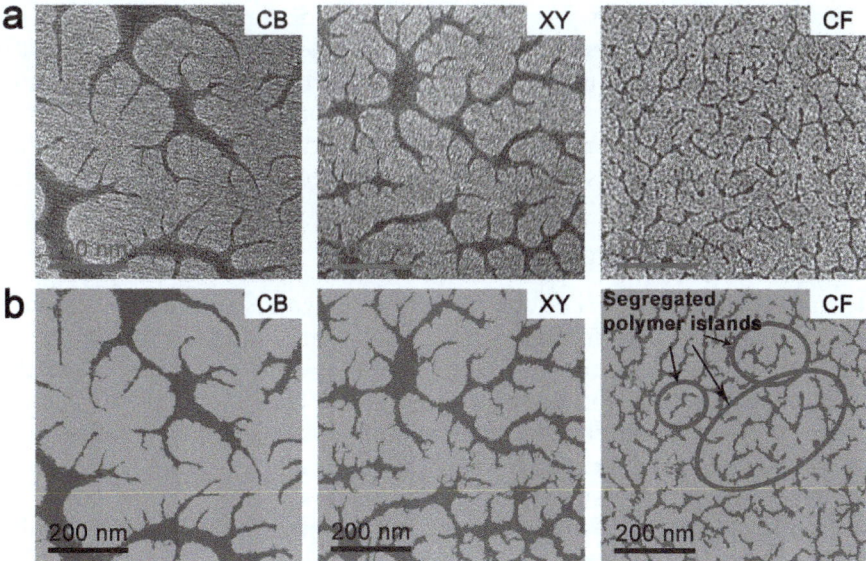

Figure 4.7 Transmission electron microscopy (TEM) blend morphology of 50:50 wt% PTB7:N2200 blend films processed from CB, XY, or CF. (a) Uncorrected TEM images and (b) false-coloured images showing the PTB7 and N2200 domains.

blends prepared from either chlorobenzene, xylene or chloroform.[72] Such films exhibit an interesting dendritic structure with differences in the characteristic length scale of morphology evident. The smallest features are observed for the case of the blends coated from chloroform, consistent with the low boiling point and high volatility of chloroform that leads to a rapid drying time. Interestingly, despite having a coarser morphology than the chloroform-processed blend, the xylene-processed blend was found to optimise device performance attributed to a better interconnection of phases.[72]

Greater flexibility in morphology control can be achieved through the use of solvent mixtures or solvent additives. The combination of a low-boiling point solvent with a high-boiling point solvent enables finer tuning of film drying time compared to use of each solvent separately.[73] Solvent choice also affects polymer aggregation and crystallization. For the P3HT:N2200 system, as identified by Neher and co-workers,[74] the acceptor polymer N2200 aggregates in solvents such as xylene and chlorobenzene that are commonly used. Using a cosolvent such as chloronaphthalene with large and highly polarizable aromatic cores, aggregation in solution was suppressed, resulting in a finer morphology and superior device performance.[74] On the other hand, use of the solvent additive diiodooctane in the PBDTTT-C-T:PPDIDTT system has been found to improve device performance by facilitating aggregation and crystallization of the donor polymer.[75] An interesting recent approach to facilitate finer morphologies is the use of low molecular weight compatibilizers. Cheng *et al.* reported the use of a conjugated molecule with a similar structure to that of the acceptor polymer as a non-volatile additive to suppress aggregation of the acceptor and to facilitate mixing with the polymer donor.[75] Slota *et al.* have also reported the use of conjugated oligomeric compatibilizers based on a diblock of donor and acceptor structures to tune interfacial area and phase purity.[76] Thus, there are many and varied approaches for influencing the morphology of a polymer blend film by varying deposition parameters. However, our present level of understanding is largely empirical and an improved understanding of the mechanisms at play will facilitate future device improvements.

4.3.2   Molecular Weight

The molecular weight of donor and acceptor polymers also plays an important role in morphology formation during solution coating. For higher molecular weight polymers the onset of polymer:polymer interactions as the solvent evaporates during casting will commence sooner, leading to a coarser degree of phase separation. Additionally, the entropy of mixing is inversely proportional to the degree of polymerization, meaning that lower molecular weight polymers will result in phases that are less pure.

Surprisingly, there are few systematic studies on the influence of molecular weight on the morphology and device performance of all-polymer solar cells. Veenstra *et al.* reported changes in the microstructure of MDMO-PPV:PC-NEPV blends as a function of the molecular weight of the acceptor polymer PCNEPV (Figure 4.8.)[77] All films in Figure 4.8 have a weight ratio of 1 : 1, with a higher molecular weight of the acceptor leading to larger-sized structures.

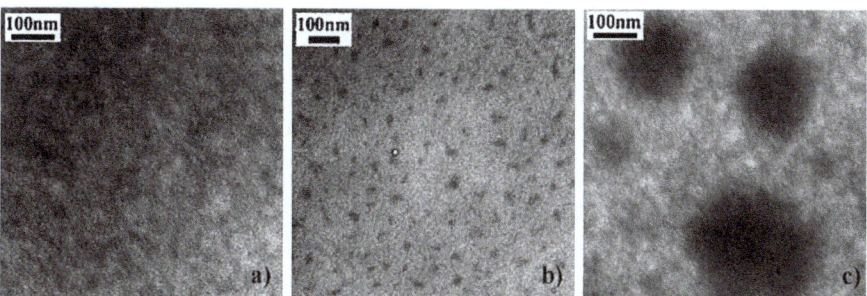

Figure 4.8 Zero-loss filtered TEM images of thin MDMO-PPV:PCNEPV blend film samples. (a) Sample 1 with low M_W (3500 g mol^{-1}) PCNEPV derivative, (b) sample 2 with medium M_W (48 000 g mol^{-1}) PCNEPV derivative, and (c) sample 3 with high M_W (113 500 g mol^{-1}) PCNEPV derivative.

The darker regions, which correspond to PCNEPV-rich domains, do not fill 50% of the images, indicating that there is significant intermixing within the observed domains, consistent with photoluminescence measurements. Interestingly, the performance of devices based on such films was found to be very similar.[77] Such an observation suggests that the nature of intermixing within these domains, rather than their observed size, is most important for MDMO-PPV:PCNEPV device performance.

More recently, Mori *et al.* have observed that the device performance of P3HT:PF12TBT was optimized with the use of a high molecular weight acceptor polymer.[78] For this system, thermal annealing of the blend is required to enhance the crystallinity of the P3HT component. While low (8500 g mol^{-1}) and medium (20 000 g mol^{-1}) molecular weights of PF12TBT resulted in better device performance at mild annealing temperatures (<120 °C), devices using the high molecular weight PF12TBT (78 000 g mol^{-1}) outperformed the other devices for higher annealing temperatures. This observation was attributed to the higher glass transition temperature of high molecular weight PF12TBT, which affords the benefits of P3HT crystallization without a significant coarsening of phase separation.

4.3.3 Crystallinity

As for the P3HT:PF12TBT system, many high efficiency all-polymer systems employ semicrystalline polymers as one or more of the components. The process of crystallization can play an important role in microstructure evolution. Sepe *et al.* have recently studied the microstructural evolution of the similar P3HT:PF8TBT6 system.[79] First, examining the phase behavior of PF8TBT6 with the amorphous regiorandom P3HT (P3HT-RA), Sepe *et al.* found that melting the P3HT-RA:PF8TBT6 film and slowly cooling (aiding the thermodynamic equilibration of the samples) led to the formation of mesoscale, lateral phase separated morphologies characteristic of amorphous–amorphous mixtures in thin films (Figure 4.9a). In contrast, films employing semicrystalline batches of P3HT did not exhibit a coarse phase-separated morphology, instead

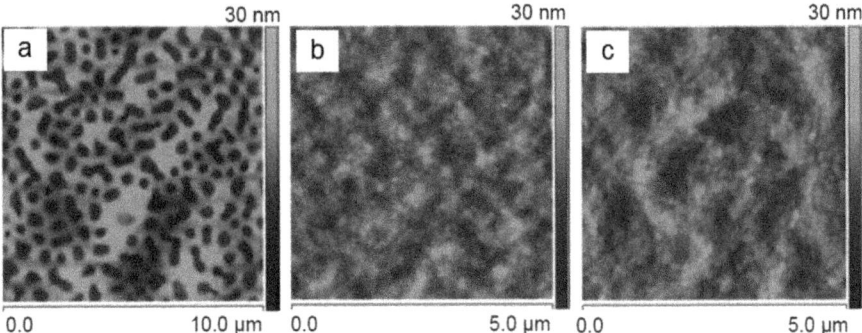

Figure 4.9 (a) AFM images of a P3HT-RA:F8TBT blend after melt annealing. (b and c) AFM images of P3HT:F8TBT blends based on two different semicrystalline P3HT batches after melt annealing.

showing a fibrillar surface texture (Figure 4.9b and c). X-ray scattering measurements also found a lack of macroscopic demixing in the blends utilizing semicrystalline P3HT, with the kinetic diffusion of PF8TBT6 limited by P3HT crystallization. A long period of 16 nm was observed by grazing incidence small-angle X-ray scattering (GISAXS), linked to the size of P3HT crystallites. Strikingly, devices fabricated using melt-annealed films exhibited reasonable efficiencies, confirming that annealing to >200 °C did not lead to gross phase-separation. The overall picture obtained from this study is that crystallization of P3HT constructs a framework on the 10 nm length scale where the structural size is determined by the lamellar P3HT crystals.[79] PF8TBT6 is enriched in the interlamellar P3HT regions, promoting charge transfer and charge transport. Thus the crystallization behavior of semicrystalline polymers can provide a mechanism for nanostructure formation on the length-scale of the exciton diffusion length, and arrest large scale phase separation.

4.3.4 Side Chains

Side-chain engineering is becoming an increasingly used approach to tuning the morphology and interfacial interactions in organic electronics. Changing polymer side-chain length can enable tuning of solubility, crystallization dynamics and the thermodynamic properties of polymers, which in turn affects morphology evolution and device performance.[80] Changing the nature of the side chain can also strongly influence material properties. For example, tuning of side-chain branching position has recently been shown to enable improved charge transport mobility[81,82] and is currently being applied to polymer:fullerene solar cells.[83] In the field of all-polymer solar cells, Hashimoto and co-workers have successfully applied the use of conjugated side-chains, as discussed above. Another interesting example of side-chain engineering is that the use of side-chains incorporating bulky phenyl groups has been shown to improve the performance of solar cells, compared with conventional linear side chains (Figure 4.10). While bulky

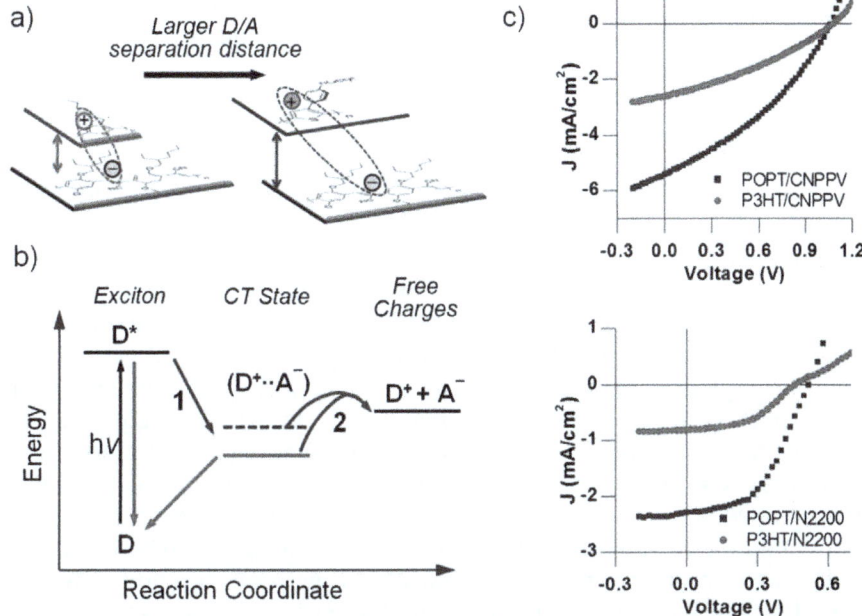

Figure 4.10   (a and b) Schematic diagrams demonstrating how steric interactions can lead to an increase in backbone spacing and destabilization of the geminate pair due to a different energy landscape. (c) Example current–voltage curves showing the enhanced performance of POPT-based devices compared to P3HT-based devices.

phenyl groups will potentially be detrimental to π–π stacking and hence charge transport in the bulk, they can be beneficial in tuning the separation of donor and acceptor species at the donor–acceptor interface. Holcombe et al.[84] compared the operation of all-polymer solar cells using P3HT and POPT as the electron donor. It was argued that the bulky side group on POPT chains leads to increased physical separation of donor:acceptor chains at the heterojunction, leading to a decrease in the energy of relaxed interfacial electron–hole pairs, reducing the barrier to charge separation.[84] Studying blends with a number of polymer acceptors, the performance of POPT-based cells was found to be consistently higher than that of P3HT-based devices, providing strong evidence for a connection between device performance and heterojunction conformation. Therefore, where device efficiency is limited by interfacial charge separation rather than bulk charge transport, the use of bulky side groups can actually be beneficial.

4.3.5   Mini-Summary

The development of higher efficiency all-polymer solar cells relies upon microstructure optimization almost as much as upon the development of new materials. Future research will benefit from a synergy among synthetic chemists, materials scientists and device physicists whereby important parameters affecting morphology, such as molecular weight, side chains and crystallinity, are well defined and controlled.

4.4 Conclusions

The development of all-PSCs in the past few years was partly driven by the development of high-performance n-type polymeric semiconductors such as NDI or PDI-based polymers. However, initial applications of high-performance n-type polymeric semiconductors to all-PSCs have produced disappointing results. Two critically important factors need to be considered in order to utilize these high performance n-type polymers effectively and to construct high-efficiency all-PSCs. First, because highly crystalline polymers often tend to form large polymer domains, it is important carefully to control the processing of the polymer:polymer solutions to obtain a morphology containing reasonably small polymer domains. On the other side, the main driving force of the all-PSC field in the past two years has been the development of donor polymers that have low optical bandgaps, high hole mobility, and, most importantly, matching LUMO levels with the acceptor polymers. The LUMO offset between the donor and acceptor polymers is one of the most important parameters determining the potential performance of all-PSCs. All-PSCs may have unique advantages over polymer:fullerene PSCs in this area, because high-efficiency all-PSCs have been demonstrated with a small LUMO offset value of 0.1 eV.

References

1. G. Yu, J. Gao, J. C. Hummelen, F. Wudl and A. J. Heeger, *Science*, 1995, **270**, 1789.
2. J. J. M. Halls, C. A. Walsh, N. C. Greenham, E. A. Marseglia, R. H. Friend, S. C. Moratti and A. B. Holmes, *Nature*, 1995, **376**, 498.
3. G. Li, V. Shrotriya, J. S. Huang, Y. Yao, T. Moriarty, K. Emery and Y. Yang, *Nat. Mater.*, 2005, **4**, 864.
4. H. Y. Chen, J. H. Hou, S. Q. Zhang, Y. Y. Liang, G. W. Yang, Y. Yang, L. P. Yu, Y. Wu and G. Li, *Nat. Photonics*, 2009, **3**, 649.
5. X. G. Guo, N. J. Zhou, S. J. Lou, J. Smith, D. B. Tice, J. W. Hennek, R. P. Ortiz, J. T. L. Navarrete, S. Y. Li, J. Strzalka, L. X. Chen, R. P. H. Chang, A. Facchetti and T. J. Marks, *Nat. Photonics*, 2013, **7**, 825.
6. Y. Y. Liang, Z. Xu, J. B. Xia, S. T. Tsai, Y. Wu, G. Li, C. Ray and L. P. Yu, *Adv. Mater.*, 2010, **22**, E135.
7. Z. C. He, C. M. Zhong, S. J. Su, M. Xu, H. B. Wu and Y. Cao, *Nat. Photonics*, 2012, **6**, 591.
8. M. M. Wienk, J. M. Kroon, W. J. H. Verhees, J. Knol, J. C. Hummelen, P. A. Van Hal and R. A. J. Janssen, *Angew. Chem., Int. Ed.*, 2003, **42**, 3371.
9. Y. J. He, H. Y. Chen, J. H. Hou and Y. F. Li, *J. Am. Chem. Soc.*, 2010, **132**, 1377.

10. P. P. Khlyabich, B. Burkhart and B. C. Thompson, *J. Am. Chem. Soc.*, 2011, **133**, 14534.
11. G. J. Zhao, Y. J. He and Y. F. Li, *Adv. Mater.*, 2010, **22**, 4355.
12. R. B. Ross, C. M. Cardona, D. M. Guldi, S. G. Sankaranarayanan, M. O. Reese, N. Kopidakis, J. Peet, B. Walker, G. C. Bazan, E. Van Keuren, B. C. Holloway and M. Drees, *Nat. Mater.*, 2009, **8**, 208.
13. C. Y. Shu, W. Xu, C. Slebodnick, H. Champion, W. J. Fu, J. E. Reid, H. Azurmendi, C. R. Wang, K. Harich, H. C. Dorn and H. W. Gibson, *Org. Lett.*, 2009, **11**, 1753.
14. R. B. Ross, C. M. Cardona, F. B. Swain, D. M. Guldi, S. G. Sankaranarayanan, E. Van Keuren, B. C. Holloway and M. Drees, *Adv. Funct. Mater.*, 2009, **19**, 2332.
15. C. Z. Li, H. L. Yip and A. K. Y. Jen, *J. Mater. Chem.*, 2012, **22**, 4161.
16. Y. F. Li, *Chem.–Asian J.*, 2013, **8**, 2316.
17. M. C. Scharber, D. Wuhlbacher, M. Koppe, P. Denk, C. Waldauf, A. J. Heeger and C. L. Brabec, *Adv. Mater.*, 2006, **18**, 789.
18. G. Dennler, M. C. Scharber, T. Ameri, P. Denk, K. Forberich, C. Waldauf and C. J. Brabec, *Adv. Mater.*, 2008, **20**, 579.
19. N. Li, D. Baran, K. Forberich, F. Machui, T. Ameri, M. Turbiez, M. Carrasco-Orozco, M. Drees, A. Facchetti, F. C. Krebs and C. J. Brabec, *Energy Environ. Sci.*, 2013, **6**, 3407.
20. J. Peet, J. Y. Kim, N. E. Coates, W. L. Ma, D. Moses, A. J. Heeger and G. C. Bazan, *Nat. Mater.*, 2007, **6**, 497.
21. Y. M. Sun, G. C. Welch, W. L. Leong, C. J. Takacs, G. C. Bazan and A. J. Heeger, *Nat. Mater.*, 2012, **11**, 44.
22. S. H. Park, A. Roy, S. Beaupre, S. Cho, N. Coates, J. S. Moon, D. Moses, M. Leclerc, K. Lee and A. J. Heeger, *Nat. Photonics*, 2009, **3**, 297.
23. J. Roncali, *Acc. Chem. Res.*, 2009, **42**, 1719.
24. T. Rousseau, A. Cravino, T. Bura, G. Ulrich, R. Ziessel and J. Roncali, *Chem. Commun.*, 2009, 1673. DOI: 10.1039/b822770e.
25. H. Shang, H. Fan, Y. Liu, W. Hu, Y. Li and X. Zhan, *Adv. Mater.*, 2011, **23**, 1554.
26. Y. Sun, C. J. Takacs, S. R. Cowan, J. H. Seo, X. Gong, A. Roy and A. J. Heeger, *Adv. Mater.*, 2011, **23**, 2226.
27. H. Yan, Z. Chen, Y. Zheng, C. Newman, J. R. Quinn, F. Dotz, M. Kastler and A. Facchetti, *Nature*, 2009, **457**, 679.
28. C. H. Guo, Y. H. Lin, M. D. Witman, K. A. Smith, C. Wang, A. Hexemer, J. Strzalka, E. D. Gomez and R. Verduzco, *Nano Lett.*, 2013, **13**, 2957.
29. T. Earmme, Y. J. Hwang, N. M. Murari, S. Subramaniyan and S. A. Jenekhe, *J. Am. Chem. Soc.*, 2013, **135**, 14960.
30. E. Zhou, J. Cong, K. Hashimoto and K. Tajima, *Adv. Mater.*, 2013, **25**, 6991.
31. D. Mori, H. Benten, I. Okada, H. Ohkita and S. Ito, *Adv. Energy Mater.*, 2014, **4**, 1301006.
32. Y. Zhou, T. Kurosawa, W. Ma, Y. Guo, L. Fang, K. Vandewal, Y. Diao, C. Wang, Q. Yan, J. Reinspach, J. Mei, A. L. Appleton, G. I. Koleilat, Y. Gao,

S. C. B. Mannsfeld, A. Salleo, H. Ade, D. Zhao and Z. Bao, *Adv. Mater.*, 2014, **26**, 3767.

33. N. Wang, Z. Chen, W. Wei and Z. H. Jiang, *J. Am. Chem. Soc.*, 2013, **135**, 17060.

34. K. H. Hendriks, G. H. L. Heintges, V. S. Gevaerts, M. M. Wienk and R. A. J. Janssen, *Angew. Chem., Int. Ed.*, 2013, **52**, 8341.

35. T. B. Yang, M. Wang, C. H. Duan, X. W. Hu, L. Huang, J. B. Peng, F. Huang and X. Gong, *Energy Environ. Sci.*, 2012, **5**, 8208.

36. F. C. Chen, Q. F. Xu and Y. Yang, *Appl. Phys. Lett.*, 2004, **84**, 3181.

37. W. R. Wu, U. S. Jeng, C. J. Su, K. H. Wei, M. S. Su, M. Y. Chiu, C. Y. Chen, W. B. Su, C. H. Su and A. C. Su, *ACS Nano*, 2011, **5**, 6233.

38. G. Yu and A. J. Heeger, *J. Appl. Phys.*, 1995, **78**, 4510.

39. A. J. Breeze, Z. Schlesinger, S. A. Carter, H. Tillmann and H. H. Horhold, *Sol. Energy Mater. Sol. Cells*, 2004, **83**, 263.

40. T. Kietzke, H. H. Horhold and D. Neher, *Chem. Mater.*, 2005, **17**, 6532.

41. A. C. Arias, J. D. Mackenzie, R. Stevenson, J. J. M. Halls, M. Inbasekaran, E. P. Woo, D. Richards and R. H. Friend, *Macromolecules*, 2001, **34**, 6005.

42. M. R. Andersson, D. Selse, M. Berggren, H. Jarvinen, T. Hjertberg, O. Inganas, O. Wennerstrom and J. E. Osterholm, *Macromolecules*, 1994, **27**, 6503.

43. Q. Pei, H. Jarvinen, J. E. Osterholm, O. Inganas and J. Laakso, *Macromolecules*, 1992, **25**, 4297.

44. T. Johansson, W. Mammo, M. Svensson, M. R. Andersson and O. Inganas, *J. Mater. Chem.*, 2003, **13**, 1316.

45. A. Gadisa, M. Svensson, M. R. Andersson and O. Inganas, *Appl. Phys. Lett.*, 2004, **84**, 1609.

46. D. M. Deleeuw, M. M. J. Simenon, A. R. Brown and R. E. F. Einerhand, *Synth. Met.*, 1997, **87**, 53.

47. K. E. Aasmundtveit, E. J. Samuelsen, W. Mammo, M. Svensson, M. R. Andersson, L. A. A. Pettersson and O. Inganas, *Macromolecules*, 2000, **33**, 5481.

48. M. R. Andersson, M. Berggren, O. Inganas, G. Gustafsson, J. C. Gustafssoncarlberg, D. Selse, T. Hjertberg and O. Wennerstrom, *Macromolecules*, 1995, **28**, 7525.

49. M. Theander, O. Inganas, W. Mammo, T. Olinga, M. Svensson and M. R. Andersson, *J. Phys. Chem. B*, 1999, **103**, 7771.

50. M. Granstrom, K. Petritsch, A. C. Arias, A. Lux, M. R. Andersson and R. H. Friend, *Nature*, 1998, **395**, 257.

51. Y. Kim, S. Cook, S. A. Choulis, J. Nelson, J. R. Durrant and D. D. C. Bradley, *Chem. Mater.*, 2004, **16**, 4812.

52. C. R. McNeill, A. Abrusci, I. Hwang, M. A. Ruderer, P. Mueller-Buschbaum and N. C. Greenham, *Adv. Funct. Mater.*, 2009, **19**, 3103.

53. X. He, F. Gao, G. Tu, D. Hasko, S. Huettner, U. Steiner, N. C. Greenham, R. H. Friend and W. T. S. Huck, *Nano Lett.*, 2010, **10**, 1302.

54. D. Mori, H. Benten, J. Kosaka, H. Ohkita, S. Ito and K. Miyake, *ACS Appl. Mater. Interfaces*, 2011, **3**, 2924.

55. B. A. Jones, A. Facchetti, M. R. Wasielewski and T. J. Marks, *J. Am. Chem. Soc.*, 2007, **129**, 15259.
56. E. J. Zhou, J. Z. Cong, Q. S. Wei, K. Tajima, C. H. Yang and K. Hashimoto, *Angew. Chem., Int. Ed.*, 2011, **50**, 2799.
57. S. Fabiano, Z. Chen, S. Vahedi, A. Facchetti, B. Pignataro and M. A. Loi, *J. Mater. Chem.*, 2011, **21**, 5891.
58. J. R. Moore, S. Albert-Seifried, A. Rao, S. Massip, B. Watts, D. J. Morgan, R. H. Friend, C. R. McNeill and H. Sirringhaus, *Adv. Energy Mater.*, 2011, **1**, 230.
59. M. Schubert, D. Dolfen, J. Frisch, S. Roland, R. Steyrleuthner, B. Stiller, Z. Chen, U. Scherf, N. Koch, A. Facchetti and D. Neher, *Adv. Energy Mater.*, 2012, **2**, 369.
60. Z. A. Tan, E. Zhou, X. Zhan, X. Wang, Y. Li, S. Barlow and S. R. Marder, *Appl. Phys. Lett.*, 2008, **93**, 073309.
61. X. Zhan, Z. A. Tan, B. Domercq, Z. An, X. Zhang, S. Barlow, Y. Li, D. Zhu, B. Kippelen and S. R. Marder, *J. Am. Chem. Soc.*, 2007, **129**, 7246.
62. E. Zhou, J. Cong, Q. Wei, K. Tajima, C. Yang and K. Hashimoto, *Angew. Chem., Int. Ed.*, 2011, **50**, 2799.
63. E. Zhou, J. Z. Cong, K. Hashimoto and K. Tajima, *Adv. Mater.*, 2013, **25**, 6991.
64. F. C. Jamieson, E. B. Domingo, T. Mccarthy-Ward, M. Heeney, N. Stingelin and J. R. Durrant, *Chem. Sci.*, 2012, **3**, 485.
65. P. Kohn, Z. X. Rong, K. H. Scherer, A. Sepe, M. Sommer, P. Muller-Buschbaum, R. H. Friend, U. Steiner and S. Huttner, *Macromolecules*, 2013, **46**, 4002.
66. B. A. Collins, J. R. Tumbleston and H. Ade, *J. Phys. Chem. Lett.*, 2011, **2**, 3135.
67. C. R. McNeill, *Energy Environ. Sci.*, 2012, **5**, 5653.
68. A. C. Arias, J. D. Mackenzie, R. Stevenson, J. J. M. Halls, M. Inbasekaran, E. P. Woo, D. Richards and R. H. Friend, *Macromolecules*, 2001, **34**, 6005.
69. C. R. McNeill, A. Abrusci, I. Hwang, M. Ruderer, P. Müller-Buschbaum and N. C. Greenham, *Adv. Funct. Mater.*, 2009, **19**, 3103.
70. S. Fabiano, Z. Chen, S. Vahedi, A. Facchetti, B. Pignataro and M. A. Loi, *J. Mater. Chem.*, 2011, **21**, 5891.
71. J. R. Moore, S. Albert-Seifried, A. Rao, S. Massip, B. Watts, D. J. Morgan, R. H. Friend, C. R. McNeill and H. Sirringhaus, *Adv. Energy Mater.*, 2011, **1**, 230.
72. N. Zhou, H. Lin, S. J. Lou, X. Yu, P. Guo, E. F. Manley, S. Loser, P. Hartnett, H. Huang, M. R. Wasielewski, L. X. Chen, R. P. H. Chang, A. Facchetti and T. J. Marks, *Adv. Energy Mater.*, 2014, **4**, 1300785.
73. A. R. Campbell, J. M. Hodgkiss, S. Westenhoff, I. A. Howard, R. A. Marsh, C. R. McNeill, R. H. Friend and N. C. Greenham, *Nano Lett.*, 2008, **8**, 3942.
74. M. Schubert, D. Dolfen, J. Frisch, S. Roland, R. Steyrleuthner, B. Stiller, Z. Chen, U. Scherf, N. Koch, A. Facchetti and D. Neher, *Adv. Energy Mater.*, 2012, **2**, 369.

75. P. Cheng, L. Ye, X. Zhao, J. Hou, Y. Li and X. Zhan, *Energy Environ. Sci.*, 2014, **7**, 1351.
76. J. E. Slota, E. Elmalem, G. Tu, B. Watts, J. Fang, P. M. Oberhumer, R. H. Friend and W. T. S. Huck, *Macromolecules*, 2012, **45**, 1468.
77. S. C. Veenstra, J. Loos and J. M. Kroon, *Prog. Photovoltaics*, 2007, **15**, 727.
78. D. Mori, H. Benten, H. Ohkita, S. Ito and K. Miyake, *ACS Appl. Mater. Interfaces*, 2012, **4**, 3325.
79. A. Sepe, Z. Rong, M. Sommer, Y. Vaynzof, X. Sheng, P. Mueller-Buschbaum, D. Smilgies, Z.-K. Tan, L. Yang, R. Friend, U. Steiner and S. Huttner, *Energy Environ. Sci.*, 2014, **7**, 1725.
80. B. Friedel, C. R. McNeill and N. C. Greenham, *Chem. Mater.*, 2010, **22**, 3389.
81. T. Lei, J.-H. Dou and J. Pei, *Adv. Mater.*, 2012, **24**, 6457.
82. F. Zhang, Y. Hu, T. Schuettfort, C.-A. Di, X. Gao, C. R. McNeill, L. Thomsen, S. C. B. Mannsfeld, W. Yuan, H. Sirringhaus and D. Zhu, *J. Am. Chem. Soc.*, 2013, **135**, 2338.
83. I. Meager, R. S. Ashraf, S. Mollinger, B. C. Schroeder, H. Bronstein, D. Beatrup, M. S. Vezie, T. Kirchartz, A. Salleo, J. Nelson and I. Mcculloch, *J. Am. Chem. Soc.*, 2013, **135**, 11537.
84. T. W. Holcombe, J. E. Norton, J. Rivnay, C. H. Woo, L. Goris, C. Piliego, G. Griffini, A. Sellinger, J.-L. Brédas, A. Salleo and J. M. J. Fréchet, *J. Am. Chem. Soc.*, 2011, **133**, 12106.

CHAPTER 5

Design and Synthesis of Small Molecule Donors for High Efficiency Solution Processed Organic Solar Cells

SETH McAFEE*[a], GREGORY C. WELCH*[a], AND COREY V. HOVEN*[b]

[a]Department of Chemistry, Dalhousie University, 6274 Coburg Road, Halifax, Nova Scotia, Canada B3H 4R2; [b]Next Energy Technologies, Inc., 5385 Hollister Avenue #115, Santa Barbara, CA 93111, USA
*E-mail: gregory.welch@dal.ca, seth.mcafee@dal.ca, corey@nextenergytech.com

5.1 Introduction

Solution processed organic photovoltaic devices (OPVs) have emerged as a promising clean energy generating technology owing to their potential to enable low-cost manufacturing *via* printing or coating techniques, the capacity for incorporation onto light-weight, flexible substrates, and their color/transparency tunability.[1-5] Considerable research and development of organic solar cells has been based on two classes of donor–acceptor active layers: (1) vapor deposited small molecule/fullerene compositions[6-8] and (2) solution-processed polymers (*large* molecules)–fullerene compositions.[9-11] Through primarily

RSC Polymer Chemistry Series No. 17
Polymer Photovoltaics: Materials, Physics, and Device Engineering
Edited by Fei Huang, Hin-Lap Yip, and Yong Cao
© The Royal Society of Chemistry 2016
Published by the Royal Society of Chemistry, www.rsc.org

donor molecule development and device optimization, power conversion efficiencies (PCE) for polymer based systems have reached 9.2% for single layer devices[12] and 10.6% for tandem cells.[13] Vapor deposited small molecule based OPV, such as those being developed by Heliatek, have demonstrated tandem cell PCEs of 12%, and extrapolated lifetimes greater than 30 years.[14] While promising, the active layer materials used within these OPVs are not without their challenges,[15] including a strict performance and stability dependence on molecular weight (MW)[16] and purity,[17] and end-capping effects,[18] respectively, for polymers, and limited processing options for insoluble small molecules.

Considering these limitations, the investigation of a third class of OPVs that make use of solution-processed small molecule–fullerene active layers has been industriously pursued.[19,20] Solution processible small molecules offer several advantages over their polymeric counterparts; issues pertaining to molecular weight are eliminated because their structures are well defined, they have non-reactive end-capping groups, and can be readily purified *via* an array of techniques including chromatography and crystallization.[20] These characteristics reduce batch-to-batch variations, and ultimately higher purity materials lead to greater performance and longer lifetimes. Furthermore, small molecule architectures are sensitive to subtle structure changes at both the core and the ends of the molecule, and thus electronic energy levels, optical absorption, and self-assembly tendencies can be systematically tuned to maximize device performance.[21-23] The absolute structure of small molecules can also be determined *via* single crystal X-ray diffraction, which has enabled unprecedented structure–property–function correlations in the field of OPVs.[24-27] In comparison to their vapor deposited small molecule counterparts, solution processible small molecules also offer advantages in that the deposition methods are more versatile, do not require high vacuum, and are expected to have much lower capital and manufacturing costs.

Very recently, several new classes of solution processible small molecule donors with favorable optical and electronic properties have been reported for use in OPV devices. When coupled with fullerene electron acceptors, such as [6,6]-phenyl-C_{71}-butyric acid methyl ester (PC$_{71}$BM), solution processible small molecule based solar cells have achieved record PCEs over 8%[28-30] for single layers devices and 10% for tandem cells.[31] This chapter provides an overview of several important classes of solution processible small molecules used as donor materials in high performance OPV devices. For greater details on materials and devices, the reader is referred to several excellent reviews by Meerholz,[32] Nguyen,[33] Bäuerle,[34] Zhan,[35] and Roncali.[36]

5.2   Device Operation

A detailed analysis of photocurrent generation in OPVs is beyond the scope of this chapter but it is important to understand that high performance OPVs operate on the principle that two materials are required to generate free charge carriers. A simplified view of standard device architecture and of device operation is shown in Figure 5.1. Most soluble small molecule donors have been employed in standard configuration solar cell devices consisting of

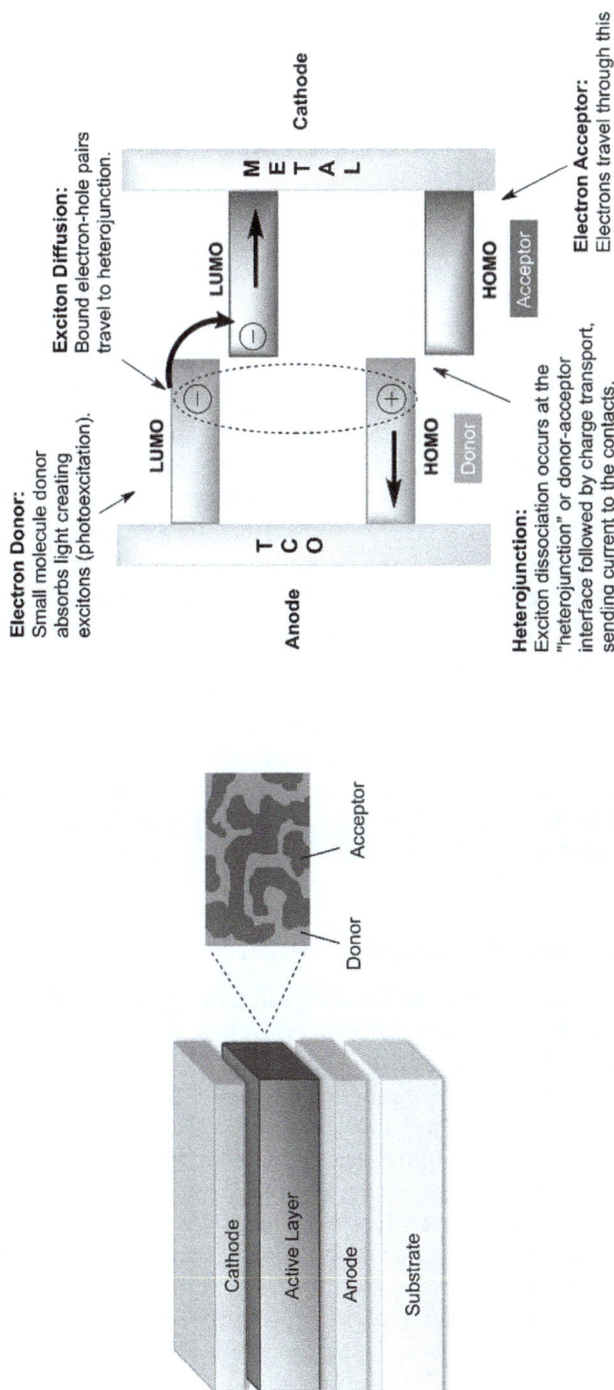

Figure 5.1 (Left) Standard device architecture. The active layer is composed of a bulk-heterojunction (BHJ) network. (Right) Simplified view of the device operation for an organic solar cell.

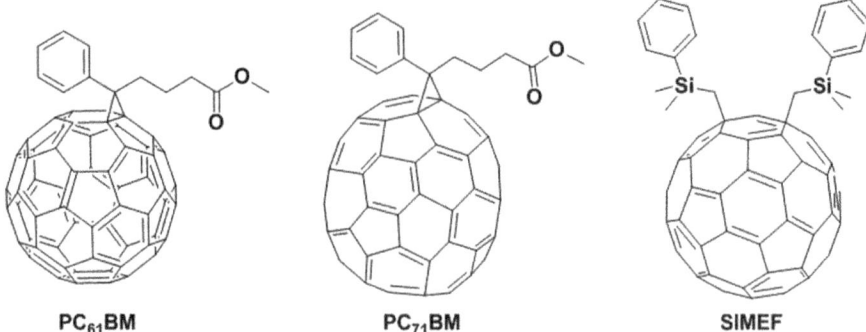

Figure 5.2   Examples of fullerene derivatives used in high performance solution processed small molecule BHJ solar cells.

a bulk-heterojunction (BHJ) active layer sandwiched between two electrodes atop a glass or plastic substrate.[31-33] Within the BHJ active layer, combinations of materials with high ionization potentials (electron *donor*) and high electron affinity (electron *acceptor*) are responsible for light absorption, exciton formation, charge separation, and charge transport. Key parameters that define PCE are the open circuit voltage (V_{OC}), short-circuit current (J_{SC}), and fill factor (FF) at a given incident light intensity.[37,38] Ubiquitous to OPV devices is the use of soluble fullerene derivatives as the electron acceptor, owing to their low-lying lowest unoccupied molecular orbital levels and isotropic electron mobility (Figure 5.2).[39-41] The vast majority of OPV performance improvements have been a result of the development of new "donor" materials that are designed to maximize light absorption and transport charge, and form ordered nanostructures when blended with fullerene acceptors.

5.3   Small Molecule Donor Design

There are several key parameters to consider when designing small molecule donors to be paired with fullerene acceptors in solution processed BHJ solar cells; these include: (1) strong optical absorption in the visible and near-infrared regions of the electromagnetic spectrum to maximize photon harvesting, (2) lowest unoccupied molecular orbital (LUMO) and highest occupied molecular orbital (HOMO) energy levels compatible with fullerene acceptors and electrode materials, (3) functional groups and planar π-conjugated backbones to promote intermolecular π–π interactions (important for achieving high charge carrier mobility) and suitable donor–acceptor phase separation, (4) sufficient solution viscosity and solubility to enable thin film formation *via* solution deposition, (5) synthetic procedures that are simple, high yielding, and highly tunable to ensure that both gram quantities can be made, and molecular libraries created.[33,34,42] With these considerations in mind, several research groups have developed a series of solution processible small molecules that challenge those of the very best polymeric and vapor-deposited molecules in solar cell performance. It is worth noting that, for commercialization of OPVs, high PCE multi-junction OPV devices are expected to be required, most likely consisting of a cell with a middle-band-gap (MBG; ~1.5–2.1 eV) organic semiconductor and a cell with a low-band-gap (LBG; ~1.2–1.5 eV) organic semiconductor where the overlap of absorption spectra between the front and back cells is minimized.[43,44] In terms of satisfying these commercialization criteria, high PCE MBG donors have been demonstrated; however, more efforts are required to develop high PCE LBG donors.

5.4   Historical Perspective

There is a long history of research associated with the development of solution processed small molecule bulk-heterojunction solar cells (SM-BHJ). Traditionally, small molecule donors have suffered from difficulties in forming uniform thin films from solution owing to low solution viscosity and a strong tendency to crystallize. Thus they have primarily been utilized in thermally evaporated devices giving high photovoltaic performace.[45] In the mid 2000's, Roncali *et al.* developed a series of tetrahedral shaped oligothiophene molecular donors that could be solution processed with $PC_{61}BM$ to give solar cells with PCE of ~0.2%.[46] Around the same time, Anthony *et al.* reported on a series of soluble acene derivatives which achieved PCEs of ~1% when paired with $PC_{61}BM$.[47] A major problem with these initial compounds was poor spectral overlap, with the solar spectrum having limited optical absorption beyond 600 nm. Nonetheless, these initial results demonstrated that small molecule donors could indeed be processed from solution to give working solar cell devices.[20]

5.5   Dye Based Molecules (BODIPY, Squaraine, and Merocyanine)

To address the need for higher optical absorption and better photon harvesting properties, a strategy emerged exploiting organic dyes to create new donor materials. Initially, a series of papers were published from 2008–2010 describing the functionalization of strongly absorbing dyes to create small molecules capable of acting as donors in BHJ solar cells. Zeisel and Roncali showed that highly absorbing ($\varepsilon > 100\,000$ M^{-1} cm^{-1}) boron–dipyrromethene (BODIPY) dyes could be made soluble and form films from solution by attaching oligooxyethylene chains to the π-conjugated backbone (**1** in Figure 5.3).[48,49] When incorporated into devices with $PC_{61}BM$, PCEs in excess of 1% were obtained and, notably, photocurrent generation beyond 750 nm was achieved. Through structural optimization of the dye component, the PCEs of BODIPY-based SM-BHJ solar cells have been improved to ~5% (**2** in Figure 5.3).[50] Most importantly, these new classes of small molecules enabled photocurrent generation beyond 900 nm, allowing for capture of near-IR photons.[51] Based on the success of this ideology, dye-based small molecule donors have been actively explored. Both the Marks and Würthner research groups have reported on functionalized squaraine dyes that exhibited broad

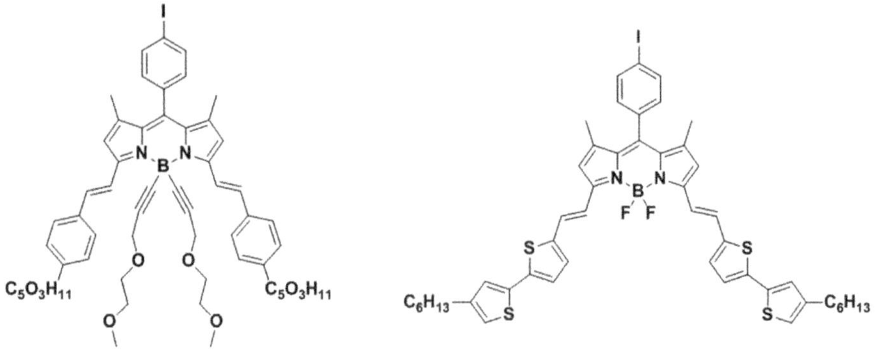

Figure 5.3 Two examples of BODIPY dye based small molecules that have been used to fabricate SM-BHJ solar cells.

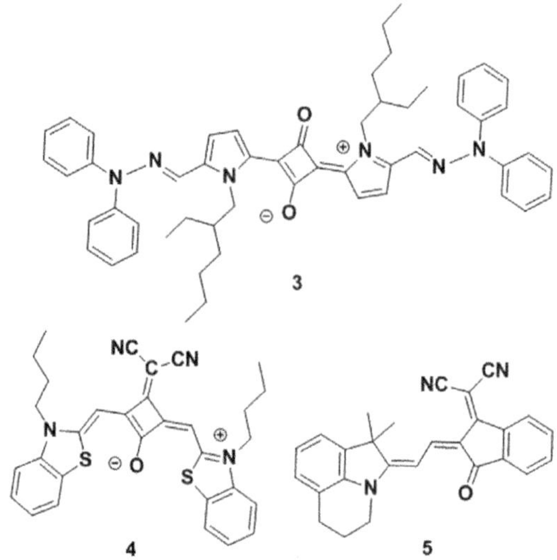

Figure 5.4 Examples of squaraine dye (**3** and **4**) and merocyanine dye (**5**) based small molecules that have been used to fabricate SM-BHJ solar cells.

and intense thin-film absorption spectra that extended well into the near IR (examples are shown in Figure 5.4, compounds **3** and **4**).[52–54] Devices using squaraine:PC$_{61}$BM active layers gave PCEs on the order of 1–2%. These devices ultimately suffered from small open circuit voltages (V_{OC}) and poor fill factors (FF), a result of the high-lying HOMO levels of squaraine dyes and poor active layer morphologies. While not using truly solution processed active layers, Forrest and co-workers have utilized soluble squaraine dyes in combination with C$_{60}$ to yield efficient OPV devices with PCEs in slight excess of 6%.[55–58] Würthner was also able successfully to incorporate merocyanine dyes into BHJ solar cells (*e.g.* compound **5** in Figure 5.4).[59–61] Such dyes are easily synthesized and exhibited strong absorption in the red region of the solar spectrum. PCEs upwards of 2.5% were obtained upon solution processing the dye with soluble fullerene derivatives. These devices had a higher V_{OC} in comparison to the squaraine-based devices, yet once more suffered from poor fill factors.

5.6 Dye Based Molecules – Diketopyrrolopyrrole

First reported in 1974 by Farnum *et al.*,[62] the diketopyrrolopyrrole (DPP) chromophore is an easily synthesized, strongly absorbing and thermally stable building block. DPP can be easily functionalized with aliphatic side chains on the amide-nitrogen atom, enabling dissolution in common organic solvents. Importantly, the presence of two electron withdrawing amide functional groups renders DPP a good acceptor within the context of donor–acceptor (D–A) organic π-conjugated materials.[63–65] The DPP core is commonly substituted in the 3 and 6 positions with thienyl or phenyl substituents [DPP(Th)$_2$ and DPP(Ph)$_2$ in Figure 5.5, respectively],[66] although furan[67,68] and selenophene[69,70] substituted DPP have also been reported. The smaller five-membered thienyl heterocycles lead to smaller dihedral angles between the DPP core and the pendant substituent, resulting in greater planarity of the molecular backbone, thus increasing π-delocalization and intermolecular π–π interactions. Additionally, thiophene is a stronger electron donor than phenyl, enhancing intramolecular charge transfer transitions. These effects result in DPP(Th)$_2$ based materials typically exhibiting small bandgaps and greater charge transport properties than DPP(Ph)$_2$ derivatives, and therefore they have extensively been used to make high performance materials for organic PV devices.[63,65]

Over the past several years, DPP-based small molecules have emerged as arguably the most studied in the area of solution processed small molecule organic PV devices (Table 5.1). Pioneering work by Nguyen and co-workers[71–73] at the University of California Santa Barbara in 2008 demonstrated the utility of DPP-based oligothiophenes as donor materials in fullerene based (bulk-heterojunction) BHJ solar cells. An early derivative with 2-hexyl bithiophene end-capping units was used as a donor molecule and resulted in device PCE over 3% (Compound **6** in Figure 5.5).[74] Subsequently, in 2009, the same group showed that the PCE could be improved to 4.4% using benzofuran end-capping units (Compound **7** in Figure 5.6) in place of 2-hexyl bithiophene, a record PCE that stood until 2012.[75] The complete four-step synthesis of **7** is outlined in Figure 5.6. The thiophene functionalized diketopyrrolopyrrole [DPP(Th)$_2$] is synthesized in high yield by reaction between commercially available dimethyl succinate and thiophene-2-carbonitrile in the presence of base. Incorporation of aliphatic side chains and reactive bromine functionalities is achieved by a S_N2 nucleophilic *N*-alkylation of the DPP core and electrophilic bromination of the thiophene 2-position, respectively. Finally, end-capping benzofuran units are installed *via* a palladium

Table 5.1 SM-BHJ solar cell data for DPP based small molecule donors.

	Onset of absorption[a] (nm)	Processing[b]	Acceptor	Weight ratio (w/w)	V_{OC} (V)	J_{SC} (mA cm^{-2})	FF (%)	PCE (%)	Ref.
6	800	Annealing	PC$_{71}$BM	5:5	0.75	9.2	41	3.0	74
7	710	Annealing	PC$_{61}$BM	6:4	0.92	10.0	48	4.4	75
8	752	Annealing	PC$_{71}$BM	6:4	0.94	8.6	50	4.0	79
9	740[c]	Annealing	PC$_{71}$BM	2:1	0.76	8.3	58	4.1	80
10	880	None	PC$_{71}$BM	1:1	0.74	2.8	58	1.2	81
10	880	DIO	PC$_{71}$BM	1:1	0.73	13.6	48	4.7	81
11	705	Annealing	PC$_{71}$BM	1:1	0.82	10.8	56	4.9	82
12	729	Annealing	PC$_{71}$BM	1:0.75	0.63	14.6	58	5.3	83
13	720	Annealing	PC$_{71}$BM	1.5:1	0.84	11.3	42	4.0	84
14	709	DIO	PC$_{71}$BM	1:1	0.87	9.5	53	4.4	86
15	755	DIO	PC$_{71}$BM	1:1	0.72	11.8	62	5.3	87
16	752	Annealing	PC$_{61}$BM	1:1	0.84	11.9	58	5.8	88
17	730	DIO	PC$_{71}$BM	1:1	0.77	11.4	63	5.5	89
18	732	Annealing	PC$_{61}$BM	1.5:1	0.76	12.0	51	4.7	90
19	755	DIO	PC$_{71}$BM	1:1	0.89	9.0	61	4.8	91
20	760	CN	PC$_{71}$BM	2:3	0.86	10.4	62	5.5	91

[a]As determined from neat thin-film optical absorption spectra.
[b]Annealing = post-deposition thermal annealing of active-layer; DIO = active layer cast from solvent mixtures containing small quantities of diiodooctane (DIO); CN = active layer cast from solvent mixtures containing small quantities of chloronapthalene (CN).
[c]Estimated by authors from thin optical absorption spectra.

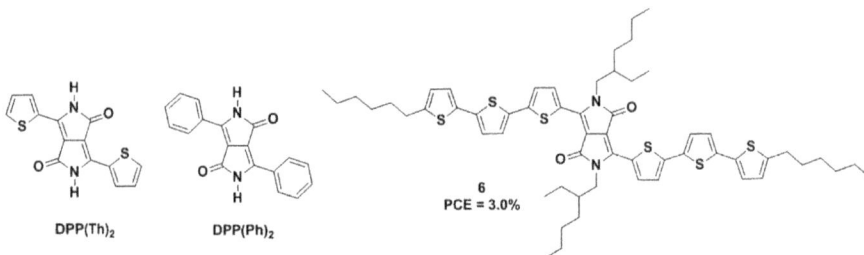

DPP(Th)$_2$ DPP(Ph)$_2$ 6 PCE = 3.0%

Figure 5.5 Chemical structures of thiophene [DPP(Th)$_2$] and phenyl [DPP(Ph)$_2$] substituted diketopyrrolopyrroles (DPP), and one of the first reported small molecules based upon DPP that was utilized as a donor molecule in SM-BHJ solar cells.

catalyzed Suzuki coupling reaction. Important properties of **7** relevant to its use as a donor material in SM-BHJ solar cells are its ability to form uniform thin films from solution, optical absorption in the solid state extending beyond 700 nm, relatively deep HOMO and LUMO energy levels that are compatible with PCBM derivatives, and a strong tendency to form highly ordered nanometer-sized domains when blended with PCBM derivatives. Compound **7** has been widely studied in SM-BHJ solar cells with consistent improvement in PCE.[76–78]

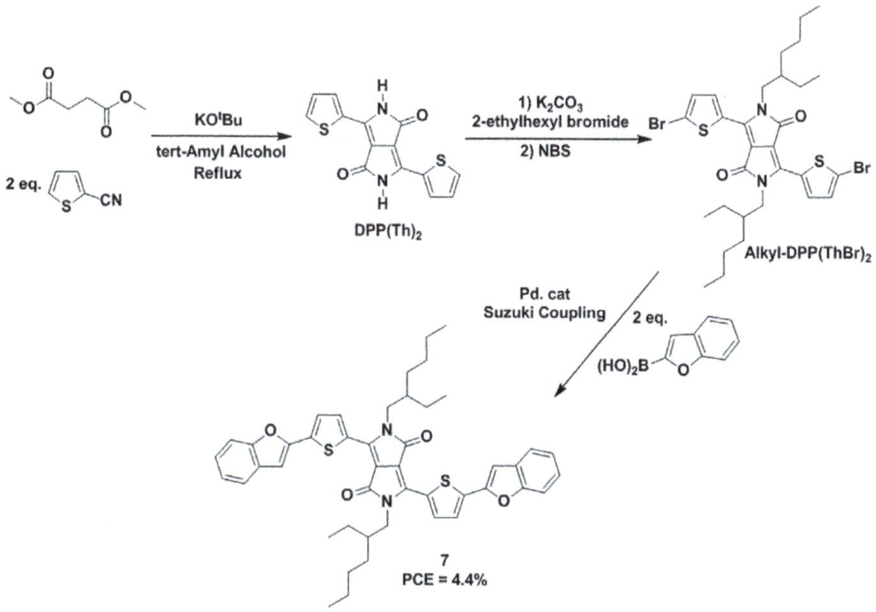

Figure 5.6 Synthesis of DPP small molecule **7**.

Exploiting the success of Nguyen and co-workers, further investigation into the functionalization of DPP small molecules found that the incorporation of electron-rich moieties exhibited narrow band-gaps; however, their HOMO levels were relatively high.[74] Alternatively, if DPP was linked to an electron-withdrawing unit the HOMO level decreased but the band-gap became undesirably wider.[75] In the pursuit of a DPP-based material that includes both a low HOMO level and a narrow band-gap, thiophene-2-carboxylate emerged as a feasible end-cap unit.[79] This compound can lower HOMO levels through its weak electron-withdrawing ester functional group, while its small size is not disruptive to the conjugation between the core and the end-cap units, supporting a narrow band-gap. This theoretical design proved fruitful; the synthesized molecule (Compound **8**, Figure 5.7) exhibited both a low-lying HOMO energy level of −5.3 eV and a narrow band-gap of 1.65 eV with broad absorption, extending to nearly 750 nm. In terms of OPV performance, a very high V_{OC} of 0.94 V and a PCE reaching 4.0% are the highlights of this molecular design.

It has been agreed that the influence of the end-cap unit on device performance of DPP-based materials is not limited to electronic properties, leading to the investigation of structure–property relationships within these materials. The introduction of π-stacking moieties onto the ends of a DPP core can facilitate favorable end-to-end π–π interactions, leading to enhanced intermolecular charge transport through directed self-assembly.[80] Compound **9** (Figure 5.7) was synthesized bearing pyrene end-caps; these were chosen because of the strong tendency for these planar moieties to π-stack.

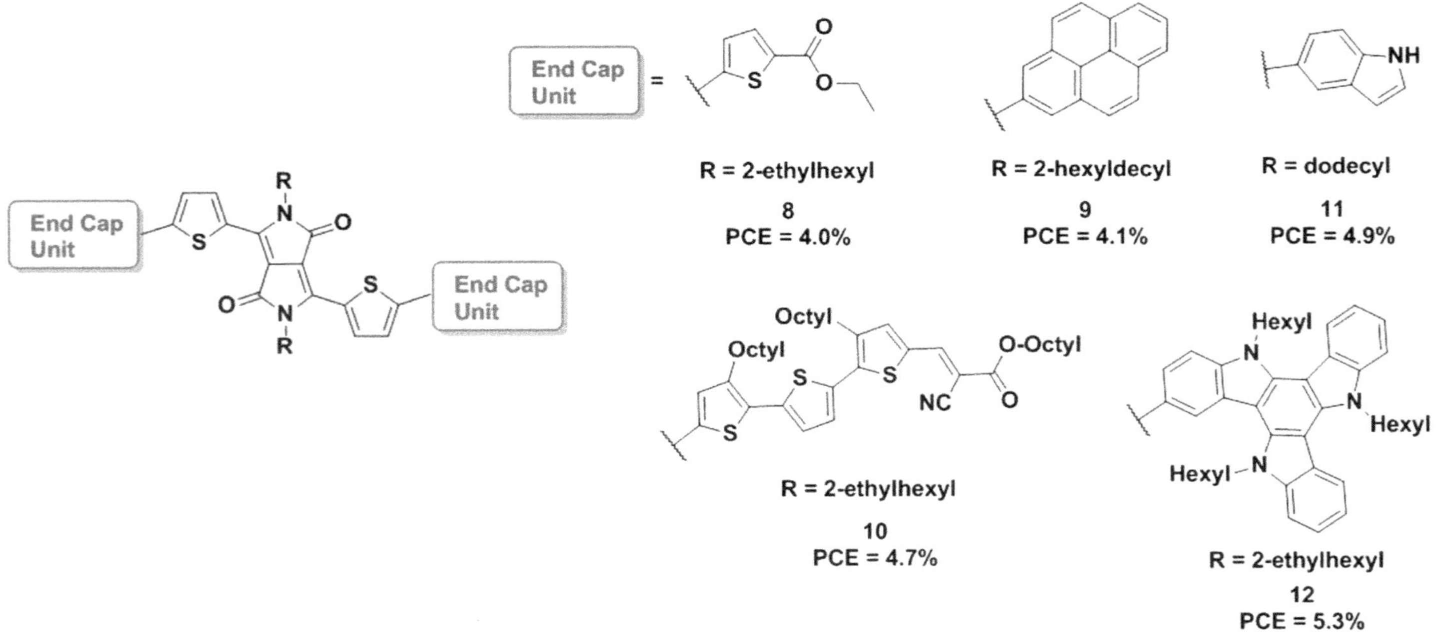

Figure 5.7 Structures of small molecules with DPP core and aromatic end-cap units that have enabled the fabrication of SM-BHJ solar cells with PCE greater than 4%.

Optimized device performance yielded high fill factors of 0.58, leading to a maximum PCE of 4.1%, indicating that the pyrene end-cap unit does in fact promote molecular packing and active layer morphology favorable for high device PCE.

The introduction of end-cap units also serves to extend the overall π-conjugated system. With this in mind, a DPP-based material flanked by two terthiophene donor units and terminated with octyl cyanoacetate units has been synthesized (Compound **10**, Figure 5.7).[81] This molecular design serves two purposes: to increase intramolecular charge transfer through the extended π-system effectively narrowing the optical band-gap, while the octyl side chains improve the solubility of the material for purification and subsequent processing. Photoelectric properties show a broad spectrum, extending to nearly 880 nm, with an optical band-gap estimated to be 1.41 eV. Device optimization of a 1:1 blend with $PC_{71}BM$ exhibited a PCE of 4.7% with a 3% (wt/v) 1,8-diiodooctane (DIO) additive to manipulate morphology. An innate low fill factor and V_{OC}; however, limit this small molecule design, and improving these properties has been emphasized as a priority for future development of these materials.

It is evident that various end-cap units can be explored as a means to tune the properties of a small molecule further, including the optical band-gap, HOMO/LUMO levels, and self-assembly. Based on these considerations, it has emerged that indole-based end-cap units have exhibited some of the highest PCE values for DPP-based materials.[82] Compound **11** (Figure 5.7) incorporates a single indole end-cap tethered at the 5-position, where the location of indole integration proved to be crucial, because substitution in the 2-position has been shown to blend too well with $PC_{71}BM$, leading to amorphous films and suppressed device performance.[26] Ultimately, the electron-rich indole proved to enhance the absorption coefficient of the compound and contributed to the broad absorption spectrum, extending to 705 nm, corresponding to an optical band-gap of 1.76 eV. In terms of device performance, the material showed a PCE of 4.9% from a 1:1 blend with $PC_{71}BM$ when thermally annealed.[82]

Building on the success of indole end-cap units, triazatruxene derivatives have also demonstrated high photovoltaic device performance. Triazatruxene consists of three fused carbazole molecules; owing to their inherent planarity these electron-donating units provide adequate π-stacking, meanwhile the indole moieties offer a site for easy alkylation to achieve high solubility. Compound **12** (Figure 5.7) incorporates two triazatruxene end-cap units; they were mono-functionalized at the 3-position and installed *via* a Suzuki coupling reaction.[83] DPP substitution can occur at either the 2 or the 3-position, with 3-position based substitution considerably easier owing to the less reactive *meta*-position of the amino group of the carbazole subunit. In terms of optoelectronic properties, the material shows a broad absorption spectrum past 700 nm, with large extinction coefficients and an optical band-gap of 1.70 eV. Device performance for a thermally annealed 1:0.75 $PC_{71}BM$ blend returned a PCE of 5.3%, which is among the highest performing systems of solution processed DPP-based materials.[83]

We have surveyed high performance compounds that employ DPP as the molecular core; however, there also exists a class of compounds that instead employ DPP as the end-cap unit. One of the first high performance molecules of this variety was based on a donor core of naphthodithiophene. This unit offers an extended π-conjugated system, affording strong intermolecular orbital overlap and electron-donating properties in view of enhanced charge separation and transport. The core was functionalized with 2-ethylhexyloxy groups for adequate solubility in common organic solvents, followed by the incorporation of two mono-substituted DPP molecules as end-caps (Figure 5.8).[84] Compound **13** (Figure 5.8) displayed an absorption spectrum that extends out to 720 nm, corresponding to a band-gap energy of 1.72 eV with a low-lying HOMO level of −5.4 eV. Solar cell device performance was optimized and it was found that a thermally annealed blend of **13**:PC$_{61}$BM exhibited a PCE of 4.0%, a promising result for the pursuit of similar molecular designs.[84]

As this class of "bis-DPP" materials was developed,[85] enforcing planarity along the π-conjugated backbone emerged as an important design strategy for targeting improved J_{SC}. This concept led to the synthesis of a naphthalene core flanked by DPP end-caps (Compound **14**, Figure 5.9).[86] The backbone planarity of this system was designed to promote efficient crystallinity, enhancing the charge carrier mobility. This design strategy proved to be successful; the material displayed appropriate optoelectronic properties consistent with DPP-based materials and gave a PCE of 4.4% with high V_{OC} and J_{SC} values when blended in a 1:1 ratio with PC$_{71}$BM.[86]

The choice of "core" unit for DPP flanked small molecules has been explored for materials with a proven track record of OPV device performance; 2-D benzodithiophene (2-D BDT) is one such example. It has attracted considerable interest with promising photovoltaic properties through structural modification to its substituents. Compounds **15** and **16** (Figure 5.9), varying only in alkyl side chains, were independently designed with this in mind and the synthesized materials showed excellent solution processability, thermal stability, broad absorption spectra, very high extinction coefficients and appropriate energy levels. Solution processed SM-BHJ OPV devices with **15**:PC$_{71}$BM active layers exhibited PCE values of 5.3%,[87] while those based upon **16**:PC$_{61}$BM active layers reached PCE values of 5.8%.[88] The latter result represents the highest PCE recorded for SM-BHJ solar cells using DPP based donors. The repeated success of A–D–A type molecular donors with DPP as the A unit has recently seen further improvement to already impressive PCEs. Modification to the substitution pattern on a benzodithiophene was the target for the synthesis of compounds **17** and **18** (Figure 5.9), leading to PCEs up to 5.5%[89] and 4.7%,[90] respectively. Compared to the more linear compounds **13–16**, the small molecules **17** and **18** adopt bent and zig-zag shapes, respectively, as drawn optimized structures. This change in molecular shape led to improved blend morphologies, resulting in higher charge carrier mobilities and increases in both short-circuit current and fill factors when compared to the parent compound **13**. These results serve to provide further evidence for the intrinsic high performance of this class of materials, reinforcing their potential in the pursuit of high performance SM-BHJ solar cells.

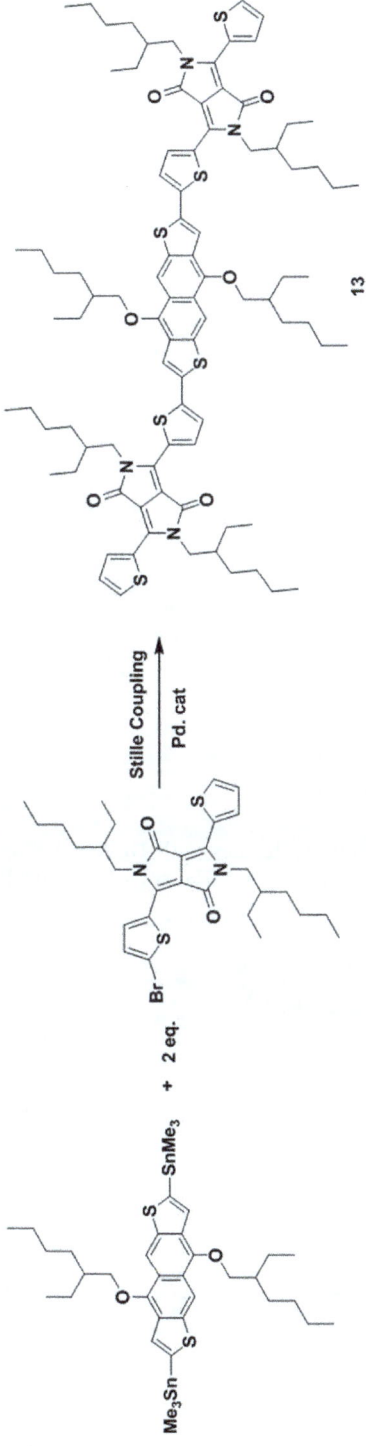

Figure 5.8 Synthesis and structure of the first reported DPP small molecule using DPP as end-capping units that gave high SM-BHJ solar cell performance.

The types of DPP molecule we have discussed are broadly classified into two groups: mono-DPP where one DPP unit serves as the molecular core, and bis-DPP where two DPP molecules are incorporated as the end-cap units. The initial success of the former was replicated by the latter and has been appropriately extended to a tris-DPP molecular system, effectively increasing the π-conjugation of the system and enhancing the push–pull characteristics of the molecule.[84,85] Compounds **19** and **20** (Figure 5.10), incorporate three DPP units, which impart a low-lying HOMO energy level of −5.4 and −5.2 eV

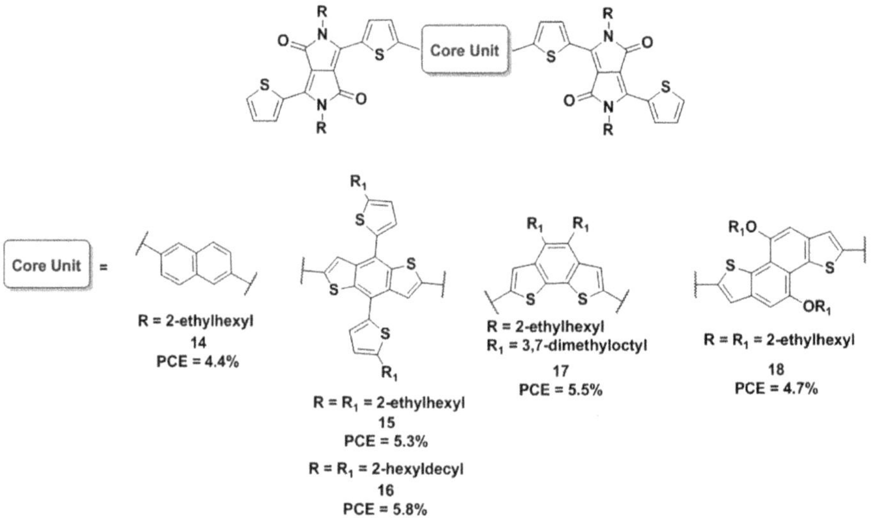

Figure 5.9   Structures of small molecules with DPP end-capping units that have enabled the fabrication of SM-BHJ solar cells with PCE greater than 4%.

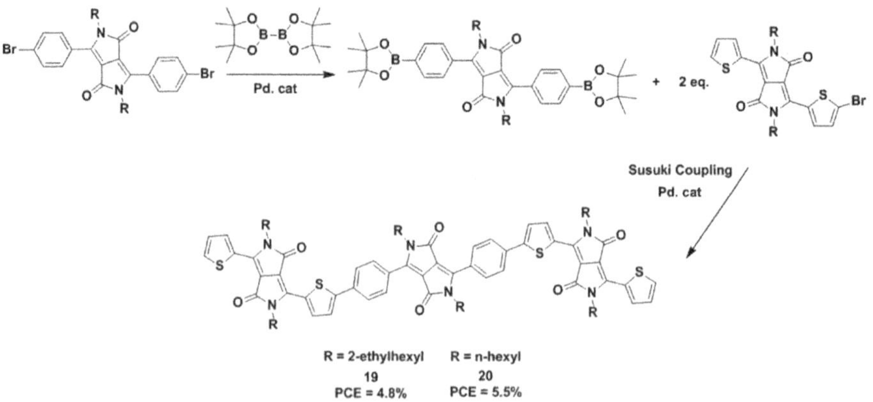

Figure 5.10   Synthesis and structure of the first reported DPP small molecules consisting of three DPP units that gave high SM-BHJ solar cell performance.

respectively, with the slight difference owing to the crystallinity of the material for the two different alkyl chains.[91] Furthermore, the nature of the phenyl–thiophene linkages induce a twist to the backbone supporting a lower HOMO energy level and improving the overall solubility in common organic solvents. When blended with $PC_{71}BM$, BHJ OPV device performance has been optimized to reach 5.5% for compound **20** and 4.8% for compound **19**.[91] The disparity in PCE corresponds to the differences in crystallinity inherent from the alkyl substitutions where the sterically demanding 2-ethylhexyl side chain has a lower tendency for crystallization. The success of this endeavor provides the motivation to investigate extending the number of core units to other high performance systems and evaluate their OPV potential.

5.7 Dye Based Molecules – Isoindigo

Isoindigo is an amide-based acceptor for organic electronics that has been gaining considerable interest; this material is an abundant naturally occurring isomer of the well-known dye indigo (Figure 5.11).[92–94] Isoindigo is composed of two symmetrical oxindole rings that are through conjugated at their 3-carbon by a site of unsaturation, effectively binding the two electron-withdrawing carbonyls and two electron-rich phenyl rings in a *trans*-conformation. This compound offers a donor–acceptor–donor type system with a fully delocalized HOMO and localized LUMO at the electron-withdrawing lactam rings. Extension of π-conjugation occurs through substitution at the 6,6' positions of the isoindigo ring. A 5,5' substitution is also known; however, this structure is not through conjugated, which restricts its potential in organic electronics.[95] In terms of electronic and structural properties, isoindigo offers an extended absorption towards the near infrared with a low-lying HOMO/LUMO energy level and high optical stability. Coupled with a planar π-conjugated symmetric structure, high charge carrier mobility and excellent film morphology without aggregation, this makes these materials ideal candidates for donor materials in OPVs.[92,96,97] These aforementioned advantages are similar to those of diketopyrrolopyrroles, encompassing strong electron-withdrawing character, low HOMO levels, relatively small bandgaps and planar structures. Isoindigo is more accessible; it can be easily

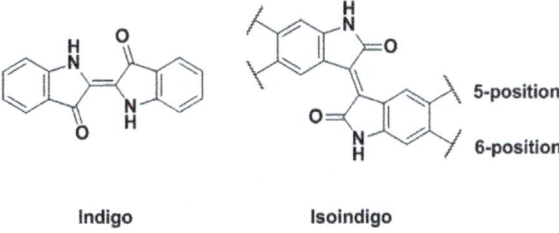

Indigo   Isoindigo

Figure 5.11 Structures of indigo and isoindigo dyes. Isoindigo can be functionalized in both the 5- and 6- positions to make strongly absorbing small molecules for use as donors in SM-BHJ solar cells.

synthesized in bulk quantities from readily available starting materials and is inherently more soluble.

The acid catalyzed condensation of isatin and oxindole has been the most direct route for synthesizing isoindigo-based materials. The commercial availability of brominated starting indoles makes this a very attractive synthesis occurring in near-quantitative yields of a brominated product, fully functionalized for subsequent reactions. Isoindigo preparation is followed by a high-yielding S_N2 nucleophilic *N*-alkylation of the isoindigo core to render the material adequately soluble for the remaining synthetic procedure, purification and solution processing. The preparation of a brominated and alkylated isoindigo core is summarized in Figure 5.12A.[92]

This molecular core offers a great deal of synthetic flexibility; its chemical structure allows for the integration of a variety of functional groups to fine-tune the properties of the material. These reactions are typically palladium-catalyzed cross-couplings such as the Stille reaction[98–101] or Suzuki[96] coupling, while a few examples of direct heteroarylation[102] have also been reported.

In addition to extending the π-conjugated structure, modifications can also be made directly within the molecular core; the incorporation of electron deficient heteroatom moieties serves to fine-tune the localized LUMO energy level. Of this variety, both thienoisoindigo[103] and azaisoindigo[100] molecular cores have been synthesized (Figure 5.12B). Alterations have also been shown for the substitution of an aromatic hydrogen with electron-withdrawing units where chlorinated,[104] fluorinated,[105] and cyanated[106] isoindigo molecular cores have been synthesized (Figure 5.12C). These electron-withdrawing species effectively lower the LUMO energy level, with fluorine highlighted for exhibiting negligible steric hindrance owing to its innate small size.

The first application of isoindigo-based materials in OPVs employed bithiophene as an electron donor, **21** and **22** (Figure 5.13).[107] The electron

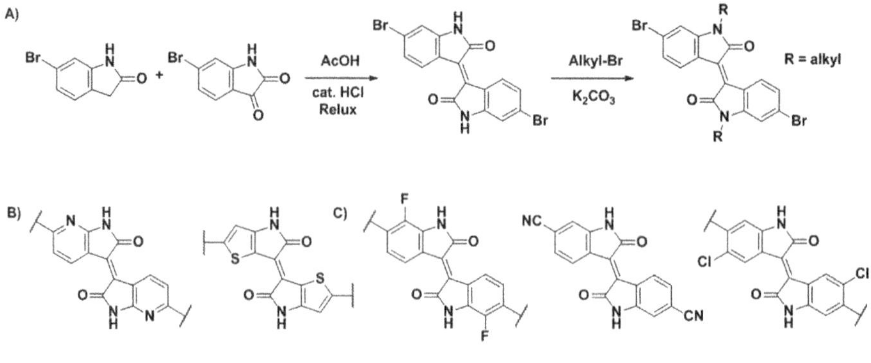

Figure 5.12 (A) General synthetic route towards isoindigo with reactive bromine groups at the 6-position. (B) Azaisoindigo and thienoisoindigo core structures. (C) Fluorine, cyano, and chlorine substituted isoindigo core structures.

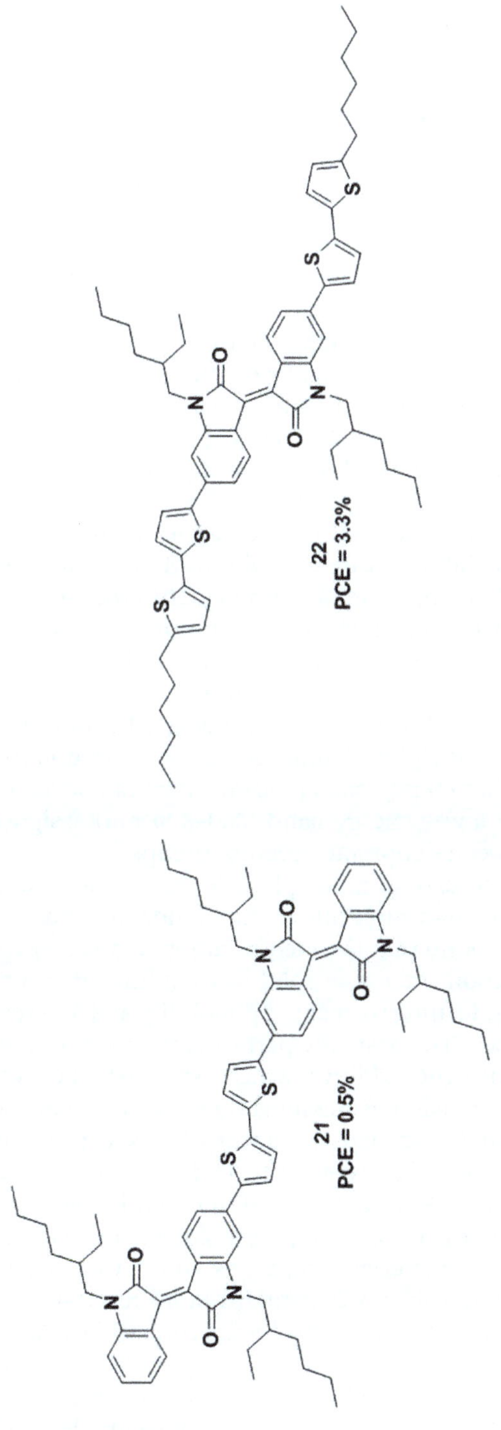

Figure 5.13 Two of the first examples of small molecules based upon the isoindigo core structure that were used in SM-BHJ solar cells.

rich bithiophene units can be incorporated *via* Suzuki coupling between mono- or bis-borylated bithiophene units and a brominated isoindigo core. Both compounds exhibited thin-film optical absorption extending beyond 700 nm, and deep HOMO levels of −5.6 eV. Initial SM-BHJ solar cell devices using $PC_{61}BM$ exhibited PCE values of 0.5% and 1.7% for **21** and **22**, respectively. The high performance of solar cells based upon **22** was due, in part, to a high degree of molecular ordering within the active layer thin-film. Upon device optimization of **22**:$PC_{61}BM$ based solar cells using a binary processing additive mixture (PDMS/TEG), the PCE was able to be improved to 3.3%.[108] These results have set the foundation on which isoindigo-based small molecules for OPVs can be built.

Inspired by the success of isoindigo-based oligothiophenes, alterations to the electron-rich donor component were investigated, leading to the synthesis of seven narrow band-gap small molecules with different electron-donating strengths (Compounds **23–29** in Figure 5.14). These materials were accessed by standard Suzuki or Stille couplings of the bis-brominated core with functionalized end-capping units.[98,99,101]

The synthesis of these materials offered a means to explore the relationship between molecular structure and photovoltaic performance, focusing on the influence of electron-donating units within the isoindigo backbone. It was found that, in general, increasing the number of thiophenes from one (**23**) to two (**24**) to three (**25**) decreased solubility; however, all small molecules exhibited good thermal stability and broad absorbance in the visible spectrum with a red-shift accompanying increased conjugation length. All spectra showed two absorption bands, consistent with donor–acceptor systems where the higher energy band (300–470 nm) can be attributed to π–π* transitions and the lower energy band (500–800 nm) assigned to internal charge transfer between donor and acceptor groups.[98]

Solution-processed BHJ organic photovoltaic devices were fabricated with each isoindigo small molecule as the donor material and $PC_{71}BM$ as the acceptor. It was found that the incorporation of an alkyl chain (**26**) onto the conjugated backbone increases solubility but hinders π–π packing of the isoindigo molecules, leading to decreased mobility and as a result poor photovoltaic performance. The best OPV performance was found for **25**:$PC_{71}BM$ blends processed with DIO additive, which exhibited a PCE of 3.2%. It was concluded that the extension of π-conjugation effectively narrowed the band-gap, red-shifted and broadened the absorption spectrum, increased the charge carrier mobility and therefore returned the highest PCE.[98]

Compounds **27** and **28** were designed and synthesized by Roncali and co-workers to explore the effect of aliphatic side-chain positioning and end-cap donor strength on materials properties and device performance. Both compounds were synthesized *via* standard Stille reactions. Compound **27** has two aliphatic side chains on the isoindigo core, while compound **28** positions two aliphatic side chains on the dithienopyrrole end-cap units. Notably **28** was found to have poor solubility in most common organic solvents except tetrahydrofuran (THF), owing to strong intermolecular H-bonding between

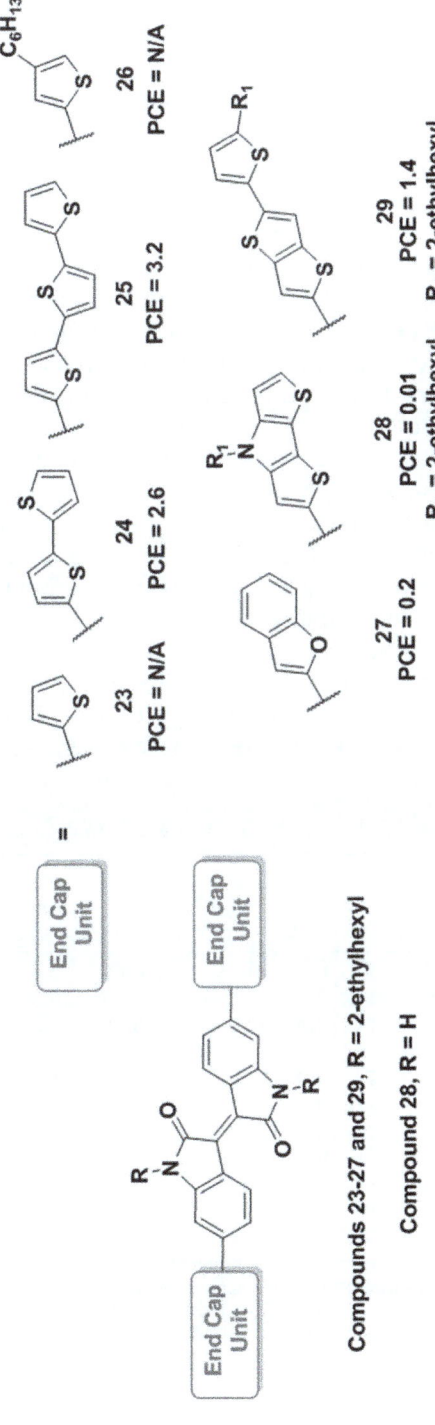

Figure 5.14 Structures of isoindigo based small molecules that have been used as donor materials in SM-BHJ solar cells.

the isoindigo core units, while **27** was found to be readily soluble in most organic solvents. Therefore, to ensure materials processability, it is important to functionalize the isoindigo building block with solubilizing side chains. Both compounds exhibited strong optical absorption in the visible region of the solar spectrum, with absorption extending beyond 700 and 800 nm for **27** and **28**, respectively. The optical band-gap for **28** is a result of a high HOMO energy level due to the stronger electron donating ability of dithieno-pyrrole *vs.* benzofuran. When incorporated into BHJ solar cells with $PC_{61}BM$ as the electron transport material, poor device performance with PCE below 1% was obtained. These results were surprising, especially for **28** which had favorable optical and electronic properties and is similar in structure to the high performance DPP derivative (**7**). The related compound **29** was designed to have ideal electronic energy levels for use as a donor in OPV application with fullerene acceptors. Indeed, this compound had HOMO/LUMO energy levels of 5.4 eV and 3.9 eV, respectively, and broad optical absorption extending to 800 nm, both attributable to the electron deficient isoindigo core and extended π-conjugation through the fused thiophene end-capping units. Despite these favorable features, low PCE values of ~1.4% were obtained for **29**:$PC_{61}BM$ based devices, largely attributed to very rough active layer, and thus through rigorous processing and device optimization it is expected that the PCE of these devices will rapidly increase.

The ease of synthesis has seen further modification of isoindigo-based small molecules with the primary focus on extending the π-conjugation of the system to control the band-gap and energy levels. This has been met with mixed results, highlighting the promise of these materials, but also the need

Table 5.2 SM-BHJ solar cell data for isoindigo based small molecule donors.

Onset of absorption[a] (nm)		Processing[b]	Acceptor	Weight ratio (w/w)	V_{OC} (V)	J_{SC} (mA cm^{-2})	FF (%)	PCE (%)	Ref.
21	743	Thermal	$PC_{61}BM$	3:2	0.66	2.4	36	0.5	107
22	705	Thermal	$PC_{61}BM$	1:1	0.75	6.3	38	1.7	107
22	705	PDMS/TEG thermal	$PC_{61}BM$	1:1	0.95	7.8	45	3.3	107
23	685	None	$PC_{71}BM$	7:3	No significant PV response				98
24	729	DIO	$PC_{71}BM$	7:3	0.93	6.6	43	2.6	98
25	775	DIO	$PC_{71}BM$	7:3	0.87	8.2	43	3.2	98
26	689	DIO	$PC_{71}BM$	7:3	No significant PV response				98
27	701	Thermal	$PC_{61}BM$	1:2	0.60	1.0	38	0.26	101
28	838	None	$PC_{61}BM$	1:2	0.35	0.15	16	0.01	101
29	804	None	$PC_{61}BM$	3:2	0.72	6.03	32	1.4	99

[a]As determined from neat thin-film optical absorption spectra.
[b]Thermal = post-deposition thermal annealing of active-layer; DIO = active layer cast from solvent mixtures containing small quantities of diiodooctane (DIO); PDMS/TEG = active layer cast from solvent mixtures containing small quantities of polydimethylsiloxane (PDMS) and tetraethylene glycol (TEG).

for more thorough investigation of their ability to form uniform thin films, electronic features, and structure–property–function relationships.

The OPV device performance of all the isoindigo-based small molecules discussed in this section has been summarized (Table 5.2). While in general the solar cell performance is low compared to other small molecule systems, we have highlighted isoindigo small molecules, because their polymer-based counterparts have been shown to exhibit excellent PCE values >7% when incorporated into BHJ solar cell devices.[109] Therefore, these results reinforce the need for further development of isoindigo-based small molecules as an organic functional material and improvements in materials processing and device fabrication.

5.8 Porphyrins

The investigation of porphyrins as the active donor component in organic solar cells is inspired by natural photosynthetic systems, specifically chlorophylls (Figure 5.15). Chlorophyll is the active component that absorbs light and carries out photochemical charge separation to store light energy from the sun, so it is conceivable that similar porphyrin-based structures could find application in organic photovoltaics.[110–112]

These analogous porphyrin structures are excellent light harvesters, reflected by their high molar absorptivity coefficients. Porphyrin-based donors possess an extensive π-conjugated system capable of fast electron transfer to acceptors, an attractive feature for all organic electronics. Furthermore, their properties can be easily tuned *via* straightforward synthetic

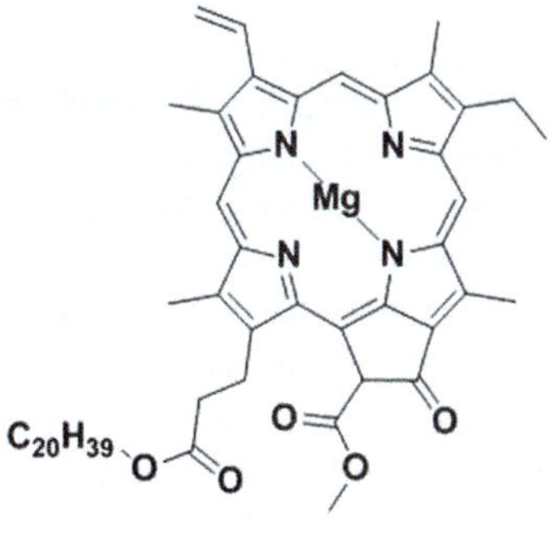

Chlorophyll *a*

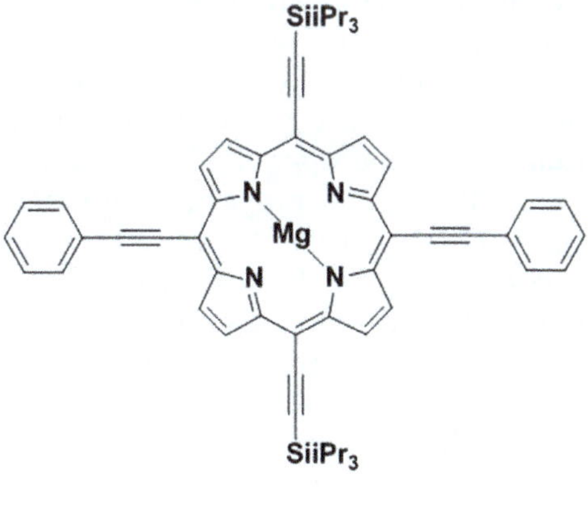

30
PCE = 1.5%

Figure 5.15 Porphyrin structures. (Left) Chlorophyll *a* and (right) early derivative used as a donor material in SM-BHJ solar cells.

modification, at the periphery structure or through the insertion of a metal into the cavity.[113] In terms of industrial applications, porphyrin-based derivatives are known to be air stable and quite robust, with recent literature reporting that devices based upon compound **30** (Figure 5.15) can maintain over 73% of its initial PCE under ambient air.[114] The outlined chemical and physical properties of porphyrins coupled with their ease of synthetic alteration make these materials suitable candidates for OPV studies.

The synthesis of a standard porphyrin structure is illustrated in Figure 5.16. The porphyrin core can be easily synthesized *via* the acid catalyzed condensation between two equivalents of an aldehyde functionalized with the desired flexible side chains and two equivalents of a dipyrrole. The synthesis of the aldehyde and dipyrrole starting materials from basic building blocks has also been reported.[115] The compound can be subsequently brominated at the available methylene units; this provides an adequate site for further reactivity. Finally, the porphyrin is equipped with a metal center (Zn, Cu, Ni, Fe, *etc.*); one of these can be incorporated into the cavity of the porphyrin by reacting with a given metal salt.[116]

Further modification to the porphyrin core becomes dependent on the desired carbon–carbon bond-forming reaction for the remaining synthetic procedure. A porphyrin core can be borylated with an organoboronic acid or ester in preparation for Suzuki couplings.[117] Alternatively, an alkyne can be installed from trimethylsilylacetylene following the removal of the trimethylsilyl functionality, for Sonogashira couplings.[111] The incorporation of these functional groups is illustrated in Figure 5.17. These functionalities provide sites of reactivity with a given transition metal catalyst for C–C bond-forming reactions, leading to porphyrins that can be flanked by common acceptor building blocks and enabling the well-known donor–acceptor approach to be extended to these materials.

In terms of application, porphyrins have been successfully incorporated as organic dyes in dye sensitized solar cells (DSSCs);[110] however, they have not been extensively studied in BHJ device architectures. One of the early examples of their potential as materials for organic electronics focused on their self-assembly properties, targeting clearly defined interdigitated structures. A highly insoluble, crystalline donor, tetrabenzoporphyrin (**31**), can be accessed from the thermal retro-Diels–Alder conversion of its precursor (**CP**) (Figure 5.18).[118] This material demonstrated the preferred columnar structure in a BHJ active layer, and with this morphology a respectable device PCE of 5.2% was obtained when using the acceptor bis(dimethylphenylsilylmethyl)[60]fullerene (SIMEF).[118,119] The controllable BHJ morphology and respectable performance indicated the need for further examination of porphyrin-based small molecules for the production of efficient solar cells.

As more porphyrin-based materials were developed, it was reported that short exciton diffusion lengths and low charge carrier mobilities were the likely culprits of poor device performance. In an attempt to enhance the overall performance of these materials, the donor–acceptor approach was investigated as a means to facilitate internal charge transfer. At the time, this

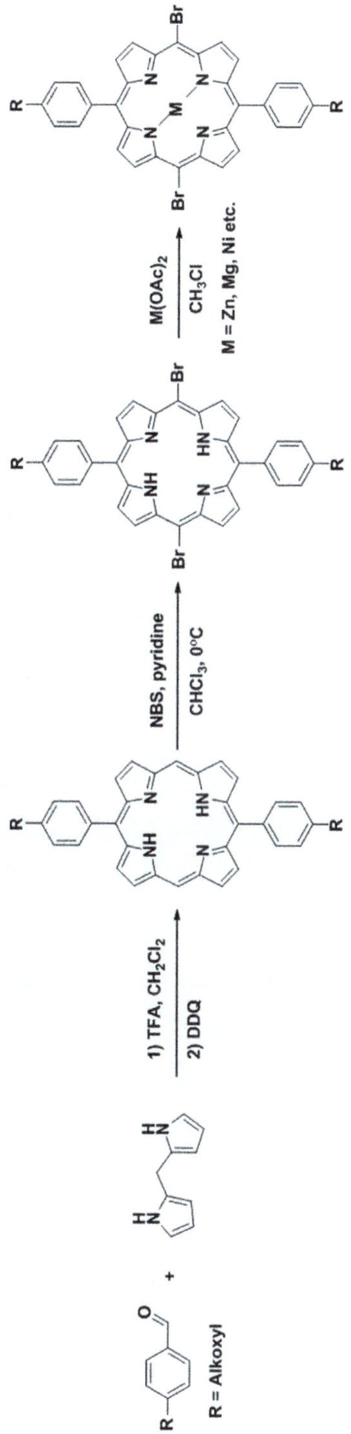

Figure 5.16 General procedure for the synthesis of porphyrins.

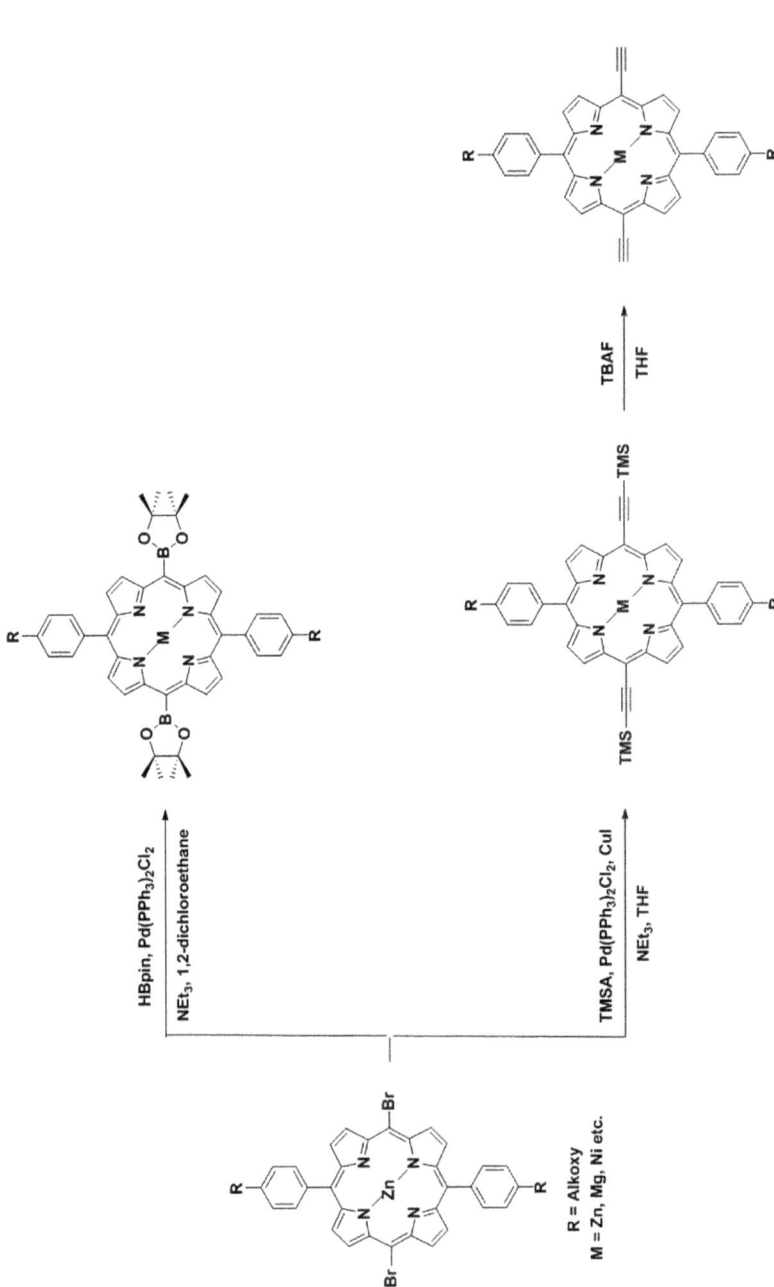

Figure 5.17 General functionalization of brominated porphyrins to yield compounds suitable for Suzuki and Sonogashira coupling reactions.

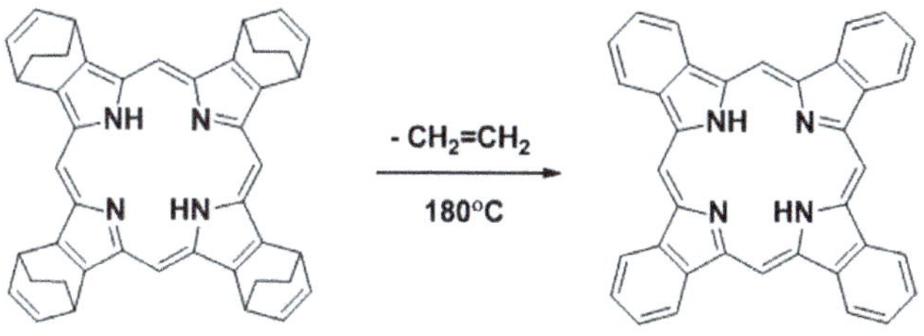

$$\text{CP} \xrightarrow[\text{180°C}]{- \text{CH}_2=\text{CH}_2} \text{31}$$

CP

31
PCE = 5.2%

Figure 5.18 Synthesis of tetrabenzoporphyrin (**31**). CP can be solution processed and, upon thermal treatment, compound **31** is generated, which can be used as a donor in SM-BHJ solar cells.

design strategy had not been previously considered for the design of porphyrin-related BHJ solar cell materials. To assess this methodology, a well-known acceptor component, 2,1,3-benzothiadiazole (**BT**), end-capped with 3-hexylthienyl (**TBT**) was linked by ethynylene bridging units to a porphyrin core to form the zinc-containing porphyrin **32**.[113] A second derivative without ethynylene was also synthesized, **33**, to assess the structure–property relationships in relation to the bridging units.[113] Both compounds (Figure 5.19) show suitable solubility in common organic solvents due to the 3,5-bis(dodecyloxy)phenyl groups at *meso*-positions of the zinc porphyrin rings. UV-vis absorption extending beyond 900 nm highlights the improved internal charge transport of **32**, which has been attributed to the co-planarity of the material. Compound **33** only exhibits weak absorption from 400 to 600 nm, due to complete localization of molecular orbitals as a result of a non-planar structure. Specifically, theoretical calculations predicted that the **BT** moieties preferred an orientation perpendicular to the porphyrin core. It was thus concluded that the ethynylenes promote planarity, leading to an entirely delocalized π-conjugated backbone.[113] This feature, coupled with the push–pull nature of the porphyrin core and **BT** moieties was found to facilitate internal charge transfer. The solution processed SM-BHJ solar cells with **32**:PC$_{71}$BM active layers exhibited a promising performance with a PCE of up to 4.0% compared to just 0.7% for **33**:PC$_{71}$BM.[113,120]

Further alterations to these materials through side chain engineering have gone on to yield unprecedented photovoltaic performance of porphyrins as the donor component in a solution processed SM-BHJ solar cell. Previously, bulky 3,5-di(dodecyloxy)-phenyl substituents have been used to ensure the solubility of the porphyrins; however, it was found that the length of these chains perturbed performance. These long flexible side chains at the 3,5-positions protrude out of the porphyrin plane and hinder intermolecular π–π stacking, culminating in a suppressed device performance. In an attempt to mitigate this influence on device performance, 4-octyloxy-phenyl groups were used in place of 3,4-di(dodecyloxy)-phenyl groups in the synthesis of a new conjugated donor–acceptor porphyrin with diketopyrrolopyrrole (**DPP**) as the acceptor component (Figure 5.20).[111] This design was shown to yield materials

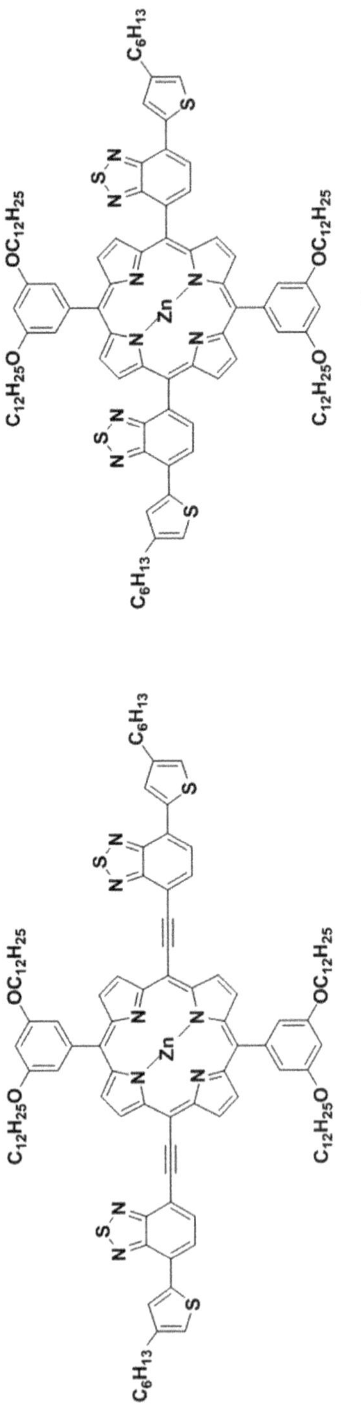

32
PCE = 4.0%

33
PCE = 0.7%

Figure 5.19 Structures of donor–acceptor type porphyrin small molecules used as donors in SM-BHJ solar cells.

34
PCE = 3.7%

35
PCE = 7.2%

R$_1$ = H
R$_2$ = OC$_{12}$H$_{25}$

R$_1$ = OC$_8$H$_{17}$
R$_2$ = H

Figure 5.20 Structures of porphyrin–DPP based small molecules used as donors in SM-BHJ solar cells.

with excellent solubility in organic solvents and broad and strong optical absorption covering the visible and near infrared region in both solution and film. Compared to the weaker absorbance peak at 790 nm for **34**, the absorbance of **35** at 813 nm is red-shifted and its intensity increased, highlighting a stronger co-facial π–π stacking than **34**. This confirmed the notion that less bulky substituents at the porphyrin periphery promote intermolecular π–π interaction in the solid state. The synthesized narrow band-gap porphyrin with less bulky substituents was able simultaneously to facilitate intramolecular charge transport and to increase the intermolecular π–π stacking in film, leading to a high PCE of up to 7.2% with PC$_{61}$BM as the acceptor component.[111]

Porphyrin-based materials represent another encouraging small molecule design for organic photovoltaics (Table 5.3). Their comparable structure to nature's own light harvesters can be accessed from straightforward organic synthesis, which represents one of the most attractive features of these materials. Furthermore, their properties can be easily tuned for a particular function *via* a wide range of synthetic modification at the periphery.

5.9 Oligothiophenes (Donor–Acceptor–Donor–Acceptor–Donor)

Building on the established literature, and taking into account the specific criteria for which to develop small molecule donors, the Bazan research group reported on a highly modular molecular framework that was found

Table 5.3 SM-BHJ solar cell data for porphyrin based small molecule donors.

	Onset of absorption[a] (nm)	Processing[b]	Acceptor	Weight ratio (w/w)	V_{OC} (V)	J_{SC} (mA cm^{-2})	FF (%)	PCE (%)	Ref.
30	No spectra	None	PC$_{61}$BM	5:1	0.86	4.3	40	1.5	114
31	720[c]	Retro Diels–Alder	PC$_{61}$BM	3:7	0.55	7.0	51	2.0	118
31	720[c]	Retro Diels–Alder	SIMEF	3:7	0.76	9.7	62	5.2	118
32	957	Pyridine	PC$_{71}$BM	1:3	0.85	9.5	50	4.0	113
33	650[c]	Pyridine	PC$_{71}$BM	1:3	0.88	2.8	29	0.7	113
34	879	None	PC$_{61}$BM	1:1	0.76	9.8	50	3.7	120
35	912	None	PC$_{61}$BM	1:1.2	0.74	15.0	53	5.8	111
35	912	DIO	PC$_{61}$BM	1:1.2	0.71	16.0	64	7.2	111

[a]As determined from neat thin-film optical absorption spectra.
[b]Retro Diels–Alder = post-deposition thermal annealing of active-layer to chemically modify the active organic components; pyridine = active layer cast from solvent mixtures containing small quantities of pyridine; DIO = active layer cast from solvent mixtures containing small quantities of diiodooctane (DIO).
[c]Estimated by authors from thin optical absorption spectra.

to yield materials with near ideal properties for use as donor molecules in BHJ solar cells.[29] The architecture consisted of an acceptor–donor–acceptor (A–D–A) core flanked with end-capping units. Two acceptors were utilized to increase the electron affinity across the conjugated backbone, ensuring deep HOMO levels, while end-capping units served to extend π-conjugation and tailor self-assembly properties. Important to this class of compounds was the use of a pyridyl[2,1,3]thiadiazole (PT) building block as the electron acceptor, which promoted strong intramolecular charge transfer when coupled with electron donors, resulting in narrow band-gaps.[42] Additionally, the unsymmetrical nature of PT imparted selective reactivity, allowing for the synthesis of mono-functionalized materials in high yield. Among the many derivatives synthesized, it was found that utilizing a 2-ethylhexyl substituted dithienosilole (DTS) donor and 2-hexylbithiophene end-capping units resulted in virtually ideal optical and electronic properties. Figure 5.21 shows the synthesis and structure of compound **36**.[42] Note that, in this case, the pyridyl N-atoms are in a position *distal* to the DTS core. Compound **36** can be considered to have a D–A–D–A–D type architecture.[22] First, the material exhibited strong optical absorption in the visible and near-IR part of the solar spectrum, with peak absorption at 700 nm in the solid state (near the region of max photon flux). Second, **36** was found to have HOMO and LUMO energy levels at −5.2 eV and −3.6 eV, respectively, making it compatible with common fullerene derivatives. Third, the molecule has a high organic solvent solubility (>20 mg mL^{-1}) as a result of alkyl side chains both perpendicular and parallel to the molecular backbone, allowing for uniform film formation. In addition, the small molecule has a planar π-conjugated backbone, important for intramolecular π-delocalization and intermolecular π-stacking, and has been show to yield high charge carrier mobilites.[121] Incorporation into BHJ devices using PC$_{71}$BM as the acceptor led to initial PCEs of 3.2% after thermal treatment of the active layer.[42] Optimization of the device architecture by utilizing a molybdenum oxide hole transport layer resulted in a PCE increase to 4.5%.[122] Subsequent optimization of the solution processing conditions [*i.e.* using diiodooctane (DIO) as a solvent additive] led to additional increases of PCE up to 5.6%.[23] To further improve device performance, in collaboration with the Heeger research group, Bazan and co-workers developed the regio-isomer **37**, where the pyridyl-N atoms are positioned proximal to the central DTS unit (Figure 5.22).[123] A slight change in the molecular geometry towards a more 'banana' shape resulted in a greater tendency for **37** to self-assemble into highly ordered nanostructures, leading to higher hole mobilities, increased light absorption, and ultimately a higher PCE of 5.7% when the devices were thermally treated.[122] Active layers processed using the solvent additive DIO gave record PCE values of 6.7%.[123] Investigation of the thin-film microstructure revealed highly crystalline "donor" domains on the order of 30–50 nm, ideal for efficient charge separation and transport. This result represented, for the first time, a solution processed small molecule BHJ device comparable in PCE to its polymer counterparts.

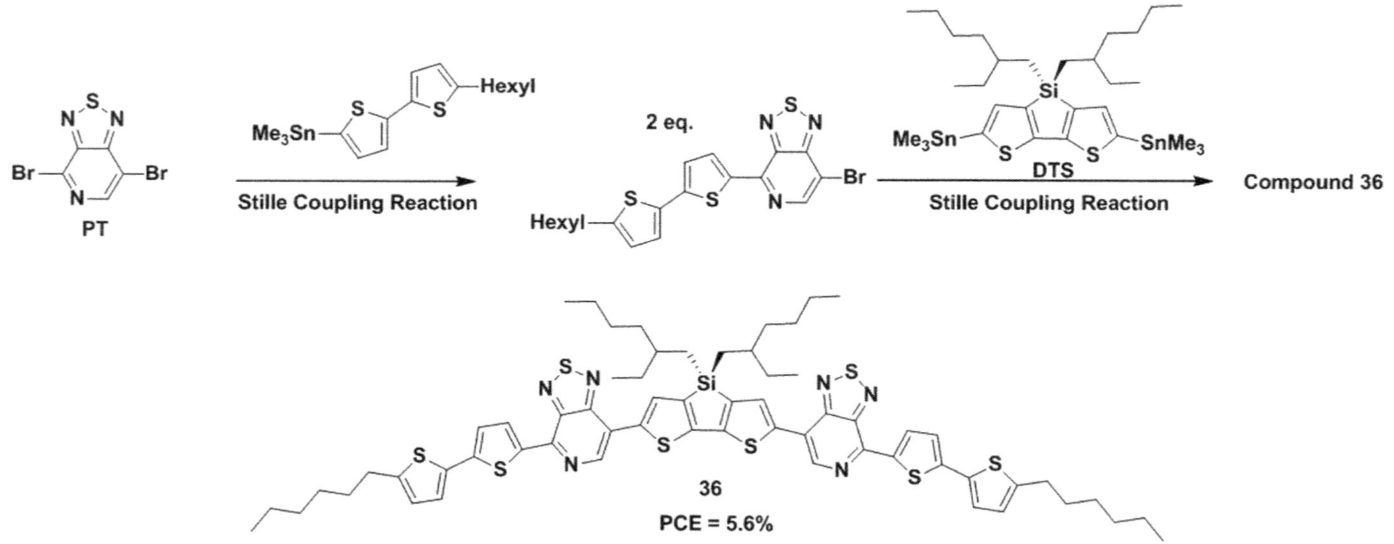

Figure 5.21 Synthesis and structure of a DTS-PT based small molecule used as donor in SM-BHJ solar cells. Compound **36** is synthesized from the outside-in, resulting in the pyridyl N-atoms in a position *distal* to the DTS core.

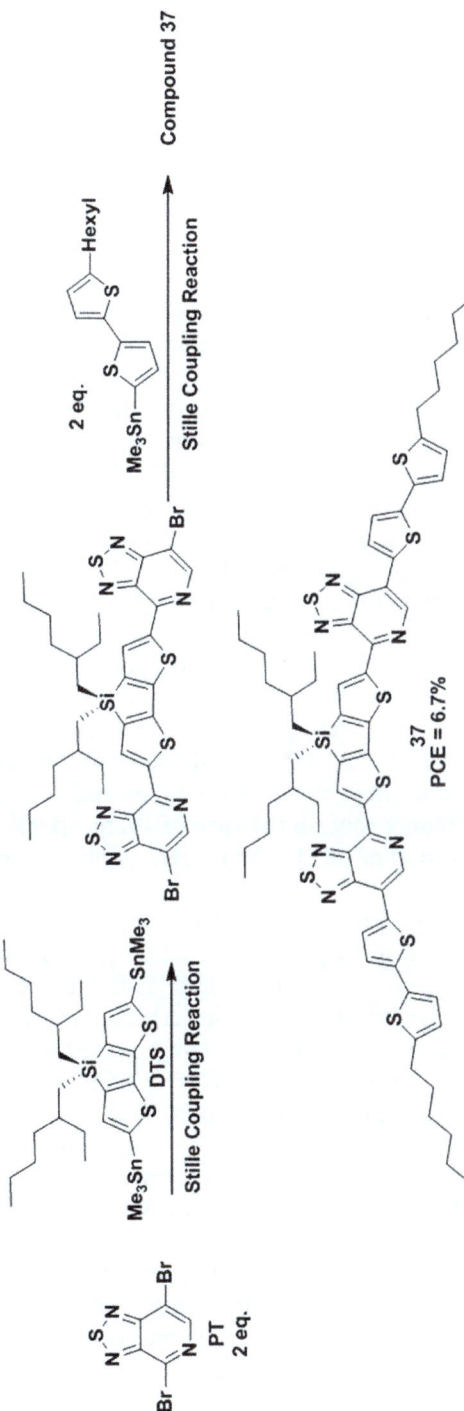

Figure 5.22 Synthesis and structure of a DTS-PT based small molecule used as donor in SM-BHJ solar cells. Compound 37 is synthesized from the inside-out, resulting in the pyridyl N-atoms in a position *proximal* to the DTS core.

To further probe the impact of pyridyl-N atom substitution, the unsymmetrical compound **38** was prepared.[23] Here, one pyridyl N-atom is in a position *distal* to the DTS core, while the other is in a position *proximal* to the DTS core. The structure and synthesis are presented in Figure 5.23. Compound **38** was synthesized in moderate yield *via* a series of Still cross-coupling reactions. Important to the synthesis was the preparation of the intermediate Sn-DTS(PTT), which was easily coupled with the 2-hexyl bithiophene substituted PT building block to yield the desired unsymmetrical small molecule. Surprisingly, when blended with PC$_{71}$BM and processed under optimized conditions similar to those for devices made with **36** and **37**, a lower PCE of 3.2% was obtained. The lower device performance was a result of a decrease in thin-film crystallinity which was attributed to the lower symmetry of **38** *vs.* **36** and **37**.[23]

Further progressing the evolution of this series of compounds, Bazan and co-workers reported on two analogues of **37** with extended π-conjugated backbones. Compounds **39** and **40** can be expressed as having D–A–D–A–D–A–D and D–A–D–A–D–A–D–A–D type architectures and were designed to explore the influence of chromophore elongation on molecular and bulk properties.[124] As shown in Figure 5.24, compounds **39** and **40** can be synthesized *via* a Stille cross-coupling reaction between the intermediate Sn-DTS(PTT) and either BTBr$_2$ or DTS(PTBr)$_2$, respectively. Compared to compound **37**, both **39** and **40** showed extended optical absorption beyond 800 nm, high melting and crystallization temperatures, and a stronger tendency in the solid state to resist structural reorganization upon thermal annealing. When incorporated into solar cell devices with PC$_{61}$BM, those based upon **39** reached a maximum PCE of 5.8%, whereas those using **40** reached a maximum PCE of 6.5%. It is important to note that, in the case of **40**, no special processing of the active layer (*i.e.* thermal annealing or solvent additives) was required to reach such high PCE values.[124]

An issue encountered with the PT based small molecules was the susceptibility of the pyridyl N-atom to protonation. Compounds **36** and **37** were shown to be readily protonated by strong acid.[125,126] This process results in a dramatic lowering of the frontier molecular orbital levels and a significant narrowing of the band-gap. Thus, when the conducting polymer PEDOT:PSS was employed as the hole transport layer (HTL) in standard BHJ solar cell devices, poor device performance was obtained as a result of the formation of interfacial traps due to interaction between acidic sites of PEDOT:PSS and the basic N-atoms of the PT unit.[125,126] To circumvent this issue, Bazan and co-workers replaced the pyridyl N-atom with a C–F moiety, compound **41** (Figure 5.25).[127] This simple substitution greatly improved stability of the small molecule in the presence of acid. It is important to note that the brominated version of the FBT acceptor (FBT-Br$_2$) exhibits different reactivity from the PT analogue. Under Stille cross-coupling conditions, substitution at the Br atom away from the fluorine substituent is more favored (Figure 5.25). Here, sterics proved to be more important than electronics. With regard to SM-BHJ solar cell devices, blends of **41**:PC$_{71}$BM gave high PCE values of

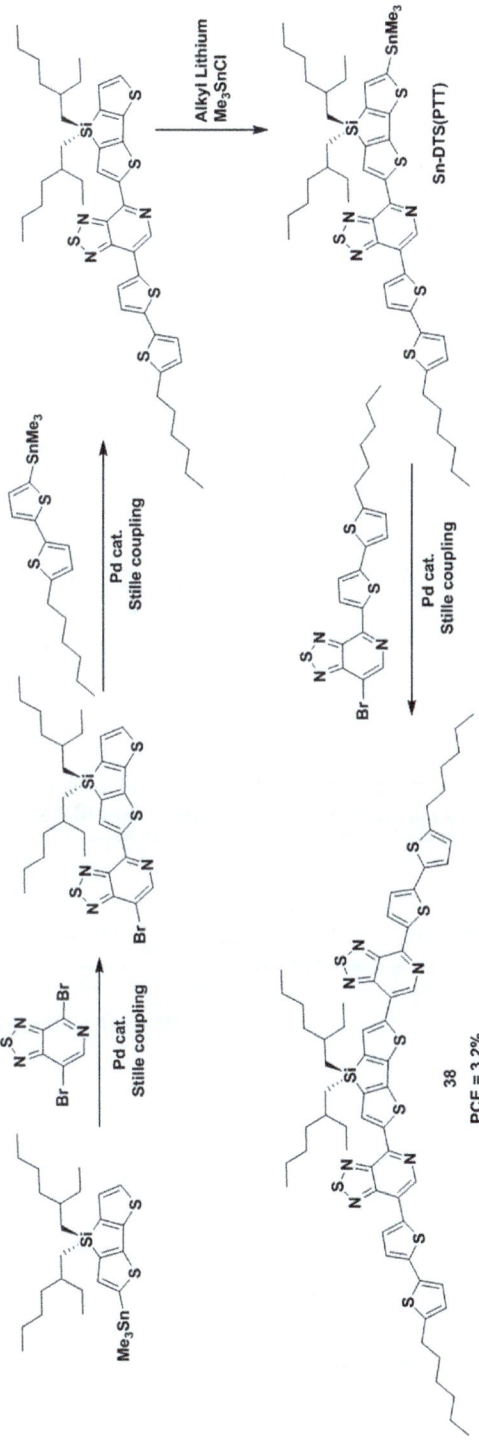

Figure 5.23 Synthesis and structure of an unsymmetrical DTS-PT based small molecule used as donor in SM-BHJ solar cells. Compound 38 has the pyridyl N-atoms in a position *proximal* and *distal* to the DTS core.

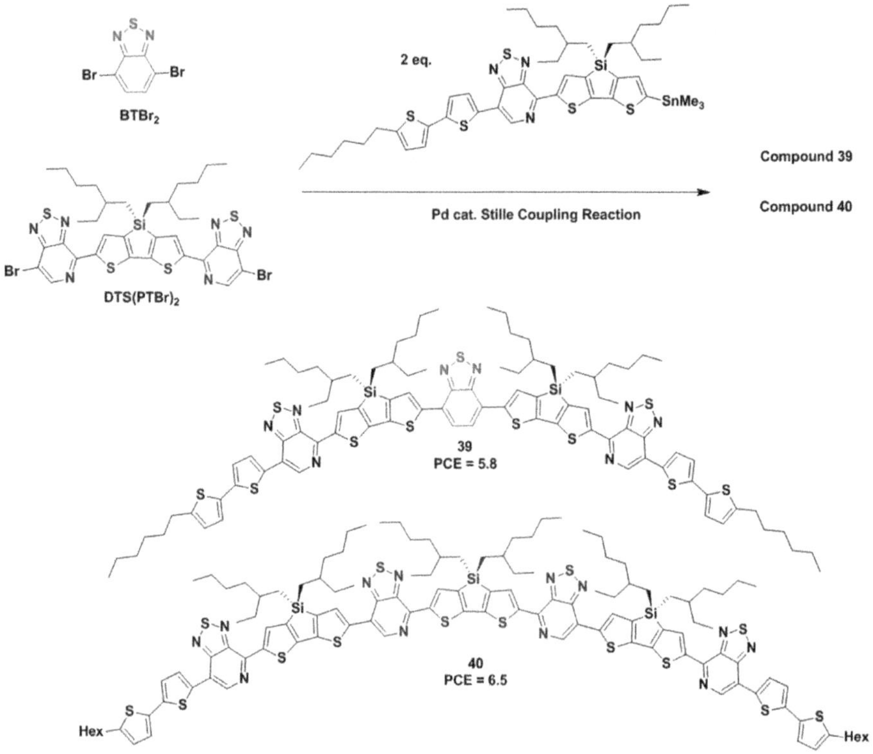

Figure 5.24   Synthesis and structure of DTS-PT small molecules with extended conjugated backbones used as donors in SM-BHJ solar cells.

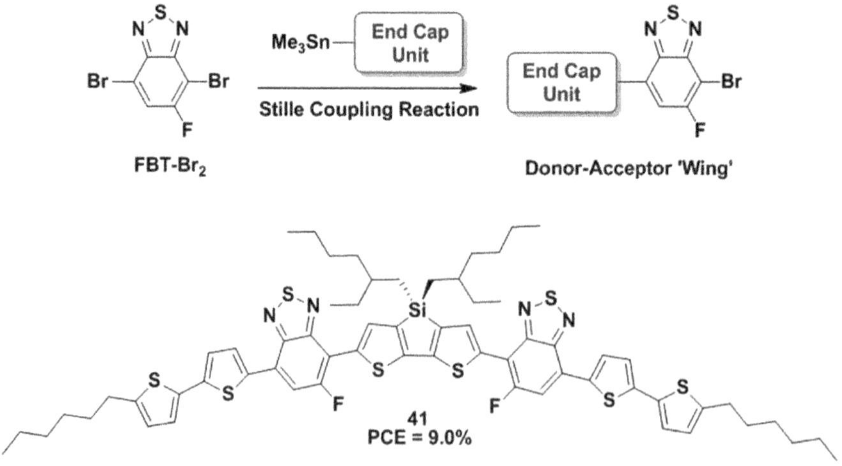

Figure 5.25   Structure of compound **41** based upon the FBT building block used as donor in SM-BHJ solar cells. Different from the PT acceptor, preferential reactivity of FBT-Br$_2$ occurs at the bromine atom away from the fluorine atom.

Table 5.4 SM-BHJ solar cell data for PT and FBT based small molecule donors.

	Onset of absorption^a (nm)	Processing^b	Acceptor	Weight ratio (w/w)	V_{OC} (V)	J_{SC} (mA cm^{-2})	FF (%)	PCE (%)	Ref.
36	820	Annealing	PC$_{71}$BM	6:4	0.70	10.9	42	3.2	42
36	820	Annealing	PC$_{61}$BM	6:4	0.71	11.7	54	4.5	122
36	820	DIO	PC$_{71}$BM	7:3	0.73	12.7	60	5.6	23
37	815	Annealing	PC$_{61}$BM	6:4	0.78	12.2	60	5.7	122
37	815	DIO	PC$_{71}$BM	7:3	0.78	14.4	59	6.7	123
38	810	DIO	PC$_{71}$BM	7:3	0.72	9.8	45	3.2	23
39	860	Annealing	PC$_{61}$BM	6:4	0.71	13.6	60	5.8	124
40	880	None	PC$_{61}$BM	1:1	0.66	15.2	65	6.5	124
41	800	DIO	PC$_{71}$BM	6:4	0.81	12.8	68	7.0	127
41	800	DIO	PC$_{71}$BM	6:4	0.73	15.2	67	7.9	128
41	800	DIO	PC$_{71}$BM	6:4	0.78	15.5	75	9.0	30

^{a}As determined from neat thin-film optical absorption spectra.
^{b}Thermal = post-deposition thermal annealing of active-layer; DIO = active layer cast from solvent mixtures containing small quantities of diiodooctane (DIO).

7% using PEDOT:PSS HTLs.[127] These results were important, because the improved material stability enabled a wider range of device and processing optimizations to be carried out to enhance the device PCE. Through these optimizations, Heeger and co-workers demonstrated 7.9% efficient **9**:PC$_{71}$BM solar cells in an inverted architecture,[128] and 9.0% efficient **9**:PC$_{71}$BM solar cells using barium cathodes.[30]

The donor–acceptor type small molecules reported on by Bazan and co-workers have proved to be excellent light-harvesting materials for use as donors in BHJ solar cells (Table 5.4). The simple, facile and reproducible synthesis, combined with excellent film forming capabilities and high device performance, make this class of materials well positioned to enable commercialization of organic solar cells.

5.10 Oligothiophenes (Acceptor–Donor–Acceptor)

In 2009, the Chen research group first reported on a series of narrow band-gap oligothiophenes that exhibited excellent performance when incorporated into SM-BHJ solar cells (Table 5.5).[28,129] They demonstrated that end-capping long septithiophenes with electron withdrawing units provided materials with absorption profiles extending into the red region of the solar spectrum, and deep HOMO levels, desired to give large V_{OC}. Critical to these materials was the incorporation of long alkyl side chains on six of the thiophene units, which rendered the compounds highly soluble in common organic solvents and enabled the formation of uniform thin films from solution.

One of the first derivatives reported was an *n*-octyl substituted septithiophene end-capped with dicyanovinyl functional groups (**42**).[129] The synthesis of compound **42** is shown in Figure 5.26. The septithiophene core (7T) is grown from the inside out *via* a series of nickel-catalyzed Kumada

Table 5.5 SM-BHJ solar cell data for end-capped oligothiophene based small molecule donors.

	Onset of absorption[a] (nm)	Processing[b]	Acceptor	Weight ratio (w/w)	V_{OC} (V)	J_{SC} (mA cm^{-2})	FF (%)	PCE (%)	Ref.
42	800		PC$_{71}$BM	1:1.4	0.82	10.2	29	2.5	130
42	800	DIO	PC$_{71}$BM	1:1.4	0.88	12.4	34	3.7	131
43	716		PC$_{61}$BM	1:0.5	0.88	9.9	51	4.5	132
44	712		PC$_{61}$BM	1:0.5	0.93	9.9	49	4.5	132
45	708		PC$_{61}$BM	1:0.5	0.86	10.7	55	5.1	132
46	832		PC$_{61}$BM	1:1	0.80	8.2	72	4.7	133
46	832	PDMS	PC$_{61}$BM	1:1	0.80	8.6	72	4.9	133
47	1033		PC$_{61}$BM	1:1	0.76	3.1	28	0.7	133
49	730		PC$_{61}$BM	1:0.5	0.92	6.8	39	2.5	134
50	742		PC$_{61}$BM	1:0.8	0.90	7.5	60	4.0	134
51	733		PC$_{61}$BM	1:0.5	0.92	14.0	47	6.1	135
52	678		PC$_{61}$BM	1:0.5	0.93	9.8	60	5.4	136
53	673		PC$_{61}$BM	1:0.5	0.95	8.0	60	4.6	137
54	717		PC$_{61}$BM	1:0.8	0.80	11.5	64	5.8	138
55	713		PC$_{61}$BM	1:0.8	0.91	10.8	65	6.4	139
55	713		PC$_{71}$BM	1:0.8	0.93	11.4	65	6.9	139
55	713	PDMS	PC$_{71}$BM	1:0.8	0.93	12.2	65	7.4	139
56	721		PC$_{71}$BM	1:0.8	0.91	13.1	63	7.5	139
56	721	PDMS	PC$_{71}$BM	1:0.8	0.93	13.2	66	8.1	139
57	705		PC$_{71}$BM	1:0.8	0.96	12.4	53	6.3	139
57	705	PDMS	PC$_{71}$BM	1:0.8	0.96	11.9	59	6.8	139
58	705		PC$_{71}$BM	1:0.8	0.90	12.0	70	7.6	139
58	705	PDMS	PC$_{71}$BM	1:0.8	0.92	12.1	72	8.0	139
59	780		PC$_{71}$BM	1.5:1	0.91	9.5	48	4.2	140
60	770		PC$_{71}$BM	1.5:1	1.03	10.1	55	5.7	140
61	775		PC$_{71}$BM	1.5:1	0.92	8.6	65	5.1	140
62	775		PC$_{71}$BM	1.5:1	0.92	11.1	66	6.8	140
62	775		PC$_{71}$BM	1:0.8	0.93	11.4	68	7.2	140
63	715	PDMS	PC$_{71}$BM	1:0.8	0.94	12.5	69	8.1	140
63	715	PDMS	PC$_{71}$BM	1:0.8	1.82	7.7	72	10.1	31

[a] As determined from neat thin-film optical absorption spectra.
[b] DIO = active layer cast from solvent mixtures containing small quantities of diiodooctane (DIO); PDMS = active layer cast from solvent mixtures containing small quantities of polydimethylsiloxane (PDMS).

cross-coupling reactions. Formyl groups were installed on the ends of 7T by utilizing the Vilsmeier–Haack reaction to give the intermediate 7T-bis-formyl. The target compound **42** was obtained by Knoevenagel condensation of the aldehyde with malononitrile. Compared to homogeneous oligothiophenes, the incorporation of the dicyano vinyl moiety served to both increase and red-shift the optical absorption. Compound **42** exhibited thin-film optical absorption extending beyond 750 nm and a relatively deep HOMO level of −5.1 eV, and therefore was used as a donor to fabricate solution processed SM-BHJ solar cells with PC$_{61}$BM. Initial devices reached a maximum PCE of 2.5%.[130] Through the use of solvent additives, the PCE was further improved

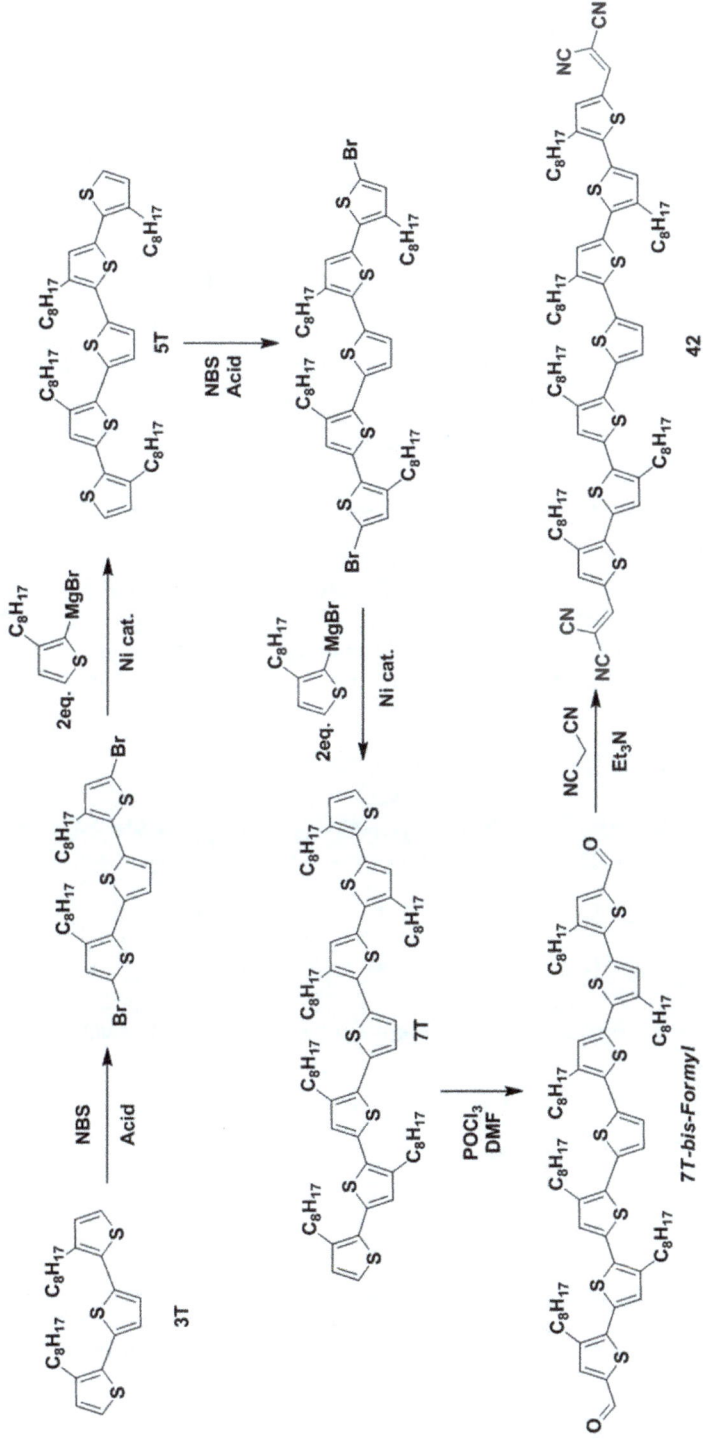

Figure 5.26 Synthesis of the septithiophene small molecule **42** used as a donor in SM-BHJ solar cells.

to 3.7%, with a high V_{OC} and J_{SC} of 0.88 V and 12.4 mA cm^{-2}, respectively, with the FF limiting.[131] The low FF was attributed to a high rigidity of the molecule, resulting in a low solubility in common processing solvent.

The intermediate bis-aldehyde compound (7T-bis-Formyl) proved to be a universal framework from which to construct a range of new narrow bandgap small molecules, for which the absorption profiles, HOMO/LUMO levels, and material solubility could be readily tuned. Figure 5.27 shows a series of nine small molecules with varying end-capping units reported by Chen

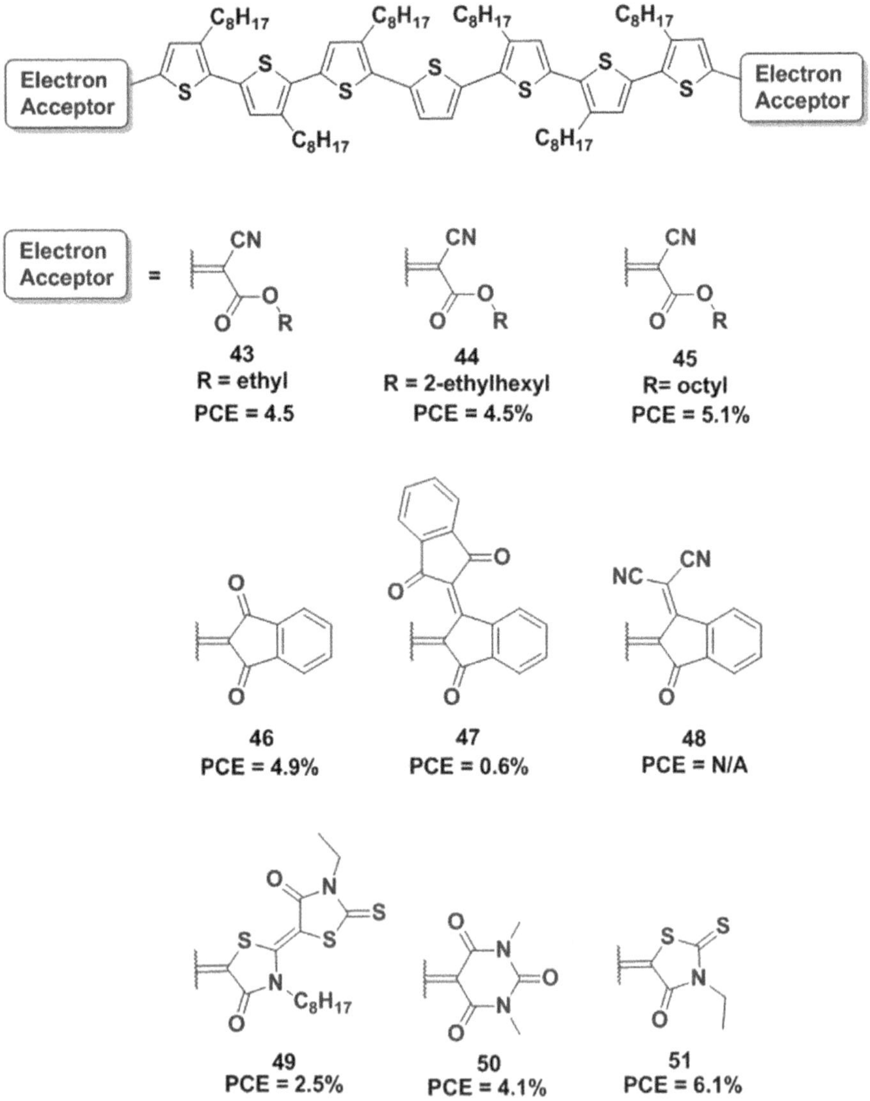

Figure 5.27   Structures of septithiophene based small molecules with varying end-capping units used as donors in SM-BHJ solar cells.

and co-workers. To overcome the solubility issues encountered with the dicy-ano vinyl compound (**42**) and improve thin-film morphology, electron with-drawing alkyl cyano acetate groups were incorporated as terminal acceptor units. Compounds with ethyl (**43**), 2-ethylhexyl (**44**), and octyl (**45**) substitu-ents were synthesized and used as donors with $PC_{61}BM$. OPV devices showed improved PCEs on the order of 4.5–5%, with notably high FF of ~50%.[132] Quite remarkably, compound **45** exhibited a high solubility in chloroform at >200 mg mL^{-1}, which is important for forming the thicker films (>300 nm) required for large area device fabrication.

Incorporation of electron deficient π-stacking end-capping units (com-pounds **46–48**) was found to lower the LUMO energy levels of the compounds relative to **43–45**, and to red-shift the optical absorption spectra.[133] Devices based upon compound **46** with indan(1,3)dione end groups gave PCE values on the order of 4.9%. These devices exhibited extremely high fill factors of 72% owing to the strong tendency for the indan(1,3)dione moiety to direct self-assembly. Use of [1,2′]biindenylidene-3,1′,3′-trione end groups (**47**) was found to inhibit self-assembly, leading to poor device performance, while use of 2-(3-oxo-2,3-dihydroinden-1-ylidene)malononitrile (**48**) rendered the material too insoluble use in solution processed OPV devices.[133]

To further improve device efficiencies, various terminal dye moieties were utilized in an effort to increase light absorption. Compounds **49–51**, bearing the (*E*)-3-ethyl-5-(3-octyl-4-oxothiazolidin-2-ylidene)-2-thioxothiazolidin-4-one, 1,3-dimethylpyrimidine-2,4,6(1*H*,3*H*,5*H*)-trione, and 3-ethylrhodanine dyes, all exhibited strong visible light absorption that was considerably red-shifted from the alkyl cyano acetate derivatives (**43–45**).[134] Compounds **49** and **50** had moderate solubility in organic solvents, whereas compound **51**, with 3-ethylrhodanine end-groups, has a high solubility of >100 mg mL^{-1} in chloroform. All compounds had similar HOMO/LUMO energy levels of approximately −5.1 and −3.4 eV, respectively. Unfortunately, compounds **49** and **50** showed poor organization in the solid state, which led to low charge carrier mobilities and only moderate OPV performance when blended with $PC_{61}BM$. On the other hand, devices based upon compound **51** yield high PCE values of 6.1%, with a high V_{OC} of 0.92 V, a J_{SC} of 14 mA cm^{-2}, and a decent FF of 47%.[135]

Based on these results, it was clear that octyl cyano acetate, 3-ethylrhodanine, and indan(1,3)dione were the best end-capping units. Therefore, these units were used to study how modification to the backbone affected mate-rials properties and device performance. The general synthesis of an oli-gothiophene with a benzodithiophene core unit is shown in Figure 5.28. The 3T-formyl wing is readily synthesized *via* a series of palladium catalyzed Stille cross-coupling reactions. This unit is easily coupled to donor cores using Stille cross-coupling conditions, followed by installation of the appro-priate end-cap *via* a Knoevenagel condensation reaction. It should be noted that Chen and co-workers performed several studies looking at the effect of conjugation length and concluded that small molecules with seven aromatic units performed the best in OPV devices.[28]

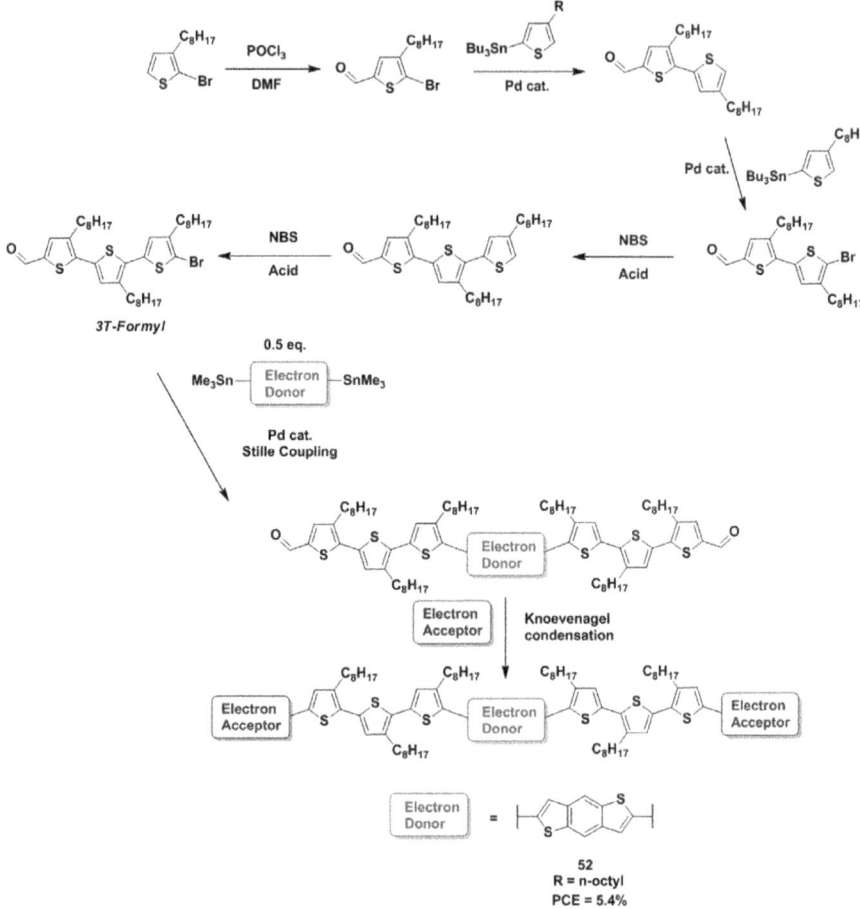

Figure 5.28 General synthesis of oligothiophenes with different electron donor "core" units and electron accepting "end-cap" units.

Figures 5.28 and 5.29 show three structures with octyl cyano end-capping units and varying cores. Compound **52**, with an unsubstituted benzodithiophene (BDT) core, had similar optical absorption and energy levels to **45**, with a thiophene core, but was found to exhibit a slightly higher charge carrier mobility.[136] Solar cell devices based on blends of **52**:PC$_{61}$BM had a better PCE, 5.4%, than those using **45** as the donor.[136] The greater PCE was a result of a slightly higher V_{OC} and improved FF. Substitution of the BDT core with alkoxyl chains resulted in a slight widening of the band-gap, and **53**:PC$_{61}$BM devices exhibited a lower PCE of 4.5%, as a result of a lower J_{SC} value.[137] Incorporation of the widely popular dithienosilole (DTS) building block (compound **54**) resulted in a significant red-shift of the optical absorption compared to those with BDT cores.[138] This was a result of an increase in the HOMO energy level. Impressively, solar devices based upon **54**:PC$_{61}$BM blends gave PCEs of 5.8%. Investigation of the thin-film microstructure

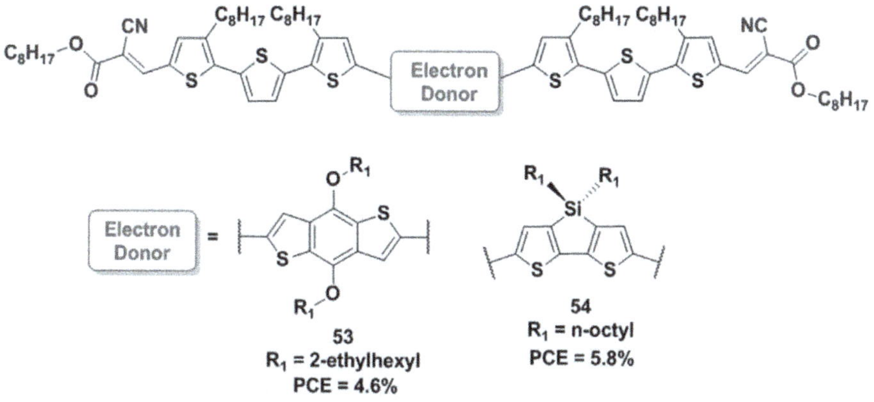

Figure 5.29 Structures of oligothiophene based small molecules with octyl cyano acetate end-capping units and different donors, used as donors in SM-BHJ solar cells.

revealed a near ideal morphology with interconnecting domains on the order of 10–20 nm.[138]

While compound **54** gave the best OPV performance, it was identified that compound **53** with the substituted BDT core gave devices with the highest V_{OC} and FF, and thus effort was made to modify these compounds to improve light absorption and subsequently J_{SC} in solar cell devices. Building on previous knowledge, the octyl cyano acetate end-groups in compound **53** were replaced with stronger absorbing 3-ethylrhodanine dyes to give compound **55** (Figure 5.30).[139] This change resulted in both a red-shift and an increase in the thin-film optical absorption. Solar cell devices based upon **55**:PC$_{61}$BM blends exhibited the expected increase in J_{SC} when compared to **53**:PC$_{61}$BM blends (from 8.0 to 10.8 mA cm^{-2}), and as a result a much higher PCE of 6.4% was obtained.[139] Further increases in PCE up to 6.9% were realized upon using PC$_{71}$BM as the electron acceptor. Amazingly, by processing the active layer with small amounts of polydimethylsiloxane (PDMS), an excellent PCE of 7.4% was obtained.[139] Further exploring this architecture, the substituents on the central BDT unit were modified. Compounds **56–58** incorporate alkyl-thiophene substituents on the core DBT unit, increasing the dimensionality of the molecules (Figure 5.30). Using PC$_{71}$BM as an electron acceptor and processing the active layer with PDMS, maximum PCEs of 6.8, 8.1, and 8.0% were obtained for solar cells using donor compounds **56**, **57**, and **58**. In all cases, the thin-films were uniform (*i.e.* low surface roughness), and exhibited continuous interpenetrating networks of evenly distrusted donor and acceptor domains on the order of tens of nanometers, properties ideal for rapid charge separation and transport. These were the first reports of solution processed SM-BHJ solar cells breaching the 8% mark.[139]

Based on the success of the BDT core framework, Li and co-workers reported on a series of small molecules with BDT cores and indan(1,3)dione

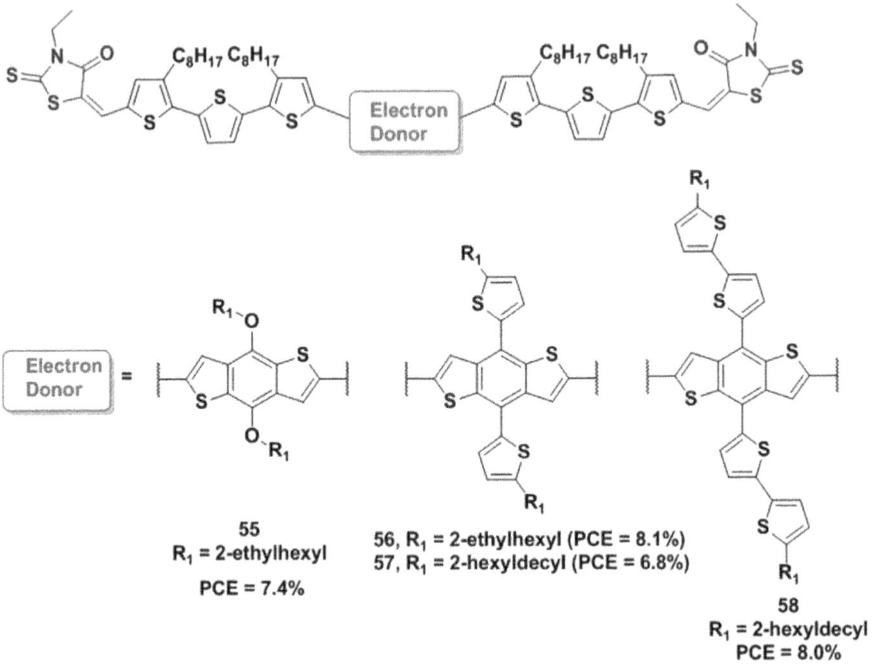

Figure 5.30 Structures of oligothiophene based small molecule with 3-ethyl-rhodanine end-capping units and various donors, used as donors in SM-BHJ solar cells.

end-capping units.[140] They explored both the substitution on the central BDT core and the effect of conjugation length (Figure 5.31). All small molecules were found to have broad optical absorption covering the 450–750 nm range and relatively deep HOMO levels of around −5.2 eV. Solar cell devices were fabricated using a 1.5:1 small molecule:$PC_{71}BM$ blend ratio. It was found that increasing the conjugation length improved device PCE, primarily as a result of an increased FF, while the use of alky-thiophene substituents on the BDT core improve all parameters. A best PCE of 6.7% was obtained when using **62** as the donor molecule. It is worth noting that devices based on **60**:$PC_{71}BM$ were found to have V_{OC} values in excess of 1 V.[140]

Finally, Yang and co-workers reported on an analogous small molecule to compound **56**, except that octyl groups replaced ethyl on the rhodanine dye terminal unit.[31] Compound **63** (Figure 5.32) was shown to have excellent organic solvent solubility and film-forming abilities, in addition to near ideal optical and electronic properties. Solar cells based upon **63**:$PC_{71}BM$ (1:0.8) active layers gave an impressive PCE value of 7.2%, which was further improved to 8.0% using the PDMS processing additive. With fabrication of tandem solar cells using **63**:$PC_{71}BM$ blends as both photoactive layers, a double digit PCE of 10.1% was recorded, making this the first report of a solution processed organic cell breaching the 10% barrier using all small molecule active layers.[31] These findings have brought great attention to the area,

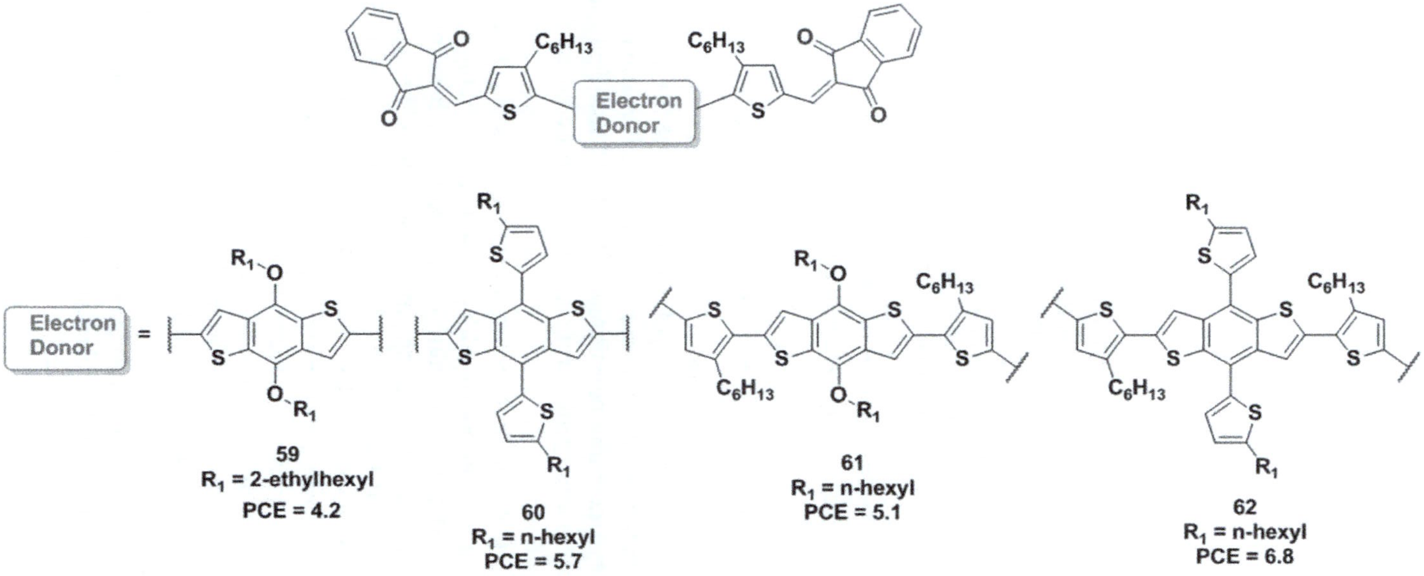

Figure 5.31 Structures of oligothiophene based small molecule with indan(1,3)dione end-capping units and various donors, used as donors in SM-BHJ solar cells.

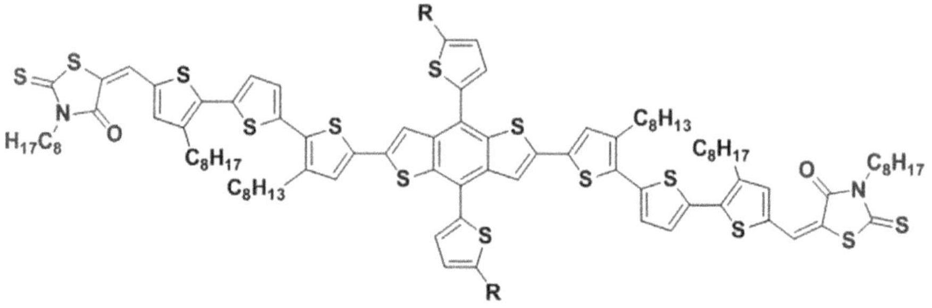

63
R = 2-ethylhexyl
PCE = 8.0% (Sinlge Junction)
PCE = 10.1% (Tandom Cell)

Figure 5.32 Structure of an oligothiophene-based small molecule with 3-ethyl-rhodanine end-capping units and a 2-D BDT core, used as a donor in both single and tandem junction SM-BHJ solar cells.

and have helped establish solution processed SM-BHJ solar cells as one of the most promising clean energy technologies.

5.11 Comments on Device Optimization

Key to the commercial potential of soluble small molecules is that they can be solution processed similarly to polymers. Like polymers, the morphology of the small molecule BHJ, and consequently the performance, is highly influenced by the processing conditions.[23,76,108,141–143] An ideal morphology would consist of an interpenetrating network of ordered small molecule and fullerene domains to allow for efficient charge separation and transport to the electrodes. The morphology of the network is determined by an exceptionally complex and poorly understood interplay of molecular design and processing.[144] Methods to obtain optimized active layer morphologies and improve the performance of polymer based OPVs through processing include: using different solvents,[76] solvent annealing,[77] and using solvent additives.[123] Meanwhile, thermal annealing and similar techniques have also now been applied to form high performance small molecule BHJs.[75,122] However, a key contrast to polymers is that soluble small molecules typically have a lower viscosity, a higher tendency to crystallize, and typically optimize at higher donor and additive concentrations.[42,75,123]

As noted, solar cells fabricated with $7:PC_{71}BM$ active layers yielded devices with a long-standing record in PCE. Top-performing devices had an active layer that was a $3:27:PC_{71}BM$ ratio cast from chloroform and were annealed at 110 °C. Chloroform is a relatively fast drying solvent and annealing was used to control domain size to lead from a PCE of 0.3% for as-cast devices to 4.4% for devices annealed at 110 °C.[75] Similar PCEs could also be achieved for annealed $7:PC_{71}BM$ devices cast from other relatively low boiling point

(<100 °C) solvents, based on the analysis of Hansen solubility parameters,[76] and through controlled solvent vapor annealing.[77,78] In all cases, as-cast devices exhibited poor PCE and thermal or solvent–vapor annealing improved the donor–acceptor phase separation and induced crystallization of the active materials, increasing light absorption, charge separation, and charge transport.

Devices made using compounds 37 and 41 have achieved some of the highest known PCEs to date for soluble small molecule based solar cells.[30,123,127] These devices also achieved the highest PCEs with a donor–acceptor ratio greater than 1. For this class of materials, it was found that a high boiling point (BP) solvent [*e.g.* diiodooctane (DIO), BP ~332 °C at 760 mmHg] could be used as a processing additive to improve the morphology and yield PCEs. Interestingly, the additive concentrations were notably lower (<1% wt/v) than typically used for polymers,[143] and the PCE was highly sensitive to concentration.[23,123] The morphology as a function of temperature and additive concentration were annualized in-depth by UV-vis, transmission electron microscopy (TEM), and grazing incidence wide angle X-ray scattering (GIWAXS), along with single crystal X-ray diffraction. It was postulated that the addition of DIO as an additive or by thermal annealing leads to a morphology including "grains" or "wires" of highly crystalline 41, but too much DIO or too high a temperature leads to overly large domains of 41, resulting in reduced PCEs.[25] It was stated that these excessively large domains lead to a reduced maximum achievable short circuit current, as demonstrated by reduced photo generated currents at negative biases and by analysis of transient photocurrent response. *In-situ* analysis of the morphological evolution *via* GIWAXS during and shortly after spincoating with a high boiling point solvent additive was also demonstrated.[145] In a related fashion, high PCE devices based upon A–D–A oligiothiophenes such as compounds 43 through 63 also typically have a donor : acceptor ratio greater than 1. They are also typically cast from chloroform, have no post-deposition treatment such as annealing, and additives such as DIO or CN were found generally to harm device performance. However, the highest efficiency A–D–A oligiothiophenes were cast with a small amount (~0.2 mg mL^{-1}) of PDMS added.[28]

It is also important to note the effect of device architecture. In the case of compounds 36 and 37, the use of the standard hole transport material (HTL) PEDOT:PSS led to devices with low PCE. Use of metal oxide HTLs (nickel oxide or molybdenum oxide) significantly improved device performance by greater than 100%, and allowed for record efficiencies to be achieved.[123,125] For compound 41, employing inverted device architectures,[128] zinc oxide optical spacers,[146] polyelectrolyte interlayers,[147] and barium cathodes,[30] all led to significant improvements in device PCE, with a record 9% being achieved.[30]

To this end, while innovative molecular design has led to the realization of small molecule donors with ideal optical and electronic properties, and the ability to form uniform thin-films from solution, these molecules must not be treated like polymers when considering the fabrication of BHJ solar cells. It almost all reports, D : A ratios greater than 1 and the use of processing additives in very small amounts (<1% wt/v) were required to form favorable thin-film morphologies and record device performance. In contrast, the highest performing polymer solar cells utilize a higher concentration of fullerene compared to polymer and are typically processed with 1–5% wt/v processing additives.

5.12   Conclusions and Future Outlook

Over the past four years, creative molecular design of small molecule donors has led to a near doubling of the power conversion efficiency of solution processed small molecule solar cells, with improvements from ~4 to ~9% (Figure 5.33) for single-junction solar cells. Record PCEs of 10% have been achieved in tandem cell devices, and are directly comparable to devices utilizing conjugated polymers. From the work of Bazan and Chen, it can be recognized that, in addition to the initial design principles, small molecule donors should be constructed with at least seven π-conjugated units in the molecular backbone and be decorated with sufficient numbers of alkyl side chains to ensure good solubility and uniform film formation from solution. In each case of high performance donors, it should be noted that not one molecule, but rather a large series of compounds were synthesized and evaluated on their structure–property–function relationships; therefore, the importance of a highly tunable synthesis cannot be overvalued. Since the report of PCEs in excess of 6% in 2012, the number of publications focused on the development of small molecule donors has increased exponentially, with now numerous reports of devices giving PCEs in the range of 5–6%. It is fully expected that the performance of solution processed SM-BHJ solar cells will continue to increase through creative molecular design and innovative device fabrication.

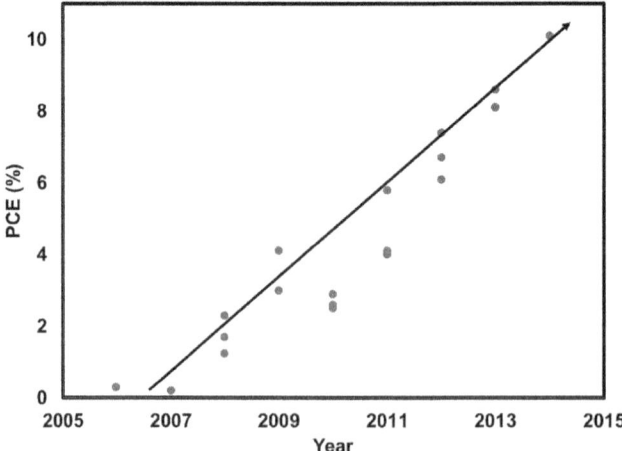

Figure 5.33   Highest power conversion efficiency values for solution processed SM-BHJ solar cells reported by year. Note the consistent increase since 2007.

References

1. S. B. Darling and F. You, *RSC Adv.*, 2013, **3**, 17633.
2. G. Li, R. Zhu and Y. Yang, *Nat. Photonics*, 2012, **6**, 153.
3. R. Søndergaard, M. Hösel, D. Angmo, T. T. Larsen-Olsen and F. C. Krebs, *Mater. Today*, 2012, **15**, 36.
4. Y.-W. Su, S.-C. Lan and K.-H. Wei, *Mater. Today*, 2012, **15**, 554.
5. B. Kippelen and J.-L. Brédas, *Energy Environ. Sci.*, 2009, **2**, 251.
6. Y.-H. Chen, L.-Y. Lin, C.-W. Lu, F. Lin, Z.-Y. Huang, H.-W. Lin, P.-H. Wang, Y.-H. Liu, K.-T. Wong, J. Wen, D. J. Miller and S. B. Darling, *J. Am. Chem. Soc.*, 2012, **134**, 13616.
7. P. Sullivan, S. Schumann, R. Da Campo, T. Howells, A. Duraud, M. Shipman, R. A. Hatton and T. S. Jones, *Adv. Energy Mater.*, 2013, **3**, 239.
8. M. Riede, T. Mueller, W. Tress, R. Schueppel and K. Leo, *Nanotechnology*, 2008, **19**, 424001.
9. H. Zhou, L. Yang and W. You, *Macromolecules*, 2012, **45**, 607.
10. G. Dennler, M. C. Scharber and C. J. Brabec, *Adv. Mater.*, 2009, **21**, 1323.
11. S. Günes, H. Neugebauer and N. S. Sariciftci, *Chem. Rev.*, 2007, **107**, 1324.
12. Z. He, C. Zhong, S. Su, M. Xu, H. Wu and Y. Cao, *Nat. Photonics*, 2012, **6**, 591.
13. J. You, L. Dou, K. Yoshimura, T. Kato, K. Ohya, T. Moriarty, K. Emery, C.-C. Chen, J. Gao, G. Li and Y. Yang, *Nat. Commun.*, 2013, **4**, 1446.
14. Heliatek, http://www.heliatek.com, (accessed Apr 23, 2014).
15. Z. B. Henson, K. Müllen and G. C. Bazan, *Nat. Chem.*, 2012, **4**, 699.
16. R. C. Coffin, J. Peet, J. Rogers and G. C. Bazan, *Nat. Chem.*, 2009, **1**, 657.
17. C. Duan, F. Huang and Y. Cao, *J. Mater. Chem.*, 2012, **22**, 10416.
18. J. K. Park, J. Jo, J. H. Seo, J. S. Moon, Y. D. Park, K. Lee, A. J. Heeger and G. C. Bazan, *Adv. Mater.*, 2011, **23**, 2430.
19. J. L. Delgado, P.-A. Bouit, S. Filippone, M. Herranz and N. Martín, *Chem. Commun.*, 2010, **46**, 4853.
20. J. Roncali, *Acc. Chem. Res.*, 2009, **42**, 1719.
21. G. C. Welch, R. C. Bakus, S. J. Teat and G. C. Bazan, *J. Am. Chem. Soc.*, 2013, **135**, 2298.
22. Z. B. Henson, G. C. Welch, T. van der Poll and G. C. Bazan, *J. Am. Chem. Soc.*, 2012, **134**, 3766.
23. C. J. Takacs, Y. Sun, G. C. Welch, L. A. Perez, X. Liu, W. Wen, G. C. Bazan and A. J. Heeger, *J. Am. Chem. Soc.*, 2012, **134**, 16597.
24. A. Zhugayevych, O. Postupna, R. C. Bakus II, G. C. Welch, G. C. Bazan and S. Tretiak, *J. Phys. Chem. C*, 2013, **117**, 4920.
25. J. A. Love, C. M. Proctor, J. Liu, C. J. Takacs, A. Sharenko, T. S. van der Poll, A. J. Heeger, G. C. Bazan and T.-Q. Nguyen, *Adv. Funct. Mater.*, 2013, **23**, 5019.
26. J. Liu, B. Walker, A. Tamayo, Y. Zhang and T.-Q. Nguyen, *Adv. Funct. Mater.*, 2013, **23**, 47.
27. C. Kim, J. Liu, J. Lin, A. B. Tamayo, B. Walker, G. Wu and T.-Q. Nguyen, *Chem. Mater.*, 2012, **24**, 1699.

28. Y. Chen, X. Wan and G. Long, *Acc. Chem. Res.*, 2013, **46**, 2645.
29. J. E. Coughlin, Z. B. Henson, G. C. Welch and G. C. Bazan, *Acc. Chem. Res.*, 2014, **47**, 257.
30. V. Gupta, A. K. K. Kyaw, D. H. Wang, S. Chand, G. C. Bazan and A. J. Heeger, *Sci. Rep.*, 2013, **3**, 1965.
31. Y. Liu, C.-C. Chen, Z. Hong, J. Gao, Y. (Michael) Yang, H. Zhou, L. Dou, G. Li and Y. Yang, *Sci. Rep.*, 2013, **3**, 3356.
32. F. Würthner and K. Meerholz, *Chem.–Eur. J.*, 2010, **16**, 9366.
33. B. Walker, C. Kim and T.-Q. Nguyen, *Chem. Mater.*, 2011, **23**, 470.
34. A. Mishra and P. Bäuerle, *Angew. Chem., Int. Ed.*, 2012, **51**, 2020.
35. Y. Lin, Y. Li and X. Zhan, *Chem. Soc. Rev.*, 2012, **41**, 4245.
36. J. Roncali, P. Leriche and P. Blanchard, *Adv. Mater.*, 2014, **26**, 3821.
37. Y.-J. Cheng, S.-H. Yang and C.-S. Hsu, *Chem. Rev.*, 2009, **109**, 5868.
38. C. J. Brabec, N. S. Sariciftci and J. C. Hummelen, *Adv. Funct. Mater.*, 2001, **11**, 15.
39. C. L. Chochos, N. Tagmatarchis and V. G. Gregoriou, *RSC Adv.*, 2013, **3**, 7160.
40. A. Montellano López, A. Mateo-Alonso and M. Prato, *J. Mater. Chem.*, 2011, **21**, 1305.
41. J. E. Anthony, A. Facchetti, M. Heeney, S. R. Marder and X. Zhan, *Adv. Mater.*, 2010, **22**, 3876.
42. G. C. Welch, L. A. Perez, C. V. Hoven, Y. Zhang, X.-D. Dang, A. Sharenko, M. F. Toney, E. J. Kramer, T.-Q. Nguyen and G. C. Bazan, *J. Mater. Chem.*, 2011, **21**, 12700.
43. J. D. Kotlarski and P. W. M. Blom, *Appl. Phys. Lett.*, 2011, **98**, 053301.
44. G. Dennler, M. C. Scharber, T. Ameri, P. Denk, K. Forberich, C. Waldauf and C. J. Brabec, *Adv. Mater.*, 2008, **20**, 579.
45. N. M. Kronenberg, V. Steinmann, H. Bürckstümmer, J. Hwang, D. Hertel, F. Würthner and K. Meerholz, *Adv. Mater.*, 2010, **22**, 4193.
46. S. Roquet, R. de Bettignies, P. Leriche, A. Cravino and J. Roncali, *J. Mater. Chem.*, 2006, **16**, 3040.
47. M. T. Lloyd, A. C. Mayer, S. Subramanian, D. A. Mourey, D. J. Herman, A. V. Bapat, J. E. Anthony and G. G. Malliaras, *J. Am. Chem. Soc.*, 2007, **129**, 9144.
48. T. Rousseau, A. Cravino, T. Bura, G. Ulrich, R. Ziessel and J. Roncali, *Chem. Commun.*, 2009, 1673.
49. T. Rousseau, A. Cravino, E. Ripaud, P. Leriche, S. Rihn, A. De Nicola, R. Ziessel and J. Roncali, *Chem. Commun.*, 2010, **46**, 5082.
50. T. Bura, N. Leclerc, S. Fall, P. Lévêque, T. Heiser, P. Retailleau, S. Rihn, A. Mirloup and R. Ziessel, *J. Am. Chem. Soc.*, 2012, **134**, 17404.
51. A. Bessette and G. S. Hanan, *Chem. Soc. Rev.*, 2014, **43**, 3342–3405.
52. F. Silvestri, M. D. Irwin, L. Beverina, A. Facchetti, G. A. Pagani and T. J. Marks, *J. Am. Chem. Soc.*, 2008, **130**, 17640.
53. U. Mayerhöffer, K. Deing, K. Gruß, H. Braunschweig, K. Meerholz and F. Würthner, *Angew. Chem., Int. Ed.*, 2009, **48**, 8776.
54. D. Bagnis, L. Beverina, H. Huang, F. Silvestri, Y. Yao, H. Yan, G. A. Pagani, T. J. Marks and A. Facchetti, *J. Am. Chem. Soc.*, 2010, **132**, 4074.

55. G. Wei, R. R. Lunt, K. Sun, S. Wang, M. E. Thompson and S. R. Forrest, *Nano Lett.*, 2010, **10**, 3555.
56. G. Wei, S. Wang, K. Renshaw, M. E. Thompson and S. R. Forrest, *ACS Nano*, 2010, **4**, 1927.
57. G. Wei, S. Wang, K. Sun, M. E. Thompson and S. R. Forrest, *Adv. Energy Mater.*, 2011, **1**, 184.
58. G. Wei, X. Xiao, S. Wang, K. Sun, K. J. Bergemann, M. E. Thompson and S. R. Forrest, *ACS Nano*, 2012, **6**, 972.
59. N. M. Kronenberg, M. Deppisch, F. Würthner, H. W. A. Lademann, K. Deing and K. Meerholz, *Chem. Commun.*, 2008, 6489.
60. H. Bürckstümmer, N. M. Kronenberg, K. Meerholz and F. Würthner, *Org. Lett.*, 2010, **12**, 3666.
61. H. Bürckstümmer, N. M. Kronenberg, M. Gsänger, M. Stolte, K. Meerholz and F. Würthner, *J. Mater. Chem.*, 2010, **20**, 240.
62. D. G. Farnum, G. Mehta, G. G. I. Moore and F. P. Siegal, *Tetrahedron Lett.*, 1974, **15**, 2549.
63. D. Chandran and K.-S. Lee, *Macromol. Res.*, 2013, **21**, 272.
64. C. B. Nielsen, M. Turbiez and I. McCulloch, *Adv. Mater.*, 2013, **25**, 1859.
65. S. Qu and H. Tian, *Chem. Commun.*, 2012, **48**, 3039.
66. B. Tieke, A. R. Rabindranath, K. Zhang and Y. Zhu, *Beilstein J. Org. Chem.*, 2010, **6**, 830.
67. P. Sonar, J.-M. Zhuo, L.-H. Zhao, K.-M. Lim, J. Chen, A. J. Rondinone, S. P. Singh, L.-L. Chua, P. K. H. Ho and A. Dodabalapur, *J. Mater. Chem.*, 2012, **22**, 17284.
68. C. H. Woo, P. M. Beaujuge, T. W. Holcombe, O. P. Lee and J. M. J. Fréchet, *J. Am. Chem. Soc.*, 2010, **132**, 15547.
69. A. J. Kronemeijer, E. Gili, M. Shahid, J. Rivnay, A. Salleo, M. Heeney and H. Sirringhaus, *Adv. Mater.*, 2012, **24**, 1558.
70. K. A. Mazzio, M. Yuan, K. Okamoto and C. K. Luscombe, *ACS Appl. Mater. Interfaces*, 2011, **3**, 271.
71. A. B. Tamayo, B. Walker and T.-Q. Nguyen, *J. Phys. Chem. C*, 2008, **112**, 11545.
72. A. B. Tamayo, M. Tantiwiwat, B. Walker and T.-Q. Nguyen, *J. Phys. Chem. C*, 2008, **112**, 15543.
73. M. Tantiwiwat, A. Tamayo, N. Luu, X.-D. Dang and T.-Q. Nguyen, *J. Phys. Chem. C*, 2008, **112**, 17402.
74. A. B. Tamayo, X.-D. Dang, B. Walker, J. Seo, T. Kent and T.-Q. Nguyen, *Appl. Phys. Lett.*, 2009, **94**, 103301.
75. B. Walker, A. B. Tamayo, X.-D. Dang, P. Zalar, J. H. Seo, A. Garcia, M. Tantiwiwat and T.-Q. Nguyen, *Adv. Funct. Mater.*, 2009, **19**, 3063.
76. B. Walker, A. Tamayo, D. T. Duong, X.-D. Dang, C. Kim, J. Granstrom and T.-Q. Nguyen, *Adv. Energy Mater.*, 2011, **1**, 221.
77. A. Viterisi, F. Gispert-Guirado, J. W. Ryan and E. Palomares, *J. Mater. Chem.*, 2012, **22**, 15175.
78. K. Sun, Z. Xiao, E. Hanssen, M. F. G. Klein, H. H. Dam, M. Pfaff, D. Gerthsen, W. W. H. Wong and D. J. Jones, *J. Mater. Chem. A*, 2014, **2**, 9048.

79. M. Chen, W. Fu, M. Shi, X. Hu, J. Pan, J. Ling, H. Li and H. Chen, *J. Mater. Chem. A*, 2012, **1**, 105.

80. O. P. Lee, A. T. Yiu, P. M. Beaujuge, C. H. Woo, T. W. Holcombe, J. E. Millstone, J. D. Douglas, M. S. Chen and J. M. J. Fréchet, *Adv. Mater.*, 2011, **23**, 5359.

81. H. Wang, F. Liu, L. Bu, J. Gao, C. Wang, W. Wei and T. P. Russell, *Adv. Mater.*, 2013, **25**, 6519–6525.

82. G. D. Sharma, M. A. Reddy, K. Ganesh, S. P. Singh and M. Chandrasekharam, *RSC Adv.*, 2013, **4**, 732.

83. T. Bura, N. Leclerc, R. Bechara, P. Lévêque, T. Heiser and R. Ziessel, *Adv. Energy Mater.*, 2013, **3**, 1118.

84. S. Loser, C. J. Bruns, H. Miyauchi, R. P. Ortiz, A. Facchetti, S. I. Stupp and T. J. Marks, *J. Am. Chem. Soc.*, 2011, **133**, 8142.

85. B. Walker, J. Liu, C. Kim, G. C. Welch, J. K. Park, J. Lin, P. Zalar, C. M. Proctor, J. H. Seo, G. C. Bazan and T.-Q. Nguyen, *Energy Environ. Sci.*, 2013, **6**, 952.

86. Y. S. Choi and W. H. Jo, *Org. Electron.*, 2013, **14**, 1621.

87. J. Huang, C. Zhan, X. Zhang, Y. Zhao, Z. Lu, H. Jia, B. Jiang, J. Ye, S. Zhang, A. Tang, Y. Liu, Q. Pei and J. Yao, *ACS Appl. Mater. Interfaces*, 2013, **5**, 2033.

88. Y. Lin, L. Ma, Y. Li, Y. Liu, D. Zhu and X. Zhan, *Adv. Energy Mater.*, 2013, **3**, 1166.

89. T. Harschneck, N. Zhou, E. F. Manley, S. J. Lou, X. Yu, M. R. Butler, A. Timalsina, R. Turrisi, M. Ratner, L. X. Chen, R. Chang, A. Facchetti and T. Marks, *Chem. Commun.*, 2014, **50**, 4099–4101.

90. S. Loser, H. Miyauchi, J. W. Hennek, J. Smith, C. Huang, A. Facchetti and T. J. Marks, *Chem. Commun.*, 2012, **48**, 8511.

91. J. Liu, Y. Sun, P. Moonsin, M. Kuik, C. M. Proctor, J. Lin, B. B. Hsu, V. Promarak, A. J. Heeger and T.-Q. Nguyen, *Adv. Mater.*, 2013, **25**, 5898–5903.

92. R. Stalder, J. Mei, K. R. Graham, L. A. Estrada and J. R. Reynolds, *Chem. Mater.*, 2014, **26**, 664.

93. P. Deng and Q. Zhang, *Polym. Chem.*, 2014, **5**, 3298.

94. T. Lei, J.-Y. Wang and J. Pei, *Acc. Chem. Res.*, 2014, **47**, 1117–1126.

95. L. A. Estrada, R. Stalder, K. A. Abboud, C. Risko, J.-L. Brédas and J. R. Reynolds, *Macromolecules*, 2013, **46**, 8832.

96. R. Stalder, J. Mei, J. Subbiah, C. Grand, L. A. Estrada, F. So and J. R. Reynolds, *Macromolecules*, 2011, **44**, 6303.

97. R. Stalder, J. Mei and J. R. Reynolds, *Macromolecules*, 2010, **43**, 8348.

98. W. Elsawy, C.-L. Lee, S. Cho, S.-H. Oh, S.-H. Moon, A. Elbarbary and J.-S. Lee, *Phys. Chem. Chem. Phys.*, 2013, **15**, 15193.

99. T. Wang, Y. Chen, X. Bao, Z. Du, J. Guo, N. Wang, M. Sun and R. Yang, *Dyes Pigm.*, 2013, **98**, 11.

100. N. M. Randell, A. F. Douglas and T. L. Kelly, *J. Mater. Chem. A*, 2013, **2**, 1085.

101. A. Yassin, P. Leriche, M. Allain and J. Roncali, *New J. Chem.*, 2013, **37**, 502.

102. L. G. Mercier and M. Leclerc, *Acc. Chem. Res.*, 2013, **46**, 1597.

103. G. W. P. V. Pruissen, F. Gholamrezaie, M. M. Wienk and R. A. J. Janssen, *J. Mater. Chem.*, 2012, **22**, 20387.
104. T. Lei, J.-H. Dou, Z.-J. Ma, C.-J. Liu, J.-Y. Wang and J. Pei, *Chem. Sci.*, 2013, **4**, 2447.
105. T. Lei, J.-H. Dou, Z.-J. Ma, C.-H. Yao, C.-J. Liu, J.-Y. Wang and J. Pei, *J. Am. Chem. Soc.*, 2012, **134**, 20025.
106. W. Yue, T. He, M. Stolte, M. Gsänger and F. Würthner, *Chem. Commun.*, 2013, **50**, 545.
107. J. Mei, K. R. Graham, R. Stalder and J. R. Reynolds, *Org. Lett.*, 2010, **12**, 660.
108. K. R. Graham, P. M. Wieruszewski, R. Stalder, M. J. Hartel, J. Mei, F. So and J. R. Reynolds, *Adv. Funct. Mater.*, 2012, **22**, 4801.
109. W.-F. Su, C.-C. Ho, C.-A. Chen, C.-Y. Chang and S. Darling, *J. Mater. Chem. A*, 2014, **2**, 8026–8032.
110. C.-L. Wang, J.-Y. Hu, C.-H. Wu, H.-H. Kuo, Y.-C. Chang, Z.-J. Lan, H.-P. Wu, E. W.-G. Diau and C.-Y. Lin, *Energy Environ. Sci.*, 2014, **7**, 1392.
111. H. Qin, L. Li, F. Guo, S. Su, J. Peng, Y. Cao and X. Peng, *Energy Environ. Sci.*, 2014, **7**, 1397.
112. J. Wei, H. Li, M. P. Barrow and P. B. O'Connor, *J. Am. Soc. Mass Spectrom.*, 2013, **24**, 753.
113. Y. Huang, L. Li, X. Peng, J. Peng and Y. Cao, *J. Mater. Chem.*, 2012, **22**, 21841.
114. T. Yamamoto, J. Hatano, T. Nakagawa, S. Yamaguchi and Y. Matsuo, *Appl. Phys. Lett.*, 2013, **102**, 013305.
115. C.-H. Lee and S. Lindsey, *J. Tetrahedron*, 1994, **50**, 11427.
116. S. Ito, T. Murashima, N. Ono and H. Uno, *Chem. Commun.*, 1998, 1661.
117. A. G. Hyslop, M. A. Kellett, P. M. Iovine and M. J. Therien, *J. Am. Chem. Soc.*, 1998, **120**, 12676.
118. Y. Matsuo, Y. Sato, T. Niinomi, I. Soga, H. Tanaka and E. Nakamura, *J. Am. Chem. Soc.*, 2009, **131**, 16048.
119. M. Guide, X.-D. Dang and T.-Q. Nguyen, *Adv. Mater.*, 2011, **23**, 2313.
120. L. Li, Y. Huang, J. Peng, Y. Cao and X. Peng, *J. Mater. Chem. A*, 2013, **1**, 2144.
121. A. Ko Ko Kyaw, D. Hwan Wang, H.-R. Tseng, J. Zhang, G. C. Bazan and A. Heeger, *J. Appl. Phys. Lett.*, 2013, **102**, 163308.
122. W. L. Leong, G. C. Welch, J. Seifter, J. H. Seo, G. C. Bazan and A. J. Heeger, *Adv. Energy Mater.*, 2013, **3**, 356.
123. Y. Sun, G. C. Welch, W. L. Leong, C. J. Takacs, G. C. Bazan and A. J. Heeger, *Nat. Mater.*, 2012, **11**, 44.
124. X. Liu, Y. Sun, L. A. Perez, W. Wen, M. F. Toney, A. J. Heeger and G. C. Bazan, *J. Am. Chem. Soc.*, 2012, **134**, 20609.
125. A. Garcia, G. C. Welch, E. L. Ratcliff, D. S. Ginley, G. C. Bazan and D. C. Olson, *Adv. Mater.*, 2012, **24**, 5368.
126. E. L. Ratcliff, R. C. B. Ii, G. C. Welch, T. S. van der Poll, A. Garcia, S. R. Cowan, B. A. MacLeod, D. S. Ginley, G. C. Bazan and D. C. Olson, *J. Mater. Chem. C*, 2013, **1**, 6223–6234.

127. T. S. Van der Poll, J. A. Love, T.-Q. Nguyen and G. C. Bazan, *Adv. Mater.*, 2012, **24**, 3646.
128. A. K. K. Kyaw, D. H. Wang, V. Gupta, J. Zhang, S. Chand, G. C. Bazan and A. J. Heeger, *Adv. Mater.*, 2013, **25**, 2397.
129. Y. Liu, J. Zhou, X. Wan and Y. Chen, *Tetrahedron*, 2009, **65**, 5209.
130. Y. Liu, X. Wan, B. Yin, J. Zhou, G. Long, S. Yin and Y. Chen, *J. Mater. Chem.*, 2010, **20**, 2464.
131. B. Yin, L. Yang, Y. Liu, Y. Chen, Q. Qi, F. Zhang and S. Yin, *Appl. Phys. Lett.*, 2010, **97**, 023303.
132. Y. Liu, X. Wan, F. Wang, J. Zhou, G. Long, J. Tian, J. You, Y. Yang and Y. Chen, *Adv. Energy Mater.*, 2011, **1**, 771.
133. G. He, Z. Li, X. Wan, J. Zhou, G. Long, S. Zhang, M. Zhang and Y. Chen, *J. Mater. Chem. A*, 2013, **1**, 1801.
134. G. He, Z. Li, X. Wan, Y. Liu, J. Zhou, G. Long, M. Zhang and Y. Chen, *J. Mater. Chem.*, 2012, **22**, 9173.
135. Z. Li, G. He, X. Wan, Y. Liu, J. Zhou, G. Long, Y. Zuo, M. Zhang and Y. Chen, *Adv. Energy Mater.*, 2012, **2**, 74.
136. Y. Liu, X. Wan, F. Wang, J. Zhou, G. Long, J. Tian and Y. Chen, *Adv. Mater.*, 2011, **23**, 5387.
137. J. Zhou, X. Wan, Y. Liu, Y. Zuo, Z. Li, G. He, G. Long, W. Ni, C. Li, X. Su and Y. Chen, *J. Am. Chem. Soc.*, 2012, **134**, 16345.
138. J. Zhou, X. Wan, Y. Liu, G. Long, F. Wang, Z. Li, Y. Zuo, C. Li and Y. Chen, *Chem. Mater.*, 2011, **23**, 4666.
139. J. Zhou, Y. Zuo, X. Wan, G. Long, Q. Zhang, W. Ni, Y. Liu, Z. Li, G. He, C. Li, B. Kan, M. Li and Y. Chen, *J. Am. Chem. Soc.*, 2013, **135**, 8484.
140. S. Shen, P. Jiang, C. He, J. Zhang, P. Shen, Y. Zhang, Y. Yi, Z. Zhang, Z. Li and Y. Li, *Chem. Mater.*, 2013, **25**, 2274.
141. J. J. Jasieniak, B. B. Y. Hsu, C. J. Takacs, G. C. Welch, G. C. Bazan, D. Moses and A. J. Heeger, *ACS Nano*, 2012, **6**, 8735.
142. L. G. Kaake, G. C. Welch, D. Moses, G. C. Bazan and A. J. Heeger, *J. Phys. Chem. Lett.*, 2012, **3**, 1253.
143. M. T. Dang and J. D. Wuest, *Chem. Soc. Rev.*, 2013, **42**, 9105–9126.
144. J. Peet, A. J. Heeger and G. C. Bazan, *Acc. Chem. Res.*, 2009, **42**, 1700.
145. L. A. Perez, K. W. Chou, J. A. Love, T. S. van der Poll, D.-M. Smilgies, T.-Q. Nguyen, E. J. Kramer, A. Amassian and G. C. Bazan, *Adv. Mater.*, 2013, **25**, 6380.
146. A. K. K. Kyaw, D. H. Wang, D. Wynands, J. Zhang, T.-Q. Nguyen, G. C. Bazan and A. J. Heeger, *Nano Lett.*, 2013, **13**, 3796.
147. H. Zhou, Y. Zhang, C.-K. Mai, S. D. Collins, T.-Q. Nguyen, G. C. Bazan and A. J. Heeger, *Adv. Mater.*, 2014, **26**, 780.

Interface Engineering of Polymer Solar Cells

KAI ZHANG[a], CHUNHUI DUAN[a], FEI HUANG*[a], AND YONG CAO[a]

[a]Institute of Polymer Optoelectronic Materials and Devices, State Key Laboratory of Luminescent Materials and Devices, South China University of Technology, Guangzhou 510640, P. R. China
*E-mail: msfhuang@scut.edu.cn

6.1 Introduction

Owing to the diminishing of fossil energy sources, the development of renewable energy technologies has attracted great attention all around the world. Photovoltaic technologies, which can directly convert solar radiation into electricity, are promising and practical approaches to utilizing solar energy. Among all kinds of photovoltaic technology, polymer solar cells (PSCs), which are based on the solution-processed physical blend of a p-type semiconducting conjugated polymer (serving as an electron donor) and an n-type fullerene derivative (serving as an electron acceptor) with a bulk-heterojunction (BHJ) structure, have received growing attention due to their immense advantages such as low cost, light weight and flexibility.[1,2] Over the past few years, appreciable achievements have been realized on improving the performance of PSCs with power conversion efficiencies (PCEs) over 9% for single-junction devices[3] and PCEs over 10% for tandem solar cells.[4] Lifetimes of more than 6 years (estimated from accelerated testing) have been

RSC Polymer Chemistry Series No. 17
Polymer Photovoltaics: Materials, Physics, and Device Engineering
Edited by Fei Huang, Hin-Lap Yip, and Yong Cao
© The Royal Society of Chemistry 2016
Published by the Royal Society of Chemistry, www.rsc.org

demonstrated as well.[5] Meanwhile, a roll-to-roll large scale manufacturing process to fabricate PSC modules has also been demonstrated.[6] Nevertheless, further progress in PCEs and their lifetimes is still urgent and challenging to push forward the practical application of PSCs. Typically, a PSC consists of a transparent electrode, a polymer:fullerene active layer and a reflective electrode to generate a so-called sandwiched configuration. The development of efficient active materials is extremely important and has also gained most attention in the research fields. It is also well recognized that the nature of the contact between the active layer and electrodes is an essential factor determining the performance of PSCs.[7,8] Interface engineering, through the incorporation of new interfacial materials with desired charge selectivity and compatibility with solution-processing of multi-layer devices, is extremely critical to further improving the performance of PSCs. In this chapter, we first discuss the functions and design criteria of the interfacial layer for PSCs. Subsequently, interface engineering of PSCs with both conventional and inverted configurations by utilization of several representative classes of interfacial materials is highlighted.

6.2   Functions and Design Criteria of the Interfacial Layer

The PCE of PSCs is determined by three parameters, *i.e.* open-circuit voltage (V_{OC}), short-circuit current density (J_{SC}) and fill factor (FF). The nature of interface contact between the active layer and the electrodes has a significant influence on all three aforementioned parameters. The efficiencies of charge transport and charge extraction at the electrodes after exciton generation and dissociation at the BHJ layer play a critical role in the value of J_{SC}. The V_{OC} is proportional to the built-in potential across the device, which is often defined as the work function (WF) difference between the hole-collecting electrode (HCE) and the electrode-collecting electrode (ECE).[9,10] This means that barrierless contacts (Ohmic contacts) between active layer and electrodes are desired to obtain as large a V_{OC} as possible. The value of FF[11–13] is directly related to the series resistance (R_s) and shunt resistance (R_{sh}) of the PSC device. An R_s as small as possible and an R_{sh} as large as possible are desired to gain large FF. The R_s is a sum of the bulk resistance of the various layers (including electrodes, BHJ layer and interfacial layers) in the device and the contact resistance between them, while R_{sh} is governed by the quality of the thin films and the nature of the electrical contact at their interfaces. In short, the optimization of the interfaces to form Ohmic contacts with an energy barrier height as low as possible is critical to realizing state-of-the-art PSCs. Interface engineering *via* the insertion of appropriate interfacial layers between the BHJ active layer and the electrodes was regarded as a feasible and effective approach. Great achievement has been made in recent years in interfacial engineering of PSCs, which has led to champion PCE values benefiting from the development of interfacial materials.[3,14]

6.2.1 Functions of Interfacial Materials

Generally, the functions of an interfacial layer include, but are not limited to:

6.2.1.1 Enhancing Charge Extraction at Electrodes and Tuning Energy Alignment

In PSCs, a barrier height of several tens of mV may cause severe charge accumulation. Ohmic contacts can be formed at both electrodes by inserting an appropriate interfacial layer between the active layer and electrodes to shift the WF of the electrodes to ideal values, from which the V_{OC} can reach the upper limit, because the magnitude of V_{OC} is limited by the difference between the effective WF of the HCE and ECE in PSCs. Under the condition of Ohmic contact, the charge extraction at both electrodes will be accelerated and the recombination loss will be reduced; benefiting from both effects, enhanced J_{SC} is achievable.

6.2.1.2 Controlling Electrode Polarity and Charge Extraction Selectivity

As the WF of both top and bottom electrodes can be altered by interfacial layers, the polarity of electrodes and charge extraction selectivity are able to be tuned by incorporating appropriate interfacial layers. Reversing the polarity of electrodes in a device with conventional configuration can lead to the fabrication of a device with inverted configuration (see Figure 6.1). Using these two basic configurations, tandem and multi-junction PSCs can be constructed. In conventional devices, a transparent interfacial layer with hole-transport ability is often deposited onto indium tin oxide (ITO) to ensure efficient hole extraction, and low WF metals are usually applied to improve electron extraction. In addition, the ideal interfacial layers should also improve the charge selectivity at electrode–active layer interface,

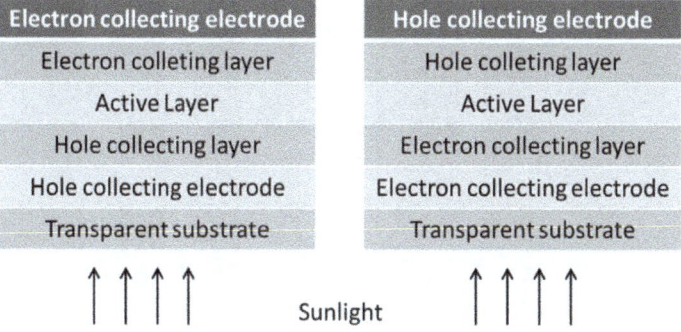

Figure 6.1 Schematics of PSCs with conventional configuration (left) and inverted configuration (right).

i.e., a hole-collecting layer (HCL) ensures efficient hole extraction and electron-blocking simultaneously at HCE–BHJ interface, while an electron-collecting layer (ECL) guarantees efficient electron extraction and hole-blocking concomitantly at the ECE–ECL interface. Highly efficient charge selectivity can reduce interfacial charge recombination and improve rectification of the *J–V* response and FF of the device.

6.2.1.3   *Improving Stability*

Ideal interfacial layers should work as protective layers to impede chemical reactions between electrodes and the active layer, and the diffusion of metal ions into organic layers, in order to improve the stability of the devices. For example, PEDOT:PSS (poly(3,4-ethylenedioxythiophene):poly(styrene sulfonate), see Scheme 6.1), a most commonly used HCL material, can etch ITO and cause the diffusion of In ions into PEDOT:PSS and the active layer and further degrade both materials owing to the acidic nature of PEDOT:PSS.[15,16] With regard to ECL, low WF metals such as Ca and Al, the most commonly used interfacial materials, are very susceptible to moisture and oxygen. Therefore, PEDOT:PSS and low WF metals are unlikely to be the ultimate interfacial layer for PSC technology owing to their stability concerns, despite the fact that they are currently extensively used. The development of interfacial materials that are intrinsically stable and can prevent the corrosion of moisture and oxygen is highly desirable for future PSC technology.

6.2.1.4   *Inducing the Formation of Optimized Active Layer Morphology*

The microstructure of the active layer has a significant influence on almost all procedures of PSC operation and thereby the overall performance of PSCs. Many approaches have been proposed to obtain an ideal nano-sized

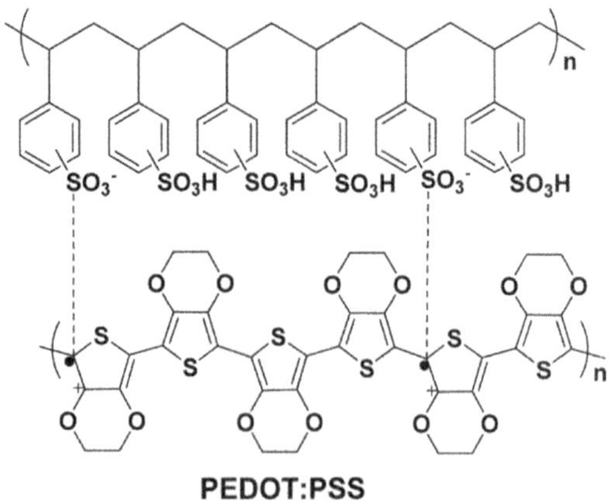

PEDOT:PSS

Scheme 6.1   Chemical structure of PEDOT:PSS.

bicontinuous penetration network structure of the BHJ active layer. It was observed that the surface energy and surface chemistry of the bottom interfacial layer exert a substantial influence on both the lateral and vertical phase separation of the active layer.[17-22] They have the potential to control the orientation of the polymer backbone (face-on or edge-on),[23,24] which greatly affects charge transport across the active layer. Further, based on the concentration gradients of the donor-phase and acceptor-phase in the vertical direction, the utilization of appropriate device configuration (conventional or inverted) is necessary to maximize the efficiency of PSCs.

6.2.1.5 Controlling Light Harvesting

Owing to the relative low charge carrier mobility in organic semiconducting materials, the thickness of the light-absorbing layer was usually limited to around 100 nm to mitigate the charge recombination loss. Generally, an active layer with greater thickness is favorable for absorbing more incident light and thereby obtaining enhanced J_{SC} but will cause more severe charge recombination loss and lead to reduced FF. As a result, achieving completely harvesting of all incident light by an active layer with a relatively small thickness is highly attractive for solving such a dilemma. Up to now, several light-trapping approaches, such as folded device configurations,[25] diffraction gratings,[26,27] aperiodic dielectric stacks,[28] texturing transparent electrodes,[29] optical spacers[30,31] and plasmonic nanostructures,[32-34] have been developed to improve the performance of thin PSC devices. Among them, the latter two light-trapping strategies rely greatly on interfacial engineering. Interfacial layers can serve as optical spacers between the electrodes and the active layer to modulate effectively the distribution of the optical field of the incident light in the active layer so as to maximize the light-harvesting by PSC devices. The inclusion of metallic nanostructures in the interfacial layers of PSCs could effectively improve light harvesting in the active layer owing to the optical enhancement by the surface plasmonic resonance effect of metallic nanostructures.[35-37] Light-trapping by plasmonic resonance could also be applied in tandem solar cells by doping metal nanoparticles into the interconnecting layer to improve light absorption in both front and rear subcells.[37]

6.2.2 Design Criteria for Interfacial Materials

Based on the discussions above, efficient interfacial materials for PSCs should: (1) facilitate Ohmic contact formation between active layer and electrodes; (2) possess appropriate energy levels to ensure efficient charge selectivity for electrodes; (3) have sufficient conductivity to minimize resistive losses; (4) have high transparency across the visible to near-infrared (vis-NIR) light region to reduce absorption losses; (5) possess large bandgap to confine excitons in the active layer; (6) have sufficient stability to retard undesirable chemical and physical reactions between active layer and electrode; (7) have good solution processability at low temperature; (8) be robust to benign multilayer device fabrication; (9) have good film-formation ability. It is worth pointing out that a single interfacial material generally cannot simultaneously have all these functions.

6.3   Interfacial Materials for Conventional Polymer Solar Cells

6.3.1   Anode Contact

For PSCs with conventional architecture, the goal of interface modification for anode contact is to enhance hole extraction selectively. Table 6.1 summarizes the performance of PSCs that apply different HCLs. PEDOT:PSS is the most widely used HCL for anode contact in PSCs with conventional structure. Structurally, PEDOT:PSS is a polyelectrolyte complex of PEDOT$^+$:PSSH$^-$ in which the PEDOT is p-doped. PEDOT:PSS possesses several advantages as a HCL material. First, it has relatively high electrical conductivity which can be further tuned by adjusting the compositional ratio between PEDOT and PSS or modifying the film morphology with additives. Second, it possesses high optical transparency in the vis-NIR range. Third, it is processed from water with good film-forming ability and can resist the erosion of common organic solvents that are used for dissolving active material. Indeed, some of the state-of-the-art PSCs with champion PCE values use PEDOT:PSS as the HCL.[38,39] However, PEDOT:PSS suffers from several serious drawbacks as an HCL in PSCs. First, PEDOT:PSS can etch ITO, and thereby reduce the chemical stability at the ITO–PEDOT:PSS interface, owing to its high acidity.[16,40,41] Second, the electrical and structural inhomogeneities of PEDOT:PSS limit its charge-collecting ability.[42] Furthermore, the electron-blocking capacity of PEDOT:PSS is questionable as well. It was reported that PEDOT:PSS can even serve as an electrode for electron-collecting.[43] Some efforts were hence devoted to modifying PEDOT:PSS towards performance enhancement in PSC devices.[44,45] For example, Xiao *et al.* reported that the fluxing of PEDOT:PSS by ethylene glycol could improve the conductivity and transparency of PEDOT:PSS, and thereby enhance charge extraction at the HCE and light-absorption by the active layer, respectively.[44] The drawbacks of PEDOT:PSS also drove vigorous development of new interfacial material. Among these new materials, semiconducting transition metal oxides (TMO) and organic interfacial materials gained remarkable success as efficient HCLs in PSCs.

Inorganic materials are extensively used as HCL in conventional PSCs, and MoO$_3$ is one of the most commonly used. Shrotriya *et al.* first reported the use of thermally evaporated MoO$_3$ as the HCL for BHJ-PSCs to replace PEDOT:PSS.[46] Later, Sun *et al.* reported more efficient and stable BHJ-PSCs in the conventional architecture based on a PCDTBT:PC$_{71}$BM active layer (see Scheme 6.2) and MoO$_3$ as the HCL.[47] The use of the MoO$_3$ improved light absorption within the active layer and thereby led to a PCE >6% at BHJ layer thickness up to 200 nm. Additionally, PSCs based on MoO$_3$ exhibit superior long-term air stability when compared to the cells fabricated with PEDOT:PSS. However, in terms of low-cost and large-scale production, the

Table 6.1 Summary of PSC devices of conventional structure employing different HCLs.

HCL	Structure	J_{SC} (mA cm^{-2})	V_{OC} (V)	FF (%)	PCE (%)	Ref.
PEDOT:PSS/ EG	ITO/PEDOT:PSS/EG/P3HT:PCBM/ Ca/Al	11.5	0.57	71	4.7	44
MoO$_3$	ITO/MoO$_3$/P3HT:PCBM/Ca/Al	8.9	0.60	62	3.3	46
MoO$_3$	ITO/MoO$_3$/PCDTBT:PC$_{71}$BM/TiO$_x$/Al	10.9	0.89	67	6.5	47
s-MoO$_3$	ITO/s-MoO$_3$/P3HT:PCBM/Al	8.4	0.64	59	3.1	48
s-MoO$_3$	ITO/s-MoO$_3$/P3HT:PCBM/Ca/Al	7.7	0.60	68	3.1	52
s-MoO$_3$	ITO/s-MoO$_3$/P3HT:PCBM/Al	9.5	0.59	68	3.8	53
V$_2$O$_5$	ITO/V$_2$O$_5$/P3HT:PCBM/Ca/Al	8.8	0.59	59	3.1	46
s-V$_2$O$_5$	ITO/s-V$_2$O$_5$/P3HT:PCBM/Al/Ag	9.6	0.53	59	3.0	58
s-V$_2$O$_5$	ITO/s-V$_2$O$_5$(210 nm)/P3HT:PCBM/ Ca/Al	8.0	0.56	51	2.3	59
s-V$_2$O$_5$	ITO/s-V$_2$O$_5$/P3HT:PCBM/Ca/Al	9.7	0.63	64	3.9	60
s-V$_2$O$_5$	ITO/s-V$_2$O$_5$/P3HT:ICBA/ZnO/Al	10.1	0.89	61	5.5	61
WO$_3$	ITO/WO$_3$/P3HT:PCBM/Ca/Al	—	—	69	3.1	54
s-WO$_3$	ITO/s-WO$_3$/P3HT:IC$_{70}$BA/Ca/Al	10.9	0.84	70	6.4	62
s-WO$_3$	ITO/s-WO$_3$/P3HT:PCBM/Al	8.5	0.59	65	3.2	63
NiO	ITO/NiO/P3HT:PCBM/LiF/Al	11.3	0.64	69	5.2	55
s-NiO	ITO/s-NiO/P3HT:PCBM/Ca/Al	8.6	0.58	71	3.6	64
s-NiO	ITO/s-NiO/PCDTBT:PC$_{71}$BM/Ca/Al	11.5	0.88	65	6.7	65
s-NiO	ITO/s-NiO/PBDT-TPD:PCBM/LiF/Al	10.4	0.85	57	5.1	29
NH$_2$-SAM	ITO/NH$_2$-SAM/P3HT:PCBM/Al	5.7	0.55	30	0.95	66
CH$_3$-SAM	ITO/CH$_3$-SAM/P3HT:PCBM/Al	6.8	0.57	32	1.2	66
CF$_3$-SAM	ITO/CF$_3$-SAM/P3HT:PCBM/Al	13.9	0.60	38	3.2	66
GO	SWNT/GO/P3HT:PCBM/Al	10.3	0.57	53	3.1	73
GO	ITO/GO/IZO/P3HT:PCBM/LiF/Al	10.0	0.63	63	3.9	74
GO	ITO/GO/P3HT:PCBM/Al	11.4	0.57	54	3.5	76
Pr-GO	ITO/pr-GO/P3HT:PCBM/Ca/Al	9.3	0.59	67	3.6	77
r-GO	ITO/r-GO/P3HT:PCBM/LiF/Al	10.2	0.59	66	4.0	79
GO-OSO$_3$H	ITO/GO-OSO$_3$H/P3HT:PCBM/Ca/Al	10.2	0.61	71	4.4	78
GOR	ITO/GOR/P3HT:PCBM/Ca/Al	10.0	0.62	67	4.1	83
GO/NiO$_x$	ITO/GO/NiO$_x$/P3HT:PCBM/LiF/Al	8.7	0.60	66	3.5	84
WO$_3$/PEDOT	ITO/WO$_3$/PEDOT/P3HT:PCBM/Al	9.3	0.61	51	2.9	86
SPDPA	ITO/SPDPA/P3HT:PCBM/Ca/Al	10.3	0.60	68	4.2	87
PSSA-*g*-PANI	ITO/PSSA-*g*-PANI//P3HT:PCBM/Al	10.9	0.59	62	4.0	88
TPDSi$_2$:TFB	ITO/TPDSi$_2$:TFB/MDMO-PPV/PCBM/Al	4.6	0.89	54	2.2	89

vacuum deposition process, which is commonly employed for MoO$_3$ fabrication, should be avoided. Developing solution processed high quality MoO$_3$ is necessary. Liu *et al.* reported the solution processed MoO$_3$ layer from an acidified aqueous (pH 1–1.5) dispersion of ammonium molybdate that was spincoated onto ITO and heat treated at 160 °C.[48] However, the resulting MoO$_3$ layer exhibited 10–100 nm scale aggregates (see Figure 6.2) of nanocrystalline MoO$_3$, which is found to reduce the shout resistance for denser MoO$_3$ layers. Meyer *et al.* reported MoO$_3$ films spincoated from MoO$_3$ nanoparticles dispersed in xylene.[49] Girotto *et al.* and Yang *et al.* developed solution-processed MoO$_3$ as HCL materials by using the sol–gel technique.[50,51] However, these processes require relatively high temperatures (>250 °C) to crystallize the MoO_3 film, rendering it incompatible with polymer substrates for future roll-to-roll manufacturing. Murase *et al.* and Jasieniak *et al.* independently reported a simple, solution-processed route from precursors to develop MoO_3 thin-films with a low annealing temperature (<100 °C) and smooth surface (see Figure 2.3; surface roughness <2 nm),[52,53] which is quite compatible with the roll-to-roll technique.

Apart from MoO_3, other TMOs such as V_2O_5,[46] WO_3,[54] and NiO[55] have also been used as HCL for PSCs. These large bandgap metal oxides possess good optical transparency in the visible and near infrared regions, which allows photons to reach the active layer, and high conductivity. In addition, the conduction band of these transition metal oxides is sufficiently higher than the lowest unoccupied molecular orbital (LUMO) of both organic donor and acceptor materials, which effectively blocks electron leakage through the anode.[56,57] A thermally evaporated thin layer of V_2O_5[46] and solution processed V_2O_5 layer prepared from a vanadium-oxytriisopropoxide–isopropyl alcohol solution as HCL in PSCs have been reported.[58] Wang *et al.* developed solution processed V_2O_5 nanoparticles as HCL.[59] Surprisingly, it was found that a

P3HT   PCDTBT   PBDT-TPD   MEMO-PPV

PTB7   PFO-DBT35   PFOTBT

m/n=65/35

PIDT-PhanQ   MEH-PPV   PSBTBT   P

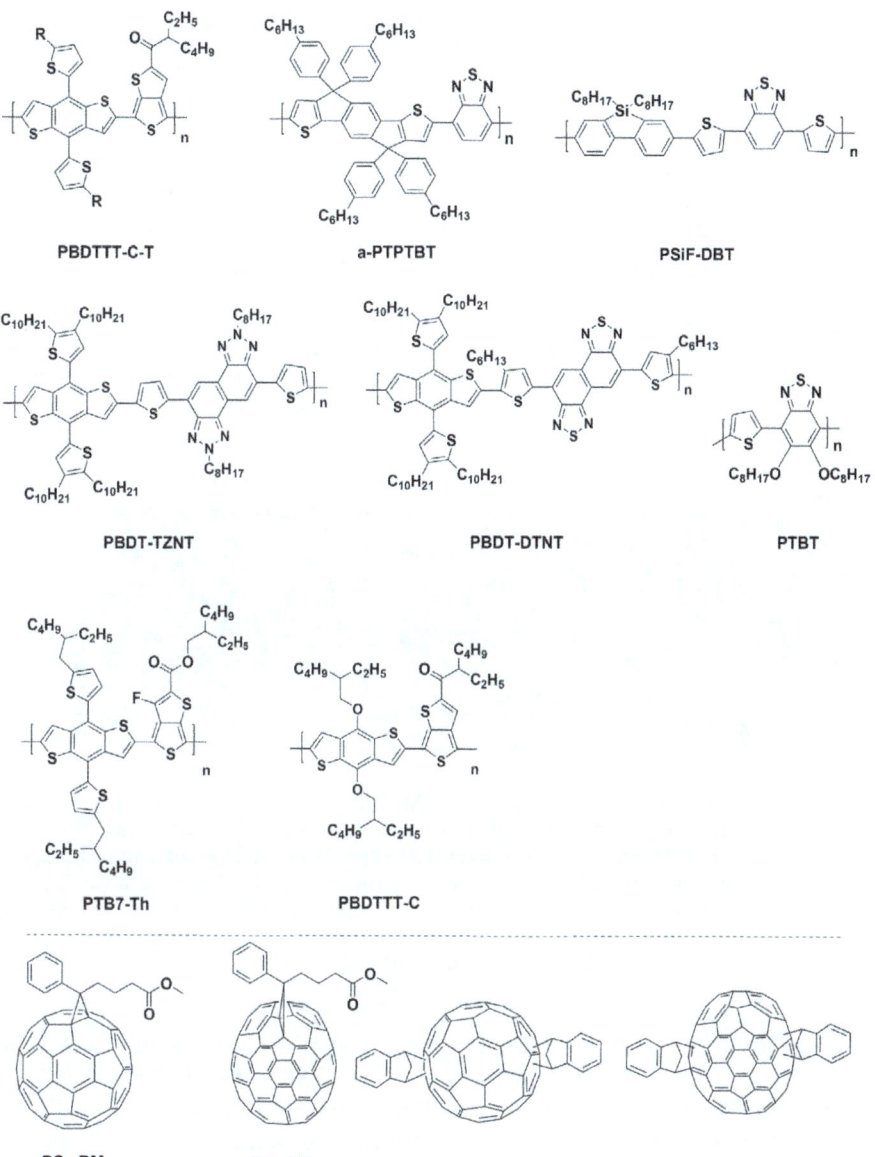

Scheme 6.2 Chemical structures of polymer donor materials and fullerene acceptors.

relatively thick V$_2$O$_5$ nanoparticle layer (*e.g.* 210 nm), which is believed good for device durability and reliability, can still work properly for PSC devices. Some other low-temperature solution-processed V$_2$O$_5$ were reported by Xie *et al.*[60] and Yusoff *et al.*,[61] respectively. In addition to V$_2$O$_5$, solution processible WO$_3$ was also reported,[62,63] where the precursor tungsten(VI) isopropoxide was

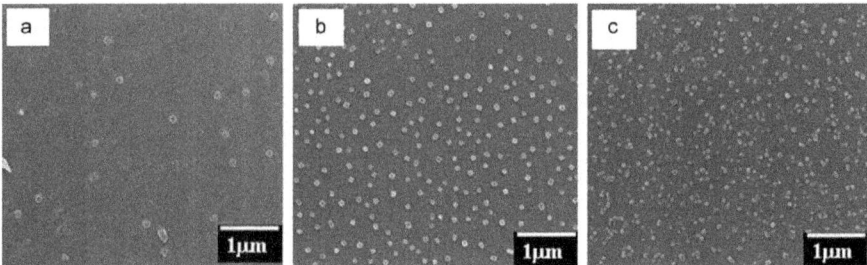

Figure 6.2 Scanning electron microscopy (SEM) images of solution-processed MoO$_3$ morphology on ITO substrate spincoated from aqueous solution at 5000 (a), 3000 (b) and 1500 rpm (c), respectively. Reproduced with permission from ref. 48. Copyright 2010 Elsevier B.V.

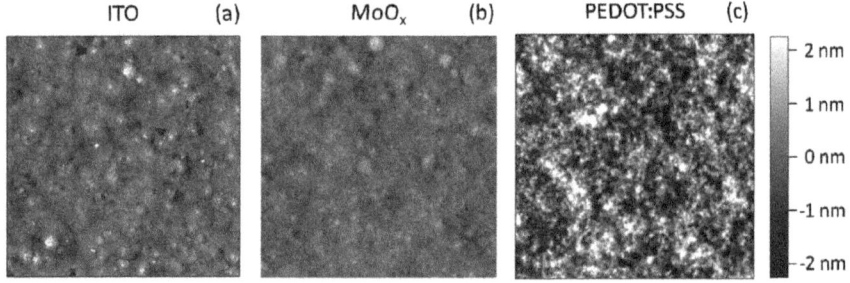

Figure 6.3 Atomic force microscopy (AFM) images showing the topological morphology of (a) bare ITO, (b) ITO with a 10 nm thick sMoO$_x$ thin film, and (c) ITO with a 40 nm thick PEDOT:PSS layer. The corresponding root mean square (rms) surface roughness of each film, across the 2 μm × 2 μm areas shown here, was calculated to be 0.448 nm, 0.354 nm, and 1.12 nm, respectively. Reproduced with permission from ref. 53. Copyright 2012 Wiley-VCH Verlag GmbH & Co. KGaA.

decomposed into WO$_3$ by thermal annealing at 150 °C for 10 min in air. The s-WO$_3$ layer shows high hole mobility and high light transmittance. The PSCs with the s-WO$_3$ HCL showed enhanced performance in comparison with the PEDOT:PSS-modified devices. Olson's group employed a nickel metal organic ink precursor to fabricate an NiO layer on an ITO anode.[64] This solution-deposited NiO, annealed at 250 °C and plasma treated, achieved similar PSC device results as reported with NiO films from pulsed laser deposition (PLD) as well as PEDOT:PSS. Later, the same group applied their solution processible NiO to a PCDTBT:PC$_{71}$BM system and achieved a PCE of 6.7%, whereas the PEDOT:PSS device under the same experimental conditions showed a PCE of 5.7%.[65] A similar phenomenon was also reported by Zhai *et al.*[29]

The modification of the electrode WF of PSC devices to adjust the barrier height between the two layers by self-assembled monolayers (SAMs) has been successfully reported in several publications.[17,66–69] The strength and

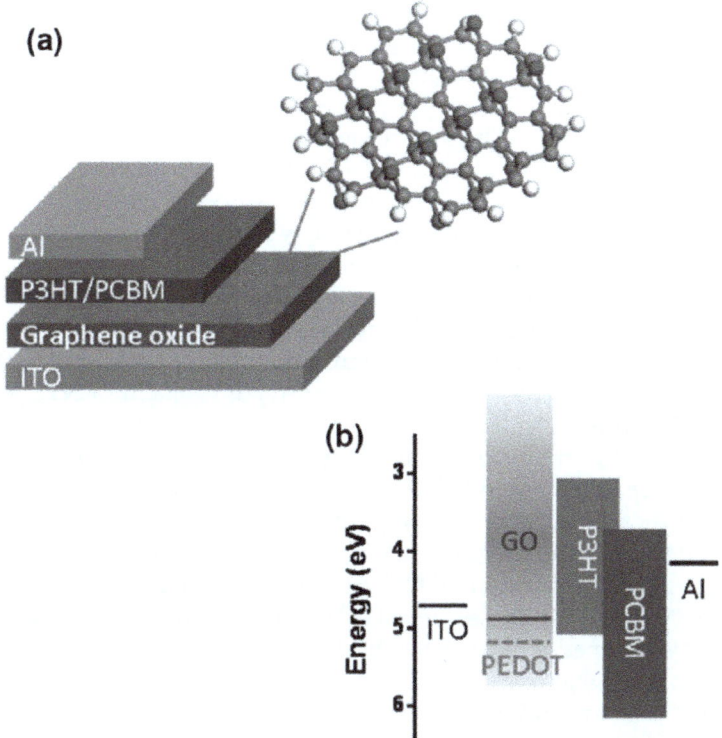

Figure 6.4 (a) Schematic of the photovoltaic device structure consisting of the following: ITO/GO/P3HT:PCBM/Al. (b) Energy level diagrams of the bottom electrode ITO, interlayer materials (PEDOT:PSS, GO), P3HT (donor), and PCBM (acceptor), and the top electrode Al. Reproduced with permission from ref. 76. Copyright 2010 American Chemical Society.

direction of the achieved dipole can be altered by changing the molecular structure containing electron-withdrawing or electron-donating moieties; thus, the surface dipoles can in turn increase or decrease the WF of these substrates.[70,71] Moreover, it was found that the surface properties of the inserted SAMs can greatly influence the morphology of the upper organic active layer.[66,22]

Graphene oxide (GO) is another very interesting material that has shown promise as an efficient HCL for PSCs.[73–75] GO is an oxidized derivative of graphene, which can be prepared by chemical oxidation of naturally abundant graphite. Moreover, GO can be produced and processed in solution at large scale with low cost, particularly attractive for massive applications. Li *et al.* reported the first use of GO (see Figure 6.4) as an efficient HCL in PSCs.[76] However, owing to the insulating nature of GO, PSCs based on a P3HT:PCBM active layer and GO HCL always exhibited a FF less than 65% whereas the typical value for high performance PEDOT:PSS-based devices was about 70%. Besides, the device performance was highly sensitive to the thickness of the GO layer.[75] Therefore, it was necessary to develop highly conductive solution processed GO. Yun *et al.* developed a solution-processible reduced graphene oxide (pr-GO) by reducing GO with *p*-toluenesulfonyl hydrazide in an aqueous solution. The pr-GO showed about 10^5 times higher conductivity than that of GO.[77] The PSC device with P3HT:PCBM active layer and pr-GO as HCL showed a PCE of 3.63% with FF of 66.7%. Liu *et al.*[78] and Jeon *et al.*[79] also reported highly conductive solution-processed GO, which resulted in PSCs based on a P3HT:PCBM active layer with high FF, up to 70%. Carbon nanotubes (CNTs) represent another all carbon-based material; they possess high electrical conductivity, a WF of ~5.0 eV which matches well with the WF of ITO and the highest occupied molecular orbital (HOMO) of most donor polymers, and outstanding optical transparency in a broad spectral range from the ultraviolet (UV) to deep infrared region, making CNTs potential HCLs in BHJ-PSCs.[80–83]

To improve the performance of PSCs further, hybrid anode interfacial materials have been developed. PSCs with different combinations of anode interfacial materials, including GO, NiO, NiO/GO and GO/NiO, were compared and the best device performance was found in the GO/NiO structure, showing major improvement of FF.[84] GO has also been used as an additive to PEDOT:PSS in P3HT:PCBM solar cells.[85] In addition to GO, Kim *et al.* demonstrated a WO_3 doped PEDOT:PSS hybrid cascade HCL for PSCs with dramatically enhanced long-term stability and slightly improved PCE from 2.80% to 2.92%.[86] The improvements resulted from the incorporation of a hybrid anode interfacial layer, which caused efficient hole extraction, enhanced photo absorption near the wavelength range of 425–575 nm and effectively blocked indium diffusion from the etched ITO glass.

Relative to transition metal oxides, the most attractive benefits of organic HCL materials are their solution processability and structural diversity. The chemical structures of some organic HCL materials are shown in Scheme 6.3.

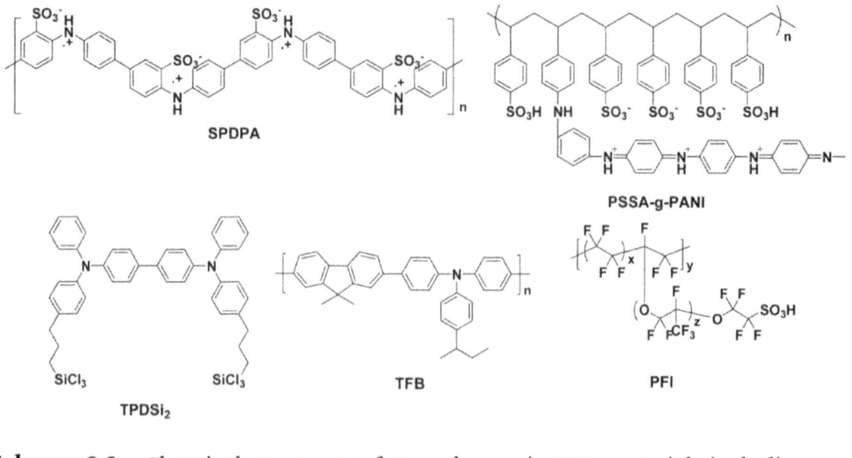

Scheme 6.3 Chemical structures of several organic HCL materials including conducting polymers and cross-linkable materials.

A self-doped conducting polymer, sulfonated poly(diphenylamine) (SPDPA), has been reported as a PEDOT:PSS alternative in PSCs. Compared with PEDOT:PSS, such a doped semiconducting HCL layer could potentially eliminate the electrical inhomogeneity due to the absence of an insulating component. Moreover, it was revealed that the SPDPA film underneath could induce an oriented arrangement of the conjugated polymer in the active layer.[87] A self-doped graft polymer, PSSA-*g*-PANI, was found to have exceptional electrochemical stability, high transparency, and high conductivity. The electrical conductivity of PSSA-*g*-PANI could be tuned by adjusting the ratio between PSSA and PANI. The PSSA-*g*-PANI with the highest conductivity outperforms PEDOT:PSS in PSC devices.[88] By adding a perfluorinated ionomer (PFI) to PSSA-g-PANI, the WF of HCL can be tuned in a wide range from 5.28 to 6.09 eV. The PSC devices based on such a composite HCL showed significantly enhanced lifetime relative to PEDOT:PSS-based devices.[89]

Cross-linkable triarylamines are another class of HCL materials. Generally, conjugated materials based on aromatic amines have good hole transporting ability and an ideal energy level to selectively collect holes, while cross-linkable groups help the formation of robust films to resist sequential solvent corrosion. Scheme 6.3 exhibited some of the cross-linkable HCL materials based on aromatic amines. For example, Hains *et al.*[90] reported the use of a new HCL based on a blend of the cross-linkable aromatic molecule TPDSi$_2$ and an amine-based polymer, TFB. Compared with PEDOT:PSS, the TPDSi$_2$:TFB blend produced considerably enhanced V_{OC}, PCE, and thermal stability. Using the bilayer HCL of PEDOT:PSS and TPDSi$_2$:TFB together in the same cell substantially reduced dark current and yielded even higher V_{OC}.

6.3.2 Cathode Contact

It is preferable for the cathode interface to have a low WF contact for efficient electron extraction. Low WF metals, such as calcium (Ca), barium (Ba) or magnesium (Mg), are usually inserted into the interface between Al and organic active layer to improve the device performance.[91–93] However, the low WF metal is vulnerable to oxidation under ambient conditions, and electrode degradation is a major concern for this type of device.[94] Therefore, the development of new interfacial materials to use as a cathode interlayer is still required.

Inorganic fluorides, such as lithium fluoride (LiF), are promising cathode interfacial materials for BHJ-PSCs. Brabec *et al.* first demonstrated that an ultrathin deposition (<1 nm) of LiF below the top Al contact in MDMO-PPV: PCBM BHJ-PPSCs led to significant improvement of both fill factor and V_{OC}, resulting in an increase in the overall PCE of over 20%.[95] Jönsson *et al.* reported the thickness-dependence of LiF where, for (sub)monolayer coverage, Al deposition decomposed LiF, causing Li-doping of the organic material to yield a low WF contact, while thicker LiF created a dipole layer that downshifts the WF.[96,97] Other metal fluorides, including NaF, KF and CsF, were also reported to exhibit similar phenomena.[98,99] Cesium carbonate

(Cs$_2$CO$_3$) is another promising ECL, which can be deposited by either thermal evaporation or solution spincoating.[100] Furthermore, the replacement of Ag with Al as the cathode did not meaningfully change the values of the photovoltaic parameters, indicating that the effect of the Cs$_2$CO$_3$ interfacial layer was insensitive to the cathode metal.[100,101] However, the actual product of thermally evaporated Cs$_2$CO$_3$ (Cs$_2$O or Cs$_2$CO$_3$) is still uncertain.[102–104] It is widely accepted that the formation of an Al–O–Cs complex yields the low WF contact and thus facilitates the electron extraction.[100]

TiO$_x$ is a well investigated n-type interfacial material for PSC devices. This material is transparent in the visible light spectrum but absorbs UV light. The layer thickness of the interfacial layer is tunable without absorption losses in visible light and therefore can additionally act as an optical spacer (see Figure 6.5).[30,105–107] TiO$_x$ possesses a conduction band edge and valence band edge of −4.4 eV and −8.1 eV, respectively, which would endow TiO$_x$ with good electron extraction ability from the active layer and outstanding hole-blocking ability.[30,108] Besides, the TiO$_x$ films not only showed good electron selectivity but also a good water and oxygen barrier, leading to significantly improved device stability.[109–111] Lee *et al.* used a solution-based sol–gel process to fabricate a TiO$_x$ layer with a thickness of around 30 nm on top of the active layer.[105,109] Because the TiO$_x$ layer was treated at a temperature of 80 °C (far below the crystallization temperatures to the anatase or rutile phases, $T_c \geq 450$ °C), the film showed an amorphous structure. Later, Heeger's group reported a promising PCE of 6.1% for BHJ-PSCs with PCDTBT:PC$_{71}$BM blend as the active layer by inserting TiO$_x$ as an ECL and optical spacer.[107] Note that the processing temperature of TiO$_x$ was 80 °C, which is compatible with most organic active layers. Besides, the internal quantum efficiency of the device was close to 100%, which implied that all absorbed photons result in a separated pair of electron and hole charge carriers. However, the electron mobility of TiO$_x$ is 1.7×10^{-4} cm^2 V^{-1} s^{-1}, which is almost two orders of magnitude lower than that of PC$_{61}$BM, and this may potentially limit the device performance. Park *et al.* fabricated

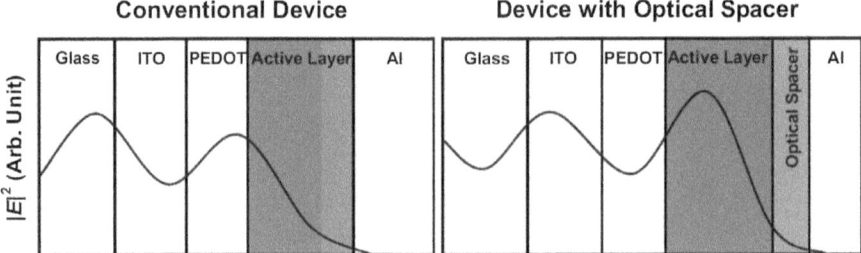

Figure 6.5   Schematic representation of the spatial distribution of the squared optical electric field strength $|E|^2$ inside the devices with a structure of ITO/PEDOT/active layer/Al (left) and ITO/PEDOT/active layer/optical spacer/Al (right). Reproduced with permission from ref. 30. Copyright 2006 Wiley-VCH Verlag GmbH & Co. KGaA.

Cs-doped film TiO_x by mixing Cs_2CO_3 solution with a nanocrystalline TiO_x solution prepared from a sol–gel process. The TiO_x layer obtained exhibited better Ohmic contact between the organic active layer and the metal electrode and also afforded a much better photovoltaic performance for a P3HT:PCBM solar cell than that of a cell using a sole TiO_x interfacial layer.[112] Zinc oxide is another inherently n-type, large bandgap semiconductor, with electronic properties similar to those of TiO_x. ZnO has a electron mobility of 6.6×10^{-2} cm^2 V^{-1} S^{-1}, which is much higher than that of TiO_x.[113] It has been reported that solution-processed ZnO nanoparticle (NP) films have an even higher electron mobility, reaching 2.5 cm^2 V^{-1} S^{-1}.[114] The high electron mobility, ideal conduction band and solution processibility of ZnO NPs indicate that ZnO would be a good ECL material to facilitate electron extraction from the active layer to the metal electrode. The direct use of ZnO as an ECL in conventional PSCs was seen by using low temperature solution-processed ZnO NPs.[61,115–117] In addition, the wide bandgap of ZnO (3.4 eV) would endow it with good hole blocking ability and good transparency. Moreover, Gilot *et al.* had proven that ZnO can also serve as an effective optical spacer for thin (<60 nm) active layers.[31]

Compared with their inorganic counterparts, organic interfacial materials, especially water/alcohol soluble conjugated polymers (WSCPs) and water/alcohol soluble small molecular (WSSMs), have attracted growing attention and exhibited more potential owing to their solution processibility from environmentally friendly solvents, excellent ECE modification ability and their tunable functions through facile modification in molecular structures.[118,119] The chemical structures of some organic ECL materials are shown in Scheme 6.4.

Cao's group reported an amino-group-functionalized polyfluorene neutral WSCP PFN that possessed good solubility in alcohol and performed very well as an electron injection layer (EIL) in polymer light emitting diodes (PLEDs).[120–123] When this material was used as an ECL in PSCs, the device performance was significantly enhanced relative to that of devices without PFN ECLs for certain donor materials.[124–126] Since then, PFN has become one of the most studied interfacial materials in the PSC field,[39,126,127] and has inspired the development of many other novel WSCPs. He *et al.* applied PFN to PCDTBT:PC_{71}BM and PTB7:PC_{71}BM solar cells.[39] The PSCs incorporating the PFN ECL showed simultaneous enhancement in J_{SC}, V_{OC} and FF, which led to PCE increases of 6.79% and 8.37% for PCDTBT- and PTB7-based devices, respectively. All of these results show that PFN is an excellent ECL material for PSCs with conventional device structures. Aside from amino-group-functionalized polyfluorene, phosphonate-functionalized polyfluorene can also be used as a cathode interlayer in PSCs. Zhao *et al.* reported enhancements in the device performance of the P3HT:PC_{61}BM solar cells by introducing a phosphonate-functionalized conjugated polymer PF-EP between the active layer and the Al electrode.[128] The PF-EP layer can be readily deposited on top of the P3HT:PC_{61}BM film by spincoating it from its ethanol solution, and it does not destroy the underlying P3HT:PC_{61}BM blend

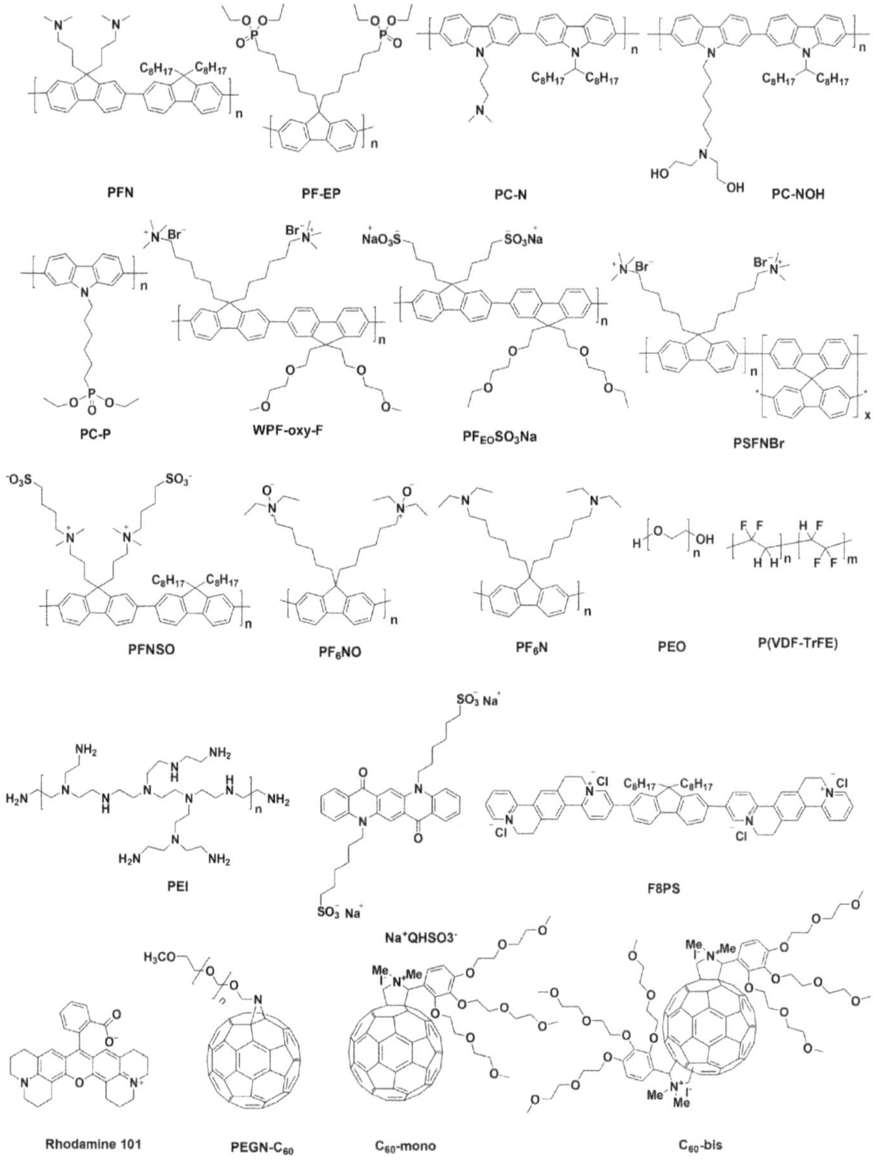

Scheme 6.4 Chemical structures of several organic ECL materials employed in conventional PSCs.

layer morphology. The group's results showed that PF-EP could effectively increase the R_{sh} and improve electron collection, because the PF-EP layer was able to restrain the penetration of Al atoms into the active layer that might have resulted in an increased leakage current and quenched the photo-generated excitons. In recent years, a large number of reports have examined the application of WSCPs with various conjugated backbones. Several novel WSCPs have been developed from polycarbazole for PSC application. Chen's group synthesized a series of water/alcohol-soluble neutral poly(2,7-carbazole) polymers, including PC-N, PC-NOH and PC-P with amino-, diethanolamino- and phosphonate-functionalized side chains, and successfully applied them in PSCs.[129]

Conjugated polyelectrolytes (CPEs) are polymers that contain π-conjugated backbones and functional groups that ionize in highly dielectric media. Compared with neutral WSCPs, CPEs have the advantages of true orthogonal solubility with common active layer materials (such as water and alcohol) and being totally insoluble in non- and low-polarity solvents (such as toluene and chlorobenzene). PSCs with different CPEs as ECL all showed enhanced device performance compared with their control devices without CPE layers.[130–133] Although CPEs have exhibited an excellent cathode modification ability, traditional CPEs have reportedly suffered from the migration of their mobile counterions under an external electric field, resulting in a delayed turn-on in PLEDs[134,135] and even a deterioration in the stability of PLEDs and PSCs.[136–139] Zwitterionic CPEs with cations and anions that covalently bond together to eliminate mobile ions have therefore been developed as ECLs in PSCs.[140,141] Duan *et al.* designed and synthesized a series of zwitterionic CPEs (PFNSO-BT, PFNSO, PFNSO-TPA) that had identical sulfobetaine zwitterionic groups on their side chains but different conjugated main chains. They found that every zwitterionic CPE could boost the photovoltaic performance of PSCs when applied as the ECL, and achieved an impressive PCE of 8.74% for the PTB7:PC$_{71}$BM composite.[140] The same group also developed amino *N*-oxide-functionalized polyfluorenes (PNOs) as ECLs in PSCs.[142] The amino *N*-oxide polymers were synthesized by oxidizing their amino-functionalized precursor polymers with hydrogen peroxide. The amine oxidized polymers displayed excellent solubility in methanol. PSCs based on a PCDTBT:PC$_{71}$BM blend and using PF$_6$NO as the ECL showed a PCE of up to 6.9%, much higher than that of a bare Al electrode device (PCE = 4.0%).

Non-conjugated water/alcohol-soluble polymers represent a different class of the solution-processible cathode interfacial materials applied in both PLEDs and PSCs.[143–145] Zhang *et al.* investigated the differences in the interfacial properties of WSCPs (PF$_6$NO and PF$_6$N) and non-conjugated water/alcohol-soluble polymers [polyethylene oxide (PEO) and polyethylene imine (PEI)]. The results imply that a conjugated backbone may improve electron transport in an ECL, and that it is therefore indispensible in cathode interfacial material design. Moreover, conjugated ECLs also showed a better device performance than non-conjugated ECLs for devices with an Au electrode, meaning that, in addition to the highly polar side chain group, the semiconducting main chains of the conjugated interlayer materials contributed significantly to their excellent interfacial modification ability.[146]

The WSSMs, which possess the advantage of batch-to-batch homogeneity in terms of molecular weight and ease of purification, have also been reported as ECLs in PSCs.[147–150] Pho *et al.* reported a solution-processed WSSM Na$^+$QHSO$_3^-$ (see Scheme 6.4) based on the commercially available pigment quinacridone as the ECL in a PSC.[147] They found that, while the PCE of PCDTBT:PC$_{71}$BM could be enhanced to 5.2%, the control device with bare Al showed a PCE of only 4.3%. The main increase, interestingly, came from the FF instead of the V_{OC} as commonly reported for typical WSCP ECLs. The authors speculated that the discrepancy in improvement may have come from the difference in conjugated motifs (fluorene in CPEs *vs.* quinacridone) and/or the difference in the pendant ionic groups (cationic tetraalkylammonium *vs.* anionic alkylsulfonate). Ye *et al.* designed and synthesized a series of electron-deficient pyridinium salts with different cores.[148] One of these salts, known as F8PS, was introduced as the ECL in a PSC owing to its ability to form a uniform film *via* spincoating. The device performance showed that the PCE of the PCDTBT:PC$_{71}$BM solar cell could be increased from 4.32% to 6.56% with the insertion of the F8PS ECL. The pyridinium rings were notably situated at the backbone of the molecules, quite different from the commonly used CPEs, in which polar groups are located at the terminal of the side chains. Solution-processed zwitterionic-conjugated small molecules have also been used as ECLs in PSCs.[150]

Fullerene-based ECL materials have shown great potential for application in PSCs. They not only match the energy level of the LUMO of a commonly used acceptor, but also possess a sufficiently deep HOMO energy level, making them energetically ideal candidates for an ECL to facilitate electron selection and hole blocking in PSCs. Jen's group developed two methanol-soluble fullerene surfactants incorporating a cationic nitrogen and polar PEO groups as interfacial layers for ECEs in PSCs. These materials possessed appropriate electron mobility and the capability of tuning electrode WFs to improve electron extraction and photocurrent generation.[151,152] Devices with either a C$_{60}$ mono- or bis-adduct inserted between PIDT-PhanQ:PC$_{71}$BM and various metal ECEs showed much improved V_{OC}, J_{SC}, FF and PCE. The PSC performance improved significantly (70% for the Al electrode and 40% for the Ag electrode). High-performing PSCs using a C$_{60}$-bis ECL-modified Ag electrode were realized, with PCEs as high as 6.63%, superior even to those of the C$_{60}$-bis/Al and Ca/Al devices (Table 6.2).

Ferroelectric-material is a newly emerged interfacial material that can be used as an ECL. Yuan *et al.* reported on the efficiency enhancement in PSCs by incorporating Langmuir–Blodgett (LB) deposition of a ferroelectric polymer monolayer P(VDF-TrFE) between the electrodes and photoactive layer.[153] The enhanced efficiency was explained by the presence of an electric filed due to the polarization of the ferroelectric layer; thus, non-radiative recombination of charges or charge transfer excitons was suppressed, leading to increased J_{SC} and V_{OC}, and therefore, an increase in PCE from 1–2% to 4–5%. This additional electric field within the active layer of the solar cell has been claimed to be large, permanent, and to eliminate the need for an external bias. Later, the same group demonstrated the synthesis of ferroelectric P(VDF-TrFE) NPs and their application in enhancing the PCE of PSCs.[154] Compared to the LB method, this method gives better control of the P(VDF-TrFE) coverage, which is critical for optimizing the performance of the

Table 6.2 Summary of PSC devices of conventional structure employing different ECLs.

ECL	Structure	J_{sc} (mA cm^{-2})	V_{OC}(V)	FF (%)	PCE (%)	Ref.
Ag	ITO/PEDOT/P3HT:PCBM/Ag	9.0	0.40	37	1.3	92
Al	ITO/PEDOT/P3HT:PCBM/Al	10.0	0.42	38	1.6	92
Mg:Ag	ITO/PEDOT/P3HT:PCBM/Mg:Ag/Ag	10.3	0.57	51	2.9	92
LiF	ITO/PEDOT/P3HT:PCBM/LiF/Al	10.4	0.58	57	3.5	92
Ca	ITO/PEDOT/P3HT:PCBM/Ca/Al	10.4	0.60	61	3.8	92
Ba	ITO/PEDOT/P3HT:PCBM/Ba/Al	10.5	0.60	62	3.9	92
Cs$_2$CO$_3$	ITO/PEDOT/P3HT:PCBM/Cs$_2$CO$_3$/Al	9.5	0.56	60	3.1	101
Cs$_2$CO$_3$	ITO/PEDOT/P3HT:PCBM/Cs$_2$CO$_3$/Ag	9.5	0.55	57	3.0	101
TiO$_x$	ITO/PEDOT/P3HT:PCBM/TiO$_x$/Al	11.1	0.61	66	4.8	30
TiO$_x$	ITO/PEDOT/P3HT:PCBM/TiO$_x$/Al	10.8	0.62	61	4.1	109
TiO$_x$	ITO/PEDOT/PCDTBT:PC$_{71}$BM/TiO$_x$/Al	10.6	0.88	66	6.1	107
TiO$_x$-Cs	ITO/PEDOT/P3HT:PCBM/TiO$_x$-Cs/Al	10.8	0.58	67	4.2	112
ZnO	ITO/PEDOT/P3HT:PCBM/ZnO/LiF/Al	9.8	0.53	56	2.9	116
ZnO	ITO/PEDOT/P3HT:PCBM/ZnO/Ag	9.3	0.53	46	2.3	116
ZnO	ITO/PEDOT/P3HT:PCBM/ZnO/Au	9.6	0.38	44	1.6	116
ZnO	ITO/PEDOT/P3HT:PCBM/ZnO/Al	7.6	0.67	61	3.1	117
ZnO	ITO/V$_2$O$_5$/P3HT:ICBA/ZnO/Al	10.1	0.89	61	5.5	61
PFN	ITO/PEDOT/PCDTBT:PC$_{71}$BM/PFN/Ca/Al	12.7	0.90	59	6.8	39
PFN	ITO/PEDOT/PTB7:PC$_{71}$BM/PFN/Ca/Al	—	—	—	8.4	39
PF-EP	ITO/PEDOT/P3HT:PCBM/PF-EP/Al	10.3	0.64	66	4.3	128
PC-N	ITO/PEDOT/PFO-DBT$_{35}$:PCBM/PC-N/Al	3.6	1.00	41	1.5	129
PC-NOH	ITO/PEDOT/PFO-DBT$_{35}$:PCBM/PC-NOH/Al	4.0	1.02	41	1.7	129
PC-P	ITO/PEDOT/PFO-DBT$_{35}$:PCBM/PC-NOH/Al	3.8	1.01	40	1.5	129
WPF-oxy-F	ITO/PEDOT/P3HT:PCBM/WPF-oxy-F/Al	9.9	0.63	61	3.8	130
PFEOSO$_3$Na	ITO/PEDOT/P3HT:PCBM/PFEOSO$_3$Na/Al	11.1	0.64	63	4.5	153
PSFNBr	ITO/PEDOT/PFOTBT:PCBM/PSFNBr/Al	9.4	1.04	48	4.3	154
PFNSO	ITO/PEDOT/PTB7:PC$_{71}$BM/PFNSO/Al	16.4	0.73	73	8.7	140
PFNSO-TPA	ITO/PEDOT/PTB7:PC$_{71}$BM/PFNSO-TPA/Al	17.1	0.71	62	7.5	140
PFNSO-BT	ITO/PEDOT/PTB7:PC$_{71}$BM/PFNSO-BT/Al	16.8	0.65	61	6.6	140
PF$_6$NO	ITO/PEDOT/PCDTBT:PC$_{71}$BM/PF$_6$NO/Al	11.6	0.91	66	6.9	142
PF$_6$N	ITO/PEDOT/PCDTBT:PC$_{71}$BM/PF$_6$N/Al	6.9	0.93	61	5.6[a]	146
PEO	ITO/PEDOT/PCDTBT:PC$_{71}$BM/PEO/Al	6.3	0.92	60	5.0[a]	146
PEI	ITO/PEDOT/PCDTBT:PC$_{71}$BM/PEI/Al	7.1	0.92	56	5.2[a]	146

(continued)

Table 6.2 (continued)

ECL	Structure	J_{sc} (mA cm^{-2})	V_{oc}(V)	FF (%)	PCE (%)	Ref.
Na$^+$QHSO$_3^-$	ITO/PEDOT/PCDTBT:PC$_{71}$BM/Na$^+$QHSO$_3^-$/Al	9.8	0.84	63	5.2	147
F8PS	ITO/PEDOT/PCDTBT:PC$_{71}$BM/F8PS/Al	11.3	0.94	62	6.6	148
Rhodamine 101	ITO/PEDOT/PCDTBT:PC$_{71}$BM/Rhodamine101/Al	11.1	0.94	59	6.2	150
C$_{60}$-mono	ITO/PEDOT/PIDT-PhanQ:PC$_{71}$BM/C$_{60}$-mono/Al	11.2	0.86	62	6.0	151
C$_{60}$-mono	ITO/PEDOT/PIDT-PhanQ:PC$_{71}$BM/C$_{60}$-mono/Ca/Al	11.1	0.87	64	6.2	151
C$_{60}$-mono	ITO/PEDOT/PIDT-PhanQ:PC$_{71}$BM/C$_{60}$-mono/Ag	11.3	0.87	64	6.3	151
C$_{60}$-bis	ITO/PEDOT/PIDT-PhanQ:PC$_{71}$BM/C$_{60}$-bis/Al	11.2	0.88	60	5.9	152
C$_{60}$-bis	ITO/PEDOT/PIDT-PhanQ:PC$_{71}$BM/C$_{60}$-bis/Cu	10.1	0.87	61	5.4	152
C$_{60}$-bis	ITO/PEDOT/PIDT-PhanQ:PC$_{71}$BM/C$_{60}$-bis/Ag	11.5	0.88	61	6.2	152
P(VDF-TrFE) LB	ITO/PEDOT/P3HT:PC$_{71}$BM/P(VDF-TrFE) LB/Al	12.8	0.59	60	4.5	155
P(VDF-TrFE) LB	ITO/PEDOT/PSBTBT:PC$_{71}$BM/P(VDF-TrFE) LB/Al	13.8	0.66	54	4.9	155
P(VDF-TrFE) NPs	ITO/PEDOT/PCDTBT:PC$_{71}$BM/P(VDF-TrFE) NPs/Al	11.3	0.91	63	6.6	156
Porphyrin-SAM	ITO/MoO$_x$/P3HT:PC$_{71}$BM/Porphyrin-SAM/Al	11.9	0.69	59	4.8	157
Porphyrin-SAM	ITO/MoO$_x$/P3HT:ICBA/Porphyrin-SAM/Al	11.0	0.88	70	6.8	157
Porphyrin-SAM	ITO/MoO$_x$/PCDTBT:PC$_{71}$BM/Porphyrin-SAM/Al	12.5	0.92	62	7.1	157
ZnO/BA-OCH$_3$	ITO/PEDOT/P3HT:PCBM/ZnO/BA-OCH$_3$/Al	11.6	0.65	55	4.2	158
ZnO/BA-CH$_3$	ITO/PEDOT/P3HT:PCBM/ZnO/BA-CH$_3$/Al	11.6	0.64	49	3.6	158
ZnO/BA-H	ITO/PEDOT/P3HT:PCBM/ZnO/BA-H/Al	11.5	0.64	48	3.5	158
ZnO	ITO/PEDOT/P3HT:PCBM/ZnO/Al	11.3	0.60	47	3.2	158
ZnO/BA-SH	ITO/PEDOT/P3HT:PCBM/ZnO/BA-SH/Al	10.4	0.45	42	2.0	158
ZnO/BA-CF$_3$	ITO/PEDOT/P3HT:PCBM/ZnO/BA-CF$_3$/Al	9.0	0.30	31	0.84	158
ZnO/BA-CN	ITO/PEDOT/P3HT:PCBM/ZnO/BA-CN/Al	8.2	0.27	28	0.62	158
ZnO:PEG	ITO/PEDOT/P3HT:PCBM/ZnO:PEG/Al	10.7	0.60	69	4.4	116
ZnO/r-GO	ITO/PEDOT/PTB7:PC$_{71}$BM/ZnO/r-GO/Al	15.2	0.72	69	7.5	159
TiO$_2$/r-GO	ITO/PEDOT/PTB7:PC$_{71}$BM/TiO$_2$/r-GO/Al	15.0	0.74	67	7.5	159
GO/TiO$_x$	ITO/PEDOT/PCDTBT:PC$_{71}$BM/GO/TiO$_x$/Al	12.4	0.88	68	7.5	160
rGO-TiO$_x$	ITO/PEDOT/P:PC$_{71}$BM/rGO-TiO$_x$/Al	11.0	0.76	64	5.3	161
GO-Cs	ITO/GO/P3HT:PCBM/GO-Cs/Al	10.3	0.61	59	3.7	162
rGO-pyrene-PCBM	ITO/PEDOT/P3HT:PCBM/rGO-pyrene-PCBM/Al	9.8	0.64	62	3.9	163
TIPD	ITO/PEDOT/MEH-PPV:PCBM/TIPD/Al	5.7	0.87	51	2.5	164
PW$_{12}$-POM	ITO/PEDOT/P3HT:PCBM/PW$_{12}$-POM/Al	10.1	0.65	41	2.7	165

^{a}Data observed under light intensity of 70 mW cm^{-2}.

PSCs. With the denser P(VDF-TrFE) nanoislands obtained by this method, a larger electric field, around 20 ± 3 V μm^{-1} in PCDTBT:PC$_{71}$BM film, can be induced by the P(VDF-TrFE) NPs, which is 50–100% higher than the electric filed generated by the 1–2 monolayer LB P(VDF-TrFE). However, Asadi *et al.* argued that ferroelectric functionalized PSC, with a "truly" ferroelectric interlayer, will be subjected to ferroelectric depolarization.[155] They demonstrated that inserting a layer of a ferroelectric polymer in the solar cell stack only leads to improved PCE for non-optimized solar cells with non-Ohmic contacts. In fact, in the best-case scenario, the performance of the "ferroelectric-material" functionalized solar cell approaches that of optimized cells with standard LiF/Al cathodes.

SAM has been also successfully used as an interfacial layer in PSCs. Vasilopoulou *et al.* reported on enhanced PSC performance resulting from the incorporation of a water/methanol-soluble porphyrin molecule as cathode interlayer. It was demonstrated that the self-organization of this porphyrin compound into aggregates in which molecules adopt a face-to-face orientation parallel to the organic semiconducting substrate induced a large local interfacial electric field that results in a significant enhancement of exciton dissociation.[156] SAMs can be used in combination with n-type metal oxides, such as TiO$_x$ and ZnO, to improve the cathode interface properties further.[157] Using SAMs that can form favorable dipole and covalent bonding between ZnO and metal devices showed significant improvement in efficiency and it also enabled high WF metals such as Ag and Au to be used as cathodes. Hybridization of ZnO with poly(ethylene glycol) can also be used to engineer the WF, morphology, refractive index and charge transporting properties of the ECL, which provided an efficient way to improve PSC performance.[116]

In terms of work on GO as an ECL, being an ambipolar material for efficient transport of both holes and electrons, GO derivatives with energy levels tunable by functionalization can also be used as ECLs for PSCs. Liu *et al.* reported the first GO-based electron extraction material, using the cesium-neutralized GO (GO-Cs) as the ECL for PSCs.[158] The GO-Cs modified electrode showed a WF of 4.0 eV, matching well with the LUMO level of the PCBM acceptor. PSC devices with GO-Cs as ECL and P3HT:PCBM active layer exhibited fairly comparable V_{OC}, J_{SC}, FF and PCE to those of the corresponding control device with LiF as ECL, indicating that GO-Cs was indeed an excellent ECL. Qu *et al.* had developed a new graphene–fullerene composite rGO-pyrene-PCBM, where PCBM as the most commonly used acceptor material in BHJ-PSCs was attached to rGO *via* the noncovalent functionalization approach with pyrene used as an anchoring bridge. PSC devices with rGO-pyrene-PCBM as ECL exhibited a V_{OC} of 0.64 V, J_{SC} of 9.07 mA cm^{-2}, FF of 62%, and PCE of 3.89%, which was higher than the 3.39% of the control device without ECL.[159] Silva's group also reported solution-processible ECL for PSC devices based on composites of metal oxides (ZnO, TiO$_x$) and reduced GO. The resulted devices based on a PTB7:PC$_{71}$BM active layer showed a PCE in the range of 7.4–7.5%, which was slightly better than the control devices with only ZnO or TiO$_x$ ECL.[160] Wang *et al.* had fabricated highly efficient PSCs with a GO/TiO$_x$ bilayer as ECL and PCDTBT:PC$_{71}$BM as the active layer. As shown in Figure 6.6, the GO layer was deposited by graphene stamping transfer from copper foil with a thermal-release tape, followed by oxidation with HNO$_3$. The PSC device thus fabricated showed a PCE as high as 7.5%, along with a V_{OC} of 0.88 V, J_{SC} of 12.4 mA cm^{-2} and FF of 68%.[161] A similar result was reported by Sharma *et al.*[162]

A solution processible titanium chelate, titanium (diisopropoxide) bis (2,4-pentanedionate) (TIPD), was used as the ECL in PSCs based on the blend of MEH-PPV:PCBM. The PCE of the PSC with TIPD reached 2.52%, which was increased by 51.8% in comparison with that (1.66%) of the device without TIPD.[163] Palilis *et al.* reported the use of a water-soluble tungsten polyoxometalate H$_3$PW$_{12}$O$_{40}$ (PW$_{12}$-POM) as an efficient ECL incorporated into P3HT:PCBM PSCs. The J_{sc} of the PW$_{12}$-POM modified device was enhanced by ~40%, the V_{oc} increased from 0.61 V to 0.65 V and the FF from 36% to 41%, resulting a PCE enhancement of ~70% (from 1.57% for the

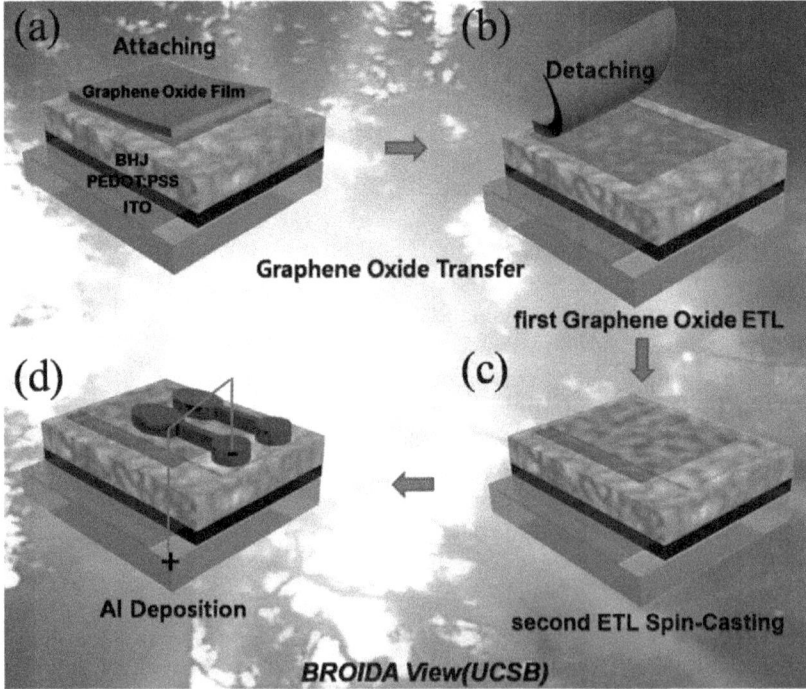

Figure 6.6   Schematic illustration and fabrication steps of BHJ solar cells with an ECL of graphene oxide (GO) applied by stamping transfer. (a) Attachment of the transfer film on top of the BHJ active layer; (b) after detachment of the film, the first ECL of GO is uniformly transferred and coated onto the BHJ layer; (c) spincasting of the second ECL of TiO$_x$ on top of GO; (d) completed device structure after Al deposition. Reproduced with permission from ref. 161. Copyright 2013 Wiley-VCH Verlag GmbH & Co. KGaA.

reference to 2.7% for the PW_{12}-POM modified device). The improvement was attributed to enhanced electron transfer/extraction at the PW_{12}-POM–Al interface as a result of the favorable interfacial energy level alignment and possible enhancement of the local electric field due to the nanoscale morphology of the PW_{12}-POM layer.[164]

6.4 Interfacial Materials for Inverted Polymer Solar Cells

6.4.1 Anode Contact

Several classes of interfacial material have been used for anode modification in the inverted cells. Thermally evaporated metal oxides, such as MoO_3,[165,166] V_2O_5,[104,167,168] NiO[169] and WO_3,[170] are efficient materials that can promote Ohmic contact between the BHJ layer and ITO electrode. Record single-junction and tandem PSCs based on thermally evaporated MoO_3 were reported by Cao's[3] and Yang's[4] groups. However, although the best inverted solar cells are fabricated with thermo-evaporated transitional metal oxides, solution processible HCL is preferred because of the vacuum evaporation energy cost and incompatibility with roll-to-roll large area device fabrication. Therefore, some efforts have been made to design a solution-processing route for metal oxide materials. Lee *et al.* reported microwave-assisted synthesis of stable suspensions of ultrasmall (<4 nm diameter) MoO_3 nanoparticles (NP MoO_x), and demonstrated room temperature solution deposition of NP MoO_x HCL for inverted PSC devices.[171] It was also found that oxidizing NP MoO_x with H_2O_2 could further increase PCE, but this was still not as good as evaporated MoO_3 and PEDOT devices. Later on, Li *et al.* reported a water-free, room-temperature and solution-processed approach to forming a MoO_x film. Devices based on a P3HT:PCBM active layer with this solution-processed MoO_x as HCL exhibited a comparable performance to the evaporated MoO_3 devices. The author further doped Ag nanoparticles into this MoO_3 solution and increased the PCE to 4.37%.[172] Huang *et al.* dispersed V_2O_5 powder in isopropanol through ultrasonic agitation and spincast the dispersion onto a P3HT:PCBM active layer to form HCL.[173] Chen *et al.* also developed solution processible V_2O_5, *via* sol–gel processes, as an HCL in inverted PSCs and studied the long-term stability.[174] Besides, it was found that NiO can also be spin casted onto an active layer to form HCL.[175,176] Stubhan *et al.* reported solution processible WO_3 as HCL, and the device showed a comparable performance to the control PEDOT device.[63]

GO can be used as an effective HCL to improve the performance of inverted PSCs as well. The thin GO layer can be deposited on top of the BHJ layer through simple spincoating from its butyl alcohol solution with controllable coverage and thickness. The optimized inverted PSCs showed performance comparable or even superior to those using PEDOT:PSS as HCL.[177] Later, the same group demonstrated that GO can be used as a new type of solution-processed dopant for efficient p-doping of the surface of conjugated polymers.[178] PSC devices based on Al, Ag and Au electrodes with GO HCL exhibited much better performance than those without GO HCL. More information about GO as interfacial materials are given in another review.[75] The data on inverted structure PSC devices employing different HCLs are summarized in Table 6.3.

6.4.2 Cathode Contact

An ITO electrode modified by a thin layer of Ca (1 nm) acted as the cathode. With an optimal MoO_3 (3 nm) as the HCL, the inverted cell had a PCE of 3.55%.[179] Alkali metal salts can also be used as an interfacial material to tune the WF of ITO for electron collection. Cs_2CO_3 is successfully used to inverted the polarity of PSCs with a P3HT:PCBM active layer, and inverted device structures with performance comparable to that exhibited by conventional solar cells have been reported,[167] where the Cs_2CO_3 cathode interfacial layer was either thermally evaporated or spincoated from its 2-ethoxyethanol solution onto ITO substrates. The PCE of inverted PSC was further increased by thermal annealing of the Cs_2CO_3 spincoated layer at 150 °C.[104] The PCE increased from 2.3%, for non-annealed Cs_2CO_3, to around 4.2% for Cs_2CO_3 annealed at 150 °C. Ultraviolet photoelectron spectroscopy (UPS) indicated a decrease of Cs_2CO_3 WF upon annealing, likely due to the decomposition into doped Cs_2O, as revealed by X-ray photoelectron spectroscopy (XPS). The same group also applied Cs_2CO_3 in inverted semitransparent PSC, where a transparency of around 70% was obtained in the wavelength range where the active layer has no absorption.[180]

Similar to the case in conventional cells, TiO_x and ZnO[181–183] are the most widely studied ECL for inverted PSCs owing to their high optical transparency in the visible and near infrared region, high carrier mobility and solution processibility. As summarized in Table 6.4, many demonstrations of the use of these n-type metal oxides as the ECL for inverted solar cells have been reported in the literature. Crystalline TiO_x have been proposed as ECL in inverted P3HT:PCBM solar cells,[184,185] deposited onto ITO substrate by liquid phase dispersion or by potentiostatic anodization of a titanium layer, followed by high temperature annealing (500 °C) to convert the initial amorphous oxide, showing a morphology of vertically oriented nanotubes, to the crystalline anatase phase. The nanostructured TiO_x led to a much higher efficiency (2.25%) of the related solar cells,[186] compared with those made of a planar TiO_x layer (0.85%), due to its enhanced ability to collect negative charges. The low temperature sol–gel process has been used to deposit TiO_x on ITO to fabricate inverted PSCs based on P3HT:PCBM PSC, showing a PCE of ~3.1%.[187] ZnO has a similar band structure to TiO_x but higher electron mobility, which is advantageous for reducing the electrical resistance in PSCs. Inverted P3HT:PCBM solar cells with a silver top anode and ZnO interlayer between ITO and the active layer showed very high diode rectification and PCE approaching 3%, indicating efficient hole-blocking and electron extraction.[188] The zinc acetate (ZnAc) sol–gel precursor was directly spincast onto ITO and then thermally annealed at 300 °C for 5 min

Table 6.3 Summary of inverted structure PSC devices employing different HCLs.

HTL	Structure	J_{SC} (mA cm^{-2})	V_{OC} (V)	FF (%)	PCE (%)	Ref.
MoO$_3$	ITO/TiO$_2$/P3HT:PCBM/MoO$_3$/Al	5.9	0.61	56	2.0	165
MoO$_3$	ITO/TiO$_2$/P3HT:PCBM/MoO$_3$/Ag	6.6	0.63	62	2.6	165
MoO$_3$	ITO/TiO$_2$/P3HT:PCBM/MoO$_3$/Au	5.6	0.61	59	2.0	165
MoO$_3$	FTO/ZnO/P3HT:PCBM/MoO$_3$/Ag	8.9	0.62	57	3.1	166
MoO$_3$	ITO/PFN/PTB7:PC$_{71}$BM/MoO$_3$/Al	17.2	0.74	72	9.2	3
V$_2$O$_5$	ITO/Cs$_2$CO$_3$/P3HT:PCBM/V$_2$O$_5$/Al	8.4	0.56	62	2.3	167
V$_2$O$_5$	ITO/Cs$_2$CO$_3$/P3HT:PCBM/V$_2$O$_5$/Al	11.1	0.59	64	4.2	104
V$_2$O$_5$	ITO/ZnO/P3HT:PCBM/V$_2$O$_5$/Ag	10.4	0.58	65	3.9	168
NiO	ITO/TiO$_2$/P3HT:PCBM/NiO/Ag	7.9	0.46	40	1.5	169
WO$_3$	ITO/TiO$_2$/P3HT:PCBM/WO$_3$/Al	6.9	0.58	—	2.4	170
WO$_3$	ITO/TiO$_2$/P3HT:PCBM/WO$_3$/Ag	7.2	0.60	60	2.6	170
WO$_3$	ITO/TiO$_2$/P3HT:PCBM/WO$_3$/Au	6.8	0.59	—	2.3	170
NP MoO$_x$	ITO/ZnO/P3HT:PCBM/NP MoO$_x$/Ag	8.7	0.50	48	2.1	171
s-MoO$_x$	ITO/TiO$_2$/P3HT:PCBM/s-MoO$_x$/Ag	9.3	0.63	65	3.8	172
s-MoO$_x$:Ag NP	ITO/TiO$_2$/P3HT:PCBM/s-MoO$_x$:Ag NP/Ag	10.4	0.63	67	4.4	172
s-MoO$_x$	ITO/TiO$_2$/PBDTTT-C-T:PC$_{71}$BM/s-MoO$_x$/Ag	16	0.77	58	7.2	172
s-MoO$_x$:Ag NP	ITO/TiO$_2$/PBDTTT-C-T:PC$_{71}$BM/s-MoO$_x$:Ag NP/Ag	16.8	0.77	62	7.9	172
s-V$_2$O$_5$	ITO/ZnO/P3HT:PCBM/s-V$_2$O$_5$/Ag	10.8	0.55	60	3.6	173
s-V$_2$O$_5$	ITO/ZnO/P3HT:PCBM/s-V$_2$O$_5$/Ag	10.1	0.57	67	3.9	174
s-V$_2$O$_5$	ITO/ZnO/a-PTPTBT:PCBM/s-V$_2$O$_5$/Ag	11.6	0.82	53	5.0	174
s-NiO	ITO/ZnO/P3HT:ICBA/s-NiO/Ag	11.3	0.79	63	5.6	176
s-WO$_3$	ITO/AZO/Si-PCPDTBT:PC$_{71}$BM/s-WO$_3$/Ag	12.8	0.62	60.4	4.8	63
GO	ITO/ZnO/C$_{60}$-SAM/P3HT:PCBM/GO/Ag	8.7	0.64	65	3.6	177

Table 6.4 Summary of inverted structure PSC devices employing different ECLs.

HTL	Structure	J_{sc} (mA cm^{-2})	V_{OC} (V)	FF (%)	PCE (%)	Ref.
Ca	ITO/Ca/P3HT:PCBM/MoO$_3$/Ag	8.3	0.65	66	3.6	179
Cs$_2$CO$_3$	ITO/Cs$_2$CO$_3$/P3HT:PCBM/V$_2$O$_5$/Al	8.4	0.56	62	2.3	167
Cs$_2$CO$_3$	ITO/Cs$_2$CO$_3$/P3HT:PCBM/V$_2$O$_5$/Al	11.1	0.59	64	4.2	104
ZnO	PEN/ITO/ZnO/P3HT:PCBM/MoO$_3$/Ag	11.3	0.59	56	3.8	181
ZnO	ITO/ZnO/PSiF-DBT:PCBM/MoO$_3$/Au	5.0	0.90	60	3.8	182
ZnO	ITO/ZnO/P3HT:PCBM/V$_2$O$_5$/Al	10.8	0.60	62	4.0	183
ZnO	ITO/ZnO/P3HT:PCBM/Ag	11.2	0.56	48	3.0	188
ZnO	ITO/ZnO/PCDTBT:PC$_{71}$BM/MoO$_3$/Ag	10.4	0.88	69	6.3	189
ZnO NPs	ITO/ZnO NPs/P3HT:PCBM/PEDOT/Ag	11.2	0.62	54	3.8	190
ZnO NPs	ITO/ZnO NPs/PIDT-PhanQ:PC$_{71}$BM/PEDOT/GO/Ag	11.6	0.86	64	6.4	191
ZnO NPs	ITO/ZnO NPs/P3HT:PCBM/V$_2$O$_5$/Al	10.7	0.61	61	4.0	192
PFN	ITO/PFN/PTB7:PC$_{71}$BM/MoO$_3$/Al	17.2	0.74	72	9.2	3
PFN-OX	ITO/PFN-OX/PBDT-TZNT:PC$_{71}$BM/MoO$_3$/Al	11.7	0.92	65	7.1	193
PFNBr	ITO/ZnO/PFNBr/PBDT-DTNT:PC$_{71}$BM/MoO$_3$/Ag	17.4	0.75	61	8.4	194
P3ImHT	ITO/P3ImHT/P3HT:PCBM/MoO$_3$/Ag	10.4	0.53	61	3.3	133
P3ImHT	ITO/P3ImHT/PCDTBT:PC$_{71}$BM/MoO$_3$/Ag	11.2	0.84	51	4.8	133
Cross-linked PCBSD	ITO/c-PCBSD/P3HT:PCBM/PEDOT:PSS/Ag	12.8	0.60	58	4.4	195
Cross-linked PCBSD	ITO/c-PCBSD/P3HT:ICBA/PEDOT:PSS/Ag	12.4	0.84	60	6.2	196
B-PCPO	ITO/B-PCPO/PCDTBT:PC$_{71}$BM/MoO$_3$/Al	9.5	0.89	62	6.2	21
Full-x	ITO/Full-x:bis-FPI(12 nm)/PIDT-PhanQ:PC$_{71}$BM/MoO$_3$/Ag	11.4	0.83	56	5.3	203
PFEN-Hg	ITO/PFEN-Hg(13 nm)/PTB7:PC$_{71}$BM/MoO$_3$/Al	17.4	0.74	71	9.1	204
ZnO/C$_{60}$-SAM	ITO/ZnO/C$_{60}$-SAM/P3HT:PCBM/PEDOT/Ag	8.7	0.64	64	3.5	177
TiO$_x$/C60-SAM	ITO/TiO$_x$/C$_{60}$-SAM/P3HT:PCBM/PEDOT/Ag	10.6	0.62	57	3.8	19
TiO$_x$/TT-SAM	ITO/TiO$_x$/TT-SAM/P3HT:PCBM/PEDOT/Ag	10.0	0.60	56	3.4	19
TiO$_x$/BA-SAM	ITO/TiO$_x$/BS-SAM/P3HT:PCBM/PEDOT/Ag	10.5	0.60	50	3.2	19
TiO$_x$/LA-SAM	ITO/TiO$_x$/LA-SAM/P3HT:PCBM/PEDOT/Ag	9.9	0.61	50	3.0	19
TiO$_x$/mono-FSAMs	ITO/TiO$_x$/mono-FSAMs/PTBT:PCBM/PEDOT/Au	8.4	0.86	60	4.3	208
TiO$_x$/bis-FSAMs	ITO/TiO$_x$/bis-FSAMs/PTBT:PCBM/PEDOT/Au	9.4	0.86	64	5.1	208
ZnO-C60	ITO/Zn-C$_{60}$/PTB7:PC$_{71}$BM/MoO$_3$/Ag	15.4	0.73	73	8.2	14
ZnO-C60	ITO/Zn-C$_{60}$/PTB7-Th:PC$_{71}$BM/MoO$_3$/Ag	15.7	0.80	74	9.4	14
GO-Cs	ITO/GO-Cs/P3HT:PCBM/GO/Al	10.7	0.51	54	3.0	158
F$_{16}$CuPc film	ITO/F$_{16}$CuPc film/ZnO/P3HT:PCBM/MoO$_3$/Ag	9.4	0.57	61	3.3	209
F$_{16}$CuPc NW	ITO/F$_{16}$CuPc NW/ZnO/P3HT:PCBM/MoO$_3$/Ag	10.7	0.57	59	3.6	209
TIPD	ITO/TIPD/PBDTTT-C:PC$_{71}$BM/MoO$_3$/Al	16.3	0.70	65	7.4	210
TOPD	ITO/TOPD/P3HT:PCBM/MoO$_3$/Al	12.8	0.58	54	4.0	211

to hydrolyze and crystallize into amorphous ZnO thin-film. Improved conductivity and mobility after annealing led to PCE of 2.97% for the inverted device. Sun *et al.* also reported that an inverted solar cell using ZnO as the ECL and a PCDTBT:PCBM active layer can yield PCE as high as 6.33%.[189] One of the problems of the sol–gel process for TiO_x and ZnO is that they require high-temperature (*e.g.* 300 °C) annealing processing conditions in order to improve the crystallinity of the material to minimize resistance. These high-temperature processing conditions are not compatible with flexible substrates, such as plastic, and industry scale roll-to-roll processes. To overcome this problem, ZnO NPs have been introduced as ECLs for inverted PSCs. The processing temperature of ZnO NPs is less than 150 °C, which is lower than required for the sol–gel precursor method.[190–192] The devices fabricated from the ZnO NPs on ITO-coated glass showed device performance very similar to that obtained from the high-temperature processed ZnO sol–gel devices.

Organic materials, as shown in Scheme 6.5, especially those that can be processed from highly polar solvents, are also promising candidates as ECLs for PSCs with inverted device architecture. For example, it was reported that PFN, a famous organic interfacial material, works also quite well as an ITO modifier to enhance electron extraction in inverted PSCs. A thin layer of PFN (5–20 nm) on ITO can not only offer Ohmic contact for enhanced electron collection due to the decrease of ITO WF but also enhance light-harvesting in the device, by which effects a record high

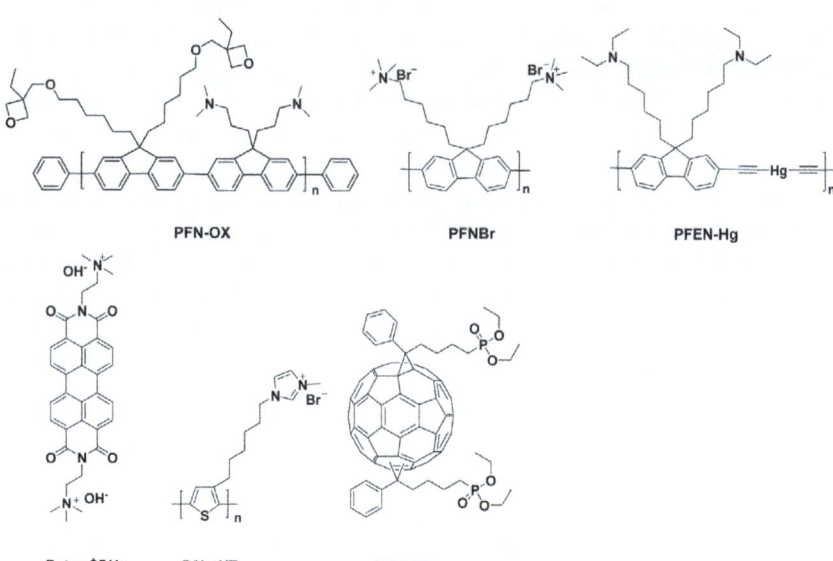

Scheme 6.5 Chemical structures of several organic ECL materials employed in inverted PSCs.

PCE value of 9.2% was realized.[3] Nevertheless, PFN would be washed away during the spincoating of the active layer. One of the approaches to overcoming such a drawback of PFN is the development of cross-linkable analogues of PFN. Huang's group reported such a new polymer, PFN-OX, which can be thermally cross-linked and can resist the corrosion of the processing solvents required for the active layer. The PFN-OX significantly enhanced the efficiency of an inverted solar cell based on a wide bandgap polymer, to an attractive level of 7.11%.[193]

Compared with organic interfacial materials based on non-ionic polar groups (such as amine, phosphorate, *etc.*), polyelectrolytes possess apparent advantages because they are almost completely insoluble in common processing solvents used for the active layer. As a result, a large number of conjugated polyelectrolytes have been reported as ideal ECLs in inverted PSCs. For example, Yang *et al.* reported the use of a bilayer ECL of ZnO and a conjugated polyelectrolyte PFNBr on top of ITO, offering a quite high PCE of 8.4% in PSCs based on a low bandgap polymer PBDT-DTNT.[194] It is worth pointing out that conjugated polyelectrolyte alone can also work well as an ECL in inverted solar cells. For example, a polythiophene-derived conjugated polyelectrolyte, P3ImHT, was successfully applied as an ECL in both P3HT- and PCDTBT-based PSCs. A more interesting finding is that the molecular weight of the interfacial polymer plays an important role in the resulted device performance, *i.e.* higher molecular weight is desired for better interfacial properties.[133] Despite exceptional interface modification ability, polyelectrolytes may suffer from their mobile counterions, which can migrate into the active layer and lead to negative effects on device performance. In addition to polyelectrolytes, small molecule salts are good ECL candidates in inverted solar cells.

Fullerene derivatives are also potentially promising ECL materials in inverted PSCs because they can form matched energy alignment with the acceptor phase in the active layer and have high electron mobility. A fullerene-based molecule, PCBSD, which bears cross-linkable groups, was found to have excellent ability to enhance electron extraction in inverted solar cells after deposition on top of ZnO. The device consisting of ITO/ZnO/cross-linked PCBSD/P3HT:PC$_{61}$BM/PEDOT:PSS/Ag greatly outperformed its control device of ITO/ZnO/P3HT:PC$_{61}$BM/PEDOT:PSS/Ag; they exhibited PCEs of 4.4% and 3.5%, respectively. Moreover, such a bilayer ECL offered exceptional device lifetime even though the device was not encapsulated.[195] Such a bilayer ECL was able to boost the efficiency of PSCs further, to 6.2%, by changing the active layer from P3HT:PC$_{61}$BM to a better-performing P3HT:ICBA.[196] Fullerene derivatives can also work well as a single ECL in inverted PSCs. For example, Duan *et al.* demonstrated the successful application of a phosphate-group-functionalized fullerene as an ECL. The efficiency of the device based on PCDTBT:PC$_{71}$BM reached 6.2%, which is much higher than that of its control device based on a bare ITO electrode. Such enhancement is attributable to the ITO modification function of the phosphate group, the good electron transport ability of fullerene and barrier-free energetic alignment at the ITO–organic interface.[21]

A fatal drawback of organic interfacial materials is their insulating nature, or low electrical conductivity, which renders the solar cell performance highly sensitive to the thickness of the ECL. In fact, a comprehensive survey informs us that insulating organic ECLs usually play a positive role within 5–10 nm.[3,21,144,194,197,198] Unfortunately, it is very challenging to produce such thin-films with good uniformity through the roll-to-roll processing applied in the manufacture of large-scale PSCs.[199] Reilly *et al.* reported the use of a dicationic perylene diimide salt, Petma⁺OH⁻, as an efficient ECL in inverted PSCs. The enhanced device performance originated from the unusual self-doping of Petma⁺OH⁻, which led to the enhancement, by 5 orders of magnitude, in the conductivity of the interfacial layer after dehydration.[200] Nevertheless, no results about the thickness dependence of device performance on ECLs were reported. Recently, Jen's group[201,202] observed the self-doping of ionic fullerene, in which electron transfer occurred from the anion of a quaternary ammonium salt to the fullerene core and thereby offered high conductivity for the fullerene films (see Figure 6.7). Employing a cross-linked self-doped fullerene robust film as the ECL in inverted solar cells offered best performance at a thickness around 10 nm. Unfortunately, further increasing the thickness of ECL to 24 nm resulted in a 40% decrease in PCE value.[203] Huang's group recently developed a metallic conjugated polymer (PFEN-Hg) with pendant amino groups which exhibited excellent electron extraction ability in inverted solar cells with a wide range of film thicknesses. When the thickness of ECL was increased from 7 nm to 19 nm, the PCEs of the resulting solar cells were almost unchanged (8.9–8.64%). PCEs of 6.82% and 4.5%, respectively, could also be obtained even if the thickness of the ECL was further increased to 29 nm and 37 nm. In stark contrast, the PCE of the control devices with non-metallic amine-functionalized interfacial material was sharply decreased from 9.03% to 1.02% and further to 0.03% when the thickness of the ECL was increased from 4 nm to 7 nm and further to 14 nm (see Figure 6.8). The extended processing window in thickness of such a metallic conjugated polymer ECL was believed to originate from the strong metal–metal interactions, which can increase the packing of the polymer film and further enhance charge carrier transport ability.[204]

The formation of SAM on ITO is a well-known approach to controlling the effective WF in OLED by shifting the vacuum level.[205] Modification with electron-accepting molecules can increase the effective WF, while modification with electron-donating molecules lowers the effective WF.[66,157] However, use of only SAM modified ITO as the cathode in inverted PSCs is rarely seen in the literature. Instead, there are many papers reporting incorporation of SAM and metal oxide as a hybrid cathode interfacial layer in inverted PSCs. The surfaces of metal oxides have hydroxyl groups that can cause charge trapping at the metal oxide–active layer interface.[206] The terminated surfaces of these hydroxyl groups lead to high-interface charge recombination due to poor

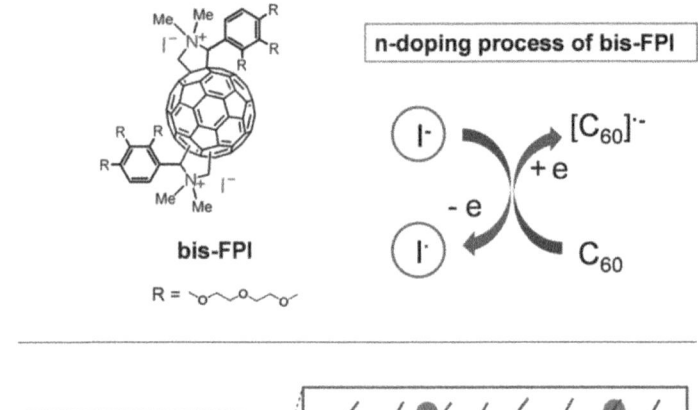

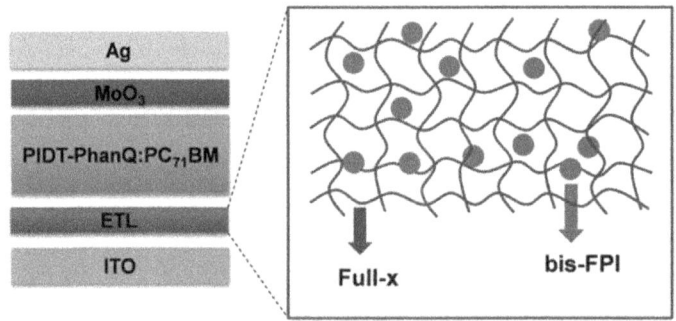

Figure 6.7   The chemical structure of bis-FPI and its doping mechanism through an anion-induced electron transfer process (top). Schematic representation of the device structure of PSCs with bis-FPI dispersed full-x layers (bottom). Reproduced with permission from ref. 203. Copyright 2014 The Royal Society of Chemistry.

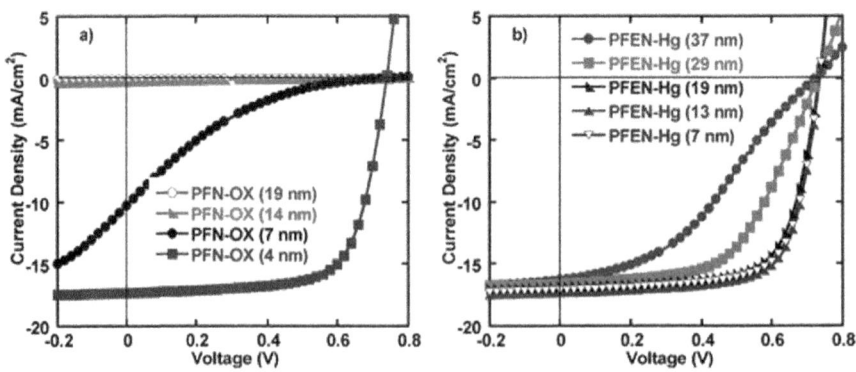

Figure 6.8   *J–V* characteristics of the inverted PSCs with PFN-OX (a) and PFEN-Hg (b) interlayers of varying thickness under AM1.5G irradiation (100 mW cm^{-2}). Reproduced with permission from ref. 204. Copyright 2013 American Chemical Society.

charge transfer. One approach that can improve the electrical and morphological properties of the metal oxide–active layer interface is to utilize a SAM between the inorganic and organic materials.[17,157,207] SAMs can be utilized to modify the interfaces of oxide and metallic surfaces significantly, to improve adhesion, compatibility, charge transfer properties, energy level alignment and affect the upper layer growth of materials. It was demonstrated that modifying the metal oxide surfaces of TiO_x, in a ZnO-based inverted PSC with a fullerene-based SAM (C_{60}-SAM), can improve the device performance, as shown in Figure 6.9. The C60-SAM affected the photo-induced charge transfer at the interface to reduce the recombination of charges, passivate inorganic surface trap states, improve the exciton dissociation efficiency at the polymer–metal oxide interface, and act as a template to influence the overlayer BHJ distribution of phases and crystallinity, leading to higher efficiency inverted PSCs.[19,177,207,208]

Recently, Chen's group presented a simple and rapid method for modification of ZnO as a cathode for inverted PSCs by doping it with a fullerene derivative (PCBE-OH, 0.5% wt% of the ZnO precursor) to give a ZnO-C_{60} nanofilm (40 nm) on ITO as the cathode. This ZnO-C_{60} cathode provided dual functionalities for enhanced electron collection, including producing a fullerene derivative-rich cathode surface and promotion of electron conductivity at the interface and bulk. For the device with ZnO-C_{60} as the ECL, the PCE was improved for PTB7-Th:PC_{71}BM from 7.64% to 9.35%, for PTB7:PC_{71}BM from 6.65% to 8.21%, and for P3HT:ICBA from 5.26% to 6.60%, compared to that with only ZnO ECL.[14] Liu reported GO-based electron extraction materials using cesium-neutralized graphene oxide (GO-Cs) as the cathode interfacial material in inverted PSCs. The inverted device, with device structure ITO/GO-Cs/P3HT:PCBM/GO/Al, showed a V_{OC} of 0.51 V, J_{SC} of 10.69 mA cm^{-2}, FF of 54% and PCE of 2.97%. The inverted devices based on GO-Cs ECL showed comparable photovoltaic performance to the corresponding standard BHJ solar cells with state-of-the-art electron-extraction layers.[158] Marks's group reported inserting nanoscopic copper hexadecafluorophthalocyanine (F_{16}CuPc) layers, as thin films for nanowires, between the ITO anode and

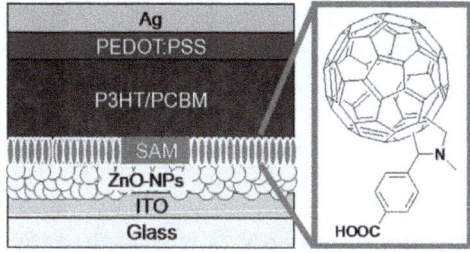

Figure 6.9 Device structure and chemical structure of an inverted solar cell with C_{60}-SAM modification. Reproduced with permission from ref. 207. Copyright 2008 American Institute of Physics.

the ZnO layer to increase PSC performance by enhancing interfacial electron transport. In inverted P3HT:PCBM cells, insertion of F_{16}CuPc nanowires increased the J_{SC} *vs.* cells with only ZnO layers, yielding an enhanced PCE of ~3.6% *vs.* ~3.0% for a control without the nanowire layer. Similar effects were observed for inverted PTB7:PC$_{71}$BM cells, where the PCE was increased from 8.1% to 8.6%.[209]

Tan *et al.* used an alcohol-soluble titanium chelate, titanium (diisopropoxide) bis (2,4-pentanedionate) (TIPD), as the ECL on ITO as a cathode in inverted PSCs with MoO$_3$/Al as anode. The active layer of the PSCs was composed of PBDTTT-C:PC$_{71}$BM blend. The buffer layer was prepared by spincoating the TIPD isopropanol solution on ITO electrode followed by thermal treatment at 60–170 °C for 10–30 min. The PCE of the inverted PSC with the TIPD buffer layer treated at 150 °C for 10 min reached 7.4%, which was increased by 16% in comparison with that (PCE of 6.4%) of the device with the conventional structure under the same experimental condition.[210] Later, the same group found that TIPD was changed to titanium(IV) oxide bis(2,4-pentanedionate) (TOPD) by thermal annealing. TOPD is also an alcohol-soluble titanium chelate. Therefore, they directly used TOPD as the ECL in the inverted PSCs based on a P3HT:PCBM active layer. The direct use of TOPD could further lower the annealing temperature and shorten the annealing time, *e.g.* when baked at 140 °C for 5 min. The PCE of the PSC with TOPD buffer layer reached 4%, which was increased by 76% in comparison with that (2.27%) of the inverted device without the TOPD layer under the same experimental conditions.[211]

6.5 Summary

Currently, efficiencies over 10% have been realized for state-of-the-art PSCs with small size. Further enhancements in both PCE and stability in large-size cells and modules are still needed to achieve commercial application of such a photovoltaic technology. The realization of such a target needs synergistic progress in the development of new active materials, the invention of new device structures, the development of efficient manufacturing processes, *etc.* Among these approaches, interface engineering plays a critical role, as we have discussed in detail above. The functions of the interlayer include many aspects, such as enhancing charge extraction, controlling electrode polarity and charge selectivity, enhancing light harvesting, tuning active layer morphology, improving device stability, and so on. In this chapter, we have also reviewed a wide range of materials employed as interfacial layers in PSCs according to the classification of device structures. All these materials show potential but none of them functions in all the above-mentioned aspects. Generally, each interfacial material has its advantages and shortcomings. Therefore, the development of new interfacial materials and further understanding of physical and chemical processes at the organic–inorganic interface are still highly critical to further enhancing the performance of PSCs.

References

1. J. G. G. Yu, J. C. Hummelen, F. Wudl and A. J. Heeger, *Science*, 1995, **270**, 1789.
2. L. Dou, J. You, Z. Hong, Z. Xu, G. Li, R. A. Street and Y. Yang, *Adv. Mater.*, 2013, **25**, 6642–6671.
3. Z. He, C. Zhong, S. Su, M. Xu, H. Wu and Y. Cao, *Nat. Photonics*, 2012, **6**, 591–595.
4. J. You, L. Dou, K. Yoshimura, T. Kato, K. Ohya, T. Moriarty, K. Emery, C.-C. Chen, J. Gao, G. Li and Y. Yang, *Nat. Commun.*, 2013, **4**, 1446.
5. C. H. Peters, I. T. Sachs-Quintana, J. P. Kastrop, S. Beaupré, M. Leclerc and M. D. McGehee, *Adv. Energy Mater.*, 2011, **1**, 491–494.
6. F. C. Krebs, S. A. Gevorgyan and J. Alstrup, *J. Mater. Chem.*, 2009, **19**, 5442–5451.
7. R. Po, C. Carbonera, A. Bernardi and N. Camaioni, *Energy Environ. Sci.*, 2011, **4**, 285–310.
8. C. Duan, K. Zhang, C. Zhong, F. Huang and Y. Cao, *Chem. Soc. Rev.*, 2013, **42**, 9071–9104.
9. E. L. Ratcliff, B. Zacher and N. R. Armstrong, *J. Phys. Chem. Lett.*, 2011, **2**, 1337–1350.
10. V. D. Mihailetchi, P. W. M. Blom, J. C. Hummelen and M. T. Rispens, *J. Appl. Phys.*, 2003, **94**, 6849–6854.
11. W. J. Potscavage, A. Sharma and B. Kippelen, *Acc. Chem. Res.*, 2009, **42**, 1758–1767.
12. P. W. M. Blom, V. D. Mihailetchi, L. J. A. Koster and D. E. Markov, *Adv. Mater.*, 2007, **19**, 1551–1566.
13. B. Qi and J. Wang, *Phys. Chem. Chem. Phys.*, 2013, **15**, 8972–8982.
14. S.-H. Liao, H.-J. Jhuo, Y.-S. Cheng and S.-A. Chen, *Adv. Mater.*, 2013, **25**, 4766–4771.
15. P. Škraba, G. Bratina, S. Igarashi, H. Nohira and K. Hirose, *Thin Solid Films*, 2011, **519**, 4216–4219.
16. K. W. Wong, H. L. Yip, Y. Luo, K. Y. Wong, W. M. Lau, K. H. Low, H. F. Chow, Z. Q. Gao, W. L. Yeung and C. C. Chang, *Appl. Phys. Lett.*, 2002, **80**, 2788–2790.
17. S. Khodabakhsh, B. M. Sanderson, J. Nelson and T. S. Jones, *Adv. Funct. Mater.*, 2006, **16**, 95–100.
18. Z. Xu, L.-M. Chen, G. Yang, C.-H. Huang, J. Hou, Y. Wu, G. Li, C.-S. Hsu and Y. Yang, *Adv. Funct. Mater.*, 2009, **19**, 1227–1234.
19. S. K. Hau, H.-L. Yip, O. Acton, N. S. Baek, H. Ma and A. K. Y. Jen, *J. Mater. Chem.*, 2008, **18**, 5113–5119.
20. X. Bulliard, S.-G. Ihn, S. Yun, Y. Kim, D. Choi, J.-Y. Choi, M. Kim, M. Sim, J.-H. Park, W. Choi and K. Cho, *Adv. Funct. Mater.*, 2010, **20**, 4381–4387.
21. C. Duan, C. Zhong, C. Liu, F. Huang and Y. Cao, *Chem. Mater.*, 2012, **24**, 1682–1689.
22. M. Campoy-Quiles, T. Ferenczi, T. Agostinelli, P. G. Etchegoin, Y. Kim, T. D. Anthopoulos, P. N. Stavrinou, D. D. Bradley and J. Nelson, *Nat. Mater.*, 2008, **7**, 158–164.

23. D. H. Kim, Y. D. Park, Y. Jang, H. Yang, Y. H. Kim, J. I. Han, D. G. Moon, S. Park, T. Chang, C. Chang, M. Joo, C. Y. Ryu and K. Cho, *Adv. Funct. Mater.*, 2005, **15**, 77–82.

24. L. Y. Park, A. M. Munro and D. S. Ginger, *J. Am. Chem. Soc.*, 2008, **130**, 15916–15926.

25. K. Tvingstedt, V. Andersson, F. Zhang and O. Inganäs, *Appl. Phys. Lett.*, 2007, **91**, 123514.

26. S.-I. Na, S.-S. Kim, J. Jo, S.-H. Oh, J. Kim and D.-Y. Kim, *Adv. Funct. Mater.*, 2008, **18**, 3956–3963.

27. M.-S. Kim, J.-S. Kim, J. C. Cho, M. Shtein, L. J. Guo and J. Kim, *Appl. Phys. Lett.*, 2007, **90**, 123113.

28. M. Agrawal and P. Peumans, *Opt. Express*, 2008, **16**, 5385–5396.

29. S. Esiner, T. Bus, M. M. Wienk, K. Hermans and R. A. J. Janssen, *Adv. Energy Mater.*, 2013, **3**, 1013–1017.

30. J. Y. Kim, S. H. Kim, H. H. Lee, K. Lee, W. Ma, X. Gong and A. J. Heeger, *Adv. Mater.*, 2006, **18**, 572–576.

31. J. Gilot, I. Barbu, M. M. Wienk and R. A. J. Janssen, *Appl. Phys. Lett.*, 2007, **91**, 113520.

32. K. Tvingstedt, N.-K. Persson, O. Inganäs, A. Rahachou and I. V. Zozoulenko, *Appl. Phys. Lett.*, 2007, **91**, 113514.

33. X. Li, W. C. H. Choy, L. Huo, F. Xie, W. E. I. Sha, B. Ding, X. Guo, Y. Li, J. Hou, J. You and Y. Yang, *Adv. Mater.*, 2012, **24**, 3046–3052.

34. K. Yao, M. Salvador, C.-C. Chueh, X.-K. Xin, Y.-X. Xu, D. W. deQuilettes, T. Hu, Y. Chen, D. S. Ginger and A. K. Y. Jen, *Adv. Energy Mater.*, 2014, **4**, 1400206.

35. A. P. Kulkarni, K. M. Noone, K. Munechika, S. R. Guyer and D. S. Ginger, *Nano Lett.*, 2010, **10**, 1501–1505.

36. J.-L. Wu, F.-C. Chen, Y.-S. Hsiao, F.-C. Chien, P. Chen, C.-H. Kuo, M. H. Huang and C.-S. Hsu, *ACS Nano*, 2011, **5**, 959–967.

37. J. Yang, J. You, C.-C. Chen, W.-C. Hsu, H.-r. Tan, X. W. Zhang, Z. Hong and Y. Yang, *ACS Nano*, 2011, **5**, 6210–6217.

38. H.-Y. Chen, J. Hou, S. Zhang, Y. Liang, G. Yang, Y. Yang, L. Yu, Y. Wu and G. Li, *Nat. Photonics*, 2009, **3**, 649–653.

39. Z. He, C. Zhong, X. Huang, W.-Y. Wong, H. Wu, L. Chen, S. Su and Y. Cao, *Adv. Mater.*, 2011, **23**, 4636–4643.

40. M. P. de Jong, L. J. van IJzendoorn and M. J. A. de Voigt, *Appl. Phys. Lett.*, 2000, **77**, 2255–2257.

41. Y.-H. Kim, S.-H. Lee, J. Noh and S.-H. Han, *Thin Solid Films*, 2006, **510**, 305–310.

42. L. S. C. Pingree, B. A. MacLeod and D. S. Ginger, *J. Phys. Chem. C*, 2008, **112**, 7922–7927.

43. S. K. Hau, H.-L. Yip, J. Zou and A. K. Y. Jen, *Org. Electron.*, 2009, **10**, 1401–1407.

44. T. Xiao, W. Cui, J. Anderegg, J. Shinar and R. Shinar, *Org. Electron.*, 2011, **12**, 257–262.

45. B. Peng, X. Guo, C. Cui, Y. Zou, C. Pan and Y. Li, *Appl. Phys. Lett.*, 2011, **98**, 243308.

46. V. Shrotriya, G. Li, Y. Yao, C.-W. Chu and Y. Yang, *Appl. Phys. Lett.*, 2006, **88**, 073508.
47. Y. Sun, C. J. Takacs, S. R. Cowan, J. H. Seo, X. Gong, A. Roy and A. J. Heeger, *Adv. Mater.*, 2011, **23**, 2226–2230.
48. F. Liu, S. Shao, X. Guo, Y. Zhao and Z. Xie, *Sol. Energy Mater. Sol. Cells*, 2010, **94**, 842–845.
49. J. Meyer, R. Khalandovsky, P. Görrn and A. Kahn, *Adv. Mater.*, 2011, **23**, 70–73.
50. C. Girotto, E. Voroshazi, D. Cheyns, P. Heremans and B. P. Rand, *ACS Appl. Mater. Interfaces*, 2011, **3**, 3244–3247.
51. T. Yang, M. Wang, Y. Cao, F. Huang, L. Huang, J. Peng, X. Gong, S. Z. D. Cheng and Y. Cao, *Adv. Energy Mater.*, 2012, **2**, 523–527.
52. S. Murase and Y. Yang, *Adv. Mater.*, 2012, **24**, 2459–2462.
53. J. J. Jasieniak, J. Seifter, J. Jo, T. Mates and A. J. Heeger, *Adv. Funct. Mater.*, 2012, **22**, 2594–2605.
54. S. Han, W. S. Shin, M. Seo, D. Gupta, S.-J. Moon and S. Yoo, *Org. Electron.*, 2009, **10**, 791–797.
55. M. D. Irwin, D. B. Buchholz, A. W. Hains, R. P. H. Chang and T. J. Marks, *Proc. Natl. Acad. Sci. U. S. A.*, 2008, **105**, 2783–2787.
56. H.-L. Yip and A. K. Y. Jen, *Energy Environ. Sci.*, 2012, **5**, 5994–6011.
57. J. Meyer, S. Hamwi, M. Kröger, W. Kowalsky, T. Riedl and A. Kahn, *Adv. Mater.*, 2012, **24**, 5408–5427.
58. K. Zilberberg, S. Trost, H. Schmidt and T. Riedl, *Adv. Energy Mater.*, 2011, **1**, 377–381.
59. H.-Q. Wang, N. Li, N. S. Guldal and C. J. Brabec, *Org. Electron.*, 2012, **13**, 3014–3021.
60. F. Xie, W. C. H. Choy, C. Wang, X. Li, S. Zhang and J. Hou, *Adv. Mater.*, 2013, **25**, 2051–2055.
61. A. R. B. Mohd Yusoff, H. P. Kim and J. Jang, *Org. Electron.*, 2013, **14**, 858–861.
62. Z. A. Tan, L. Li, C. Cui, Y. Ding, Q. Xu, S. Li, D. Qian and Y. Li, *J. Phys. Chem. C*, 2012, **116**, 18626–18632.
63. T. Stubhan, N. Li, N. A. Luechinger, S. C. Halim, G. J. Matt and C. J. Brabec, *Adv. Energy Mater.*, 2012, **2**, 1433–1438.
64. K. X. Steirer, J. P. Chesin, N. E. Widjonarko, J. J. Berry, A. Miedaner, D. S. Ginley and D. C. Olson, *Org. Electron.*, 2010, **11**, 1414–1418.
65. K. X. Steirer, P. F. Ndione, N. E. Widjonarko, M. T. Lloyd, J. Meyer, E. L. Ratcliff, A. Kahn, N. R. Armstrong, C. J. Curtis, D. S. Ginley, J. J. Berry and D. C. Olson, *Adv. Energy Mater.*, 2011, **1**, 813–820.
66. J. S. Kim, J. H. Park, J. H. Lee, J. Jo, D.-Y. Kim and K. Cho, *Appl. Phys. Lett.*, 2007, **91**, 11211.
67. G. Heimel, L. Romaner, E. Zojer and J.-L. Bredas, *Acc. Chem. Res.*, 2008, **41**, 721–729.
68. N. Beaumont, I. Hancox, P. Sullivan, R. A. Hatton and T. S. Jones, *Energy Environ. Sci.*, 2011, **4**, 1708–1711.
69. H. Wang, E. D. Gomez, Z. Guan, C. Jaye, M. F. Toney, D. A. Fischer, A. Kahn and Y.-L. Loo, *J. Phys. Chem. C*, 2013, **117**, 20474–20484.

70. C.-L. Liu, X.-Y. Liu and L. J. Borucki, *Appl. Phys. Lett.*, 1999, **74**, 34–36.

71. R. Steim, F. R. Kogler and C. J. Brabec, *J. Mater. Chem.*, 2010, **20**, 2499–2512.

72. M. Campoy-Quiles, T. Ferenczi, T. Agostinelli, P. G. Etchegoin, Y. Kim, T. D. Anthopoulos, P. N. Stavrinou, D. D. C. Bradley and J. Nelson, *Nat. Mater.*, 2008, **7**, 158–164.

73. K.-H. Tu, S.-S. Li, W.-C. Li, D.-Y. Wang, J.-R. Yang and C.-W. Chen, *Energy Environ. Sci.*, 2011, **4**, 3521–3526.

74. H. P. Kim, A. R. B. Mohd Yusoff, M. S. Ryu and J. Jang, *Org. Electron.*, 2012, **13**, 3195–3202.

75. J. Liu, M. Durstock and L. Dai, *Energy Environ. Sci.*, 2014, **7**, 1297–1306.

76. S.-S. Li, K.-H. Tu, C.-C. Lin, C.-W. Chen and M. Chhowalla, *ACS Nano*, 2010, **4**, 3169–3174.

77. J.-M. Yun, J.-S. Yeo, J. Kim, H.-G. Jeong, D.-Y. Kim, Y.-J. Noh, S.-S. Kim, B.-C. Ku and S.-I. Na, *Adv. Mater.*, 2011, **23**, 4923–4928.

78. J. Liu, Y. Xue and L. Dai, *J. Phys. Chem. Lett.*, 2012, **3**, 1928–1933.

79. Y.-J. Jeon, J.-M. Yun, D.-Y. Kim, S.-I. Na and S.-S. Kim, *Sol. Energy Mater. Sol. Cells*, 2012, **105**, 96–102.

80. V. Sgobba and D. M. Guldi, *J. Mater. Chem.*, 2008, **18**, 153–157.

81. S. Chaudhary, H. Lu, A. M. Müller, C. J. Bardeen and M. Ozkan, *Nano Lett.*, 2007, **7**, 1973–1979.

82. R. A. Hatton, N. P. Blanchard, L. W. Tan, G. Latini, F. Cacialli and S. R. P. Silva, *Org. Electron.*, 2009, **10**, 388–395.

83. J. Liu, G.-H. Kim, Y. Xue, J. Y. Kim, J.-B. Baek, M. Durstock and L. Dai, *Adv. Mater.*, 2014, **26**, 786–790.

84. M. S. Ryu and J. Jang, *Sol. Energy Mater. Sol. Cells*, 2011, **95**, 2893–2896.

85. B. Yin, Q. Liu, L. Yang, X. Wu, Z. Liu, Y. Hua, S. Yin and Y. Chen, *J. Nanosci. Nanotechnol.*, 2010, **10**, 1934–1938.

86. W. Kim, J. Kyu Kim, Y. Lim, I. Park, Y. Suk Choi and J. Hyeok Park, *Sol. Energy Mater. Sol. Cells*, 2014, **122**, 24–30.

87. C.-Y. Li, T.-C. Wen and T.-F. Guo, *J. Mater. Chem.*, 2008, **18**, 4478–4482.

88. J. W. Jung, J. U. Lee and W. H. Jo, *J. Phys. Chem. C*, 2009, **114**, 633–637.

89. M.-R. Choi, T.-H. Han, K.-G. Lim, S.-H. Woo, D. H. Huh and T.-W. Lee, *Angew. Chem., Int. Ed.*, 2011, **50**, 6274–6277.

90. A. W. Hains, J. Liu, A. B. F. Martinson, M. D. Irwin and T. J. Marks, *Adv. Funct. Mater.*, 2010, **20**, 595–606.

91. D. Gupta, M. Bag and K. S. Narayan, *Appl. Phys. Lett.*, 2008, **92**, 093301.

92. M. O. Reese, M. S. White, G. Rumbles, D. S. Ginley and S. E. Shaheen, *Appl. Phys. Lett.*, 2008, **92**, 053307.

93. H.-W. Lin, H.-W. Kang, Z.-Y. Huang, C.-W. Chen, Y.-H. Chen, L.-Y. Lin, F. Lin and K.-T. Wong, *Org. Electron.*, 2012, **13**, 1925–1929.

94. M. Jørgensen, K. Norrman and F. C. Krebs, *Sol. Energy Mater. Sol. Cells*, 2008, **92**, 686–714.

95. C. J. Brabec, S. E. Shaheen, C. Winder, N. S. Sariciftci and P. Denk, *Appl. Phys. Lett.*, 2002, **80**, 1288–1290.

96. S. K. M. Jönsson, E. Carlegrim, F. Zhang, W. R. Salaneck and M. Fahlman, *Jpn. J. Appl. Phys.*, 2005, **44**, 3695–3701.

97. B. N. Limketkai and M. A. Baldo, *Phys. Rev. B*, 2005, **71**, 085207.

98. X. Jiang, H. Xu, L. Yang, M. Shi, M. Wang and H. Chen, *Sol. Energy Mater. Sol. Cells*, 2009, **93**, 650–653.

99. E. Ahlswede, J. Hanisch and M. Powalla, *Appl. Phys. Lett.*, 2007, **90**, 163504.

100. J. Huang, Z. Xu and Y. Yang, *Adv. Funct. Mater.*, 2007, **17**, 1966–1973.

101. F.-C. Chen, J.-L. Wu, S. S. Yang, K.-H. Hsieh and W.-C. Chen, *J. Appl. Phys.*, 2008, **103**, 103721.

102. T. R. Briere and A. H. Sommer, *J. Appl. Phys.*, 1977, **48**, 3547–3550.

103. M.-H. Chen and C.-I. Wu, *J. Appl. Phys.*, 2008, **104**, 113713.

104. H.-H. Liao, L.-M. Chen, Z. Xu, G. Li and Y. Yang, *Appl. Phys. Lett.*, 2008, **92**, 173303.

105. J. K. Lee, N. E. Coates, S. Cho, N. S. Cho, D. Moses, G. C. Bazan, K. Lee and A. J. Heeger, *Appl. Phys. Lett.*, 2008, **92**, 243308.

106. A. Roy, S. H. Park, S. Cowan, M. H. Tong, S. Cho, K. Lee and A. J. Heeger, *Appl. Phys. Lett.*, 2009, **95**, 013302.

107. S. H. Park, A. Roy, S. Beaupre, S. Cho, N. Coates, J. S. Moon, D. Moses, M. Leclerc, K. Lee and A. J. Heeger, *Nat. Photonics*, 2009, **3**, 297–302.

108. J. H. Lee, S. Cho, A. Roy, H.-T. Jung and A. J. Heeger, *Appl. Phys. Lett.*, 2010, **96**, 163303.

109. K. Lee, J. Y. Kim, S. H. Park, S. H. Kim, S. Cho and A. J. Heeger, *Adv. Mater.*, 2007, **19**, 2445–2449.

110. A. Hayakawa, O. Yoshikawa, T. Fujieda, K. Uehara and S. Yoshikawa, *Appl. Phys. Lett.*, 2007, **90**, 163517.

111. D. H. Wang, S. H. Im, H. K. Lee, O. O. Park and J. H. Park, *J. Phys. Chem. C*, 2009, **113**, 17268–17273.

112. M.-H. Park, J.-H. Li, A. Kumar, G. Li and Y. Yang, *Adv. Funct. Mater.*, 2009, **19**, 1241–1246.

113. A. L. Roest, J. J. Kelly, D. Vanmaekelbergh and E. A. Meulenkamp, *Phys. Rev. Lett.*, 2002, **89**, 036801.

114. H. Faber, M. Burkhardt, A. Jedaa, D. Kälblein, H. Klauk and M. Halik, *Adv. Mater.*, 2009, **21**, 3099–3104.

115. J. Gilot, M. M. Wienk and R. A. J. Janssen, *Appl. Phys. Lett.*, 2007, **90**, 143512.

116. S. B. Jo, J. H. Lee, M. Sim, M. Kim, J. H. Park, Y. S. Choi, Y. Kim, S.-G. Ihn and K. Cho, *Adv. Energy Mater.*, 2011, **1**, 690–698.

117. P. S. Mbule, T. H. Kim, B. S. Kim, H. C. Swart and O. M. Ntwaeaborwa, *Sol. Energy Mater. Sol. Cells*, 2013, **112**, 6–12.

118. F. Huang, H. Wu and Y. Cao, *Chem. Soc. Rev.*, 2010, **39**, 2500–2521.

119. C. Zhong, C. Duan, F. Huang, H. Wu and Y. Cao, *Chem. Mater.*, 2010, **23**, 326–340.

120. F. Huang, H. Wu, D. Wang, W. Yang and Y. Cao, *Chem. Mater.*, 2004, **16**, 708–716.

121. H. Wu, F. Huang, Y. Mo, W. Yang, D. Wang, J. Peng and Y. Cao, *Adv. Mater.*, 2004, **16**, 1826–1830.
122. H. Wu, F. Huang, J. Peng and Y. Cao, *Org. Electron.*, 2005, **6**, 118–128.
123. H. Wu, F. Huang, J. Peng and Y. Cao, *Synth. Met.*, 2005, **153**, 197–200.
124. C. He, C. Zhong, H. Wu, R. Yang, W. Yang, F. Huang, G. C. Bazan and Y. Cao, *J. Mater. Chem.*, 2010, **20**, 2617–2622.
125. L. Zhang, C. He, J. Chen, P. Yuan, L. Huang, C. Zhang, W. Cai, Z. Liu and Y. Cao, *Macromolecules*, 2010, **43**, 9771–9778.
126. Z. He, C. Zhang, X. Xu, L. Zhang, L. Huang, J. Chen, H. Wu and Y. Cao, *Adv. Mater.*, 2011, **23**, 3086–3089.
127. W. Li, A. Furlan, K. H. Hendriks, M. M. Wienk and R. A. J. Janssen, *J. Am. Chem. Soc.*, 2013, **135**, 5529–5532.
128. Y. Zhao, Z. Xie, C. Qin, Y. Qu, Y. Geng and L. Wang, *Sol. Energy Mater. Sol. Cells*, 2009, **93**, 604–608.
129. X. Xu, W. Cai, J. Chen and Y. Cao, *J. Polym. Sci., Part A: Polym. Chem.*, 2011, **49**, 1263–1272.
130. S.-I. Na, S.-H. Oh, S.-S. Kim and D.-Y. Kim, *Org. Electron.*, 2009, **10**, 496–500.
131. S.-H. Oh, S.-I. Na, J. Jo, B. Lim, D. Vak and D.-Y. Kim, *Adv. Funct. Mater.*, 2010, **20**, 1977–1983.
132. J. H. Seo, A. Gutacker, Y. Sun, H. Wu, F. Huang, Y. Cao, U. Scherf, A. J. Heeger and G. C. Bazan, *J. Am. Chem. Soc.*, 2011, **133**, 8416–8419.
133. J. Kesters, T. Ghoos, H. Penxten, J. Drijkoningen, T. Vangerven, D. M. Lyons, B. Verreet, T. Aernouts, L. Lutsen, D. Vanderzande, J. Manca and W. Maes, *Adv. Energy Mater.*, 2013, **3**, 1180–1185.
134. C. Hoven, R. Yang, A. Garcia, A. J. Heeger, T.-Q. Nguyen and G. C. Bazan, *J. Am. Chem. Soc.*, 2007, **129**, 10976–10977.
135. A. Garcia, R. C. Bakus Ii, P. Zalar, C. V. Hoven, J. Z. Brzezinski and T.-Q. Nguyen, *J. Am. Chem. Soc.*, 2011, **133**, 2492–2498.
136. K. Meerholz, *Nature*, 2005, **437**, 327–328.
137. D. A. Rider, B. J. Worfolk, K. D. Harris, A. Lalany, K. Shahbazi, M. D. Fleischauer, M. J. Brett and J. M. Buriak, *Adv. Funct. Mater.*, 2010, **20**, 2404–2415.
138. J. M. Hodgkiss, G. Tu, S. Albert-Seifried, W. T. S. Huck and R. H. Friend, *J. Am. Chem. Soc.*, 2009, **131**, 8913–8921.
139. Y.-M. Chang, R. Zhu, E. Richard, C.-C. Chen, G. Li and Y. Yang, *Adv. Funct. Mater.*, 2012, **22**, 3284–3289.
140. C. Duan, K. Zhang, X. Guan, C. Zhong, H. Xie, F. Huang, J. Chen, J. Peng and Y. Cao, *Chem. Sci.*, 2013, **4**, 1298–1307.
141. F. Liu, Z. A. Page, V. V. Duzhko, T. P. Russell and T. Emrick, *Adv. Mater.*, 2013, **25**, 6868–6873.
142. X. Guan, K. Zhang, F. Huang, G. C. Bazan and Y. Cao, *Adv. Funct. Mater.*, 2012, **22**, 2846–2854.
143. F. Zhang, M. Ceder and O. Inganäs, *Adv. Mater.*, 2007, **19**, 1835–1838.
144. Y. Zhou, C. Fuentes-Hernandez, J. Shim, J. Meyer, A. J. Giordano, H. Li, P. Winget, T. Papadopoulos, H. Cheun, J. Kim, M. Fenoll, A. Dindar, W. Haske, E. Najafabadi, T. M. Khan, H. Sojoudi, S. Barlow, S. Graham, J.

L. Bredas, S. R. Marder, A. Kahn and B. Kippelen, *Science*, 2012, **336**, 327–332.

145. Y. Zhou, C. Fuentes-Hernandez, J. W. Shim, T. M. Khan and B. Kippelen, *Energy Environ. Sci.*, 2012, **5**, 9827–9832.

146. K. Zhang, X. Guan, F. Huang and Y. Cao, *Acta Chim. Sin.*, 2012, **70**, 2489.

147. T. V. Pho, H. Kim, J. H. Seo, A. J. Heeger and F. Wudl, *Adv. Funct. Mater.*, 2011, **21**, 4338–4341.

148. H. Ye, X. Hu, Z. Jiang, D. Chen, X. Liu, H. Nie, S.-J. Su, X. Gong and Y. Cao, *J. Mater. Chem. A*, 2013, **1**, 3387–3394.

149. D. Chen, H. Zhou, M. Liu, W.-M. Zhao, S.-J. Su and Y. Cao, *Macromol. Rapid Commun.*, 2013, **34**, 595–603.

150. K. Sun, B. Zhao, V. Murugesan, A. Kumar, K. Zeng, J. Subbiah, W. W. H. Wong, D. J. Jones and J. Ouyang, *J. Mater. Chem.*, 2012, **22**, 24155–24165.

151. C.-Z. Li, C.-C. Chueh, H.-L. Yip, K. M. O'Malley, W.-C. Chen and A. K. Y. Jen, *J. Mater. Chem.*, 2012, **22**, 8574–8578.

152. K. M. O'Malley, C.-Z. Li, H.-L. Yip and A. K. Y. Jen, *Adv. Energy Mater.*, 2012, **2**, 82–86.

153. Y. Yuan, T. J. Reece, P. Sharma, S. Poddar, S. Ducharme, A. Gruverman, Y. Yang and J. Huang, *Nat. Mater.*, 2011, **10**, 296–302.

154. Z. Xiao, Q. Dong, P. Sharma, Y. Yuan, B. Mao, W. Tian, A. Gruverman and J. Huang, *Adv. Energy Mater.*, 2013, **3**, 1581–1588.

155. K. Asadi, P. de Bruyn, P. W. M. Blom and D. M. de Leeuw, *Appl. Phys. Lett.*, 2011, **98**, 183301.

156. M. Vasilopoulou, D. G. Georgiadou, A. M. Douvas, A. Soultati, V. Constantoudis, D. Davazoglou, S. Gardelis, L. C. Palilis, M. Fakis, S. Kennou, T. Lazarides, A. G. Coutsolelos and P. Argitis, *J. Mater. Chem. A*, 2014, **2**, 182–192.

157. H.-L. Yip, S. K. Hau, N. S. Baek, H. Ma and A. K. Y. Jen, *Adv. Mater.*, 2008, **20**, 2376–2382.

158. J. Liu, Y. Xue, Y. Gao, D. Yu, M. Durstock and L. Dai, *Adv. Mater.*, 2012, **24**, 2228–2233.

159. S. Qu, M. Li, L. Xie, X. Huang, J. Yang, N. Wang and S. Yang, *ACS Nano*, 2013, **7**, 4070–4081.

160. K. D. G. I. Jayawardena, R. Rhodes, K. K. Gandhi, M. R. R. Prabhath, G. D. M. R. Dabera, M. J. Beliatis, L. J. Rozanski, S. J. Henley and S. R. P. Silva, *J. Mater. Chem. A*, 2013, **1**, 9922–9927.

161. D. H. Wang, J. K. Kim, J. H. Seo, I. Park, B. H. Hong, J. H. Park and A. J. Heeger, *Angew. Chem., Int. Ed.*, 2013, **52**, 2874–2880.

162. G. D. Sharma, M. L. Keshtov, A. R. Khokhlov, D. Tasis and C. Galiotis, *Org. Electron.*, 2014, **15**, 348–355.

163. Z. A. Tan, C. Yang, E. Zhou, X. Wang and Y. Li, *Appl. Phys. Lett.*, 2007, **91**, 023509.

164. L. C. Palilis, M. Vasilopoulou, A. M. Douvas, D. G. Georgiadou, S. Kennou, N. A. Stathopoulos, V. Constantoudis and P. Argitis, *Sol. Energy Mater. Sol. Cells*, 2013, **114**, 205–213.

165. C. Tao, S. Ruan, X. Zhang, G. Xie, L. Shen, X. Kong, W. Dong, C. Liu and W. Chen, *Appl. Phys. Lett.*, 2008, **93**, 193307.

166. A. K. K. Kyaw, X. W. Sun, C. Y. Jiang, G. Q. Lo, D. W. Zhao and D. L. Kwong, *Appl. Phys. Lett.*, 2008, **93**, 221107.

167. G. Li, C.-W. Chu, V. Shrotriya, J. Huang and Y. Yang, *Appl. Phys. Lett.*, 2006, **88**, 253503.

168. K. Takanezawa, K. Tajima and K. Hashimoto, *Appl. Phys. Lett.*, 2008, **93**, 063308.

169. W. Yu, L. Shen, S. Ruan, F. Meng, J. Wang, E. Zhang and W. Chen, *Sol. Energy Mater. Sol. Cells*, 2012, **98**, 212–215.

170. C. Tao, S. Ruan, G. Xie, X. Kong, L. Shen, F. Meng, C. Liu, X. Zhang, W. Dong and W. Chen, *Appl. Phys. Lett.*, 2009, **94**, 043311.

171. Y.-J. Lee, J. Yi, G. F. Gao, H. Koerner, K. Park, J. Wang, K. Luo, R. A. Vaia and J. W. P. Hsu, *Adv. Energy Mater.*, 2012, **2**, 1193–1197.

172. X. Li, W. C. H. Choy, F. Xie, S. Zhang and J. Hou, *J. Mater. Chem. A*, 2013, **1**, 6614–6621.

173. J.-S. Huang, C.-Y. Chou, M.-Y. Liu, K.-H. Tsai, W.-H. Lin and C.-F. Lin, *Org. Electron.*, 2009, **10**, 1060–1065.

174. C.-P. Chen, Y.-D. Chen and S.-C. Chuang, *Adv. Mater.*, 2011, **23**, 3859–3863.

175. D.-C. Lim, Y.-T. Kim, W.-H. Shim, A. Y. Jang, J.-H. Lim, Y.-D. Kim, Y.-S. Jeong, Y.-D. Kim and K.-H. Lee, *Bull. Korean Chem. Soc.*, 2011, **32**, 1067–1070.

176. Y.-H. Lin, P.-C. Yang, J.-S. Huang, G.-D. Huang, I.-J. Wang, W.-H. Wu, M.-Y. Lin, W.-F. Su and C.-F. Lin, *Sol. Energy Mater. Sol. Cells*, 2011, **95**, 2511–2515.

177. Y. Gao, H.-L. Yip, S. K. Hau, K. M. O'Malley, N. C. Cho, H. Chen and A. K.-Y. Jen, *Appl. Phys. Lett.*, 2010, **97**, 203306.

178. Y. Gao, H.-L. Yip, K.-S. Chen, K. M. O'Malley, O. Acton, Y. Sun, G. Ting, H. Chen and A. K. Y. Jen, *Adv. Mater.*, 2011, **23**, 1903–1908.

179. D. W. Zhao, P. Liu, X. W. Sun, S. T. Tan, L. Ke and A. K. K. Kyaw, *Appl. Phys. Lett.*, 2009, **95**, 153304.

180. J. Huang, G. Li and Y. Yang, *Adv. Mater.*, 2008, **20**, 415–419.

181. J.-C. Wang, W.-T. Weng, M.-Y. Tsai, M.-K. Lee, S.-F. Horng, T.-P. Perng, C.-C. Kei, C.-C. Yu and H.-F. Meng, *J. Mater. Chem.*, 2010, **20**, 862–866.

182. T. Yang, W. Cai, D. Qin, E. Wang, L. Lan, X. Gong, J. Peng and Y. Cao, *J. Phys. Chem. C*, 2010, **114**, 6849–6853.

183. N. Sekine, C.-H. Chou, W. L. Kwan and Y. Yang, *Org. Electron.*, 2009, **10**, 1473–1477.

184. C.-H. Huang, C.-H. Huang, T.-P. Nguyen and C.-S. Hsu, *Thin Solid Films*, 2007, **515**, 6493–6496.

185. C.-Y. Li, T.-C. Wen, T.-H. Lee, T.-F. Guo, J.-C.-A. Huang, Y.-C. Lin and Y.-J. Hsu, *J. Mater. Chem.*, 2009, **19**, 1643–1647.

186. B. Y. Yu, A. Tsai, S. P. Tsai, K. T. Wong, Y. Yang, C. W. Chu and J. J. Shyue, *Nanotechnology*, 2008, **19**, 255202.

187. C. Waldauf, M. Morana, P. Denk, P. Schilinsky, K. Coakley, S. A. Choulis and C. J. Brabec, *Appl. Phys. Lett.*, 2006, **89**, 233517.

188. M. S. White, D. C. Olson, S. E. Shaheen, N. Kopidakis and D. S. Ginley, *Appl. Phys. Lett.*, 2006, **89**, 143517.

189. Y. Sun, J. H. Seo, C. J. Takacs, J. Seifter and A. J. Heeger, *Adv. Mater.*, 2011, **23**, 1679–1683.

190. S. K. Hau, H.-L. Yip, N. S. Baek, J. Zou, K. O'Malley and A. K.-Y. Jen, *Appl. Phys. Lett.*, 2008, **92**, 253301.

191. J. Zou, H.-L. Yip, Y. Zhang, Y. Gao, S.-C. Chien, K. O'Malley, C.-C. Chueh, H. Chen and A. K. Y. Jen, *Adv. Funct. Mater.*, 2012, **22**, 2804–2811.

192. M. A. Ibrahem, H.-Y. Wei, M.-H. Tsai, K.-C. Ho, J.-J. Shyue and C. W. Chu, *Sol. Energy Mater. Sol. Cells*, 2013, **108**, 156–163.

193. Y. Dong, X. Hu, C. Duan, P. Liu, S. Liu, L. Lan, D. Chen, L. Ying, S. Su, X. Gong, F. Huang and Y. Cao, *Adv. Mater.*, 2013, **25**, 3683–3688.

194. T. Yang, M. Wang, C. Duan, X. Hu, L. Huang, J. Peng, F. Huang and X. Gong, *Energy Environ. Sci.*, 2012, **5**, 8208–8214.

195. C.-H. Hsieh, Y.-J. Cheng, P.-J. Li, C.-H. Chen, M. Dubosc, R.-M. Liang and C.-S. Hsu, *J. Am. Chem. Soc.*, 2010, **132**, 4887–4893.

196. Y.-J. Cheng, C.-H. Hsieh, Y. He, C.-S. Hsu and Y. Li, *J. Am. Chem. Soc.*, 2010, **132**, 17381–17383.

197. Y. Zhou, F. Li, S. Barrau, W. Tian, O. Inganäs and F. Zhang, *Sol. Energy Mater. Sol. Cells*, 2009, **93**, 497–500.

198. Y. Zhu, X. Xu, L. Zhang, J. Chen and Y. Cao, *Sol. Energy Mater. Sol. Cells*, 2012, **97**, 83–88.

199. F. C. Krebs, *Sol. Energy Mater. Sol. Cells*, 2009, **93**, 465–475.

200. T. H. Reilly, A. W. Hains, H.-Y. Chen and B. A. Gregg, *Adv. Energy Mater.*, 2012, **2**, 455–460.

201. C.-Z. Li, C.-C. Chueh, H.-L. Yip, F. Ding, X. Li and A. K. Y. Jen, *Adv. Mater.*, 2013, **25**, 2457–2461.

202. C.-Z. Li, C.-C. Chueh, F. Ding, H.-L. Yip, P.-W. Liang, X. Li and A. K. Y. Jen, *Adv. Mater.*, 2013, **25**, 4425–4430.

203. N. Cho, C.-Z. Li, H.-L. Yip and A. K. Y. Jen, *Energy Environ. Sci.*, 2014, **7**, 638–643.

204. S. Liu, K. Zhang, J. Lu, J. Zhang, H.-L. Yip, F. Huang and Y. Cao, *J. Am. Chem. Soc.*, 2013, **135**, 15326–15329.

205. B. de Boer, A. Hadipour, M. M. Mandoc, T. van Woudenbergh and P. W. M. Blom, *Adv. Mater.*, 2005, **17**, 621–625.

206. L.-L. Chua, J. Zaumseil, J.-F. Chang, E. C. W. Ou, P. K. H. Ho, H. Sirringhaus and R. H. Friend, *Nature*, 2005, **434**, 194–199.

207. S. K. Hau, H.-L. Yip, H. Ma and A. K.-Y. Jen, *Appl. Phys. Lett.*, 2008, **93**, 233304.

208. H. Choi, J. Lee, W. Lee, S.-J. Ko, R. Yang, J. C. Lee, H. Y. Woo, C. Yang and J. Y. Kim, *Org. Electron.*, 2013, **14**, 3138–3145.

209. S. M. Yoon, S. J. Lou, S. Loser, J. Smith, L. X. Chen, A. Facchetti and T. Marks, *Nano Lett.*, 2012, **12**, 6315–6321.

210. Z. A. Tan, W. Zhang, Z. Zhang, D. Qian, Y. Huang, J. Hou and Y. Li, *Adv. Mater.*, 2012, **24**, 1476–1481.

211. F. Wang, Q. Xu, Z. A. Tan, D. Qian, Y. Ding, L. Li, S. Li and Y. Li, *Org. Electron.*, 2012, **13**, 2429–2435.

CHAPTER 7

Solution Processed Metal Oxides and Hybrid Metal Oxides as Efficient Carrier Transport Layers of Organic Optoelectronic Devices

WALLACE C. H. CHOY*[a]

[a]Department of Electrical and Electronic Engineering, The University of Hong Kong, Pokfulam Road, Hong Kong
*E-mail: chchoy@eee.hku.hk

7.1   Introduction

Organic/polymer solar cells (OSCs) have now come to the stage of photo-voltaics to rival the conventional silicon based solar cell. The power conversion efficiency of single/bulk heterojunction OSCs can reach 9% or above.[1–4] Although the organic/polymer solar cells at present have lower power conversion efficiency and shorter lifetime than silicon based solar cells, the OSCs have several advantages including cost-effective production by printing and coating technologies, flexible structures, light weight and semi-transparency.[5]

The performance of OSCs strongly depends on the extraction of charges (electrons and holes) to respective electrodes. In order to achieve efficient

RSC Polymer Chemistry Series No. 17
Polymer Photovoltaics: Materials, Physics, and Device Engineering
Edited by Fei Huang, Hin-Lap Yip, and Yong Cao
© The Royal Society of Chemistry 2016
Published by the Royal Society of Chemistry, www.rsc.org

charge carrier extraction, the hurdle between electrode and active layer should be eliminated. The insertion of a carrier transport layer (CTL) between the active layer and electrodes can offer efficient carrier extraction.[6–8] A CTL consists of an electron transport layer (ETL) and a hole transport layer (HTL), depending on which active layer and electrode interface it is placed at. An ideal CTL material will have several desired properties:

(i) The ETL should have a low work function (WF) that matches the lowest unoccupied molecular orbital (LUMO) of the acceptor organic material of the active layer. The HTL should have a high work function to match the highest occupied molecular orbital (HOMO) of the donor organic material of the active layer.

(ii) The ETL should have good electron transport (mobility) and hole blocking properties. The HTL should have good hole transport (mobility) and electron blocking properties.

(iii) The CTL should have a wide bandgap to reduce light absorption and keep the exciton in the active layer.

(iv) The CTL should be chemically and thermally stable to avoid unexpected reactions and assure reliable performance.

(v) The CTL should have satisfactory film morphology and surface energy for the formation of a film on top.

(vi) The CTL should be produced at low cost.

Indeed, it is practically mandatory to use ETL and HTL in organic optoelectronic devices because the interlayer can enhance the charge extraction and enable the use of a wider range of materials as electrodes.[9,10] In this chapter, I will first discuss solution-processed metal oxides as ETL and HTL and then various hybrid metal oxide systems used to improve the performance of CTL further.

7.2 Solution-Processed Metal Oxides as Electron Transport Layer (ETL)

The electron transporting layer, also called the electron collection layer or cathode buffer layer, is inserted between the active layer and the cathode to transport the electron selectively from the active layer to the cathode. The cathode normally uses metallic materials which have high defect density that causes recombination of excitons and generates dipoles at the cathode–active surface and are detrimental to electron extraction efficiency.[6] Therefore, the use of a suitable ETL can benefit the overall device efficiency. Besides establishing ohmic contact between the electrode and active layer, the ETL is also responsible for other parameters of OPV devices, including the internal electric field, film morphology, and charge recombination rate.[11] The focus of the following section is to analyze how the ETL will affect the performance of OSCs in terms of the use of different materials and treatments of each material.

However, the conventional OSCs have inherent stability issues, and the low work function metal cathode can easily become oxidized in the ambient environment. Hence the inverted configuration for OSCs has emerged.[12] Transition from standard to inverted architecture is an essential progression in OSC design. As can be seen from Figure 7.2 (below), in a standard OSC the electron is extracted from the top cathode electrode and the hole is extracted from the bottom anode electrode. For an inverted OSC, the positioning of cathode and anode is the opposite: the electrons exit from the bottom cathode electrode and the holes exit from the top anode electrode.[7]

7.2.1   Zinc Oxide (ZnO)

ZnO is one of the materials most commonly used as the ETL in OSCs owing to its attractive electrical properties, low work function, ease of preparation and good air stability.[13] The low work function of ZnO allows the ETL to form ohmic contact with the active layers and ZnO is compatible with simple chemical bath methods such as roll-to-roll coating.[13] It is a non-toxic semiconductor and has a wide bandgap of 3.37 eV.[14]

Thin ZnO layers can be deposited *via* several solution-based techniques such as spray-coating, sol–gel, and nanoparticle deposition.[13,15,16] The overall power conversion efficiency (PCE) of OSCs equipped with ZnO based ETL is typically from 2% to around 7% for poly(3-hexylthiophene) (P3HT):phenyl-C_{61}-butyric acid methyl ester (PCBM) based OSCs.[13,15–17] Several treatments for ZnO can be used to improve the efficiency. For example, ultraviolet treatment of the ZnO exhibits higher conductivity for the ETL film,[18,19] and post thermal treatment of ZnO films is reported to enhance the morphology.[20,21] The ZnO layer morphology, or the roughness, has a major influence on the efficiency of the OSC.[13] This is because the rough ZnO surface can create a light trapping phenomenon which will result in better light absorption. The improved crystallization of the ZnO layer can enhance the electron extraction efficiency and hole blocking abilities.[13] However, excessive roughness of the ZnO film will have an adverse effect on the overall PCE of the OSC.[13] This is because too rough an interface will reduce the fill factor (FF), owing to the introduction of trap states at the interface with the organic active layer which will cause trap-assisted recombination and render the photocurrent dependent upon the electric field.[22] In addition, the roughness of the ZnO film will change the surface energy[23] and morphology of the active layer on top.[24] When the roughness of ZnO increases, the short circuit current (J_{SC}) will decrease as a result of the reduction of the donor–acceptor interfacial area in the active layer deposited on top.[25]

7.2.1.1   *Solution-Processed ZnO ETL*

Method I – ZnO nanoparticles. Nanostructured ZnO film is a booster of PCE for OSCs. A nanostructured ZnO layer can be solution processed and form uniform surface morphology in low annealing temperatures (≤150 °C).[17] Meanwhile, one-dimensional (1-D), 2-D and 3-D ZnO nanostructures such as nanoparticles, nanowires, nanorods, and nanowalls have been synthesized and some of them have been used as ETL for OSC ETLs.[26-28] It should be noted that ZnO nanoparticles (NPs) such as those synthesized by hydrolysis and condensation of zinc acetate dihydrate by potassium hydroxide in methanol[29-31] are not very stable in water and a ligand is usually required for stabilization.[32]

In 2012, Chen *et al.*[28] reported a solution-processed nanocrystalline ZnO ETL in an inverted OSC which exhibited a PCE of 3.7% with a device structure of ITO/nanocrystalline ZnO/P3HT:PCBM/MoO$_3$/Ag. The crystalline ZnO NPs were synthesized using zinc acetate dehydrate (ZnAc), ethanol, and potassium hydroxide (KOH). The ZnO NP solution was spincoated over the pre-cleaned ITO substrate and annealed at 120 °C for 10 minutes. The crystalline ZnO NPs provided ideal energy level alignment to PCBM, and high conductivity.

Method II – sol–gel-derived ZnO. In 2011, Heeger *et al.*[33] fabricated an inverted OSC with sol–gel-derived ZnO prepared using zinc acetate in 2-methoxyethanol with post annealing temperature below 200 °C. The device structure of glass/ITO/ZnO/PCDTBT:PC$_{70}$BM/MoO$_x$/Ag achieved average PCE of 6.08%, 5.86% and 5.10% for post annealing temperatures of 200 °C, 150 °C and 130 °C respectively. This method used relatively low annealing temperatures for the ETL and demonstrated that the ETL produced can function efficiently in inverted OSCs.

Method III – PEO-modified ZnO. In 2012, Shao *et al.*[34] reported a PSC using poly(ethylene oxide) (PEO) modified ZnO as the ETL in an inverted structure with an active layer of TQ1:PC$_{71}$BM and this exhibited a PCE improvement from 5.39% to 6.59% based on a reference device using ZnO ETL. The PEO modification is achieved by mixing ZnO CB (chlorobenzene) solution with PEO solution. The PEO shares the lone electron pair of oxygen with the ZnO and the surface traps of ZnO NP are reduced as a result. It also leads to suppression of exciton recombination, reduction of series resistance, and improved electrical coupling between ZnO and the active layer. Figure 7.1 shows the device structure and chemical composition of the active layer and PEO.

Method IV – ZnO·xH$_2$O. Wang *et al.*,[35] in 2014, developed an aqueous zinc oxide hydrate (ZnO·xH$_2$O) deposited ZnO ETL in inverted OSCs. With the PCDTBT:PC$_{71}$BM as the active layer, the average PCE of the cell was 6.48%, for which the device was annealed at a low annealing temperature of 80 °C. The low annealing temperature was achieved as a result of the low energy metal-ammine dissociation and hydroxide condensation. The group compared the device with a sol–gel processed ZnO ETL with an annealing temperature of 200 °C, and the average PCE was 5.53%, due to lower electron mobility and the rough morphology of the ZnO film. Therefore, the ZnO·xH$_2$O deposited ZnO ETL at low temperature is a low-cost solution-processed method that has great potential for commercialization.

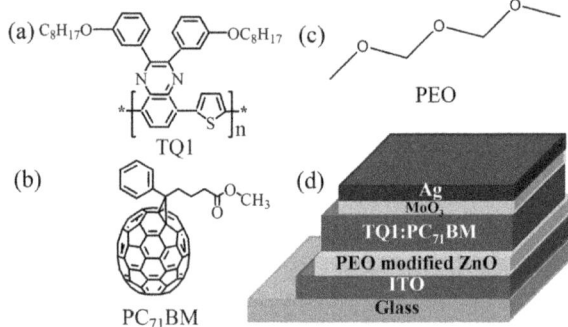

Figure 7.1   Chemical structure of (a) TQ1, (b) PC$_{71}$BM, (c) PEO, and (d) the device
structure. (Reprinted from ref. 34, with permission from Elsevier, http://
www.sciencedirect.com/science/article/pii/S0142961200001150.)

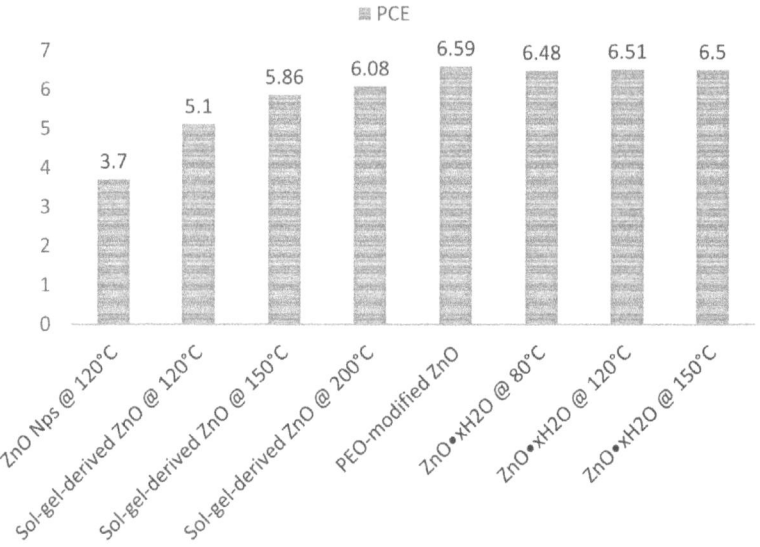

Figure 7.2   The device PCE of different ZnO ETL process methods.

7.2.1.2   ZnO ETL Performance Analysis

The device performance varies for the above-mentioned ZnO ETL process
methods. Figure 7.2 shows the PCE of OSCs and Table 7.1 shows the summarized device performances.

  From the trend of PCE shown in Table 7.1, the primary reasons for
improved PCE may include: the active layer, HTL/anode structure, annealing condition, and the processing method of ZnO ETL. For instance, the
large increase of PCE from 3.7% to 5.1% for ZnO NPs (120 °C) and sol–gel-
synthesized ZnO (120 °C), respectively, may be attributed mainly to the use
of different active layers. The donor material PCDTBT has lower HOMO level

Table 7.1 Summary of the performance of devices using different ZnO ETL process methods.

ETL process method	Cathode/ETL	Active layer	HTL/ anode	Annealing	PCE	Ref.
ZnO NPs	ITO/ZnO NP	P3HT:PCBM	MoO_3/Ag	120 °C	3.7%	28
Sol–gel-derived	ITO/ZnO	PCDTBT:$PC_{70}BM$	MoO_x/Ag	120 °C	5.1%	33
Sol–gel-derived	ITO/ZnO	PCDTBT:$PC_{70}BM$	MoO_x/Ag	150 °C	5.86%	33
Sol–gel-derived	ITO/ZnO	PCDTBT:$PC_{70}BM$	MoO_x/Ag	200 °C	6.08%	33
PEO modified	ITO/PEO-ZnO	TQ1:$PC_{71}BM$	MoO_3/Ag	—	6.59%	34
ZnO·xH_2O	ITO/ZnO·xH_2O	PCDTBT:$PC_{71}BM$	MoO_3/Al	80 °C	6.48%	35
ZnO·xH_2O	ITO/ZnO·xH_2O	PCDTBT:$PC_{71}BM$	MoO_3/Al	120 °C	6.51%	35
ZnO·xH_2O	ITO/ZnO·xH_2O	PCDTBT:$PC_{71}BM$	MoO_3/Al	150 °C	6.50%	35

than P3HT and this leads to higher device PCEs.[36,37] As mentioned earlier, the annealing temperature has a strong influence on device PCE because different annealing temperatures of ZnO ETL can alter the surface morphology, which in turn will affect the device PCE. As can be seen from Table 7.1, increasing annealing temperature results in improved PCEs.

By using PEO to modify ZnO NPs, surface traps, charge carrier recombination losses, and series resistance can be reduced, and the electrical coupling between ZnO NPs and the active layer is improved.[34] A control OSC using ZnO as the ETL showed a PCE of 4.5%. When different concentrations of PEO-modified ZnO were used and the optimal device was achieved by using 0.05% PEO-modified ZnO as ETL, the resulting PCE was 5.64%.[34] Thus 25.3% enhancement in PCE was achieved by using an optimal concentration of PEO-modified ZnO for a given device structure.[34] For the device using ZnO·xH_2O as ETL, the aqueous solution process method can enhance the charge transportation of ZnO ETL and the surface smoothness.[35] Different annealing temperatures were used and the optimal temperature was found to be 120 °C, where the device PCE (6.51%) and FF (62.4%) were at maximum.[35] The reference OSCs using ZnO ETL prepared by zinc acetate and post-annealed at 200 °C for the same structure exhibited 5.53% PCE. Therefore, the ZnO ETL prepared by aqueous solution and a post-annealing temperature of 120 °C resulted in 17.7% enhancement in PCE for the given device structure.

7.2.2 Titanium Oxide (TiO_x)

TiO_x is widely used as the ETL in OSCs because of its low work function, transparency, non-toxicity, and stability in terms of film morphology and electronic properties.[6,38] The overall PCE of the OSC is highly dependent on the crystalline structure and morphology of the TiO_2.[39] Depending on the preparation method of TiO_x ETL, such as thermal evaporation, sputtering, sol–gel pyrolysis, *etc.*, the stability varies in each specific case.[40] Amorphous TiO_x is also used as the ETL for inverted architecture owing to its suitable bandgap and energy levels.[41] By using TiO_x as the ETL and P3HT:PCBM as the active layer, the resulting PCEs are in the range of 2.0–4.0%.[42] However, by controlling the morphology and electric resistance for amorphous TiO_x, the performance can be greatly enhanced.[43] Also, the TiO_x layer can be solution-deposited at room temperature with no post-annealing treatment and exhibit high air stability, which demonstrates possible roll-to-roll manufacturing potential.[44]

The device PCE can be further improved by inserting a buffer layer between the ITO substrate and the TiO_x layer. For instance, by using bis(2-(trichlorosilyl)propyl)-malonate C_{60} (TSMC), a new cross-linkable fullerene material, on aTiO_x surface, the TSMC can passivate the hydroxyl groups on the surface of TiO_x, hence enhancing the performance significantly.[45] Meanwhile two self-assembled fullerene monolayers, 4-(2-ethylhexyloxy)-[6,6]-phenyl C_{61}-butyric acid (*p*-EHO-PCBA) and bis-4-(2-ethylhexyloxy)-[6,6]-phenyl C_{61}-butyric acid (bis-*p*-EHO-PCBA), were used on top of the TiO_x layer and exhibited a substantial resistance drop and enhanced hydrophobicity, which resulted in significant OSC PCE improvement, up to 5.13%.[46]

7.2.2.1 Solution-Processed TiO_x ETL

Method I – chemical bath deposition. In 2010, Takahashi *et al.*[47] introduced a chemical bath deposited TiO_x ETL in an inverted BHJ OSC with P3HT:PCBM and bis-P3HT:PCBM as the photoactive layer. The cell achieved a highest PCE of 3.8%. The device structure is ITO/TiO_x/P3HT:PCBM/PEDOT:PSS/Au and the deposition mechanism of the TiO_x layer can be concluded from the following equations:

$$Tio^{2+} + H_2O_2 \rightarrow Ti(O)_2^{2+} + H_2O \tag{7.1}$$

$$mTi(O)_2^{2+} + nH_2O_2 \rightarrow [Ti(O)_p(OH)_q]_m \cdot \gamma H_2O + 2mH^+ \tag{7.2}$$

$$[Ti(O)_p(OH)_q]_m \cdot \gamma H_2O \rightarrow mTiO_x + sH_2O \tag{7.3}$$

The solution of $TiOSO_4$ was added in a diluted H_2O_2 aqueous solution and the concentrations of $TiOSO_4$ and H_2O_2 were adjusted to 0.03 M. The ITO substrate was immersed in the chemical bath at 80 °C and after a certain time the deposited TiO_x layer on the ITO substrate was heated at 150 °C for 1 hour in air. The chemical bath deposition method is simple, low cost, and can be applied to large-area devices.

Method II – self-assembly TiO_2. In 2012, Choy *et al.*[48] demonstrated a self-assembly method using TiO_2 nanoparticles to form large-area solution-processed ETL. The method can produce films with enhanced surface morphology and charge extraction properties. Ethanol was used as the solvent of ligand free TiO_2 nanocrystal. The orderly alignment of TiO_2 nanoparticles is the result of hydrogen bonding in ethanol. The solution is distributed on the ITO substrate and then covered by a Petri dish for about 30 minutes so that the solution can spread to cover the substrate and deposit TiO_2 after evaporation of the ethanol. After the self-assembly process, the film is annealed for 10 minutes in air at 150 °C. The device, with the structure ITO/TiO_2/P3HT:PCBM/MoO_3/Al, exhibited average PCE of 3.91 ± 0.1%. The TiO$_2$ was deposited on a 44-inch ITO glass substrate, which showed the capability of its large area application as shown in Figure 7.3.

Method III – TiO$_x$ precursor solution. In 2013, Hadipour *et al.*[44] developed a repeatable and stable precursor for solution-processed TiO$_x$ based ETL at room temperature in a controlled atmosphere without initial or post anneal-ing. Titanium(IV) isopropoxide with acetic acid was used, and the reaction involves hydrolysis, condensation and stabilization during the formation of the precursor solution in a nitrogen-filled glove box. In addition, ethanol-amine is used in combination with ethanol to interact with the metal cations. The precursor solution is then deposited to form TiO$_x$ layer by spincoating. The TiO$_x$ layer developed was tested with conventional and inverted OSCs to show high chemical compatibility with many types of active layer blend and satisfactory adhesion to both active layer and ITO substrate. Comparison was made of small particle sized TiO$_x$ solution and larger particle sized TiO$_x$, which led to clear and hazy solutions, respectively. The optimum electrical performance was achieved by ETL using clear TiO$_x$ precursor solution. The method showed great potential for cost-effective, energy efficient and roll-to-roll mass production for the OPV.

Method IV – amorphous TiO$_x$. In 2014, Bao *et al.*[41] presented inverted PSCs with an ETL that is one-step solution-processed using amorphous TiO$_x$ (s-TiO$_x$) without an annealing process. Isopropoxide isopropanol solution was used to spincoat s-TiO$_x$ on ITO/glass substrate in an inert environment and this was dried for 60 minutes. The s-TiO$_x$ film exhibited good light transmittance and stability in a humid atmosphere. In contrast to the method developed by Hadipour,[44] the s-TiO$_x$ film is formed *via* direct hydrolysis of titanium(IV) iso-propoxide (TTIP) in air. The TTIP isopropanol solution is first spincoated onto ITO substrate in a N$_2$-filled glove box and then dried in air to allow hydrolysis

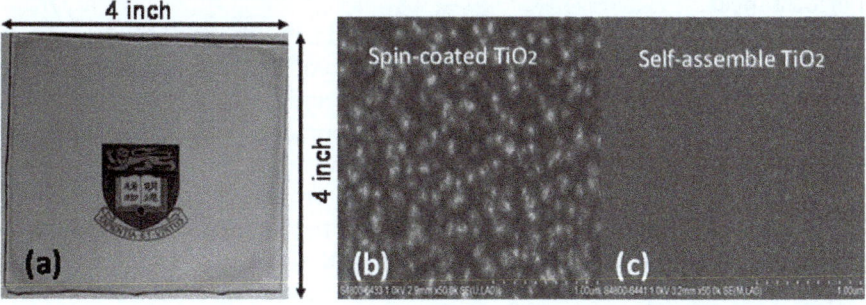

Figure 7.3 (a) Large-area 4 × 4-inch self-assembled TiO$_2$ film. (b) Atomic force microscopy (AFM) image of spincoated TiO$_2$ with rms roughness of ~8.1 nm. (c) AFM image of self-assembled TiO$_2$ with rms roughness of ~1.8 nm. (Reprinted from ref. 48, with permission from Elsevier, http://www.sciencedirect.com/science/article/pii/S1566119912002674.)

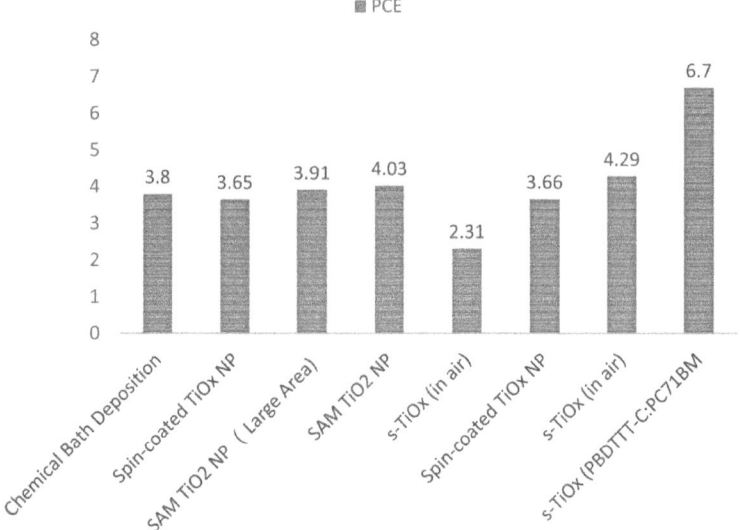

Figure 7.4   The device PCEs of different TiO$_x$ ETL process methods.

to take place, upon which the TTIP will decompose into amorphous TiO$_x$. The group fabricated two PSC devices using the s-TiO$_x$ as ETL with an active blend of P3HT:PC$_{61}$BM and PBDTTT-C:PC$_{71}$BM and reached PCEs of 4.29% and 6.7% respectively. The result further demonstrates the promising large-scale production of s-TiO$_x$ ETL in flexible high-performance PSCs.

7.2.2.2   TiO$_x$ ETL Performance Analysis

The device performance varies for the above-mentioned ZnO ETL process methods, as shown in Figure 7.4. Table 7.2 shows the summarized device performances.

The devices using TiO$_x$ as a solution-processed ETL exhibit performance similar to devices using ZnO as the solution-processed ETL. The overall PCE of the device is strongly related to the film morphology and crystalline structure of the TiO$_x$.[41,47,48] By using the self-assembly TiO$_2$ NP approach and solution-processed amorphous TiO$_x$ ETL, OSC performance improves, which can be attributed to increased J_{SC} resulting from enhanced surface morphology or roughness.[41,48]

However, the influence of the annealing temperature of TiO$_x$ ETLs on PCE is not as strong as for ZnO ETLs. As studied by Bao *et al.*, the effect of increasing annealing temperature from 25 °C to 200 °C was tested for amorphous TiO$_x$ ETL with the structure of ITO/s-TiO$_x$/P3HT:PC$_{61}$BM/MoO$_3$/Ag. The result is shown in Table 7.3.

From Table 7.3 we can see that the annealing temperature of ETL and device performance do not have a clear proportional relationship. The current density does not show significant change, indicating that the surface

Table 7.2 Summary of the performance of devices using different TiO_x ETL process methods.

ETL process method	Cathode/ ETL	Active layer	HTL/anode	Annealing	PCE	Ref.
Chemical bath deposition	ITO/TiO$_x$	P3HT:PCBM	PEDOT:PSS/ Au	150 °C	3.8%	47
Spin-coated TiO$_x$ NP	ITO/TiO$_2$	P3HT:PCBM	MoO$_3$/Al	—	3.65%	48
SAM TiO$_2$ NP (large area)	ITO/SAM TiO$_2$	P3HT:PCBM	MoO$_3$/Al	150 °C	3.91%	48
SAM TiO$_2$ NP	ITO/SAM TiO$_2$	P3HT:PCBM	MoO$_3$/Al	150 °C	4.03%	48
Solution-processed amorphous TiO$_x$ (in air)	ITO/s-TiO$_x$ (in air)	P3HT:PC$_{61}$BM	MoO$_3$/Ag	150 °C	2.31%	44
Spincoated TiO$_x$ NP	ITO/TiO$_x$ NP	P3HT:PC$_{61}$BM	MoO$_3$/Ag	—	3.66%	44

Table 7.3 Device performance using different ETL annealing temperatures for 60 minutes.

ETL	Temperature (°C)	V_{OC} (V)	J_{SC} (mA cm^{-2})	FF (%)	Avg. PCE (%)
s-TiO$_x$	25	0.62	10.81	63.91	4.23
s-TiO$_x$	90	0.62	10.67	63.40	4.15
s-TiO$_x$	160	0.62	10.78	62.97	4.20
s-TiO$_x$	200	0.62	10.63	63.64	4.17

morphology or roughness was not much changed by variation in annealing temperature. The reason is that the TTIP films can be easily decomposed to s-TiO$_x$ and form very smooth films.[41]

7.2.3 Cs$_2$CO$_3$

The Cs$_2$CO$_3$ is recognized as one of the best ETL materials.[49,50] For example, in 2006,[51] Cs$_2$CO$_3$ was used as the ETL in a P3HT-PCBM OSC with V$_2$O$_5$ as the HTL. The OSC gave an overall PCE of 2.25% and it was shown that the realization of Cs$_2$CO$_3$ would not affect the device performances because of the favorable ohmic contact formed. The insensitivity of its contacting metal electrode gives Cs$_2$CO$_3$ advantages over other ETL materials. Several treatments can tune up performance of the Cs$_2$CO$_3$ layer. Zhang et al.[52] demonstrated that annealing and UV-ozone treatment of the Cs$_2$CO$_3$ layer can enhance the electron extraction by varying the UV-ozone treatment time. Using annealing and 15-minute UV-ozone treatment, an inverted OSC with the structure ITO/Cs$_2$CO$_3$/P3HT:PCBM/MoO$_3$/Al exhibited 1% PCE improvement. Barbot et al.[53] incorporated Cs$_2$CO$_3$ and fullerene C$_{60}$ by co-sublimation and observed a reduction in ohmic losses at active layer–electrode interfaces. The resulting

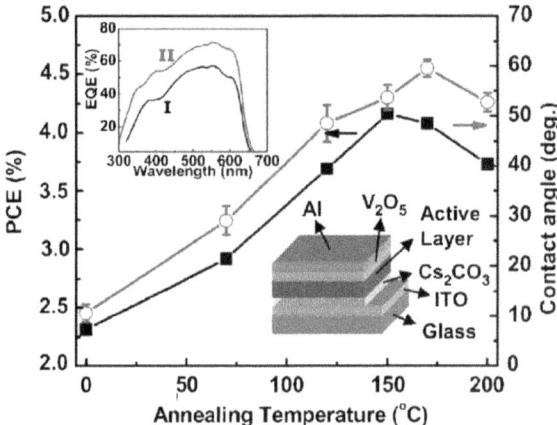

Figure 7.5   Device structure and PCE of the Cs_2CO_3 layer as a function of annealing temperature. (Reprinted from ref. 54, with permission from AIP Publishing LLC, http://scitation.aip.org/content/aip/journal/apl/92/17/10.1063/1.2918983.)

PCE enhancement achieved by the use of a Cs_2CO_3 doped C_{60} layer for the inverted OSC was 3.79% when compared with OSC using pure C_{60} as ETL.

7.2.3.1   Solution-Processed Cs_2CO_3 as ETL

Spin-coated Cs_2CO_3 with low temp annealing. In 2008, Yang *et al.*[54] demonstrated a solution-processed Cs_2CO_3 ETL in a P3HT:PCBM inverted PSC. The Cs_2CO_3 annealed at low temperature (<200 °C) showed altered work function from 3.45 to 3.06 eV and the PCE was improved from 2.31% to 4.19% by 150 °C post annealing. The Cs_2CO_3 was dissolved in 2-ethoxyethanol and spincoated on ITO substrate. The deposited substrate was thermally treated in a glove box for 20 minutes at different temperatures. The maximum FF and PCE of the cell were found at an annealing temperature of 150 °C. Figure 7.5 shows the device structure and PCE as a function of annealing temperature. The thermal annealing decreased the work function of the Cs_2CO_3 so that the Cs_2CO_3 intrinsically decomposed into a doped n-type semiconductor, Cs_2O_2, and showed a lower contact resistance.

7.2.4   Other Metal Oxide Based ETLs

Besides TiO_x and ZnO, other metal oxides such as Al_2O_3 and Nb_2O_5 are also used as ETL in OSCs.[55,56] The inverted OSCs with an ultrathin Al_2O_3 electron buffer layer showed S-shape current density (*J*)–voltage (*V*) characteristics before UV exposure or electrical bias and reappeared upon air exposure. After eliminating the S-shape electrical properties, a PCE of 2.8% was obtained in a P3HT:$PC_{60}BM$ cell.[55] As for Nb_2O_5, its conduction band is higher than the LUMO of the PCBM polymer, which means that electron transport from

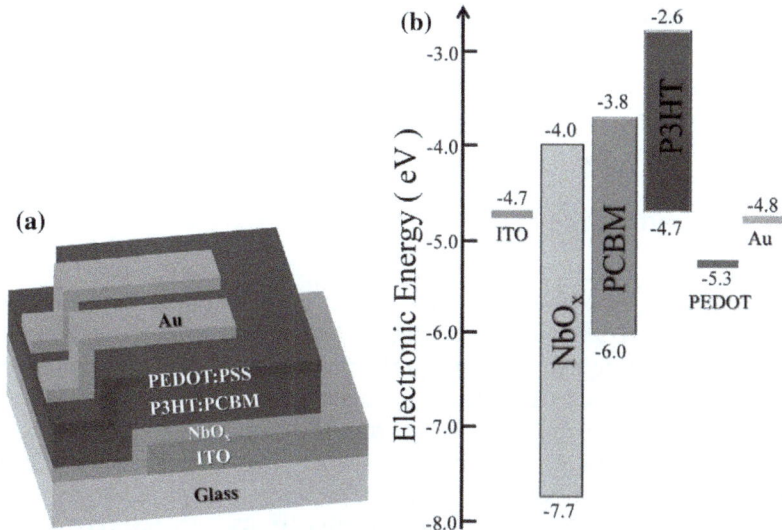

Figure 7.6 (a) Energy level diagram of NbO$_x$ and other components, (b) schematic of PSC using NbO$_x$ as ETL. (Reprinted from ref. 57, with permission from Elsevier, http://www.sciencedirect.com/science/article/pii/S0009261413010233.)

the LUMO of PCBM into the conduction band of Nb$_2$O$_5$ is not possible. However, a PCE of 2.7% was indeed achieved when Nb$_2$O$_5$ served as ETL, which suggested that electrons were transferred from the PCBM to the collecting electrode through a possible tunneling process.[56] The use of amorphous niobium oxide (NbO$_x$) as ETL *via* solution processing was demonstrated by Ohno *et al.*[57] in 2013 in a P3HT:PCBM based inverted OSC with PCE of 2.22%, as shown in Figure 7.6(a). NbO$_x$ is a metal oxide which can be used both as an ETL and a HTL owing to its appropriate energy level as shown in Figure 7.6(b). The NbO$_x$ was spincoated on ITO substrates followed by 1 hour post annealing at 150 °C. The stability of NbO$_x$ is comparable to that of TiO$_x$ ETL.

In 2013, Tan *et al.*[58] demonstrated a P3HT:indene-C$_{60}$ bisadduct (IC$_{60}$BA) based OSC using solution-processed titanium(IV) oxide bis(2,4-pentanedionate) (TOPD) as the ETL. The TOPD was dissolved in isopropanol and was deposited on the active layer by spincoating with post-annealing at 80 °C for 15 minutes. The PCE of the device reached 5%. A higher PCE of 5.59% was achieved by replacing the IC$_{60}$BA by indene-C$_{70}$ bisadduct (IC$_{70}$BA). The solution processability and low temperature annealing make it a promising ETL for PSCs.

Other low work function metal oxides such as Cu$_2$O, Ta$_2$O$_5$ and ZrO$_2$ are also reported in dye-sensitized solar cells (DSSCs), although there are relatively few reports that these metal oxides can be utilized in OSCs. For example, Ta$_2$O$_5$–ZnO film was demonstrated to be an ETL buffer layer that increased the PCE of a P3HT:PC$_{60}$BM cell from 3.7% to 4.12%.[59] It was suggested that the Ta–O–Zn bonds formed in the ETL were beneficial to the overall PCE, and that the high dielectric constant of Ta_2O_5 may also be beneficial in reducing electron recombination.

7.3 Doped and Hybrid Metal Oxides for Enhanced Electron Transport of ETL

7.3.1 Doped and Hybrid TiO_x

Doping of TiO_x has been achieved by adding Cs_2CO_3 to the precursor solution to obtain morphologically stable TiO_x:Cs layers, and an increased electrical conductivity was demonstrated by X-ray photoelectron spectroscopy (XPS), which resulted in efficient performance (PCE of 4.2%) for inverted single-junction OSCs.[60] In addition, Cs in TiO_x also gives rise to a more desirable work function for charge transport, which makes the doped TiO_x layer a promising ETL in OPVs. A further improvement of devices with TiO_x:Cs as ETL was made by energy level alignment, and a high PCE over 7% was reported for thienothiophene-based polymer (PTB7):$PC_{71}BM$ based OSCs.[61] It is also confirmed that the Sn-doped TiO_x film has improved morphology and electrical properties, and the shunt loss and interfacial charge recombination were reduced accordingly.[62] The improved device performance indicates that the properties of metal oxides can be enhanced and tuned through the addition of various dopants. However, various methods for Zn, Ta or Nb doped TiO_x aimed at increasing the electrical conductivity are limited to the application in OPVs because high processing temperatures are required.[63,64] For example, the Zn-doped TiO_x ETL has some merits due to its high transparency, greater stability and small lattice mismatch; with polyethylene oxide modification of the doped ETL surface, a significantly increased PCE of 8.1% was obtained for PTB7:$PC_{71}BM$ cell. However, to prepare such a Zn-doped TiO_x ETL, a 500 °C thermal annealing process was needed.[64]

For other low temperature approaches to improve the electrical properties of TiO_x ETL, the incorporation of metal nanoparticles (NPs) such as Au and Ag NPs into TiO_x has been demonstrated to be a successful strategy, through charge accumulation[65] and plasmonic-electrical effects.[66] The reported inverted P3HT:PCBM cells have enhanced electron extraction and increased photocurrent when NP-embedded TiO_x ETLs are used, and this results in an improved PCE of 8.2% for poly{[4,8-bis-(2-ethyl-hexyl-thiophene-5-yl)-benzo[1,2-b:4,5-b0]dithiophene-2,6-diyl]-alt-[2-(20-ethylhexanoylethylhexanoyl)-thieno[3,4-b]thiophen-4,6-diyl]} (PBDTTT-C-T):$PC_{71}BM$ based OSCs.[65,67] The main reason for the enhancement is not the optical plasmonic effects in enhancing the light absorption of the active polymer in OSCs, but the enhancement in charge extraction induced by charge accumulation effects of these metal NPs under solar illumination. Specifically, the higher energy UV light within the solar light excited electrons from the TiO_x ETL to the metal NPs and resulted in accumulation of the electrons in metal NP–TiO_x composites. The electron accumulation reduces the work function of the ETL, as illustrated in Figure 7.7, which can assist the charge extraction in OPVs.

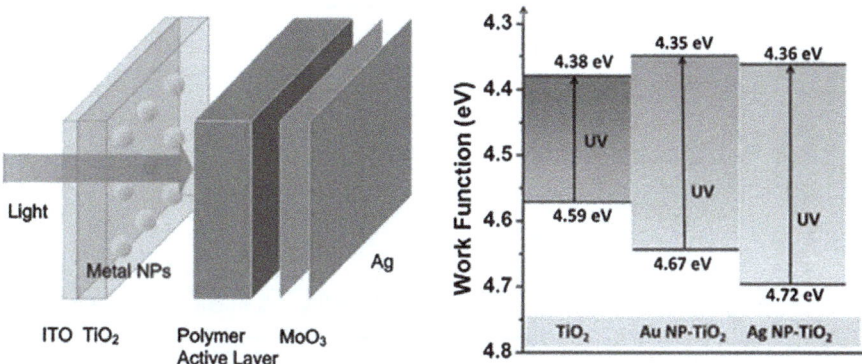

Figure 7.7 Schematic of the inverted OPV device incorporating a NP-embedded TiO$_x$ layer and corresponding WF changes of TiO$_2$ and Au/Ag NP–TiO$_2$ films from dark to UV illumination. (Reprinted from ref. 65, with permission from the Royal Society of Chemistry, http://pubs.rsc.org/en/Content/ArticleLanding/2013/EE/c3ee42440e#!divAbstract.)

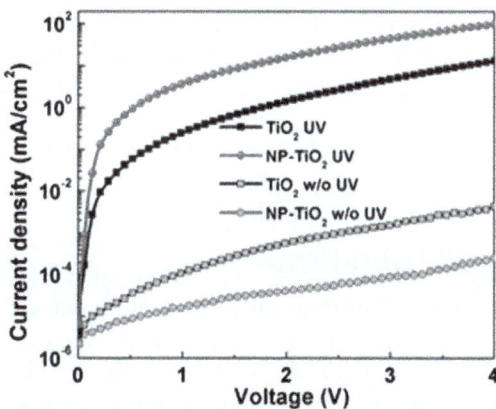

Figure 7.8 *J–V* characteristics of electron-only devices using TiO$_2$ and optimized NP–TiO$_2$ as ETLs measured with and without (w/o) UV excitation. (Reprinted from ref. 65, with permission from the Royal Society of Chemistry, http://pubs.rsc.org/en/Content/ArticleLanding/2013/EE/c3ee42440e#!divAbstract.)

As a result, the photocurrent and thus the device performance of the OPVs are increased considerably.

Another interesting property of the metal NP-incorporated TiO$_2$ is that the dark current of the electron-only device reduces when the current is enhanced under UV excitation when compared to that of pristine TiO$_2$ devices, as shown in Figure 7.8.[65] This can enhance the dynamic range of the OSCs and shows potential for photodetector applications.

By optimizing the concentration ratio of the Au NPs in the NP–TiO$_x$, the J_{SC} and FF of OSCs with various polymer active layers are notably increased, and

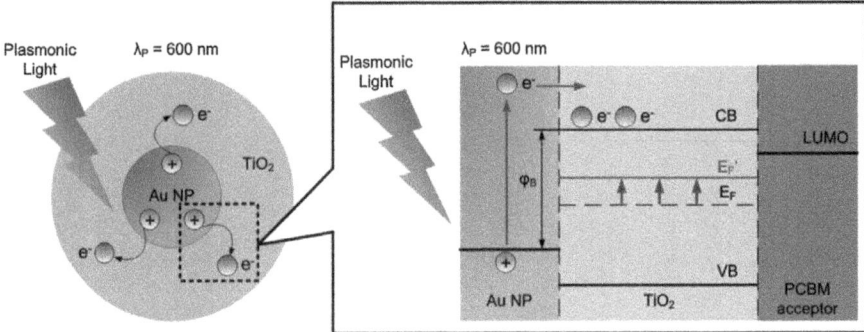

Figure 7.9   Schematic diagram of the process for plasmonic-induced charge injection. (Reprinted from ref. 66, with permission from John Wiley and Sons, http://onlinelibrary.wiley.com/doi/10.1002/adfm.201203776/abstract.)

a high PCE of 8.74% was reached in a PTB7:PC$_{71}$BM cell.[66] The experimental and theoretical results demonstrated that the Au NPs in the TiO$_x$ layer are the plasmonically generated hot carriers which can fill the trap states in TiO$_x$ and lower the effective extraction barrier. As illustrated in Figure 7.9, the improvement in device performance was induced by the charge transfer of plasmonically excited electrons from Au NPs to TiO$_x$ that enhanced the electron extraction and transportation. Moreover, the highly efficient OPVs using this NP–TiO$_x$ can operate by optical activation at a plasmonic wavelength in the visible region, rather than UV light.

7.3.2   Doped and Hybrid ZnO

Similar to the TiO$_x$ ETL, doping of ZnO films can be an effective strategy to improve device performance. It is known that the conductivity of ZnO can be significantly increased by doping with Al, Ga, or In, which are well established n-type dopants for ZnO.[68] Besides, dopants such as Sr and Ba can be doped in ZnO films to form new species; the resulting mixed metal oxides ZnSrO and ZnBaO can lead to suppression or reduction of the oxygen absorption at mobile oxygen vacancy sites on the metal oxide surface, which will improve the performance and stability of OPVs.[69] Specifically, a dual doped ZnO, by indium and fullerene derivative (InZnO-BisC$_{60}$), thin-film used as the ETL in an inverted PTB7-Th:PC$_{71}$BM single junction OPV, provided improved surface conductivity and enhanced electron mobility from 8.25×10^{-5} to 1.09×10^{-2} cm^2 V^{-1} s^{-1}.[70] This dual doped ZnO film showed the opposite gradient dopant concentration profiles, see Figure 7.10, being rich in fullerene derivate at the cathode surface in contact with the active layer and rich in indium substance in contact with the ITO surface. The resulting OPV has an improved PCE of 10.31% relative to that with ZnO without doping (8.25%), which is among the best reported for single junction OPVs.

Since both TiO$_x$ and ZnO ETLs are applicable in OSC fabrication and beneficial to device performance improvement, the combination of these two

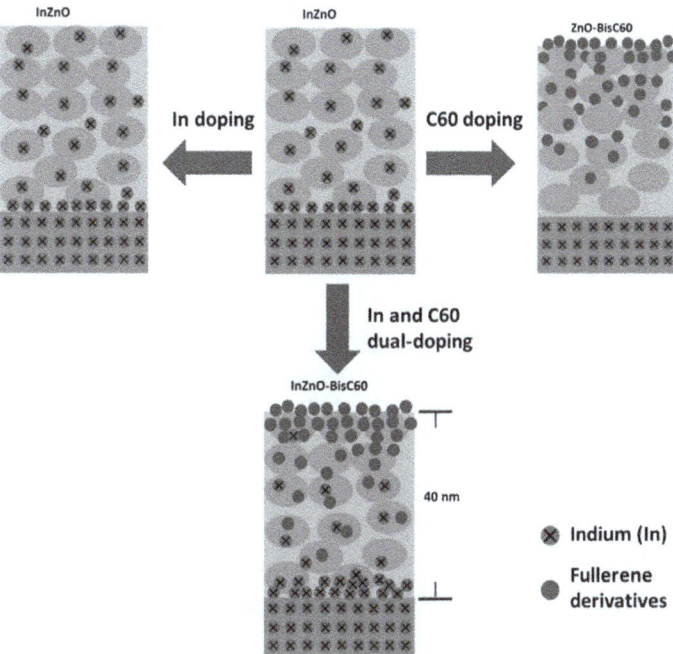

Figure 7.10 Schematic illustration of the proposed dual doped ZnO electron transporting layer. (Reprinted from ref. 70, with permission from Nature Publishing Group, http://www.nature.com/srep/2014/141029/srep06813/full/srep06813.html.)

kinds of ETL in a single cell seems a feasible strategy to take advantage of their respective features. It was demonstrated that TiO_x thin films deposited by atomic layer deposition onto ZnO surfaces can improve the efficiency of the inverted OPV when the TiO_x layer is less than 3 nm.[71] Co-sputtering of TiO_x and ZnO at room temperature was also demonstrated to produce an integrated electrode with ITO cathode for cost-efficient OPVs. However, the enhancement of efficiencies is not as great as expected, and only slightly increased or comparable efficiencies are found in these studies. Importantly, a thicker layer or higher ratio of TiO_x in these mixed ETLs will lead to reduced photovoltaic performance. Therefore, further development is needed in the concept of combining various metal oxides for functioning as ETLs.

7.4 Solution-Processed Metal Oxides Functioning as Hole Transport Layers (HTLs)

7.4.1 Solution-Processed Molybdenum Oxide (MoO$_x$) as HTLs

Molybdenum oxide (MoO_x) is typically an n-type semiconductor metal oxide with a high work function of over 5.0 eV. After metal oxide was introduced by the evaporation method to function as HTLs in OSCs,[51] more and more

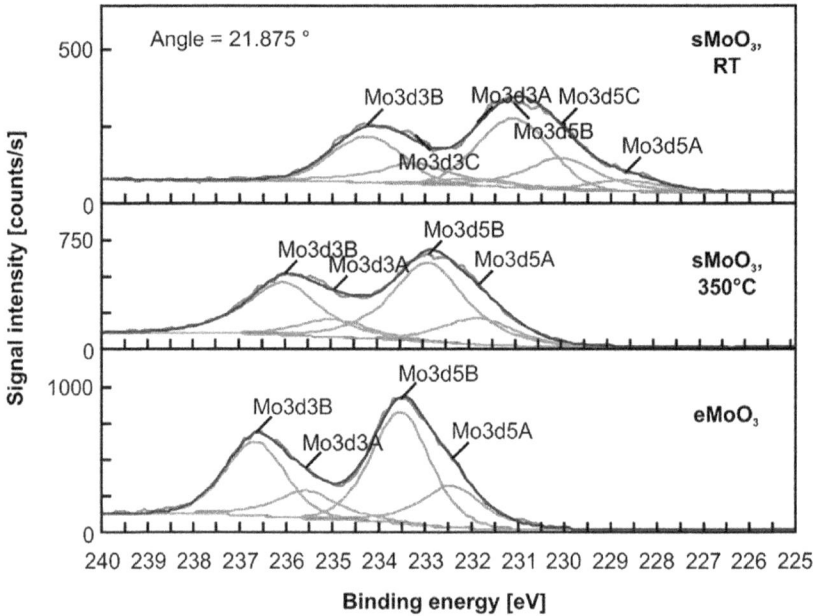

Figure 7.11 XPS spectra and deconvolutions of various MoO₃ layers. (Reprinted from ref. 72, with permission from the American Chemical Society, http://pubs.acs.org/doi/abs/10.1021/am200729k.)

attention has been attracted to developing solution-processed metal oxides to simplify device fabrication. OSCs have been demonstrated using solution-processed MoO$_x$ with a variety of process methods by various groups. Both normal and inverted architectures of OSCs have been realized which yield high efficiency and alternative device architectures.

Method I – MoO$_x$ film using sol–gel technique. Girotto *et al.* demonstrated a solution-processed MoO₃ using the sol–gel technique to work as HTL in OSCs.[72] By refluxing MoO₃ powder in H₂O₂ at 80 °C for 2 hours, a clear yellow liquid will be obtained with the addition of polyethylene glycol and 2-methoxyethanol. After film formation by spincoating on the ITO substrate, an annealing treatment over 275 °C can thermally convert the precursor film into MoO₃. As shown in Figure 7.11, the X-ray photoelectron spectroscopy (XPS) measurements of MoO₃ film before and after the conversion indicated several oxidation states existing in this layer. An ultra-smooth surface of the metal oxide was measured by atomic force microscopy (AFM) with a surface roughness root mean squared (rms) of 0.36 nm. Compared to OSCs fabricated with traditional methods of evaporated MoO₃ and spincoated PEDOT:PSS, the devices have equal performance characteristics. The MoO₃-based OSCs also provide remarkable device stability compared with OSCs with PEDOT:PSS as HTL.

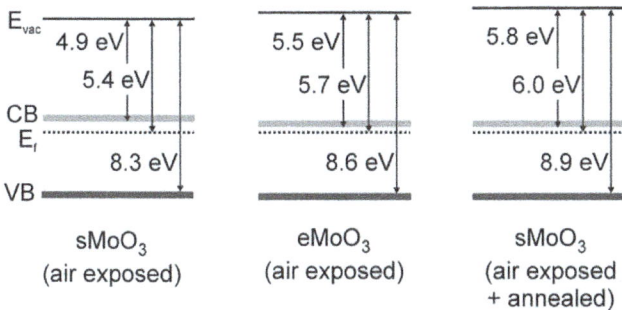

Figure 7.12 Energy levels of solution-based MoO₃ (left), vacuum evaporated MoO₃ (center) and solution-based MoO₃ after annealing to 200 °C (right). All samples were air-exposed for a short time before measuring. (Reprinted from ref. 73, with permission from John Wiley and Sons, http://onlinelibrary.wiley.com/doi/10.1002/adma.201003065/ abstract.)

Method II – the approach of crystalline molybdenum oxide nanoparticles (MoO₃ NPs). Another film formation technique of solution processed MoO₃ was prepared by Meyer *et al.* by using crystalline molybdenum oxide nanoparticles (MoO₃ NPs).[73] A suspension containing 2.5 wt% of 15 nm MoO₃ NPs and 1 wt% of a block copolymer dispersing agent in xylene was spincoated onto ITO with O₂-plasma treatment to remove the dispersing agent. As shown in Figure 7.12, a thermal annealing process at 100 °C and 200 °C was introduced to increase the work function of the MoO₃ NPs film. The study of hole-only devices showed that solution-processed MoO₃ NPs films exhibited improved properties on a low interface barrier and low leakage-current compared to the commonly used PEDOT:PSS layer.

Method III – molybdenum oxide bis(2,4-pentanedionate) (MoO₂(acac)₂) as precursor for MoOₓ HTL. A synthetic route of solution-processed MoOₓ anode buffer layers was demonstrated by Jasieniak *et al.*[74] The precursor solution in methanol can be prepared using an anhydrous and an ambient synthetic route. Both methods can be used to form ultra-smooth MoOₓ films with surface roughness rms of 0.35 ± 0.06 nm. As shown in Figure 7.13, the study of OSCs based on the solution-processed MoOₓ indicated that the performance depended strongly on the density of oxidation states within the MoOₓ film, which was verified by the XPS and ultraviolet photoelectron spectroscopy (UPS) measurements. By using another thermal decomposition method, with (NH₄)₆Mo₇O₂₄·4H₂O as precursor in deionized water, Murase *et al.* reported solution-processed MoO₃ with heat treatment at 80 °C for 1 hour in air.[75] The film formed from a 2 wt% solution was found to have a surface roughness rms of 0.8 nm. The XPS results indicated a nearly stoichiometric MoO₃ film after the decomposition. Organic photovoltaic devices using the solution-processed MoO₃ showed a high fill factor (FF) of 69.3% and gave a comparable power conversion efficiency to the PEDOT:PSS based devices.

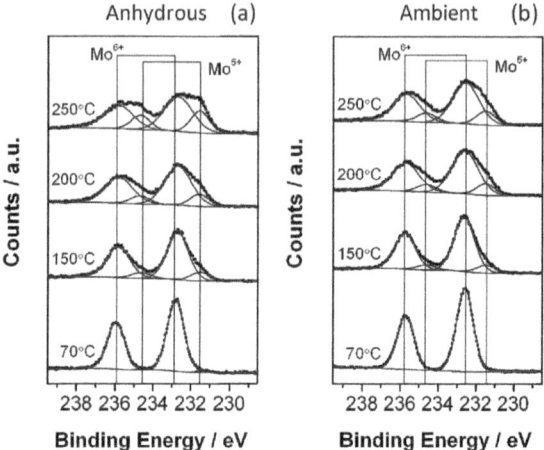

Figure 7.13   High-resolution XPS data of the Mo 3-D core level peaks observed for molybdenum oxide thin films prepared from (a) anhydrous and (b) ambient-based precursor solutions as a function of annealing temperature in a nitrogen atmosphere. (Reprinted from ref. 74, with permission from John Wiley and Sons, http://onlinelibrary.wiley.com/doi/10.1002/adma.201003065/abstract.)

Method IV – low-temperature-annealed sol–gel-derived MoO_x. This was demonstrated by Yang *et al.* as solution-processed HTL in OSCs.[76] By oxidizing the molybdenum using H_2O_2 (30%) in a cool water bath, the precursor solution was spincoated onto ITO glass and thermally annealed at 250 °C for 30 minutes to form a metal oxide film. A good surface morphology was verified by transmission electron microscopy (TEM) without pinholes, and AFM results gave a low roughness rms of 0.457 nm. As shown in Figure 7.14, the OSCs gave the same performance as the commonly used HTL of PEDOT:PSS. In a similar process, Lee *et al.* fabricated inverted OSCs using an H_2O_2 oxidized MoO_x NPs suspension assisted by microwave in *n*-butanol.[77] However, the use of air exposure or H_2O_2 oxidation to increase the Mo^{6+} fraction of MoO_x NPs may lead to an OSC hypersensitive to oxidizing treatment agencies.

Method V – hydrogen molybdenum oxide bronze (H_xMoO_3). This was demonstrated by Xie *et al.* to realize a one-step method to synthesize low-temperature solution-processed MoO_x.[78] The ethanol used in synthesizing the metal oxide based solution can control and alleviate the oxidation reaction between the metal powder and the H_2O_2. Based on this point, the oxidation states of Mo^{5+} and Mo^{6+} in the final solution product can be finely adjusted by controlling the amount of the oxidizing agent. More importantly, the XPS and UPS results indicated that the existence of a minimal amount of Mo^{5+} provided an n-doping of MoO_x after film formation, which is shown in Figure 7.15. In contrast with MoO_3 with full oxidation state of Mo^{6+}, oxygen vacancies in the film of MoO_x were essential in allowing the transition metal oxide

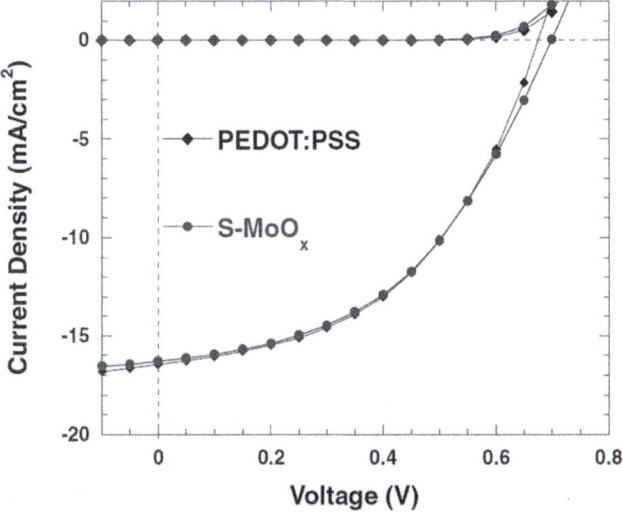

Figure 7.14 *J–V* characteristics of PSCs with s-MoO$_x$ and a PEDOT:PSS layer as the buffer layer. (Reprinted from ref. 76, with permission from John Wiley and Sons, http://onlinelibrary.wiley.com/doi/10.1002/aenm.201100598/abstract.)

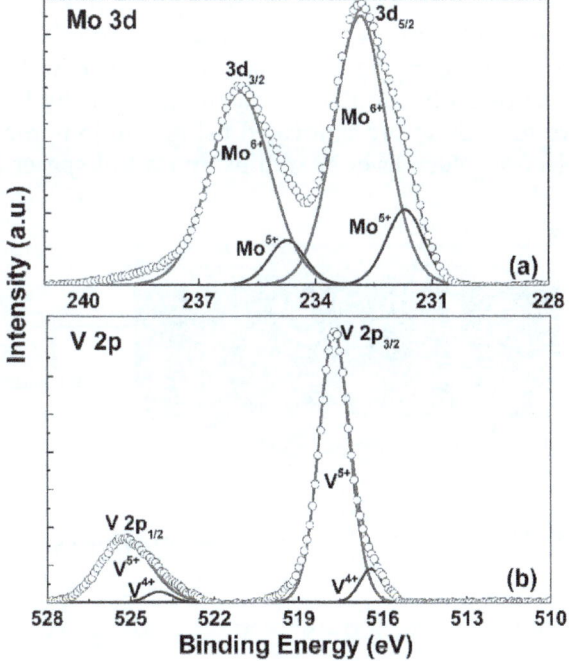

Figure 7.15 XPS spectra of: (a) Mo 3-D core level in molybdenum oxide (MoO$_3^-$); (b) V 2p core level in vanadium oxide (V$_2$O$_5^-$). The circles represent the experimental XPS spectra; the solid lines are decomposed XPS spectra. (Reprinted from ref. 78, with permission from John Wiley and Sons, http://onlinelibrary.wiley.com/doi/10.1002/adma.201204425/abstract.)

to function as an effective HTL.[79] By using the smooth layer of MoO$_x$, with a roughness rms of 1.33 nm, OSCs can reach a high PCE, over 7.7%, compared with the PCE of 7.2% from the commonly used HTL material PEDOT:PSS.

Later, Soultati *et al.* demonstrated that solution-processed hydrogen molybdenum bronze (s-H$_x$MoO$_{2.75}$) can function as a highly conductive anode interlayer in OSCs with high power conversion efficiencies.[80] After obtaining hydrated molybdenum oxide (H$_2$O·MoO$_3$) by dissolving molybdenum oxide powder in H$_2$O$_2$ with the additive polyethylene glycol, the stoichiometry of the metal oxide can be adjusted by using different annealing temperatures of the formed film. The hydrogen molybdenum bronze with a 190 °C annealing temperature, with the stoichiometry of H$_x$MoO$_{2.75}$, gave the optimal OSCs.

7.4.2   Solution-Processed Vanadium Oxide (V$_2$O$_x$) as HTL

With a quite similar energy band structure to MoO$_3$, V$_2$O$_5$ has been widely used in fabricating organic optoelectronic devices. The method most commonly used to form a V$_2$O$_5$ layer utilizes thermal evaporation to realize efficient OSCs. By introduction of solution-processed V$_2$O$_5$, both normal and inverted OSCs have been demonstrated with high efficiency and simple fabrication processes.

Method I – ultrasonic agitation. In 2009, Huang *et al.* introduced a colloidal method to realize solution-processed V$_2$O$_5$ in isopropanol (IPA) by ultrasonic agitation.[81] A roughness rms of 18.4 nm of the active layer was found after film formation from the IPA V$_2$O$_5$ solution. By hybridizing with ZnO nanorods, inverted OSCs were fabricated using solution-processed V$_2$O$_5$ to serve as an electron block layer as well as an optical spacer (Figure 7.16).

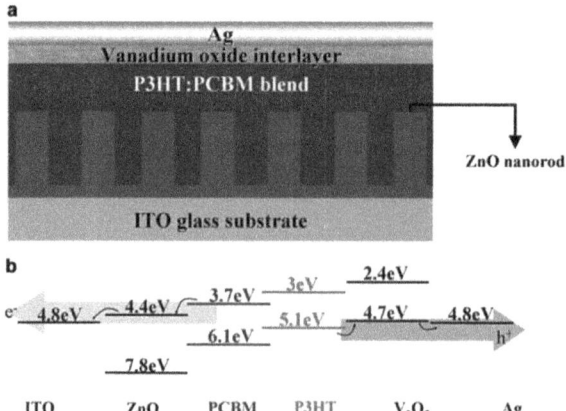

Figure 7.16   (a) Device structure of the OSCs. (b) Energy band diagram for the photovoltaic cells in this study. (Reprinted from ref. 81, with permission from Elsevier, http://www.sciencedirect.com/science/article/pii/S1566119909001396.)

A very similar method was realized later by using wet-milled transition metal oxide nanoparticles as buffer layers in OSCs.[82] A wet grinding method was used to disperse the V_2O_5 into IPA. With the grinding process, the sizes of the metal oxide particles can be effectively reduced to obtain a stable colloidal solution which features a surfactant-free and large-scale solution process.

Method II – sol–gel approach. In 2011, a sol–gel precursor method for V_2O_5 was reported by Zilberberg *et al.* to function as HTL for both normal and inverted OSCs.[83] As given in Table 7.4, by using vanadium-oxytriisopropoxide/isopropyl alcohol solution, the solution-processed V_2O_5 layer can exhibit a high work function of 5.6 eV with plasma treatment or even without any post-treatment. With a similar electrical energy band structure to thermally deposited V_2O_5 film, OSCs with solution-based V_2O_5 yielded comparable efficiency as well as very stable device performance in ambient air conditions.

Another sol–gel method for solution processed V_2O_x was reported by Tan *et al.*, who used vanadyl acetylacetonate [$VO(acac)_2$] isopropyl alcohol solution.[84] After a thermal treatment at 150 °C for 10 minutes of the formed film on ITO substrate, a V_2O_x layer with 2.4 nm surface roughness rms can be obtained. With the annealing process in air, XPS results indicated that V^{4+} and V^{5+} oxidation states existed in the composite metal oxide film, which is shown in Figure 7.17. The OSCs fabricated using the solution-processed V_2O_5 show enhanced performance in comparison with PEDOT:PSS devices. OSCs based on P3HT:$IC_{70}BA$ with solution-processed V_2O_5 as HTL gave an enhanced J_{SC}, a relatively high FF of over 70%, and a PCE of 6.35%.

Method III – hydrogen vanadium oxide bronze. In a similar work to that on H_xMoO_3, hydrogen vanadium oxide bronze ($H_xV_2O_5$) was also demonstrated by Xie *et al.* to serve as a low-temperature solution-processed water-free V_2O_x HTL in OSCs.[78] After the oxidation of vanadium powder with H_2O_2 in ethanol solution, a brown solution could be obtained. As measured by XPS, a small amount of V^{4+} oxidation state existed in the film formed after a 100 °C annealing process, which was shown in Figure 7.14. The oxygen vacancies provided an n-type dopant within the metal oxide film and gave an overall work function of 5.5 eV. OSCs using solution-processed V_2O_x as HTL showed a PCE of over 7.6%, which was better than devices with the commonly used HTL PEDOT:PSS.

Method IV – V_2O_5 hydrate method. Escobar *et al.* reported a low-temperature solution-processed V_2O_5 hydrate method from an aqueous solution of

Table 7.4 WF of pristine and thermally annealed (1 h, 110 °C in ambient air) V_2O_5 layers obtained by sol–gel processing and thermal evaporation in high-vacuum, respectively.

Sample layer	WF pristine [eV]	WF annealed [eV]
V_2O_5 sol–gel processed	5.6	5.3
V_2O_5 thermally evaporated	5.6	5.4

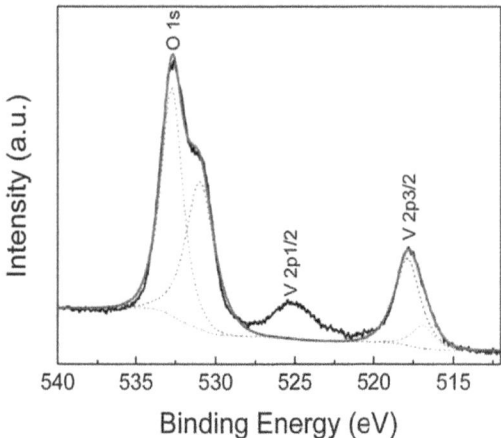

Figure 7.17   X-ray photoelectron spectra of O 1s, V 2p1/2 and V 2p3/2 for the s-VO$_x$ layer on the silicon substrate. (Reprinted from ref. 84, with permission from the Royal Society of Chemistry, http://pubs.rsc.org/en/content/articlelanding/2012/cp/c2cp43125d#!divAbstract.)

sodium metavanadate (NaVO$_3$).[85] The V$_2$O$_5$ hydrate gel can be obtained by cation-exchange from the NaVO$_3$ water solution to metavanadic acid (HVO$_3$). The dehydration process was processed with temperature annealing at 120 °C for 5 minutes after formation of the film of V$_2$O$_5$·nH$_2$O on the substrate. The XPS and UPS findings indicated that the work function, as well as the energy band structure, can be modified during the process, which is given in Figure 7.18. The outdoor stability test indicated the long lifetime of OSCs with metal oxide interfacial layers for both normal and inverted device architectures.

7.4.3   Solution-Processed Tungsten Oxide (WO$_x$) as HTL

Tungsten oxide (WO$_x$) has been applied in organic light emitting diodes (OLEDs) and OSCs to function as a hole injection or extraction layer because of its high work function when incorporated with metal or transparent electrodes.[86] In an early study of solution-processed WO$_3$ film, a sol–gel depositing technique from the hydrolysis of WOClO$_4$ precursor solution was applied by Ozer to characterize the optical and electrochemical properties of the film.[87]

Uniform solution-processed WO$_3$ film was demonstrated by Choy *et al.* to replace PEDOT:PSS as HTL in OSCs.[78] The solution of tungsten(VI) ethoxide in ethanol was spincoated on ITO substrate and stored in air overnight to complete the hydrolysis and condensation reactions. As shown in Figure 7.19, the OSC performance depended on the variation in the thickness of the WO$_3$ layer. With a similar solution-processing method, Tan *et al.* reported a precursor method for forming WO$_3$ film by using tungsten(VI) isopropoxide solution in IPA.[88] A thermal annealing process of 150 °C for 10 minutes in air was applied to decompose the precursor into WO$_3$ on ITO substrate. A small surface roughness rms of 2.6 nm was obtained on ITO substrate. OSCs with

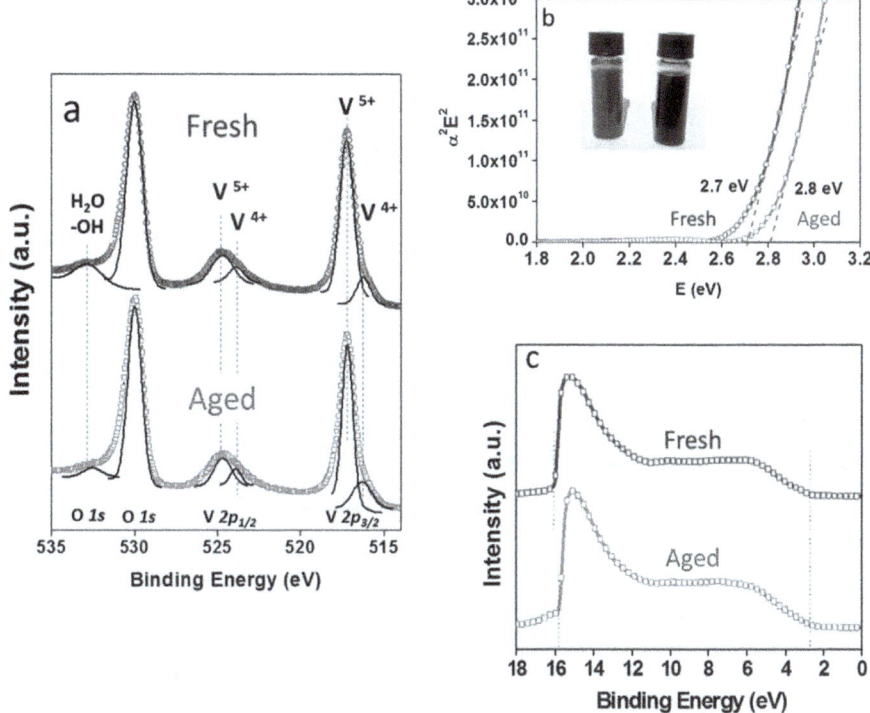

Figure 7.18   XPS (a), band gap (b) and UPS (c) spectra of the $V_2O_5 \cdot 0.5H_2O$ thin-film obtained from freshly prepared (red, above) or 24 h old (green, below) V_2O_5–IPA solution. (Reprinted from ref. 85, with permission from the Royal Society of Chemistry, http://pubs.rsc.org/en/Content/ArticleLanding/2013/EE/c3ee42204f#!divAbstract.)

solution-processed WO_3 as HTL showed enhancement of J_{SC} compared to the commonly used PEDOT:PSS.

In the same period, an alcohol-based WO_3 nanoparticle suspension method was demonstrated by Stubhan *et al.* to function as an HTL in OSCs with both normal and inverted device architectures.[89] The suspension of 7 nm particle size WO_3 can be formed on different substrates by a doctor blading method with a surface roughness rms around 6 nm. The WO_3 NPs HTL was treated by a temperature annealing process at 80 °C for 5 minutes and provided a work function of 5.35 eV. As shown in Figure 7.20, OSCs using WO_3 NPs as HTL can provide comparable performance for both the normal and inverted device architectures.

7.4.4   Doped and Hybrid Metal Oxides as HTL

Based on the hydrogen molybdenum oxide bronze, Li *et al.* demonstrated high performance inverted OSCs by using solution-processed MoO_x with unique features of room-temperature and water-free processing.[90] By introducing the vacuum treatment after the deposition of the MoO_x film on the

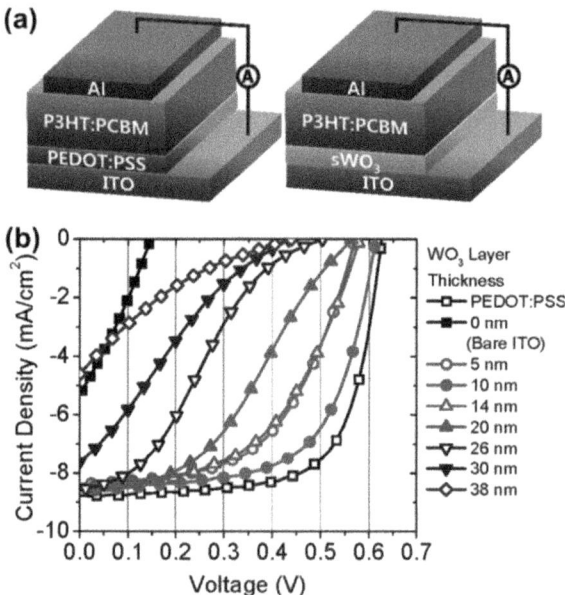

Figure 7.19 (a) Schematic diagram of P3HT:PCBM bulk heterojunction cell structures using ITO/PEDOT:PSS and ITO/sWO$_3$ substrates. (b) Current
density–voltage curves of solar cells employing various thickness of
sWO$_3$ layers and PEDOT:PSS layer as hole extraction layers. Solar cells
were measured under 1 sun illumination (AM1.5G, 100 mW cm^{-2}).
(Reprinted from ref. 88, with permission from the American Chemical
Society, http://pubs.acs.org/doi/abs/10.1021/jp304878u.)

bulk heterojunction active layer of OSCs, the Fermi level of the MoO$_x$ film can
be satisfied to realize energy level alignment to the highest occupied molecular orbital (HOMO) of the light absorption polymer material. Moreover, to
improve the performance of OSCs further, a silver NPs–molybdenum oxide
(Ag NP–MoO$_x$) composite film was prepared by incorporating 4 nm Ag NPs
into the metal oxide based solution. As shown in Figure 7.21, a wide wavelength band increment of incident photon-to-current conversion efficiency
(IPCE) spectra of OSCs indicated the enhancement of the electrical properties
of the composite HTL. By using a low bandgap polymer as donor, OSCs with
Ag NP–MoO$_x$ composite film as HTL gave a relatively high PCE of around 8%.

More recently, Li *et al.* demonstrated solution-processed cesium intercalated MoO$_3$ and V$_2$O$_5$ to realize over 1.1 eV work function tuning.[91] This
method featured a room temperature water-free solution process. The work
function of MoO$_3$ and V$_2$O$_5$ can be continuously tuned by the Cs intercalation
from 5.32 eV (MoO$_3$) and 5.41 eV (V$_2$O$_5$) to 4.17 eV (Cs$_x$MoO$_3$) and 4.12 eV
(Cs$_x$V$_2$O$_5$). As shown in Figure 7.22, the energy band structures and electrical
properties of metal oxides can be effectively modified by the Cs intercalation process. The OSCs and OLEDs using pristine and Cs-intercalated metal
oxides to function as HTL and ETL can achieve good performance in both
normal and inverted device architectures.

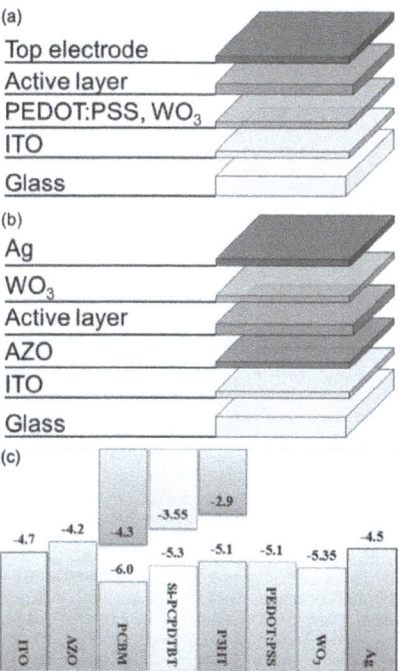

Figure 7.20 Layer stacks of the investigated devices in (a) the normal architecture, and (b) the inverted architecture. (c) Energy levels of the materials investigated in this work. (Reprinted from ref. 89, with permission from John Wiley and Sons, http://onlinelibrary.wiley.com/doi/10.1002/aenm.201200330/abstract.)

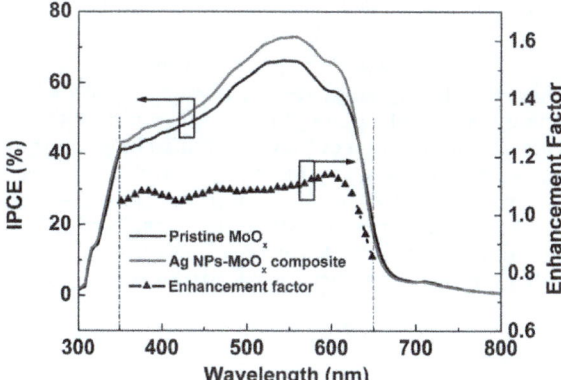

Figure 7.21 IPCE spectra (left axis) of P3HT:PCBM OSCs with pristine MoO_x and Ag NP–MoO_x composite layers and the enhancement factor (right axis) in the core absorption region from 350 nm to 650 nm, *i.e.* (IPCE of OSCs with Ag NP–MoO_x composite)/(IPCE of OSCs with pristine MoO_x). (Reprinted from ref. 90, with permission from the Royal Society of Chemistry, http://pubs.rsc.org/en/Content/ArticleLanding/2013/TA/c3ta10531h#!divAbstract.)

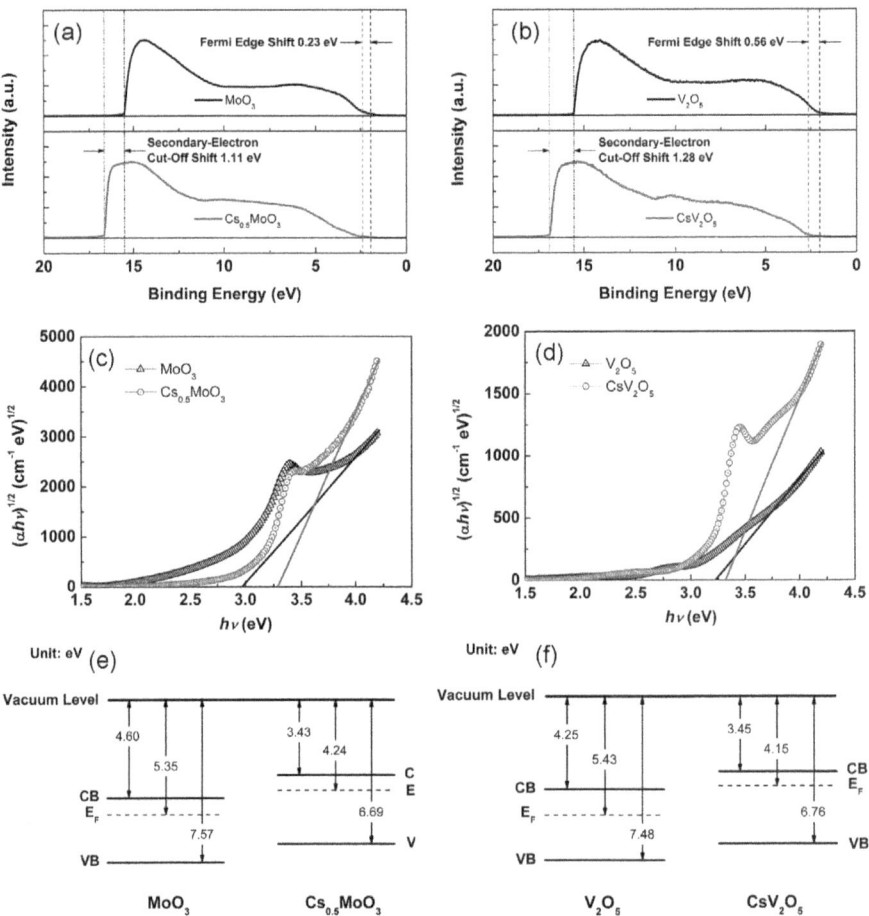

Figure 7.22 (a and b) UPS results of MoO_3, $Cs_{0.5}MoO_3$, V_2O_5 and CsV_2O_5. (c and d) Optical bandgap (E_{opt}) results from the plot of $(ahv)^{1/2}$ vs. photon energy (hv) of MoO_3, $Cs_{0.5}MoO_3$, V_2O_5 and CsV_2O_5. E_{opt} can be taken from the intersection of the extrapolated line with the energy axis. The E_{opt} values of MoO_3, $Cs_{0.5}MoO_3$, V_2O_5, and CsV_2O_5 are 2.97 eV, 3.26 eV, 3.23 eV, and 3.31 eV respectively. (e and f) Detailed energy band structures calculated from the UPS and E_{opt} results of different metal oxides. (Reprinted from ref. 91, with permission from John Wiley and Sons, http://onlinelibrary.wiley.com/doi/10.1002/adfm.201401969/abstract.)

7.5 Summary

Efficiency and stability have been the major concerns for OSCs and OLEDs when compared with the conventional semiconductor-based optoelectronic device counterpart. Great progress in metal oxide-based CTLs has been made in recent decades to improve the efficiency of both normal and inverted organic optoelectronic devices. The single junction bulk-heterojunction OSCs have now achieved PCEs of 9% or above and are continuously evolving.

Moreover, the stability has experienced significant improvement such that the device lifetime of organic optoelectronic devices has increased from "minutes" to "years" owing to the adoption of new materials and fabrication methods. Also, the invention of inverted architecture has brought substantial enhancement in the stability.

The metal oxide based CTLs (including ETLs and HTLs) play an important role in overall device performance for bulk-heterojunction OSCs. In this chapter, I have explained the functions and effects of the CTLs and have focused on CLTs using solution-processible materials in organic optoelectronic devices. Examples of different solution process methods of CTLs were presented. For various ETL and HTL materials, performance comparisons of each solution process method were discussed. By doping and introduction of metal NPs, the electrical properties of CTLs can be further improved by different concepts such as plasmonic-electrical effects, carrier accumulation, *etc.*

The processibility of the CTL material and low temperature annealing is of importance because it is required for mass production methods such as roll-to-roll printing. Traditionally, thermal deposition methods for CTL, such as thermal evaporation and sputtering, are not only energy intensive but also detrimental to overall device performance because high temperature processing can cause undesired chemical reaction or metal diffusion into the active layer film. Various devices using different solution-processed ETL have been reported for each material. These methods all help the device achieve better PCE when compared with a reference cell. The contributions of the solution-processed CTL in each case described in the review can be summarized as providing: (i) adjustment to the energy level alignment at the cathode–active layer interface; (ii) enhancement of the morphology at the ETL–active layer and cathode–ETL interfaces; (iii) improvement of electrical properties and interfacial contact; (iv) formation of an interfacial dipole that reduces work function; and (v) increase of the device stability.

Although all the different solution-processed metal oxide based materials exhibit great potential for performance optimization and commercialization, there are still some limitations regarding optical, electrical, chemical and mechanical properties for each solution-process method and materials. Therefore additional research is required in the field of CTL development to improve the performance of the OSCs and OLEDs further.

Acknowledgments

This study is supported by the University Grant Council of the University of Hong Kong (grants 10401466 and 201111159062), the General Research Fund (grants HKU711813 and HKU711612E) from the Research Grants Council of Hong Kong Special Administrative Region, China. This work is also supported by a grant from the SRFDP and RGC ERG Joint Research Scheme sponsored by the RGC of Hong Kong and the Ministry of Education (M-HKU703/12) and grant CAS14601 from CAS-Croucher Funding Scheme for Joint Laboratories.

References

1. M. Scharber and N. S. Sariciftci, Efficiency of bulk-heterojunction OSCs, *Prog. Polym. Sci.*, 2013, **38**, 1929–1940.
2. Y. H. Liu, J. B. Zhao, Z. K. Li, C. Mu, W. Ma, H. W. Hu, K. Jiang, H. R. Lin, H. Ade and H. Yan, Aggregation and morphology control enables multiple cases of high-efficiency polymer solar cells, *Nat. Commun.*, 2014, **5**, 5293.
3. X. H. Li, W. C. H. Choy, L. J. Huo, F. X. Xie, W. Sha, B. F. Ding, X. Guo, Y. F. Li, J. H. Hou, J. B. You and Y. Yang, Dual Plasmonic Nanostructures for High Performance Inverted Organic Solar Cells, *Adv. Mater.*, 2012, **24**, 3046–3052.
4. W. C. H. Choy, W. K. Chan and Y. P. Yuan, Recent Advances in Transition Metal Complexes and Light Management Engineering in Organic Optoelectronic Devices, *Adv. Mater.*, 2014, **26**, 5368–5399.
5. R. F. Service, Outlook brightens for plastic solar cells, *Science*, 2011, **332**, 293.
6. L. T. Dou, J. B. You, Z. R. Hong, Z. Xu, G. Li, R. A. Street and Y. Yang, 25th Anniversary Article: A Decade of Organic/Polymeric Photovoltaic Research, *Adv. Mater.*, 2013, **25**(46), 6642–6671.
7. C. H. Duan, C. M. Zhong, F. Huang and Y. Cao, Interface Engineering for High Performance Bulk-Heterojunction Polymeric Solar Cells, in *Organic Solar Cells: Materials and Device Physics*, ed. W.C.H. Choy, Springer, 2013, pp. 43–79.
8. H. L. Yip and K. Y. Jen, Recent advances in solution-processed interfacial materials for efficient and stable polymer solar cells, *Energy Environ. Sci.*, 2012, **5**(3), 5994–6011.
9. D. S. Fung and W. C. H. Choy, Introduction to Organic Solar Cells, in *Organic Solar Cells: Materials and Device Physics*, ed. W.C.H. Choy, Springer, 2013, pp. 1–16.
10. I. Litzov and C. Brabec, Development of Efficient and Stable Inverted Bulk Heterojunction (BHJ) Solar Cells Using Different Metal Oxide Interfaces, *Materials*, 2013, **6**, 5796–5820.
11. T. H. Lai, S. W. Tsang, J. R. Manders, S. Chen and F. So, Properties of interlayer for organic photovoltaics, *Mater. Today*, 2013, **16**, 424–432.
12. H. Ma, H. L. Yip, F. Huang and K. Y. Jen, Interface Engineering for Organic Electronics, *Adv. Funct. Mater.*, 2010, **20**, 1371–1388.
13. Z. F. Ma, Z. Tang, E. G. Wang, M. R. Andersson, O. Inganäs and F. L. Zhang, Influences of Surface Roughness of ZnO Electron Transport Layer on the Photovoltaic Performance of Organic Inverted Solar Cells, *J. Phys. Chem. C*, 2012, **116**(46), 24462–24468.
14. J. S. Bhat, A. S. Patil, N. Swami, B. G. Mulimani, B. R. Gayathri, N. G. Deshpande, G. H. Kim, M. S. Seo and Y. P. Lee, Electron irradiation effects on electrical and optical properties of sol-gel prepared ZnO films, *J. Appl. Phys.*, 2010, **108**, 043513.
15. Y. J. Kang, K. Lim, S. H. Jung, D. G. Kim, J. K. Kim, C. S. Kim, S. H. Kim and J. W. Kang, Spray-coated ZnO electron transport layer for air-stable inverted OSCs, *Sol. Energy Mater. Sol. Cells*, 2012, **96**, 137–140.

16. Y. Jouane, S. Colis, G. Schmerber, C. Leuvrey, A. Dinia, P. Leveque, T. Heiser and Y. A. Chapuis, Annealing treatment for restoring and controlling the interface morphology of organic photovoltaic cells with interfacial sputtered ZnO films on P3HT:PCBM active layers, *J. Mater. Chem.*, 2012, **22**, 1606–1612.

17. M. A. Ibrahem, H. Y. Wei, M. H. Tsai, K. C. Ho, J. J. Shyue and C. W. Chu, Solution-processed zinc oxide nanoparticles as interlayer materials for inverted OSCs, *Sol. Energy Mater. Sol. Cells*, 2013, **108**, 156–163.

18. T. Kuwabara, Y. Kawahara, T. Yamaguchi and K. Takahashi, Characterization of Inverted-Type Organic Solar Cells with a ZnO Layer as the Electron Collection Electrode by ac Impedance Spectroscopy, *ACS Appl. Mater. Interfaces*, 2009, **1**, 2107–2110.

19. H. S. Cheun, C. F. Hernandez, Y. H. Zhou, W. J. Potscavage Jr, S. J. Kim, J. W. Shim, A. Dindar and B. Kippelen, Electrical and Optical Properties of ZnO Processed by Atomic Layer Deposition in Inverted Polymer Solar Cells, *J. Phys. Chem. C*, 2010, **114**, 20713–20718.

20. N. K. Elumalai, C. Vijila, R. Jose, K. Z. Ming, A. Saha and S. Ramakrishna, Simultaneous improvements in power conversion efficiency and operational stability of polymer solar cells by interfacial engineering, *Phys. Chem. Chem. Phys.*, 2013, **15**, 19057–19064.

21. M. R. Lilliedal, A. J. Medford, M. V. Madsen, K. Norrman and F. C. Krebs, The effect of post-processing treatments on inflection points in current–voltage curves of roll-to-roll processed polymer photovoltaics, *Sol. Energy Mater. Sol. Cells*, 2010, **94**, 2018–2031.

22. W. G. Tress, K. Leo and M. Riede, Optimum mobility, contact properties, and open-circuit voltage of OSCs: A drift-diffusion simulation study, *Phys. Rev. B*, 2012, **85**, 155201.

23. A. Marmur, Wetting on Hydrophobic Rough Surfaces: To Be Heterogeneous or Not To Be, *Langmuir*, 2003, **19**, 8343–8348.

24. X. Bulliard, S. Ihn, S. Yun, Y. Kim, D. Choi, J. Y. Choi, M. Kim, M. Sim, J. H. Park, W. Choi and K. Cho, Enhanced Performance in Polymer Solar Cells by Surface Energy Control, *Adv. Funct. Mater.*, 2010, **20**, 4381–4387.

25. F. Nickel, C. Sprau, M. F. G. Klein, P. Kapetana, N. Christ, X. Liu, S. Klinkhammer, U. Lemmer and A. Colsmann, Spatial mapping of photocurrents in OSCs comprising wedge-shaped absorber layers for an efficient material screening, *Sol. Energy Mater. Sol. Cells*, 2012, **104**, 18–22.

26. Z. Q. Liang, R. Gao, J. L. Lan, O. Wiranwetchayan, Q. F. Zhang, C. D. Li and G. Z. Cao, Growth of vertically aligned ZnO nanowalls for inverted polymer solar cells, *Sol. Energy Mater. Sol. Cells*, 2013, **117**, 34–40.

27. Z. F. Shi, Y. T. Zhang, S. W. Zhuang, L. Yan, B. Wu, X. J. Cui, Z. Huang, X. Dong, B. L. Zhang and G. T. Du, Vertically aligned two-dimensional ZnO nanowall networks: Controllable catalyst-free growth and optical properties, *J. Alloys Compd.*, 2015, **620**, 299–307.

28. M. J. Tan, S. Zhong, J. Li, Z. K. Chen and W. Chen, Air-Stable Efficient Inverted Polymer Solar Cells Using Solution-Processed Nanocrystalline ZnO Interfacial Layer, *ACS Appl. Mater. Interfaces*, 2013, **5**, 4696–4701.

29. H. Womelsdorf, W. Hoheisel and G. Passing, Nanopartikulares, redispergierbares Fallungsoxid German Patent Specification 19907704 A1 (filing date: 23.02.1999).

30. H. Womelsdorf, W. Hoheisel and G. Passing, Process for producing nanoparticulate, redispersible zinc oxide gels European Patent Specification 1157064 B1 (filing date: 11.02.2000).

31. H. Womelsdorf, W. Hoheisel and G. Passing, Nanoparticulate, redispersible zinc oxide gels US Patent Specification 6,710,091 B1 (patent date: 23.03.2004).

32. F. C. Krebs, Y. Thomann, R. Thomann and J. W. Andreasen, A simple nanostructured polymer/ZnO hybrid solar cell-preparation and operation in air, *Nanotechnology*, 2008, **19**, 424013.

33. Y. M. Sun, J. H. Seo, C. J. Takacs, J. Seifter and A. J. Heeger, Inverted Polymer Solar Cells Integrated with a Low-Temperature-Annealed Sol-Gel-Derived ZnO Film as an Electron Transport Layer, *Adv. Mater.*, 2011, **23**, 1679–1683.

34. S. Y. Shao, K. B. Zheng, T. Pullerits and F. L. Zhang, Enhanced Performance of Inverted Polymer Solar Cells by Using Poly(ethylene oxide)-Modified ZnO as an Electron Transport Layer, *ACS Appl. Mater. Interfaces*, 2012, **5**, 380–385.

35. Y. W. Chen, Z. H. Hu, Z. M. Zhong, W. Shi, J. B. Peng, J. Wang and Y. Cao, Aqueous Solution Processed, Ultrathin ZnO Film with Low Conversion Temperature as the Electron Transport Layer in the Inverted Polymer Solar Cells, *J. Phys. Chem. C*, 2014, **118**, 21819–21825.

36. N. Blouin, A. Michaud, D. Gendron, S. Wakim, E. Blair, R. Neagu-Plesu, M. Belletête, G. Durocher, Y. Tao and M. Leclerc, Toward a Rational Design of Poly(2,7-Carbazole) Derivatives for Solar Cells, *J. Am. Chem. Soc.*, 2007, **130**, 732–742.

37. J. H. Hou and X. Guo, Active Layer Materials for Organic Solar Cells, in *Organic Solar Cells: Materials and Device Physics*, ed. W.C.H. Choy, Springer, 2013, pp. 17–42.

38. J. Y. Kim, S. H. Kim, H. H. Lee, K. Lee, W. Ma, X. Gong and A. J. Heeger, New Architecture for High-Efficiency Polymer Photovoltaic Cells Using Solution-Based Titanium Oxide as an Optical Spacer, *Adv. Mater.*, 2006, **18**, 572–576.

39. R. X. Peng, F. Yang, X. H. Ouyang, Y. Liu, Y. S. Kim and Z. Y. Ge, Enhanced photovoltaic performance of inverted polymer solar cells by tuning the structures of titanium dioxide, *Thin Solid Films*, 2013, **545**, 424–428.

40. E. Voroshazi, I. Cardinaletti, G. Uytterhoeven, L. Shan, M. Empl, T. Aernouts and B. P. Rand, Role of Electron- and Hole-Collecting Buffer Layers on the Stability of Inverted Polymer: Fullerene Photovoltaic Devices, *IEEE J. Photovoltaics*, 2014, **4**, 265–270.

41. L. Sun, W. F. Shen, W. C. Chen, X. C. Bao, N. Wang, X. W. Dou, L. L. Han and S. G. Wen, Simple solution-processed titanium oxide electron transport layer for efficient inverted polymer solar cells, *Thin Solid Films*, 2014, **573**, 134–139.

42. J. Xiong, J. L. Yang, B. C. Yang, C. H. Zhou, X. Hu, H. P. Xie, H. Huang and Y. L. Gao, Efficient and stable inverted polymer solar cells using TiO$_2$ nanoparticles and analysized by Mott-Schottky capacitance, *Org. Electron.*, 2014, **15**, 1745–1752.

43. T. Kuwabara, H. Sugiyama, T. Yamaguchi and K. Takahashi, Inverted type bulk-heterojunction OSC using electrodeposited titanium oxide thin films as electron collector electrode, *Thin Solid Films*, 2009, **517**, 3766–3769.

44. A. Hadipour, R. Müller and P. Heremans, Room temperature solution-processed electron transport layer for OSCs, *Org. Electron.*, 2013, **14**, 2379–2386.

45. W. W. Liang, C. Y. Chang, Y. Y. Lai, S. W. Cheng, H. H. Chang, Y. Y. Lai, Y. J. Cheng, C. L. Wang and C. S. Hsu, Formation of Nanostructured Fullerene Interlayer through Accelerated Self-Assembly and Cross-Linking of Trichlorosilane Moieties Leading to Enhanced Efficiency of Photovoltaic Cells, *Macromolecules*, 2013, **46**, 4781–4789.

46. H. S. Choi, J. H. Lee, W. H. Lee, S. J. Ko, R. Q. Yang, J. C. Lee, H. Y. Woo, C. D. Yang and J. Y. Kim, Acid-functionalized fullerenes used as interfacial layer materials in inverted polymer solar cells, *Org. Electron.*, 2013, **14**, 3138–3145.

47. T. Kuwabara, H. Sugiyama, M. Kuzuba, T. Yamaguchi and K. Takahashi, Inverted bulk-heterojunction OSC using chemical bath deposited titanium oxide as electron collection layer, *Org. Electron.*, 2010, **11**(6), 1136–1140.

48. W. C. H. Choy, D. Zhang, F. X. Xie and X. C. Li, Large-area, high-quality self-assembly electron transport layer for organic optoelectronic devices, *Org. Electron.*, 2012, **13**, 2042–2046.

49. F. C. Chen, J. L. Wu, S. S. Yang, K. H. Hsieh and W. C. Chen, Cesium carbonate as a functional interlayer for polymer photovoltaic devices, *J. Appl. Phys.*, 2008, **103**, 103721.

50. N. Li, T. Stubhan, D. Baran, J. Min, H. Q. Wang, T. Ameri and C. J. Brabec, Design of the Solution-Processed Intermediate Layer by Engineering for Inverted Organic Multi junction Solar Cells, *Adv. Energy Mater.*, 2013, **3**, 301–307.

51. G. Li, C. W. Chu, V. Shrotriya, J. S. Huang and Y. Yang, Efficient inverted polymer solar cells, *Appl. Phys. Lett.*, 2006, **88**, 253503.

52. Y. S. Xin, Z. X. Wang, L. Xu, X. W. Xu, Y. Liu and F. J. Zhang, UV-Ozone Treatment on Cs$_2$CO$_3$ Interfacial Layer for the Improvement of Inverted Polymer Solar Cells, *J. Nanomater.*, 2013, **2013**, 104825.

53. A. Barbot, B. Lucas, C. Di Bin, B. Ratier and M. Aldissi, Optimized inverted polymer solar cells incorporating Cs$_2$CO$_3$-doped C$_{60}$ as electron transport layer, *Appl. Phys. Lett.*, 2013, **102**, 193305.

54. H. H. Liao, L. M. Chen, Z. Xu, G. Li and Y. Yang, Highly efficient inverted polymer solar cell by low temperature annealing of Cs$_2$CO$_3$ interlayer, *Appl. Phys. Lett.*, 2008, **92**, 173303.

55. Y. H. Zhou, H. S. Cheun, W. J. Potscavage Jr, C. Fuentes-Hernandez, S. J. Kim and B. Kippelen, Inverted OSCs with ITO electrodes modified with an ultrathin Al$_2$O$_3$ buffer layer deposited by atomic layer deposition, *J. Mater. Chem.*, 2010, **20**, 6189–6194.

56. O. Wiranwetchayan, Z. Q. Liang, Q. F. Zhang, G. Z. Cao and P. Singjai, The Role of Oxide Thin Layer in Inverted Structure Polymer Solar Cells, *Mater. Sci. Appl.*, 2011, **2**, 1697–1701.

57. K. Hamada, N. Murakami, T. Tsubota and T. Ohno, Solution-processed amorphous niobium oxide as a novel electron collection layer for inverted polymer solar cells, *Chem. Phys. Lett.*, 2013, **586**, 81–84.

58. F. Z. Wang, L. J. Li, Q. Xu, D. P. Qian, S. S. Li and Z. A. Tan, Improved performance of polymer solar cells based on P3HT and ICBA using alcohol soluble titanium chelate as electron collection layer, *Org. Electron.*, 2013, **14**, 845–851.

59. J. L. Lan, S. J. Cherng, Y. H. Yang, Q. F. Zhang, S. Subramaniyan, F. S. Ohuchi, S. A. Jenekhe and G. Z. Cao, The effects of Ta_2O_5–ZnO films as cathodic buffer layers in inverted polymer solar cells, *J. Mater. Chem. A*, 2014, **2**, 9361–9370.

60. M. H. Park, J. H. Li, A. Kumar, G. Li and Y. Yang, Doping of the Metal Oxide Nanostructure and its Influence in Organic Electronics, *Adv. Funct. Mater.*, 2009, **19**, 1241–1246.

61. J. B. You, C. C. Chen, L. T. Dou, S. Murase, H. S. Duan, S. A. Hawks, T. Xu, H. J. Son, L. P. Yu, G. Li and Y. Yang, Metal Oxide Nanoparticles as an Electron-Transport Layer in High-Performance and Stable Inverted Polymer Solar Cells, *Adv. Mater.*, 2012, **24**, 5267–5272.

62. M. Thambidurai, J. Y. Kim, H. J. Song, Y. J. Ko, N. Muthukumarasamy, D. Velauthapillai, V. W. Bergmann, S. A. L. Weberd and C. H. Lee, Enhanced power conversion efficiency of inverted OSCs by using solution processed Sn-doped TiO_2 as an electron transport layer, *J. Mater. Chem. A*, 2014, **2**, 11426–11431.

63. R. M. Pasquarelli, D. S. Ginley and R. O. Hayre, Solution processing of transparent conductors: from flask to film, *Chem. Soc. Rev.*, 2011, **40**, 5406–5441.

64. M. Thambidurai, J. Y. Kim, Y. J. Ko, H. J. Song, H. W. Shin, J. Y. Song, Y. K. Lee, N. Muthukumarasamy, D. Velauthapillai and C. H. Lee, High-efficiency inverted OSCs with polyethylene oxide-modified Zn-doped TiO_2 as an interfacial electron transport layer, *Nanoscale*, 2014, **6**, 8585–8589.

65. F. X. Xie, W. C. H. Choy, W. Sha, D. Zhang, S. Q. Zhang, X. C. Li, C. W. Leung and J. H. Hou, Enhanced charge extraction in OSCs through electron accumulation effects induced by metal nanoparticles, *Energy Environ. Sci.*, 2013, **6**, 3372.

66. D. Zhang, W. C. H. Choy, F. X. Xie, W. Sha, X. C. Li, B. F. Ding, K. Zhang, F. Huang and Y. Cao, Plasmonic Electrically Functionalized TiO_2 for High-Performance Organic Solar Cells, *Adv. Funct. Mater.*, 2013, **23**, 4255–4261.

67. W. C. H. Choy, The Emerging Multiple Metal Nanostructures for Enhancing the Light Trapping of Thin Film Organic Photovoltaics, *Chem. Commun.*, 2014, **50**, 11984–11993.

68. D. P. Norton, Y. W. Heo, M. P. Ivill, K. Ip, S. J. Pearton, M. F. Chisholm and T. Steiner, ZnO: growth, doping & processing, *Mater. Today*, 2004, **7**, 34–40.

69. O. Pachoumi, C. Li, Y. Vaynzof, K. K. Banger and H. N. Sirringhaus, Improved Performance and Stability of Inverted Organic Solar Cells with Sol-Gel Processed, Amorphous Mixed Metal Oxide Electron Extraction Layers Comprising Alkaline Earth Metals, *Adv. Energy Mater.*, 2013, 3(11), 1428–1436.

70. S. H. Liao, H. J. Jhuo, P. N. Yeh, Y. S. Cheng, Y. L. Li, Y. H. Lee, S. Sharma and S. A. Chen, Single Junction Inverted Polymer Solar Cell Reaching Power Conversion Efficiency 10.31% by Employing Dual-Doped Zinc Oxide Nano-Film as Cathode Interlayer, *Sci. Rep.*, 2014, 4, 6813.

71. H. O. Seo, S. Y. Park, W. H. Shim, K. D. Kim, K. H. Lee, M. Y. Jo, J. H. Kim, E. S. Y. Lee, D. W. Kim, Y. D. Kim and D. C. Lim, Ultrathin TiO_2 Films on ZnO Electron-Collecting Layers of Inverted Organic Solar Cell, *J. Phys. Chem. C*, 2011, **115**, 21517–21520.

72. C. Girotto, E. Voroshazi, D. Cheyns, P. Heremans and B. P. Rand, Solution-processed MoO_3 thin films as a hole-injection layer for OSCs, *ACS Appl. Mater. Interfaces*, 2011, 3, 3244–3247.

73. J. Meyer, R. Khalandovsky, P. Görrn and A. Kahn, MoO_3 Films Spin-Coated from a Nanoparticle Suspension for Effcient Hole-Injection in Organic Electronics, *Adv. Mater.*, 2011, **23**, 70–73.

74. J. J. Jasieniak, J. Seifter, J. Jo, T. Mates and A. J. Heeger, A Solution-Processed MoO_x Anode Interlayer for Use within Organic Photovoltaic Devices, *Adv. Funct. Mater.*, 2012, **22**, 2594–2605.

75. S. Murase and Y. Yang, Solution Processed MoO_3 Interfacial Layer for Organic Photovoltaics Prepared by a Facile Synthesis Method, *Adv. Mater.*, 2012, **24**, 2459–2462.

76. T. B. Yang, M. Wang, Y. Cao, F. Huang, L. Huang, J. B. Peng, X. Gong, S. Z. D. Cheng and Y. Cao, Polymer Solar Cells with a Low-Temperature-Annealed Sol–Gel-Derived MoO_x Film as a Hole Extraction Layer, *Adv. Energy Mater.*, 2012, **2**, 523–527.

77. Y. J. Lee, J. Yi, G. F. Gao, H. Koerner, K. W. Park, J. Wang, K. Y. Luo, R. A. Vaia and J. W. P. Hsu, Low-Temperature Solution-Processed Molybdenum Oxide Nanoparticle Hole Transport Layers for Organic Photovoltaic Devices, *Adv. Energy Mater.*, 2012, **2**, 1193–1197.

78. F. X. Xie, W. C. H. Choy, C. D. Wang, X. C. Li, S. Q. Zhang and J. H. Hou, Low-Temperature Solution-Processed Hydrogen Molybdenum and Vanadium Bronzes for an Efficient Hole-Transport Layer in Organic Electronics, *Adv. Mater.*, 2013, **25**, 2051–2055.

79. M. T. Greiner, L. Chai, M. G. Helander, W. M. Tang and Z. H. Lu, Transition Metal Oxide Work Functions: The Influence of Cation Oxidation State and Oxygen Vacancies, *Adv. Funct. Mater.*, 2012, **22**, 4557–4568.

80. A. Soultati, A. M. Douvas, D. G. Georgiadou, L. C. Palilis, J. M. Feckl, S. Gardelis, M. Fakis, S. Kennou, P. Falaras, T. Stergiopoulos, N. A. Stathopoulos, D. Davazoglou, P. Argitis and M. Vasilopoulou, Solution processed hydrogen molybdenum bronzes as highly conductive anode interlayers in efficient organic photovoltaics, *Adv. Energy Mater.*, 2014, **4**, 1300896.

81. J. S. Huang, C. Y. Chou, M. Y. Liu, K. H. Tsai, W. H. Lin and C. F. Lin, Solution-processed vanadium oxide as an anode interlayer for inverted polymer solar cells hybridized with ZnO nanorods, *Org. Electron.*, 2009, **10**, 1060–1065.

82. J. H. Huang, T. Y. Huang, H. Y. Wei, K. C. Ho and C. W. Chu, Wet-milled transition metal oxide nanoparticles as buffer layers for bulk heterojunction solar cells, *RSC Adv.*, 2012, **2**, 7487–7491.

83. K. Zilberberg, S. Trost, J. Meyer, A. Kahn, A. Behrendt, D. Lützenkirchen-Hecht, R. Frahm and T. Riedl, Inverted Organic Solar Cells with Sol-Gel Processed High Work-Function Vanadium Oxide Hole-Extraction Layers, *Adv. Funct. Mater.*, 2011, **21**, 4776–4783.

84. Z. A. Tan, W. Q. Zhang, C. H. Cui, Y. Q. Ding, D. P. Qian, Q. X., L. J. Li, S. S. Li and Y. F. Li, Solution-processed vanadium oxide as a hole collection layer on an ITO electrode for high-performance polymer solar cells, *Phys. Chem. Chem. Phys.*, 2012, **14**, 14589–14595.

85. G. T. Escobar, J. Pampel, J. M. Caicedo and M. L. Cantu, Low-temperature, solution-processed, layered V_2O_5 hydrate as the hole-transport layer for stable OSCs, *Energy Environ. Sci.*, 2013, **6**, 3088–3098.

86. J. Meyer, S. Hamwi, T. Bülow, H.-H. Johannes, T. Riedl and W. Kowalsky, Highly efficient simplified organic light emitting diodes, *Appl. Phys. Lett.*, 2007, **91**, 113506S. Schumann, R. Da Campo, B. Illy, A. C. Cruickshank, M. A. McLachlan, M. P. Ryan, D. J. Riley, D. W. McComb and T. S. Jones, Inverted organic photovoltaic devices with high efficiency and stability based on metal oxide charge extraction layers, *J. Mater. Chem.*, 2011, **21**, 2381–2386.

87. N. Özer, Optical and electrochemical characteristics of sol-gel deposited tungsten oxide films: a comparison, *Thin Solid Films*, 1997, **304**, 310–314.

88. Z. A. Tan, L. J. Li, C. H. Cui, Y. Q. Ding, Q. Xu, S. S. Li, D. P. Qian and Y. F. Li, Solution-Processed Tungsten Oxide as an Effective Anode Buffer Layer for High-Performance Polymer Solar Cells, *J. Phys. Chem. C*, 2012, **116**, 18626–18632.

89. T. Stubhan, N. Li, N. A. Luechinger, S. C. Halim, G. J. Matt and C. J. Brabec, High Fill Factor Polymer Solar Cells Incorporating a Low Temperature Solution Processed WO_3 Hole Extraction Layer, *Adv. Energy Mater.*, 2012, **2**(12), 1433–1438.

90. X. C. Li, W. C. H. Choy, F. X. Xie, S. Q. Zhang and J. H. Hou, Room-temperature solution-processed molybdenum oxide as a hole transport layer with Ag nanoparticles for highly efficient inverted OSCs, *J. Mater. Chem. A*, 2013, **1**, 6614–6621.

91. X. C. Li, F. X. Xie, S. Q. Zhang, J. H. Hou and W. C. H. Choy, Over 1.1 eV Workfunction Tuning of Cesium Intercalated Metal Oxides for Functioning as Both Electron and Hole Transport Layers in Organic Optoelectronic Devices, *Adv. Funct. Mater.*, 2014, **24**, 7348–7356.

CHAPTER 8

New Science and New Technology in Semiconducting Polymers

L. KAAKE[a], D. MOSES[a], C. LUO[a], A. K. K. KYAW[a], L. A. PEREZ[a], S. PATEL[a], M. WANG[a], B. GRIMM[a], Y. SUN[a], G. C. BAZAN[a], E. J. KRAMER[a], AND ALAN J. HEEGER*[a]

[a]University of California, Santa Barbara, California, CA 93106, USA
*E-mail: ajhe@physics.ucsb.edu

8.1 Coherence and Uncertainty in Nanostructured Organic Photovoltaic Materials

8.1.1 The Mechanism for Ultrafast Electron Transfer

The field of bulk heterojunction (BHJ) solar cells was created as a result of the demonstration of ultrafast charge transfer.[1] Ultrafast observations of photoinduced infrared active vibrational modes (IRAV) associated with polaron formation unambiguously established the ultrafast photogeneration of charge carriers in BHJ materials.[2]

Our initial discovery of ultrafast electron transfer occurred in late 1992. It was a discovery based purely on curiosity. At that time, we had been working on the optical properties of semiconducting polymers for many years. Then, the fullerenes were discovered. Prof. Serdar Sariciftci was a Post-doc.

RSC Polymer Chemistry Series No. 17
Polymer Photovoltaics: Materials, Physics, and Device Engineering
Edited by Fei Huang, Hin-Lap Yip, and Yong Cao
© The Royal Society of Chemistry 2016
Published by the Royal Society of Chemistry, www.rsc.org

in our group at that time. The first evidence of charge transfer came from electron spin resonance (ESR) studies; two ESR signals were observed; one with *g*-value identical to that of the polymer and a second with *g*-value identical to that of the fullerene, unambiguously implying charge transfer.[3] We had no idea, however, of the timescale of the electron transfer. The additional fact that the luminescence of the polymer was heavily quenched by the addition of fullerenes indicated that the electron transfer must occur on a timescale significantly faster than the decay time of the photoluminescence; *i.e.* at least in the picosecond timescale.[1] Thus, we decided to measure the electron transfer time directly using ultrafast pulsed laser techniques. The result of these initial ultrafast experiments was reported in *Chem. Phys. Lett.* 1993, **213**, 389.[1] The entire field of bulk heterojunction solar cells was created as a result of this demonstration of ultrafast charge transfer. Given that the electron transfer rate was orders of magnitude faster than any competing process, we inferred that the efficiency of photoinduced charge generation must be high, implying the possibility of high efficiency solar cells. The ultrafast charge transfer was time resolved by Brabec *et al.* several years later.[4] Nevertheless, the mechanism for the ultrafast charge transfer has remained a mystery for over 20 years. Only recently has the mechanism become clear.[5]

Photogenerated electron transfer reactions in chemistry typically occur on a timescale significantly greater than the decay timescale of coherent processes. Electron transfer reactions fail this criterion in poly(phenylenevinylene) based polymers, for which the photogenerated coherent wavefunction persists for 25 fs (in solution),[6] and electron transfer occurs in ≈45 fs in the solid BHJ material.[4] In P3HT, the coherent wavefunction persists for 100 fs in solution,[7] while the electron transfer timescale is <100 fs in the BHJ material.[1,8,9] It is therefore necessary to develop testable hypotheses regarding the mechanism of electron transfer in situations where a coherent photoexcited state is involved in the ultrafast electron transfer in bulk heterojunction solar cells.

We estimated the length scale for the effective volume of the photon (and, importantly, the wavefunction describing the probability amplitude for finding a photoexcitation at a particular point in space) using position–momentum uncertainty and assuming the momentum of a single photon as Δp in the expression for the uncertainty principle.[10] The problem can also be considered semi-classically and a very similar estimate of the length scale of the wavefunction describing the photoexcitation is obtained *via* the resolution limit of a microscope; $(\lambda/2\pi n)$ where n is the index of refraction.[11] Finally, the effective volume of a photon is a non-trivial quantity which, for example, plays an important role in calculations of the refractive index of gasses from first principles.[12] In a similar manner, an organic bulk heterojunction solar cell should be especially sensitive to the effective interaction volume of the photon because it is a densely packed collection of strong absorbers. As a result, the photoexcitation process generates a delocalized coherent superposition of the eigenfunctions of the

Schrodinger equation that describes the nanostructured organic photo-voltaic blend. Thus, there is an immediate probability amplitude for finding a photoexcitation near a heterojunction boundary, thereby enabling ultrafast charge transfer (in the femtosecond regime) over relatively long distances.

If the delocalized coherent wavefunction is important for ultrafast charge transfer, the electron transfer rates must outcompete decoherence. We would therefore expect charge generation rates to be correlated with device efficiencies. Figure 8.1a shows the early timescale charge generation dynamics in a solution processed molecular BHJ material.[10] The rate of carrier generation is affected by the amount of diiodooctane (DIO) included in the solution from which the film was cast. To follow the overall trend more easily, Figure 8.1b displays the rate *vs.* additive concentration. Although the estimates of charge transfer timescale are shorter than the pulse width and, as such, are not quantitatively accurate (see ref. 10 for details), the trend is clear.

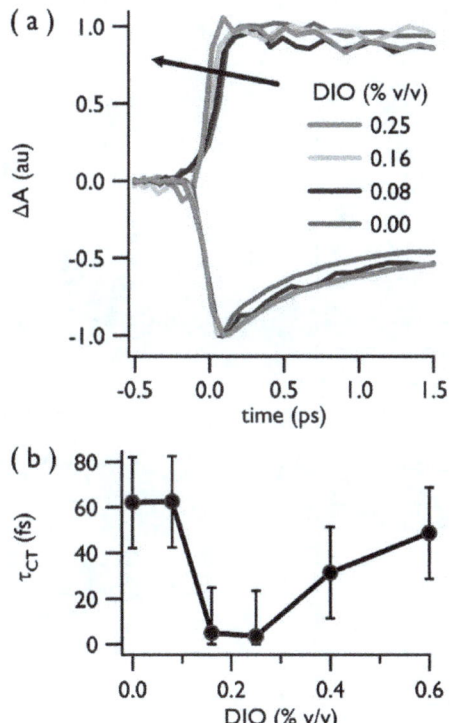

Figure 8.1   Ultrafast transient absorption dynamics. (a) Positive-going signal is the absorption of charge carriers which is affected by the amount of solvent additive (DIO). Negative-going features are the photobleaching signals of the neutral donor molecule, and are unaffected by solvent additives. (b) Qualitative charge transfer timescales obtained by fitting the traces of panel (a) with a model which accounts for a pulse width of 100 fs. Reproduced from ref. 10. Copyright 2012 American Chemical Society.

8.1.2   Ultrafast Experimental Results

Transient absorption measurements were performed on a variety of organic bulk heterojunction nano-structured materials to investigate the photogenerated charge transfer dynamics.[5] A startling generality emerged which implies that the lifetime of the delocalized coherent state produced at ultrafast timescales is sufficiently long that it plays an important role during the charge and energy transfer processes critical to ultrafast charge photogeneration.

Figure 8.2 displays difference spectra, *i.e.* changes in the absorption spectrum of each of the bulk heterojunction samples (composition specified in the figure) as a result of illumination by the pump pulse. All the spectra have common features: a negative pointing feature at shorter wavelengths, and a positive photoinduced absorption at longer wavelengths. The negative features are caused by the photobleaching of the neutral ground state of the electron donors. The positive photoinduced absorptions arise from products of the photoexcitation, associated with charge carriers and/or singlet excitons. Stimulated emission signals are also found in transient absorption spectra, and provide a negative-going contribution spectrally similar to the fluorescence.

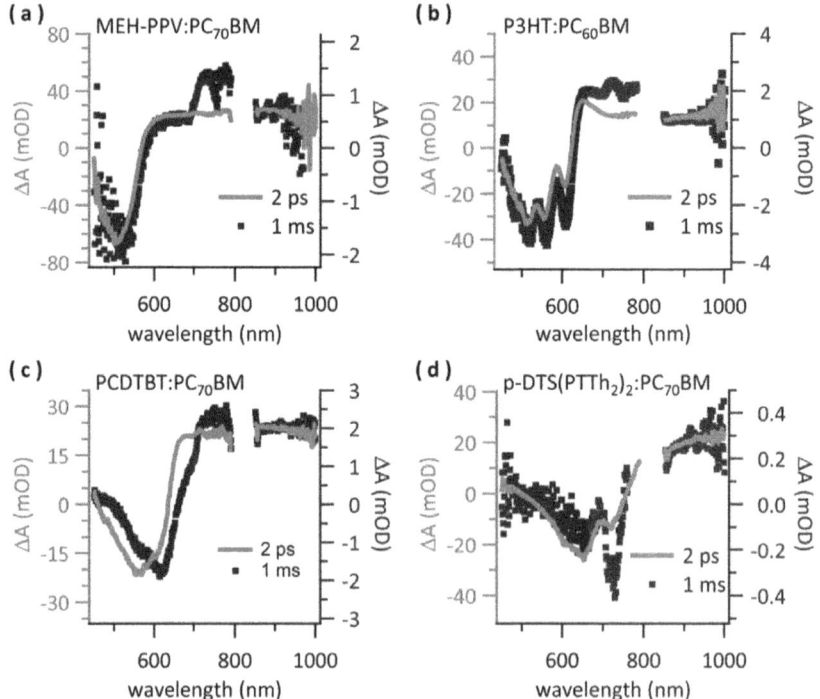

Figure 8.2   Transient absorption spectra at 2 ps and 1 ms. (a) MEH-PPV:PC$_{70}$BM bulk heterojunction, (b) P3HT:PC$_{60}$BM bulk heterojunction, (c) PCDTBT:PC$_{70}$BM bulk heterojunction, (d) *p*-DTS(PTTh$_2$)$_2$:PC$_{70}$BM bulk heterojunction. Reproduced with permission from ref. 5. Copyright 2011 American Chemical Society.

The charge carrier signal was extracted by integrating the spectral region at 850–900 nm (the strength of the signal in this spectral region is proportional to the number of mobile polarons; see ref. 5 for details). This 850–900 nm spectral region largely avoids transient absorption signals from photoexcitation of excitons. By coincidence, in each of the bulk heterojunction samples, the spectral region from 850 to 900 nm could be integrated to provide information regarding the population of charges within the heterojunction films. These spectral assignments are in agreement with those deduced by other research groups from similar measurements on MEH-PPV, P3HT, and PCDTBT (see ref. 5 for details).

Figure 8.3 shows the normalized intensity of the transient absorption signal associated with carriers produced as a result of charge transfer in a bulk heterojunction film, plotted as a function of the time delay between pump and probe pulses. The dynamics display universal behavior, independent of the materials, and correspondingly, independent of the fine details of sample morphology. The data are characterized by a component which rises on a timescale less than 100 fs and a second, slower component which rises on a timescale of approximately 50 ps. In some cases, a decrease is observed at longer timescales ($t \sim 500$ ps) indicative of carrier loss by recombination. In all cases, the maximum value of the charge carrier signal is approximately 1.5 times its value at 1 ps.

The rapidly rising component of the charge carrier dynamics in Figure 8.3 is known to arise from ultrafast electron transfer over distances of 10–20 nm between the electron donor and the fullerene acceptor. The slower rising dynamics occurs after the collapse of the initially delocalized coherent state. As a result, the remaining excitons diffuse to a heterojunction interface

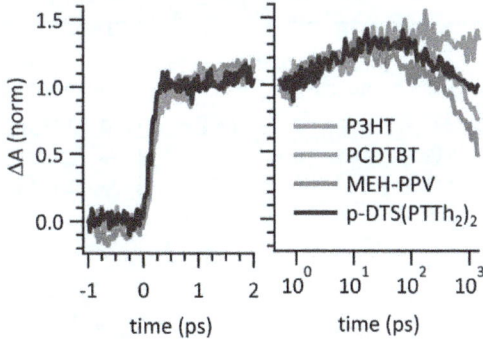

Figure 8.3   Transient absorption of bulk heterojunction materials (*e.g.* P3HT: PC$_{60}$BM, PCDTBT:PC$_{60}$BM, *etc.*). (Left) Integrated spectral intensity associated with mobile carriers, normalized to the intensity at 100 fs and plotted on a linear scale near zero time delay. (Right) Semi-log plot of the integrated spectral intensity associated with the slower component of the mobile carrier generation process, normalized to the intensity at 100 fs. Dynamics are representative of the limit of low pump intensity. Reproduced with permission from ref. 5. Copyright 2013 American Chemical Society.

where they are split, forming charge carriers with holes on the donor side of the heterojunction and electrons on the acceptor side. The observed timescale for the increase in carrier density (≈ 50 ps) is consistent with typical transport distances ~10 nm and known diffusion constants of excitons in molecular solids.[13]

These conclusions are further supported by the intensity dependence of the magnitude of the transient absorption signal at 300 fs and 50 ps, shown in Figure 8.4. Red crosses (longer, top, line) indicate the signal level at 300 fs, assigned as the ultrafast charge transfer yield. This component is linear over 2 orders of magnitude in pump intensity. Blue crosses (shorter, lower line) indicate the signal level at 50 ps minus that at 300 fs, associated with charge transfer occurring after the diffusion of excitons to a heterojunction. The latter component is strongly non-linear at pump intensities greater than ~10 μJ cm^{-2}. This behavior is the result of effects such as exciton–exciton annihilation and exciton–charge annihilation,[13] which destroy diffusing excitons prior to reaching a charge transfer interface.

The data in Figure 8.4 can be used to determine more precisely the ratio between the ultrafast charge transfer component and the charge transfer

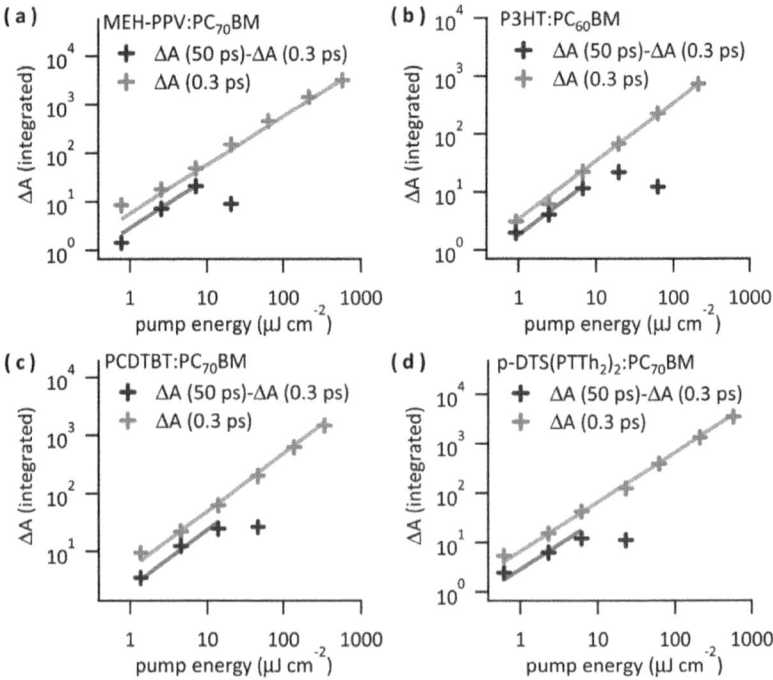

Figure 8.4   Power dependence of the integrated transient absorption signal associated with the two pathways of mobile carrier generation. (a) MEH-PPV:PC$_{70}$BM, (b) P3HT:PC$_{60}$BM, (c) p-DTS(PTTh$_2$)$_2$:PC$_{70}$BM and (d) PCDTBT:PC$_{70}$BM. Reproduced with permission from ref. 5. Copyright 2011 American Chemical Society.

following exciton diffusion. This is done by comparing the magnitude of the ultrafast component to that of the exciton diffusion component in the regime where both are linear; the ratio of diffusive carrier generation to ultrafast carrier generation across samples is 0.31 ± 0.02. The generality of the result (the ratio, 0.31 ± 0.02) is possible only if the photoexcited state which exists at the shortest timescales interacts with approximately the same volume of material as the subsequently diffusing excitons.

Despite observations that fullerenes possess non-zero solubility in several popular semiconducting polymers, accounting for the charge transfer dynamics wholly in terms of morphology is implausible given the sample to sample variations in fullerene concentration. For example, heterojunctions of p-DTS(PTTh$_2$)$_2$ contain 30% fullerene by weight, while heterojunctions of MEH-PPV are 80% fullerene. Moreover, the need for purity within the donor and acceptor domains to obtain high performance solar cells has been established through direct experimental studies.[14–16]

Delocalization *via* this very general quantum mechanical formation of the coherent wavefunction for describing the photoexcited state can be understood as the result of fundamental quantum uncertainty as described by the uncertainty principle. The collapse of this initial state is complex, but we emphasize that, even in solution, coherence in MEH-PPV has been observed to persist to ~25 fs,[6] and in P3HT to ~100 fs.[7] This implies that a coherent state of sufficient lifetime exists in BHJ materials such that it can enable ultrafast electron transfer reactions and the ultrafast generation of mobile charges. Thus, prior to collapse into the wavefunction expected for a disordered nanostructured material, electron transfer can occur from phase separated domains of electron donating materials (polymers or small molecules) to domains comprised of electron accepting materials (typically substituted fullerenes). This explains the equivalent sampling volumes, along with providing an understanding of why the power dependence in the charge generation dynamics becomes different at long timescales; the initially generated coherent state has collapsed.

8.2 High Mobility Thin-Film Transistors (TFTs) Fabricated from Semiconducting Polymers

Solution processible semiconducting polymers with excellent film forming capacity and mechanical flexibility are considered among the most progressive alternatives to conventional inorganic semiconductors. However, the random packing of polymer chains and the disorder of the polymer matrix typically result in low charge transport mobility (10^{-5}–10^{-2} cm^2 V^{-1} s^{-1}), which compromises their performance and development. Here, we summarize a general strategy – by utilizing capillary action – to mediate polymer chain self-assembly and unidirectional alignment on nano-grooved substrates. We designed a sandwich tunnel system separated by functionalized glass spacers to induce capillary action for controlling the polymer nanostructure, crystallinity, and charge transport properties. Using this capillary action, we demonstrate the achievement of saturation mobilities with average values of 21.3 cm^2 V^{-1} s^{-1} and 18.5 cm^2 V^{-1} s^{-1} on two different semiconducting polymers at a transistor channel length of 80 μm. These values are limited by the source–drain contact resistance, R_c. Using a longer channel length of 140 μm where the contact resistance is less important, we measured μ = 36.3 cm^2 V^{-1} s^{-1}. Extrapolating to infinite channel length where R_c is unimportant, we obtain the intrinsic mobility for PCDTBT at this degree of chain alignment and structural order: μ = 45 cm^2 V^{-1} s^{-1}. Our results create a promising pathway towards high performance, solution processible, and low-cost organic electronics.

Initial progress toward this goal has been reported using nano-grooved substrates[17,18] to obtain chain alignment with resulting thin-film mobilities of μ ≈ 20 cm^2 V^{-1} s^{-1}. Although promising, still higher mobilities and a general strategy for achieving such high mobilities are needed for the continued development of "plastic" electronics.[19]

As shown in Figure 8.5a, the main constituents of the sandwich casting system are comprised of two Si/SiO$_2$ substrates, separated by two glass spacers on both short sides. The polymer solution is easily trapped in the tunnel by surface tension, even if the system is slightly tilted along the longitudinal direction. Capillarity-mediated film deposition was then achieved through slow solvent drying. For 75 μL dilute polymer solution (0.25 mg mL^{-1}) prepared in chlorobenzene, five hours are required for solvent evaporation in a closed petri dish at room temperature under N$_2$ atmosphere. In general, a slow film forming process facilitates disentanglement of the macromolecules and self-assembly into crystalline nanostructures.[20–22] Aligning semiconducting polymers with highly parallel spatial orientation is still a formidable challenge, owing to the lack of external directional guidance to the complex macromolecules with multiple degrees of conformational freedom. Here, the capillary action, generated by the glass spacers, can be utilized to promote polymer chain self-alignment along the uniaxial nanogrooves on the textured substrate. The strength of the capillary action can be readily tailored by surface treatment and functionalization with self-assembled monolayers (SAMs) of selected organosilanes (Figure 8.5b and c), such as perfluorodecyltrichlorosilane (FDTS), n-decyltrichlorosilane (DTS) and 6-phenylhexyltrichlorosilane (PTS). Because of the surface attraction and capillary action generated by the PTS treatment, the solution is drawn and flows towards the spacer. The flow direction is the same as that of the uniaxial nanogrooves on the textured substrate and thus favorable for polymer alignment during solution drying (Figure 8.5d). On the other hand, the surface repulsion against the solution caused by the FDTS treatment suppresses such flow towards the spacer; a situation unfavourable to polymer alignment along the uniaxial nano-grooves (Figure 8.5d). To investigate how the strength of capillary action can affect the polymer alignment on the textured substrate, atomic force microscopy (AFM) in tapping mode was performed to examine and compare the nanomorphology on the bottom surfaces of

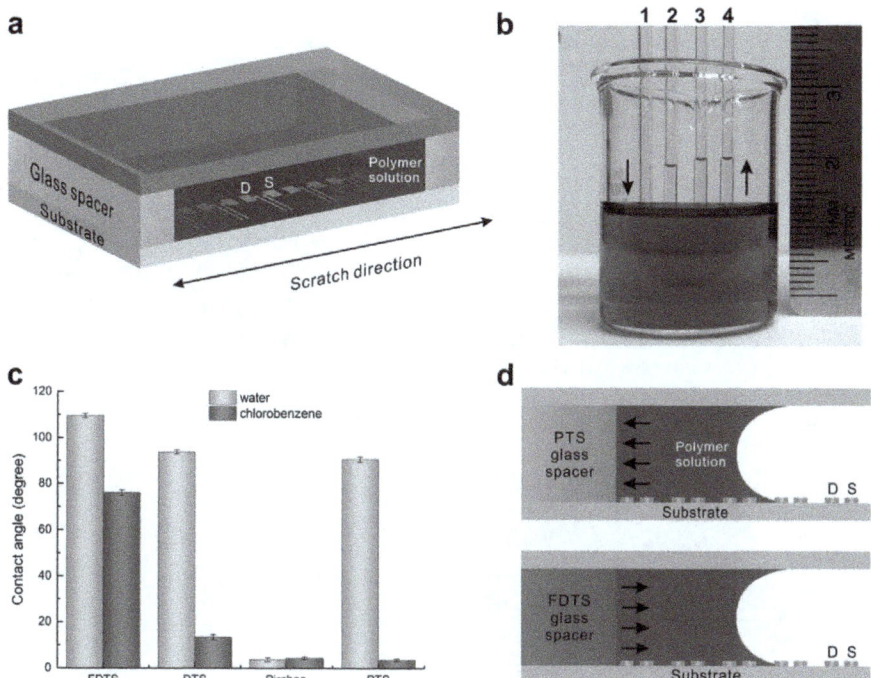

Figure 8.5 (a) Schematic illustration of the sandwich tunnel system consisting of two silicon substrates (12.2 × 7.7 × 0.5 mm) with a pair of glass spacers (7.7 × 2.0 × 1.0 mm) inserted at two sides. The two sides without spacers loaded are open. The direction of the uniaxial nano-grooves on the substrate is along the longitude and perpendicular to the spacers with various surface treatments. (b) Capillary height tests on polymer solution prepared in chlorobenzene using glass capillary tubes (inner diameter: *ca.* 1.2 mm) with various surface treatments. Tube 1, treated by FDTS; tube 2, DTS; tube 3, piranha solution; tube 4, PTS. (c) Contact angle measurements on the plain glass slides with the four different treatments using water and chlorobenzene as the testing media, respectively. (d) Schematic diagram (side view) showing the different actions on the polymer solution exerted by the spacers with two opposite surface treatments, PTS (top) and FDTS (bottom), respectively.

polymer films, prepared using two opposite treatments (FDTS and PTS) to the spacers in the sandwich systems, composed of identically scratched Si/SiO$_2$ substrates. A donor–acceptor copolymer, poly[4-(4,4-dihexadecyl-4*H*-cyclopenta[1,2-*b*:5,4-*b'*]dithiophen-2-yl)-*alt*-[1,2,5]thiadiazolo[3,4-*c*]pyridine] (PCDTPT, M_n = 140 kDa), was used to prepare the solution.[23]

Figure 8.6a and b show the AFM topographic images of the bottom surfaces (taken from areas approaching to the spacers) of two deposited films prepared using the spacers with FDTS treatment and PTS treatment, respectively. The film prepared using FDTS-treated spacers exhibits featureless surfaces and is apparently amorphous. By comparison, the film prepared using

a   **b**

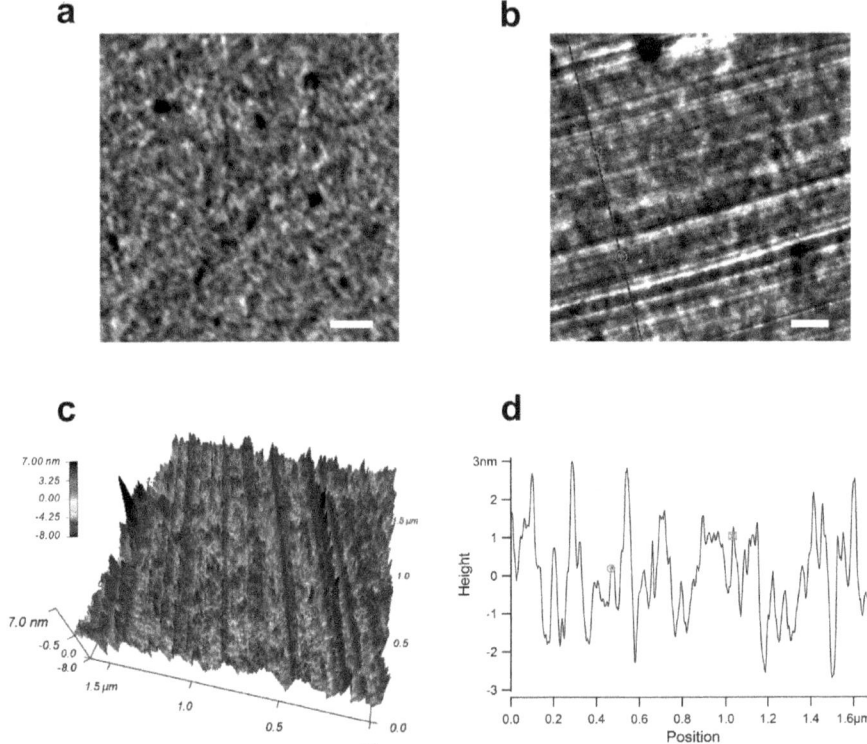

Figure 8.6   Topographic images of the bottom surfaces of two deposited films approaching FDTS-treated spacer (a) and PTS-treated spacer (b). Scale bars represent the length of 200 nm. (c) Three-dimensional (3D) AFM topography of image (b). Demonstration of the unidirectional orientation in a large domain. (d) The section analysis of a height profile traced along the direction perpendicular to the alignment in image (b). The two markers on the height profile indicate the two corresponding points in (b).

PTS-treated spacers shows groove/ridge-like nanostructures, aligned parallel to the uniaxial nano-grooves. These oriented nanostructures are present throughout the bottom surface with widths of 50–100 nm and heights of ~6 nm (Figure 8.6c and d), comparable to the dimensions of the uniaxial nano-grooves produced on the substrate. The root-mean-square (RMS) surface roughness of the FDTS-treated film (σ_{RMS} = 0.64 nm) is significantly lower than that of the PTS-treated film (σ_{RMS} = 1.55 nm) and the textured Si/SiO$_2$ substrate (σ_{RMS} = 1.57 nm). The relatively smooth surface suggests that the repulsion effect from FDTS treatment renders the polymer chains in random orientation; the deposited film could not recognize the uniaxial nano-grooves on the substrate. In contrast, the surface attraction/capillary action from PTS treatment resulted in a more structured topography with roughness comparable to that of the substrate, implying that the polymer chains have diffused, nucleated and grown along the uniaxial nano-grooves on the patterned sub-strate. This leads to a significantly improved nanomorphology with finer, ordered, and regularly connected grains, as observed in Figure 8.6b and c. It is expected that this kind of nanomorphology should result in continuous pathways necessary for efficient charge transport and impart superior aniso-tropic charge transport properties to the bottom surface of the film. On the other hand, the topography on the top surface of the deposited film using PTS-treated spacers appears essentially amorphous, indicating that only the first few monolayers of polymer are aligned upon solution drying.

As the AFM analysis is limited to the surface morphology, to gain deeper insight into the molecular packing and crystallinity in the deposited films, measurements by grazing incidence wide angle X-ray scattering (GIWAXS) were conducted on the deposited films (Figure 8.7a–d). The observation of high order reflections accompanied by strong intermolecular π–π stacking strongly supports the assertion that the macromolecules adopt an edge-on arrangement with a lamella thickness of 2.47 nm and a π-stacking distance of 0.35 nm (Figure 8.7e).[24] These results obtained from GIWAXS measurements are in good agreement with the surface information observed from AFM, thereby confirming that capillary action substantially promotes the behavior of unidirectional self-assembly and anisotropic alignment on the textured substrates.

By implementing the sandwich casting with various surface treatments over the inserted spacers, we fabricated polymer TFT devices with the pre-deposited Ni/Au electrodes as the source and drain on the SiO_2/Si substrates to define the geometry of bottom-contact and bottom-gate. In addition to PCDTPT, we extended our investigation on TFT performance to another donor–acceptor copolymer, poly[2,6-(4,4-bis-alkyl-4*H*-cyclopenta-[2,1-*b*;3,4-*b*0]-dithiophene)-*alt*-4,7-(2,1,3-enzothiadiazole)], (cyclopenta-dithiophene-ben-zothiadiazole) (CDTBTZ, M_n = 120 kDa).[25] See Figure 8.8a and b for the molecular structures. The optimal performance was obtained from devices by processing the source-to-drain (S–D) orientation identical with the procedures described above for PCDTPT; the PTS-treated spacers were again used in a tilted sand-wich system at ~12.5° inclination. Transfer and output characteristics of the best fabricated TFT are shown in Figure 8.8c and d, respectively. The mobility is extracted from the saturated region at low gate voltages (Figure 8.8c; V_G: −4 V to −13 V), where there is a good saturation in the corresponding out-put curves (Figure 8.8d). The output data at high V_G do not show saturation, therefore one can obtain accurate mobilities only in the low V_G regime (V_G ≤ 20 V). The result of the non-saturation can be seen in the change in slope of the transconductance data at V_G > 20 V. Low hysteresis is observed with multi-cycles of forward and reverse sweeping of gate–source voltage, suggest-ing that a low density of shallow traps is present at the polymer–dielectric interface consistent with the small value of the turn-on voltage (V_{th} ~ 0). The superior TFT characteristics offered by the aligned polymers are competitive with TFTs made using inorganic analogues, such as amorphous silicon and indium gallium zinc oxide.[26,27]

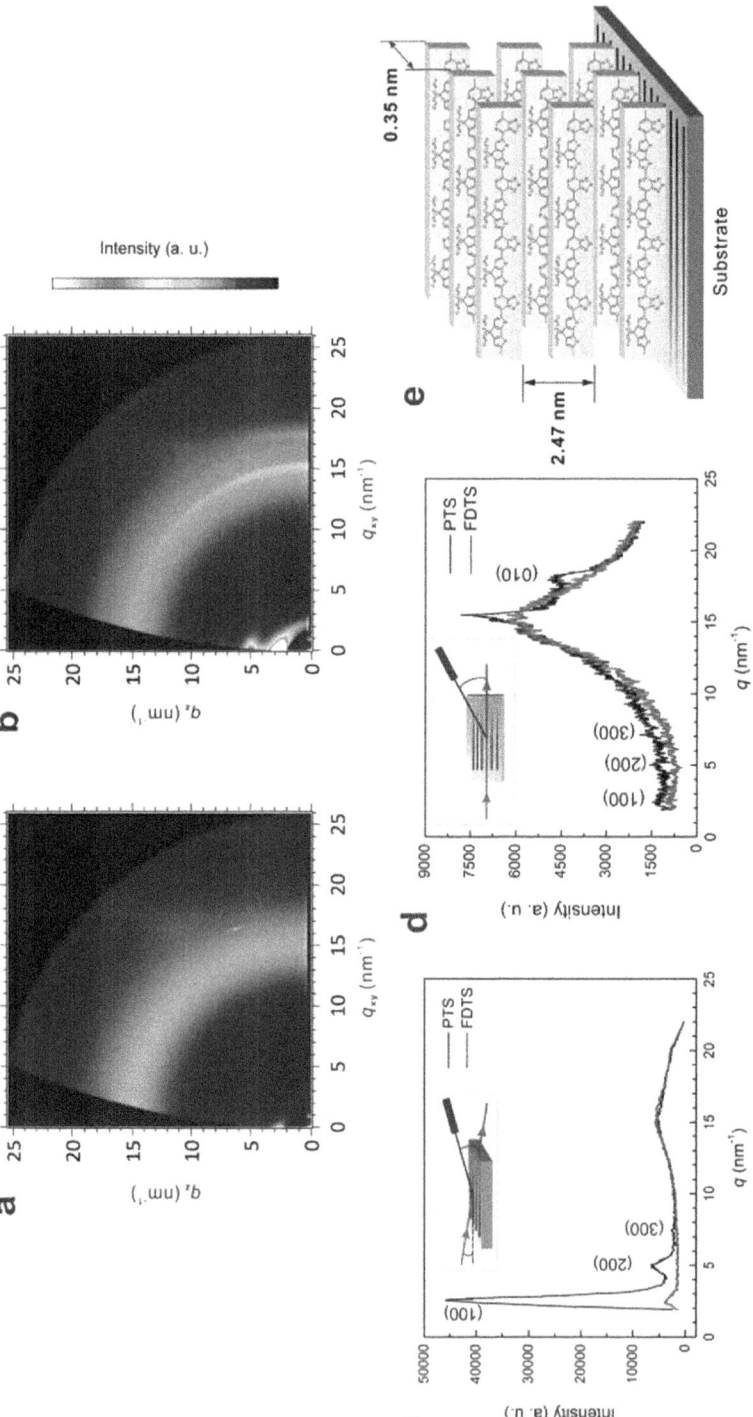

Figure 8.7 GIWAXS patterns of the films fabricated with FDTS-treated spacer (a) and PTS-treated spacer (b). GIWAXS line profiles of the two films using constant, grazing incident angle with out-of-plane (c) and in-plane (d) scattering geometry; q denotes the scattering vector. The insets show the illustration of the measurements for out-of-plane and in-plane geometries. (e) The proposed model, showing the preferential molecule orientations along the nano-groove on the textured substrate, and the calculated spacing between individual planes.

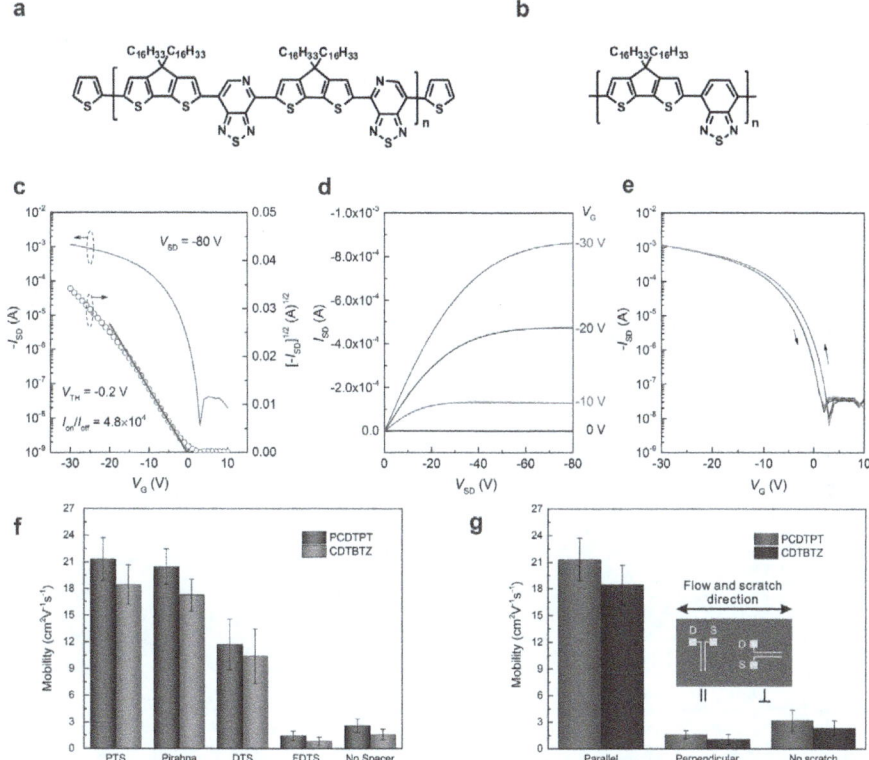

Figure 8.8 The molecular structures of polymers PCDTPT (a) and CDTBTZ (b). Transfer (c) and output (d) characteristics of the device showing a saturation mobility $\mu = 25.4$ cm^2 V^{-1} s^{-1}, fabricated from PCDTPT by treating the pair of spacers with PTS. (e) Hysteresis characterization with four cycles of sweeping of gate–source voltage. (f) Comparison of the average saturation mobility of the devices prepared from PCDTPT and CDTBTZ with various surface treatments on the pair of spacers. The S–D orientation of all devices is parallel to the solution flow and the nano-groove on the textured substrate. (g) Dependence of average saturation mobility on the S–D orientation of devices obtained from spacers using PTS treatment. The average saturation mobility in each processing condition was obtained from 20 devices close to the low-lying spacer (distance: *ca.* 0.5 mm).

The hole mobility obtained from these devices is particularly sensitive to the strength of capillary action generated by the spacers. To verify the sensitivity, we maintained the S–D orientation of all devices parallel to the uni-axial nano-groove on the textured substrate, while altering the treatments to the pair of spacers. Using spacers through PTS and piranha treatments, which exhibit similar strength of capillary action and contact angles, results in comparable average mobilities (obtained with 20 independent devices) of 21.3 cm^2 V^{-1} s^{-1} and 18.5 cm^2 V^{-1} s^{-1} for PCDTPT and CDTBTZ, respectively, as shown in Figure 8.8f. The highest values were 25.4 cm^2 V^{-1} s^{-1} and 22.2 cm^2

V^{-1} s^{-1} for the two semiconducting polymers as the channel length of 80 μm was applied. When the surface treatment applied to the spacers changes to increase contact angle or attenuate capillary action, such as using DTS and FDTS, the average saturation mobility notably decreases for devices prepared from both polymers. Specifically, the average saturation mobility obtained from PTS treatment is up to 14.9- and 22.8-fold higher than those from FDTS treatment for PCDTPT and CDTBTZ, respectively, in good agreement with AFM observation and GIWAXS data on films prepared by the two treatments. In polymer films having a high degree of alignment and crystallinity, charge carriers can travel more efficiently along the conjugated backbone in one dimension, with lower probability of hopping and trapping at grain

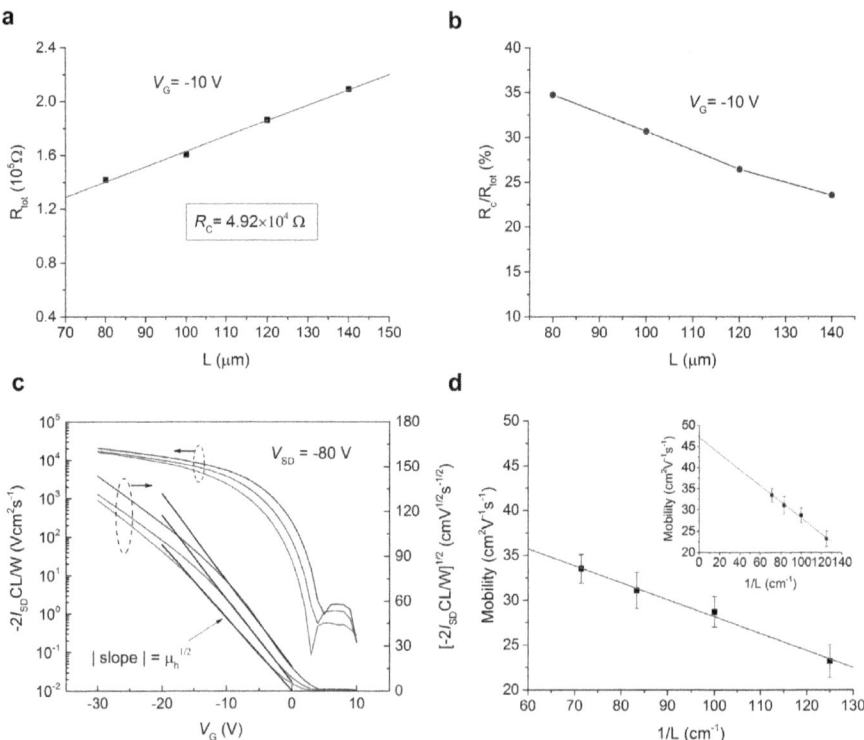

Figure 8.9   (a) Measurement of the channel resistance as a function of the channel length. The contact resistance, R_c is the value extrapolated to $L = 0$. (b) Representative transfer characteristics of devices with channel lengths of 80, 100 and 140 μm, fabricated with PTS-treated spacers. Data obtained from the saturated regime were fitted to obtain the hole mobilities. (c) Transconductance curves for the various TFT devices. (d) Channel length related average saturation mobility of PCDTBT. The average and error bars were obtained from the results of eight devices. The inset shows the mobility extrapolation to infinite channel length. The lines are linear fits to the data. In order to minimize parasitic effects, prior to each measurement the polymer surrounding the specific TFT to be measured was removed.

boundaries and structural imperfections. Moreover, comparing the devices fabricated on the same textured substrates by conventional dip coating (no spacer used), it is evident that this sandwich casting with strong capillary action enhances the charge transport along the conjugated chain. We examined the dependence of mobility on the S–D orientation relative to polymer alignment while retaining PTS treatment to all the spacers. As displayed in Figure 8.8g, the anisotropy between the average saturation mobility of devices with the S–D oriented parallel to polymer alignment *vs.* devices with the S–D perpendicular to polymer alignment was 13.6 and 17.6 for PCDTPT and CDTBTZ, respectively, demonstrating that intramolecular charge transport along the π-conjugated backbones is dominant over the transport by intermolecular charge hopping through π–π stacking. The performance of devices prepared on the native substrates without nanostructure but with the S–D parallel to the solution flow drawn by capillary action remained only slightly higher than that of devices with the S–D perpendicular to the alignment, implying that both capillary action and the uniaxial nano-grooves on the substrate have caused the polymer to self-organize with unidirectional alignment and improved charge transport along the linear backbones.

The mobility values in Figure 8.9 assume that there is no contact resistance. We have measured the contact resistance by fabricating a series of TFTs with different channel lengths by extrapolating the measured S–D channel resistance as a function of the channel length;[28,29] the contact resistance: $R_c = 4.5 \times 10^4\ \Omega$ for PCDTPT.

8.3 Conclusion

The creation of long range coherent superposition states enables the photoexcitation process to generate a delocalized coherent superposition of the eigenfunctions of the Schrodinger equation that describes the nanostructured organic photovoltaic blend. As a result, there is an immediate probability amplitude for finding a photoexcitation near a heterojunction boundary, thereby enabling ultrafast charge transfer (in the femtosecond regime) over relatively long distances.

Our studies have revealed a general and effective strategy to realizing unidirectional alignment and efficient charge transport for semiconducting polymer films deposited on textured Si/SiO$_2$ substrates. By employing sandwich casting in a tilted tunnel system, we are able effectively to utilize capillary action, generated by the functionalized spacer, to render self-assembly of semiconducting polymers along the uniaxial nano-grooves on the substrate. The strength of capillary action can be readily tailored by different surface treatments and functionalizations over the glass spacers. Charge transport in the polymer films prepared by this method is sensitive to the strength of capillary action induced by the functionalized spacers. The capillary action from PTS functionalization enables the achievement of highly oriented crystalline films with compact lamella structure, leading to the superior saturation hole mobilities of 25.4 cm^2 V^{-1} s^{-1} and 22.2 cm^2 V^{-1} s^{-1} for the two semiconducting polymers, PCDTPT and CDTBTZ, respectively. These values are limited by the S–D contact resistance, R_c. Using longer channel lengths where R_c is less important (significantly less than the actual channel resistance) we obtain a mobility of $\mu = 36.3$ cm^2 V^{-1} s^{-1}. Extrapolating to infinite channel length, we obtain the intrinsic mobility for PCDTPT at this degree of chain alignment and structural order: $\mu = 47$ cm^2 V^{-1} s^{-1}. The charge transport in the aligned films exhibits strong anisotropy, showing 13.6- and 17.6-fold higher mobility along the direction of alignment than perpendicular to the alignment for the two polymers, respectively. The presented methodology has clear potential to be applied to a broader range of semiconducting polymers for diverse optoelectronic applications. The concept of capillarity-mediated self-assembly and alignment opens up the possibility of enhancing anisotropic charge transport to create high mobility solution processible TFTs for low-cost organic electronics.

Acknowledgements

Support for these ultrafast studies was provided by the Center for Energy Efficient Materials, an Energy Frontier Research Center funded by the Office of Basic Energy Sciences of the US Department of Energy (DE-DC0001009). The high mobility TFT research was supported by the MC-CAM Program sponsored by Mitsubishi Chemical Corporation (Japan). L. A. P. acknowledges support from the ConvEne IGERT Program (NSF-DGE 0801627) and a Graduate Research Fellowship from the National Science Foundation (GRFP).

References

1. B. Kraabel, C. H. Lee, D. McBranch, D. Moses, N. S. Sariciftci and A. J. Heeger, *Chem. Phys. Lett.*, 1993, **213**, 389.
2. P. B. Miranda, D. Moses and A. J. Heeger, *Phys. Rev. B*, 2001, **64**, 081201.
3. N. S. Sariciftci, L. Smilowitz, A. J. Heeger and F. Wudl, *Science*, 1992, **258**, 1474.
4. C. J. Brabec, G. Zerza, G. Cerullo, S. De Silvestri, S. Luzzati, J. C. Hummelen and S. Sariciftci, *Chem. Phys. Lett.*, 2001, **340**, 232.
5. L. G. Kaake, D. Moses and A. J. Heeger, *J. Phys. Chem. Lett.*, 2013, **4**, 2264.
6. X. J. Yang, T. E. Dykstra and G. D. Scholes, *Phys. Rev. B*, 2005, **71**, 045203.
7. N. P. Wells and D. A. Blank, *Phys. Rev. Lett.*, 2008, **100**, 086403.
8. I. W. Hwang, D. Moses and A. J. Heeger, *J. Phys. Chem. C*, 2008, **112**, 4350.
9. J. M. Guo, H. Ohkita, H. Benten and S. Ito, *J. Am. Chem. Soc.*, 2010, **132**, 6154.
10. L. G. Kaake, G. C. Welch, D. Moses, G. C. Bazan and A. J. Heeger, *J. Phys. Chem. Lett.*, 2012, **3**, 1253.
11. L. Novotny and B. Hecht, *Principles of nano-optics*, Cambridge University Press, New York, 2006.
12. D. L. Andrews, *J. Phys. Chem. Lett.*, 2013, **4**, 3878.

13. M. Pope and C. E. Swenberg, *Electronic processes in organic crystals and polymers*, Oxford University Press, New York, 1999.
14. S. R. Cowan, W. L. Leong, N. Banerji, G. Dennler and A. J. Heeger, *Adv. Funct. Mater.*, 2011, **21**, 3083.
15. W. L. Leong, G. C. Welch, L. G. Kaake, C. J. Takacs, Y. M. Sun, G. C. Bazan and A. J. Heeger, *Chem. Sci.*, 2012, **3**, 2103.
16. Y. M. Sun, G. C. Welch, W. L. Leong, C. J. Takacs, G. C. Bazan and A. J. Heeger, *Nat. Mater.*, 2012, **11**, 44.
17. H. R. Tseng, L. Ying, B. B. Y. Hsu, L. A. Perez, C. J. Takacs, G. C. Bazan and A. J. Heeger, *Nano Lett.*, 2012, **12**, 6353.
18. H. R. Tseng, H. Phan, C. Luo, M. Wang, L. A. Perez, L. Ying, E. J. Kramer, T.-Q. Nguyen, G. C. Bazan and A. J. Heeger, *Adv. Mater.*, 2014, **26**, 2993–2998.
19. H. Yan, Z. H. Chen, Y. Zheng, C. Newman, J. R. Quinn, F. Dotz, M. Kastler and A. Facchetti, *Nature*, 2009, **457**, 679.
20. L. Song, R. K. Bly, J. N. Wilson, S. Bakbak, J. O. Park, M. Srinivasarao and U. H. F. Bunz, *Adv. Mater.*, 2004, **16**, 115.
21. M. Campoy-Quiles, T. Ferenczi, T. Agostinelli, P. G. Etchegoin, Y. Kim, T. D. Anthopoulos, P. N. Stavrinou, D. D. C. Bradley and J. Nelson, *Nat. Mater.*, 2008, **7**, 158.
22. J. Smith, R. Hamilton, I. McCulloch, N. Stingelin-Stutzmann, M. Heeney, D. D. C. Bradley and T. D. Anthopoulos, *J. Mater. Chem.*, 2010, **20**, 2562.
23. L. Ying, B. B. Y. Hsu, H. M. Zhan, G. C. Welch, P. Zalar, L. A. Perez, E. J. Kramer, T. Q. Nguyen, A. J. Heeger, W. Y. Wong and G. C. Bazan, *J. Am. Chem. Soc.*, 2011, **133**, 18538.
24. H. Sirringhaus, P. J. Brown, R. H. Friend, M. M. Nielsen, K. Bechgaard, B. M. W. Langeveld-Voss, A. J. H. Spiering, R. A. J. Janssen, E. W. Meijer, P. Herwig and D. M. de Leeuw, *Nature*, 1999, **401**, 685.
25. H. N. Tsao, D. M. Cho, I. Park, M. R. Hansen, A. Mavrinskiy, D. Y. Yoon, R. Graf, W. Pisula, H. W. Spiess and K. Mullen, *J. Am. Chem. Soc.*, 2011, **133**, 2605.
26. H. Gleskova, P. I. Hsu, Z. Xi, J. C. Sturm, Z. Suo and S. Wagner, *J. Non-Cryst. Solids*, 2004, **338**, 732.
27. K. Nomura, H. Ohta, A. Takagi, T. Kamiya, M. Hirano and H. Hosono, *Nature*, 2004, **432**, 488.
28. T. Matsumoto, W. Ou-Yang, K. Miyake, T. Uemura and J. Takeya, *Org. Electron.*, 2013, **14**, 2590.
29. Y. Xu, R. Gwoziecki, I. Chartier, R. Coppard, F. Balestra and G. Ghibaudo, *Appl. Phys. Lett.*, 2010, **97**, 063302.

CHAPTER 9

Morphology of Bulk Heterojunction Polymer Solar Cells

FENG LIU[a], YAO LIU[a], AND THOMAS P. RUSSELL*[a,b]

[a]Department of Polymer Science and Engineering, University of Massachusetts, Amherst, Massachusetts 01003, USA; [b]WPI-Advanced Institute for Materials Research, Tohoku University, Sendai 980-8577, Japan
*E-mail: tom.p.russell@gmail.com

9.1   Introduction

Organic photovoltaics (OPV) are thought as one of the more promising renewable energy sources for the near future.[1-4] The cost-effective solution-processing of thin-film polymer-based devices, their light weight and flexibility enable them to be used in small appliances, printed onto flexible substrates or integrated into building materials. In the past two decades, there have been significant efforts in both academic and industrial laboratories devoted to this rapidly growing area, so as to enhance device power conversion efficiencies.[5-9] Up to now, a record efficiency of 10.6% has been obtained from polymer-based bulk heterojunction (BHJ) tandem devices.[10] While the chemical structure of the polymer can be modified to enhance the performance, there is much to be gained by a quantitative understanding of the morphology, the parameters influencing the formation of the morphology, and routes

RSC Polymer Chemistry Series No. 17
Polymer Photovoltaics: Materials, Physics, and Device Engineering
Edited by Fei Huang, Hin-Lap Yip, and Yong Cao
© The Royal Society of Chemistry 2016
Published by the Royal Society of Chemistry, www.rsc.org

by which the morphology can be tailored to meet specific objectives. This will require a multi-disciplinary effort that integrates materials science, device physics, interface science and optics. Invariably, this will involve the utilization of more advanced characterization tools to reveal the subtle electronic process in the blended zone.[11-13] The commercialization of this technology calls for input from manufacturing engineering, so as to optimize device fabrication and processing in an industrial setting. This calls for the development of *in situ*, on-line characterization tools that can rapidly decipher the development of morphology over multiple length scales rapidly.[14-16]

This chapter deals with the morphology and related characterization of bulk heterojunction photovoltaics. The fundamental physics underpinning organic photovoltaics, OPVs, can be found in other chapters of this book or from reviews that have appeared.[1,4,17-20] Rather than a single-junction bilayer device, the successful implementation of the BHJ concept, where polymer–polymer and polymer–small molecule blends are used to generate a three-dimensional (3D), bicontinuous morphology having domains of tens of nanometers in size, led to a significant increase in the interfacial area between the donor and acceptor, which translated into a largely improved short circuit current in devices. This breakthrough was a milestone in OPV research, setting a standard for efficient device design and optimization. This advance, though, came with some caveats. In particular, to optimize efficiency, the morphology of a BHJ layer must have bicontinuous domains of the electron and hole conducting materials where the domain sizes are on the order of tens of nanometers. For polymer mixtures and mixtures of a polymer with a small molecule, this mandates a kinetically trapped morphology, because the equilibrium size scale of the domains is macroscopic. Consequently, the morphology will be strongly influenced by the conditions under which the BHJ is prepared, *i.e.* solvents (single or multiple), additives, rates of solvent evaporation, thermal annealing and kinetics of ordering. It remains a challenge to generate optimal BHJ morphologies.

Simply understanding the structural details of the BHJ active layer is challenging. If we add to this the additional challenge of determining the function of components in the blends and a solid structure–property relationship that covers materials, processing and function, the challenge is even more daunting, in fact impossible, without a substantive body of structural data coupled directly with performance. Not only is the bulk morphology of the active layer of importance but also the orientation of the components within the domains, the orientation of the components with respect to each other, the distribution of the components, both in and out of the plane of the films, and the nature of the active layer immediately adjacent to the two electrodes. No one measurement technique can provide sufficient information on all of these fronts and, as such, multiple techniques must be brought in to play. The open circuit voltage (V_{OC}), the short circuit current (J_{SC}) and the fill factor (FF) determine the power conversion efficiency of a PV device.[21] While the V_{OC} is more related to intrinsic properties of active layer materials and electrodes,[22-25] the J_{SC} and FF can be influenced by the morphology and, hence, by the processing conditions and subsequent thermal treatments. Previous studies have shown that chemical structure, molecular weight, blending ratio, solvents, additives, solvent removal rate and annealing conditions can fundamentally change the morphology of the BHJ active layer and, therefore, its performance.

BHJ solar cells are currently the most efficient type of OPV devices. The fundamental benefit is the increased interfacial area between the hole and electron-conducting domains that enhance charge separation.[26,27] State-of-the-art BHJ active layers consist of bicontinuous donor and acceptor domains having length scales commensurate with the exciton diffusion length. Devices based on this concept intrinsically show a larger J_{SC}. The J_{SC} and FF are strongly affected by the details of this phase-separated morphology, which, in turn, calls for a systematic study of the morphology–property relationship to enhance performance and reproducibly to generate a specific morphology.[28–31] Despite numerous efforts from many laboratories worldwide, there is still no consistent, quantitative description of the morphology.[13] It was shown that, in polymer-based BHJ blends, structural order, miscibility of the components, preferential segregation of the components at the electrode interfaces, the distribution of the components within the active layer, and solvent polarity determine the morphology of blends with or without subsequent thermal treatment.[30] In this chapter, we will discuss fundamental aspects of structural characterization, morphology, and the relationship of morphology with device performance.

9.2   Characterization Methods

The BHJ active layer is typically a 100–200 nm thick film where the morphology is a multi-phased system that can be different in the plane of and normal to the surface of the film. Consequently, a quantitative characterization of the morphology requires a three-dimensional description. In general, the morphology of BHJ thin films can be separated into several sub-categories: (i) the lateral phase separation, *i.e.* density correlations in the plane of the film, (ii) concentration variations normal to the film surface (vertical segregation), and (iii) the structure and morphology at the surface. Aside from the local structure, *e.g.* crystalline ordering, crystal size and crystal orientation, the orientation of the components is important for the absorption of light and the transport of excitons to the donor–acceptor interface, hole transport to the anode and electron transport to the cathode. Scheme 9.1 summarizes common methods currently used for the characterization of the morphology. We will briefly discuss these techniques, highlighting the information that can be derived, and the limitations of the technique.

9.2.1   Lateral Morphology Characterizations

The most commonly used real space method is transmission electron microscopy (TEM), which can be used for imaging, diffraction or spectroscopy, with spatial resolution on the nanometer to sub-nanometer scale.

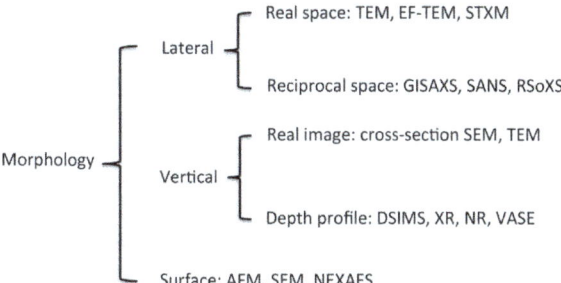

Scheme 9.1 Methods in morphology characterization. TEM: transmission electron microscopy; EF-TEM: energy filter transmission electron microscopy; STXM: scan transmission X-ray microscopy; GISAXS: grazing incidence small-angle X-ray scattering; SANS: small-angle neutron scattering; RSoXS: resonant soft X-ray scattering; SEM: scanning electron microscopy; DSIMS: dynamic secondary ion mass spectroscopy; XR: X-ray reflectivity; NR: neutron reflectivity; VASE: variable angle spectroscopic ellipsometry.

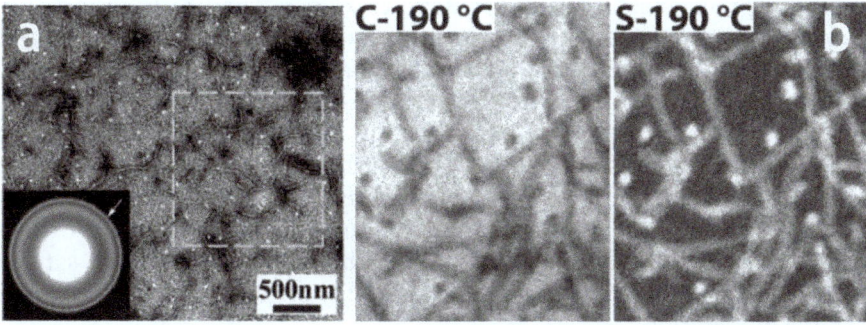

Figure 9.1 (a) TEM image of thermal annealed P3HT:PCBM blends; the clear polymer fibril structure can be seen in the image. The insert is the selected area electron diffraction. (b) EFTEM (carbon map and sulfur map) of annealed P3HT:PCBM blends. Reprinted from ref. 33 and 36 with permission from the American Chemical Society.

Owing to the large depth of field, the elastically scattered electrons produce a planar, two-dimensional projected view of the morphology of the film.[32–34] Consequently, the thickness of the film or cross-section plays a critical role in the image produced. Bright field TEM imaging is the most common and widely used morphology characterization method in OPV research. It provides an image that summarizes the overall features of the blends. Figure 9.1a shows a bright field TEM image of thermally annealed P3HT:PCBM film.[33] In the TEM image, thin fibrils (several tens of nanometers in thickness) form a network with a mesh size of several hundred nanometers. Selected area electron diffraction (SAED) patterns are shown in the insets. The scattering rings and halos in these patterns can be assigned to different structures arising from different components. More advanced TEM techniques such as tomography, where micrographs obtained at different tilt angles of the sample with respect to the incident electron beam are digitally merged to reconstruct a three dimensional image of the morphology, have been gaining usage.[35] It should be noted that the electron densities of each component in the BHJ layer are similar, therefore suitable contrast is difficult to obtain in some cases and selective staining must be used. Energy filtered TEM (EF-TEM) is another useful TEM method where the elemental composition of the active layer can be spatially mapped by taking advantage of inelastically scattered electrons that lose energy as a result of interactions with the elements in the sample. Figure 9.1b shows the carbon and sulfur elemental map images of P3HT:PCBM films annealed at 190 °C.[36] Because PCBM does not contain sulfur, significant enhancement in contrast can be observed by taking an image windowed at the sulfur absorption edge. Similar contrast can be observed for images windowed around the carbon absorption edge. Based on this information, different phases can be identified in the active layer.

Scanning transmission X-ray microscopy (STXM) is another real space morphology characterization technique that uses X-rays rather than electrons to image. The material, depending on its elemental composition, will have characteristic X-ray absorption edges that can be used to tune the contrast of the different components in the sample. Consequently, a chemical mapping of the sample can be obtained. In BHJ OPVs, the most common acceptor, PCBM, shows a distinctive absorption edge at 284.5 eV that can be used to generate contrast between donor materials and PCBM.[37] The two major disadvantages of STXM are the current spatial resolution, which is ~30–50 nm, and, because of X-ray absorption, the necessity for the sample to be very thin, which introduces the possibility of generating artifact in the sample preparation.

In addition to the real space imaging methods, reciprocal space methods, such as grazing incidence X-ray scattering (GISAXS),[38] small angle neutron scattering (SANS)[39,40] and resonant soft X-ray scattering (RSoXS), are also effective in characterizing the morphology. These scattering methods provide information on correlations over length scales ranging from the tenths to hundreds of nanometers in a plane defined by the diffraction vector. The contrast in GISAXS arises from the electron density variations, therefore any structure that changes the electron density of a domain, such as polymer crystallization or PCBM aggregation, can be observed by GISAXS. SANS uses neutron scattering length density (SLD) as the contrast. For PCBM, the high carbon concentration imparts a high SLD that can be easily distinguished from hydrogen rich donor materials, because the neutron scattering length of the proton is negative. Neutron sources are orders of magnitude less intense than synchrotron X-ray sources. To circumvent this, thick sample are usually required. However, neutrons are highly penetrating, whereas substrates such as silicon or ITO glass, metal electrodes, and uniform conducting and interlayers are essentially transparent to neutrons. Consequently, an active layer that contains fullerene-based acceptors or deuterium-labeled components can be characterized by neutrons in the device under operating conditions.

With RSoXS, contrast arises from electron density and absorption differences between the components near the absorption or resonance edges of the constituent functional moieties. RSoXS inherently is sensitive to the chemical composition and bonding of the components and, as such, can be used to investigate multi-component, multi-phase systems, where the contrast can be varied by changing the energy of the incident X-rays. With the use of soft X-rays having longer wavelengths, there is the added benefit that very small scattering vectors can be attained and, as such, much larger structural features can be probed.[41] RSoXS is usually performed in a transmission geometry. Owing to the low energy of the X-rays, absorption is an issue, but with typical BHJ thin films that are ~100 nm in thickness, the scattering volume is sufficiently large to generate sufficient signal to be observed. Shown in Figure 9.2 is the RSoXS profile of a series of DPP based low bandgap polymer BHJ blends.[42] The scattering peak location clearly showed a systematic change in the length scale of phase separation. The length scale of phase separation translates to differences in J_{SC}.

It should be noted that the different scattering and microscopy methods discussed are complementary. Although TEM, SEM and STXM provide exquisite detail on the structure of materials, the information gained is only local. Scattering methods, on the other hand, provide statistically average

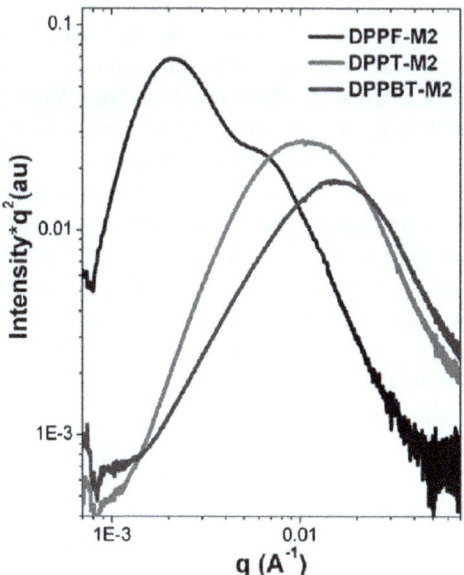

Figure 9.2   RSoXS of DPP based low bandgap polymer and PC71BM blends. The scattering peak indicates a length scale of phase separation inside the BHJ thin-film. Reprinted from ref. 42 with permission from the American Chemical Society.

information over large areas but phase information is lost. This loss in phase can actually be overcome by recent developments in X-ray ptychography, where quantitative, real space images of a sample can be obtained from the inversion of scattering in two separate, but overlapping, sample areas.

9.2.2 Vertical Morphology Characterizations

The distribution of components in BHJ thin-films normal to the surface of the film is also critical in defining the performance of the material. In the simplest extreme, if there is a preferential segregation of one component to an electrode interface, device performance can be poor even with an idealized morphology in the remaining part of the BHJ active layer. Given that charge transfer and collection occur at the electrode, transport to these interfaces requires transport normal to the surface of the film. Normally, the BHJ layer is ~100–200 nm in thickness. Consequently, methods are needed to assess composition profiles normal to the film surface or techniques must be used to examine cross-sections normal to the film surface. In the latter case, cryo-microtoming or focused ion beam (FIB) are required.

The variation in the concentration of the components normal to the film surface, *i.e.* a depth profile, can be obtained by a variety of ion beam methods or reflectivity methods. Dynamic secondary ion mass spectroscopy (DSIMS) is a commonly used technique to obtain a depth profile of specific elements in a sample. Here, an ion beam is rastered over the film surface, etching a crater into the film, and the ion fragments produced are passed into a mass spectrometer.[43] Consequently, DSIMS can be atom or isotope specific (deuterium is twice as heavy as a proton, for example). OPV polymers usually contain sulfur that can be used as a label for the donor polymer, while deuterium labeling is routinely used to locate the PCBM.[39,44,45] Figure 9.3a shows DSIMS profiles of P3HT:deuterated-PCBM blend films under different annealing conditions.[39] It is clear that pre-annealing and post-annealing treatments strongly affect the vertical segregation of PCBM, thus leading to differences in device performances.

Measurements of reflectivity, either neutron reflectivity (NR) or X-ray reflectivity (XR), are other useful tools to study the vertical composition of the film. Here, the X-rays or neutrons are incident on the surface of a thin-film at an angle θ. The specularly reflected (reflected at an angle θ from the surface plane) X-rays or neutrons are measured. Under this condition, the diffraction vector is oriented normal to the surface of the film and, as such, variations in the electron density or neutron scattering length density normal to the surface of the film are measured. A model is then used to calculate the reflectivity profile, which is compared to the experimental data. In NR the intrinsic difference in the scattering length density between a protonated conjugated polymer and PCBM gives sufficient contrast for the measurements. The contrast in XR arises from the electron density difference and it usually has a large accessible q range.[46] In both cases, it must be kept in mind that the depth profiles obtained by NR and XR are averaged

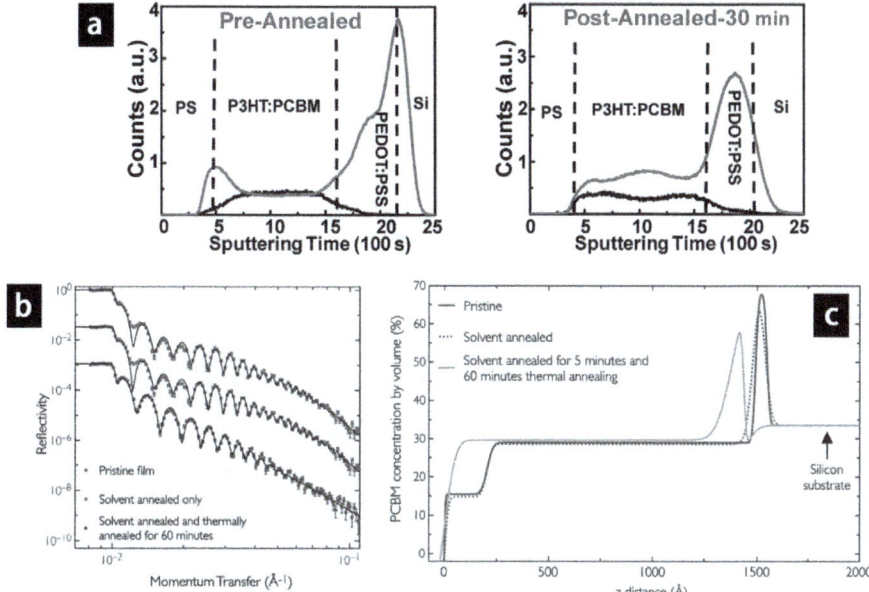

Figure 9.3   Vertical morphology characterizations. (a) DSIMS of pre-annealed P3HT:PCBM blends and post-annealed P3HT:PCBM blends. The red (upper) curve is sulfur element distribution; the black (lower) curve is D atom distribution. A clear difference in vertical segregation is seen for the two different annealing conditions. (b) NR profiles and (c) calculated vertical segregation of P3HT:PCBM blends processed using different methods. Reprinted from ref. 39 and 47 with permission from the American Chemical Society and Wiley-VCH.

(in-plane) over the coherence length of the neutrons and X-rays. Figure 9.3b shows the reflectivity profiles of PCBM distribution in P3HT:PCBM blends.[47] Figure 9.3c is the PCBM concentration profile obtained by model fitting.

Other methods, such as variable-angle spectroscopic ellipsometry (VASE), can be used to obtain a depth profile of the BHJ active layers in thin-films. However, the proper modeling of the refractive indices of BHJ components and fitting is challenging, and one must contend with the fact that most polymers are highly absorbing for wavelengths in the visible spectrum.[48–50]

9.2.3   Surface Morphology Characterization

The surface morphology of a spin-cast thin-film is also of interest in OPV research. It has been shown that surface topography is an important factor influencing the device performance, especially in cases of surface modifications using polymer electrolytes.[51] To assess the surface topography and roughness, scanning electron microscopy (SEM) and atomic force microscopy (AFM) are the most commonly used. Near edge X-ray absorption fine structure (NEXAFS), which uses polarized soft X-rays to probe the elemental composition

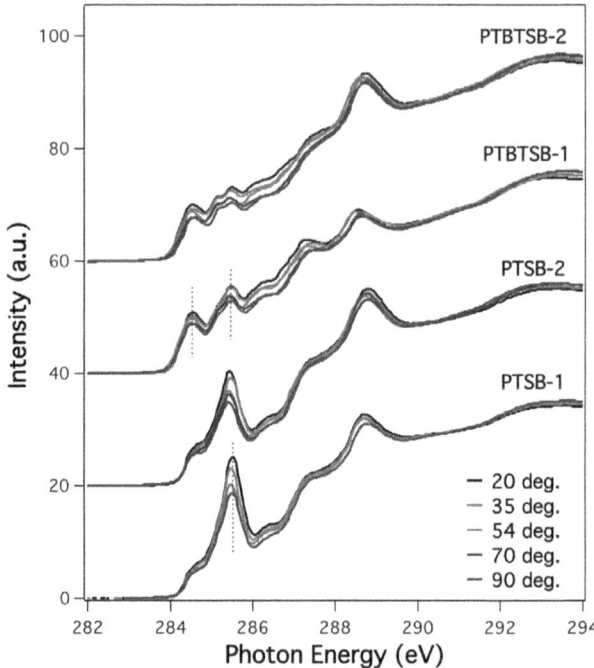

Figure 9.4   NEXAFS of conjugated polymeric zwitterions. The major peak intensity change reflects the preferred orientation of the conjugated backbone of the polymer. Reprinted from ref. 51 with permission from Wiley-VCH.

and bond orientations at the surface, is emerging as an important technique to characterize the surface of the active layer films. It is an element specific technique, because the K-shell absorption edges of different elements appear at different energies. NEXAFS can be obtained by the collection of Auger electrons, yielding information on the first nanometer of the film at the surface, while use of total electrons affords information over the first 10 nm from the surface. Since the absorption of the X-rays will depend on the orientation of dipoles (and consequently bonds), the polarized X-ray beam can be used to measure effective dichroism, *i.e.* the difference in absorption of the X-rays with their polarization direction oriented parallel to and normal to the film surface.[51] Shown in Figure 9.4 are the NEXAFS spectra for conjugated polymeric zwitterions. The peak at 285.5 eV is characteristic of the thiophene carbon=carbon (double bond) π* absorption. By changing the incidence angle of the X-rays, the aromatic backbone orientation can also be assessed.

9.2.4   Crystalline Structure Characterization

The BHJ thin-film blends normally contain a donor polymer or small molecule and PCBM. It has been shown that the miscibility of the donor material and PCBM is key in determining the morphology of the blends.[52] It has been

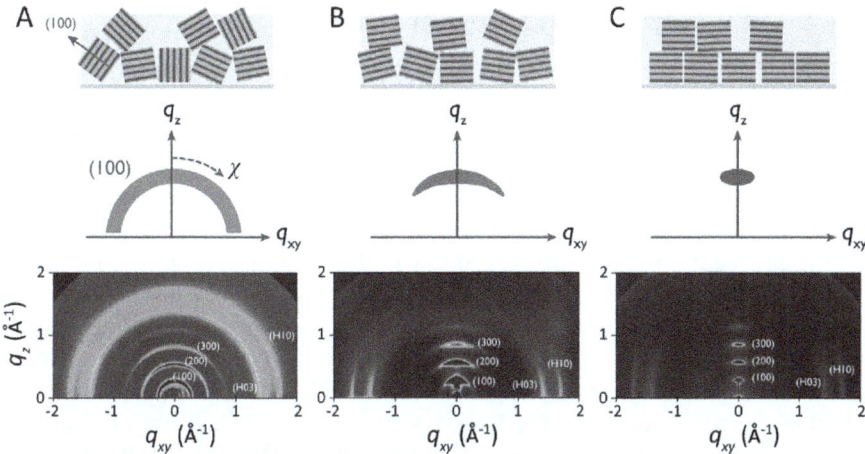

Figure 9.5 GIXD used to study the crystalline information in thin films. (A) Randomly oriented (similar to powder) arrangements of crystallites, with no preference for a specific crystallographic orientation (100) with respect to the substrate normal, produces rings in the diffraction patterns. (B) Textured or oriented films with a distribution of crystallite orientations produce arcs of diffracted intensity. (C) Highly oriented films produce spots or ellipses. The corresponding 2D GIXD patterns for PBTTT that are solid state pressed (A), as spun from solution (B) and annealed (C), are used as examples at bottom. The pressed sample (A) is mostly-not completely-randomly oriented. Reprinted from ref. 54 with permission from the American Chemical Society.

shown that mixing conjugated polymers with PCBM lead to electron mobility;[53] thus, the mixing region showed a similar function to that of PCBM aggregation. The purity of the donor part, for example in crystalline form, is one of the key important factors in BHJ blends, because they are the major conducting channels for positive carriers. Grazing incidence X-ray diffraction (GIXD) is the standard characterization technique used to study the crystal structure in thin-films and widely used for OPV BHJ characterization.[54] Shown in Figure 9.5 is an example of the use of GIXD to measure the crystal structure and orientation of conjugated materials. The Bragg spacing d can be calculated by using $d = 2\pi/q$ and the crystal size can be calculated by using Scherrer equation. The azimuthal spread of the diffraction ring is related to the orientation of the lattice planes in the thin-film. For a highly oriented crystalline thin-film, only strong diffraction spots should be observed.

9.3 Important Morphology Observations

In this part, we will discuss important observations made in OPV. Different material systems will be discussed and important processing methods and related morphology optimization will be compared and analyzed.

9.3.1 PPV Polymers and Solvent Effect

Poly(2-methoxy-5-{3′,7′-dimethyloctyloxy}-*p*-phenylene vinylene) (MDMO-PPV), which gives a power conversion efficiency of 3.0%,[55] was an important material in early OPV research. It is also one of the few conjugated polymers that have been well studied. In 2001, it was noticed that the processing solvent strongly influenced the device efficiency. A 2.5% efficiency was achieved by using chlorobenzene rather than toluene as the processing solvent.[56] The main reason for this improvement is that the macroscopic phase separation of BHJ blends cast from toluene was avoided. As a result, the J_{SC} of such devices increased significantly. Detailed morphological characterization of this system was performed using high-resolution scanning electron microscopy (HR-SEM) and TEM techniques.[57–59] It was seen that the toluene-processed thin-films had large PCBM aggregates, which limited the charge carrier generation efficiency (Figure 9.6a). In fact this feature is similar to observations in single solvent processed low bandgap polymer BHJ blends. When chlorobenzene is used as the processing solvent, the large PCBM clusters disappeared (Figure 9.6b). It was seen that the polymer formed nanoscopic ordered domains, ~20–30 nm in size, and created percolated pathways for both electrons and holes leading to a three-fold enhancement in the efficiency.[57] An 80% loading of PCBM was needed to form a good percolated morphology.[58]

9.3.2 P3HT and Thermal Annealing

Regioregular poly(3-hexylthiophene) (P3HT) is the most popular polythiophene based material in OPV research, owing to its good performance and richness in morphology.[60,61] P3T crystallizes, as evidenced by the higher order reflections in the GIXD profiles. The morphology of P3HT:PCBM blended thin-films is quite complicated,[62] depending strongly on the quality of the polymer itself as well as the processing conditions.[30] It has been continuously reported that regioregularity,[63,64] blend ratio,[65–67] molecular

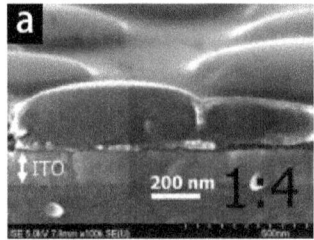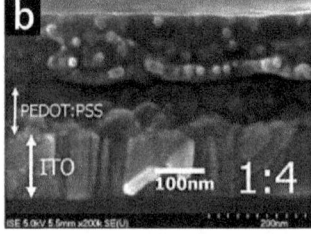

Figure 9.6   Morphology of MDMO-PPV:PCBM blends. (a) BHJ thin-film processed from toluene solution; (b) BHJ thin-film processed from chlorobenzene solution. A clear difference in morphology is seen. Reprinted from ref. 57 with permission from Wiley-VCH.

weight,[68-71] solvent,[72] annealing conditions[73-81] and additives[82-84] all strongly influence the final morphology of the blend thin-films. The most important discovery of this system is the structure–property relationship under thermal annealing. In experiments, two different thermal annealing methods have been used. Pre-annealing is where the blend film is annealed before the deposition of cathode, and post-annealing is where cathode is evaporated onto the blend film prior to thermal annealing.[39] When P3HT:PCBM blends were directly spincoated onto substrates from a commonly used chlorobenzene solution, no obvious phase separation was observed. Post-annealing (usually at 150 °C) of the thin film was proven to be effective in increasing the device performance.[33,39,79,80] This simple thermal treatment can increase the crystallinity of P3HT, which can be directly observed from the increased UV absorption at ~605 nm.[79] At the same time, thermal annealing drives the separation of the P3HT and fullerenes and produces a bi-continuous morphology.[33] In devices, it was seen that post-annealing increased both the J_{SC} and the FF.[80] It was further confirmed that the increase in the hole-mobility of the P3HT was the single most important factor leading to the marked enhancement of the efficiency.[79] It was also found that thermal annealing increased the length and connectivity of the P3HT fibrils embedded in the mixture matrix.[33] Such a thermal treatment also led to a coarsening of the PCBM phase, with increased PCBM agglomerate size, as evidenced by SANS.[40] These results, plus more detailed morphological investigations using different combined techniques and scrutinizing thermal annealed samples with different annealing time, unraveled the morphology of P3HT:PCBM blends.[39] It was seen that within a few seconds of annealing at 150 °C, a bi-continuous network morphology was formed with a characteristic length scale of ~10–15 nm, as evidenced by SANS and high resolution TEM. With further annealing, the phase separation remained relatively stable. The correlation length increased from 0.5 nm to 4.6 nm after 5 s of annealing, and reached 5.3 nm after 30 min of annealing as observed in SANS. Using GIXD, the crystal sizes of P3HT along (100) and (010) crystal planes in the direction normal to the film surface were determined to be ~23 nm and ~12 nm respectively, which is consistent with SANS and TEM results. These results, coupled with electron energy loss spectroscopy, indicated that one of the domains observed could be assigned to the P3HT crystals. Consequently, a more reasonable explanation of the observed morphology is that P3HT nucleates in the homogeneous mixture and, with time, P3HT crystals grow, literally pushing the PCBM away from the growth front(s). From classic arguments of Keith and Padden,[85] if the crystals can grow at a rate of G (in units of cm s^{-1}) and the PCBM can diffuse in the mixture with a diffusion coefficient D (in units of cm^2 s^{-1}), then the ratio D/G will yield a length scale, δ, that is characteristic of an instability at the crystal growth front.

The vertical segregation of the components in the active layer also strongly affects the device performance. It has been reported that post-annealing leads to much better performances than pre-annealing, and this was related to the structure of the film normal to the surface of the film. For pre-annealed samples, P3HT having the lower surface energy will preferentially segregate to the surface.[47,48,86] However, the difference in interfacial energies, rather than the surface energies of the components, plays a more important role for the post-annealed samples, where thermal annealing was performed in the presence of cathode. Consequently, even though the bulk morphologies of the pre-annealed and post-annealed films may be the same, the concentration of components near the cathode interface will be different for the two annealing conditions. This difference in the distribution of components normal to the film surface was revealed by DSIMS (see Figure 9.3a). For the pre-annealed sample, a significant enhancement in the P3HT concentration was observed near the surface, whereas for the post-annealed sample an increase in PCBM concentration was observed near the cathode surface. It is this slight difference in the concentration of the components at the cathode interface that can give rise to the observed differences in the efficiencies.[39]

9.3.3   PCPDTBT and Chemical Additives

PCPDTBT is one of the most well-studied low bandgap polymers that shows a good efficiency. PCPDTBT has a broad absorption up to 800 nm with a bandgap of 1.4 eV.[87,88] However, when PCPDTBT was made, devices using PCPDTBT:$PC_{71}BM$ blends as the active layer and spincoated from chlorobenzene only exhibited a moderate efficiency of 2.8%,[89] which was even lower than that of the P3HT:PCBM system. Various methods had been used to optimize the morphology of PCPDTBT:$PC_{71}BM$ blends and it was found that, by adding a small amount of processing additives, such as 1,8-octancedithiol (ODT) or 1,8-diiodooctane (DIO), a significant improvement in the device efficiency was achieved, up to 5.5%.[89,90] This approach has proven to be effective with many other low bandgap polymers and has become a standard device preparation strategy for low bandgap OPV fabrication. The additives characteristically have a high boiling point, much higher than the major solvent. The additives are commonly poor solvents for the conjugated polymer but good solvents for PCBM. Consequently, during the solution processing of the film (spincoating or solvent coating), the major solvent evaporates initially, increasing the concentration of the additive and making the solvent mixture poorer for the low bandgap polymer but retaining good solubility of the PCBM. The ramification of this is that the low bandgap polymer will precipitate, order or crystallize initially, while the PCBM remains solubilized. This establishes a network of ordered low bandgap polymer within which the remaining low bandgap polymer mixture with PCBM is deposited. This network, by default, suppresses the formation of macroscopic phase separated domains.[91]

It was noted that PCPDTBT has a relatively low crystallinity, if processed without additives.[92-94] Blending PCBM with PCPDTBT further suppresses the ordering of the PCPDTBT, resulting in a loss of percolation network for charge transport. The crystallinity of PCPDTBT is enhanced with the use of the processing additives, as evidenced by the appearance of diffraction

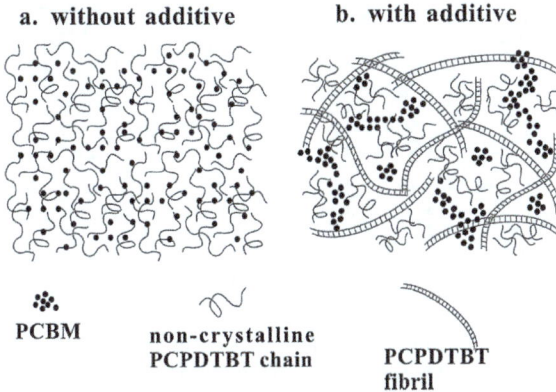

Figure 9.7 Morphology of PCPDTBT:PCBM blends processed (a) without and (b) with additives. Additive processing leads to polymer crystallization and phase separation in the blends. Reprinted from ref. 92 with permission from Wiley-VCH.

peaks in the GIXD profiles along with a red-shift of the UV-vis absorption. Moreover, PCPDTBT fibrils are observed in TEM images. The suppression of the macroscopic phase separation of the components is evidenced in the RSoXS profiles, where a distinct interference maximum is seen in the data that corresponds to the mesh size of the fibrillar network observed by TEM. SANS showed a sharp increase in scattering intensity corresponding to further phase separation in the inter-fibrillar region. These combined scattering results point to a multi-length scale morphology (Figure 9.7) that correlates well with enhanced device efficiency.[92] It is suggested that the use of additives enhanced PCPDTBT chain ordering, the formation of a bicontinuous morphology, and the degree of demixing (phase purity), leading to improved electron and hole mobility through the percolated network with reduced carrier loss.[95-98] Summaries of the resultant device performances with different additives can be found in the literature.[90,92] The additive approach is now a standard method for low bandgap polymer processing and has been extended to small molecule systems.[99]

9.3.4 PTB7 and Hierarchical Structure

In more recent low bandgap polymer developments, materials composed of benzo[1,2-b:4,5-b']dithiophene and thieno[3,4-b]thiophene units are emerging as the most promising systems that show good V_{OC}, J_{SC} and PCE.[6,100,101] Among them, PTB7 has received the most attention, showing a PCE of 7.4% in a standard device configuration[102,103] and 9.2% in an inverted device configuration.[104] PTB7 shows a face-on orientation of the crystallites, which is different from P3HT:PCBM blends. Directly casting PTB7:PCBM blends from a single solvent, such as chlorobenzene, yields a morphology consisting of large domains of PCBM, limiting the J_{SC}. As for other low bandgap polymers,

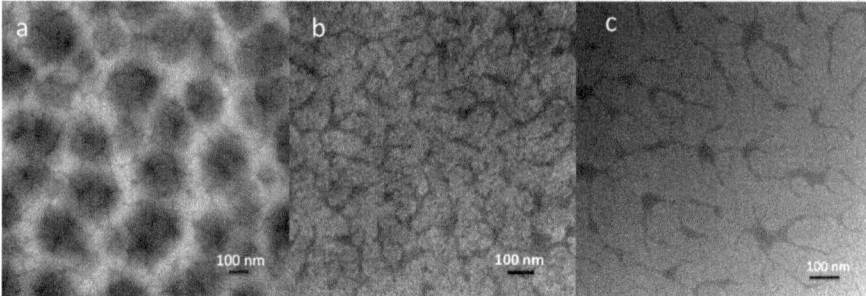

Figure 9.8   Morphology of the PTB7:PCBM system. (a) Blended thin-film processed from chlorobenzene. Two obvious length scales are seen. (b) Blended thin-film processed from chlorobenzene–DIO mixture. Large size PCBM aggregation is suppressed and a new small-length-scaled morphology (~20 nm) shows up. (c) TEM image of pure PTB7 thin-film processed from chlorobenzene. The PTB7 polymer itself forms aggregations that lead to a length scale of phase separation. Reprinted from ref. 106 with permission from Wiley-VCH.

with the addition of a processing additive, *e.g.* diiodooctane, the size scale of the morphological features is markedly reduced. While the crystallinity and orientation of the PTB7 do not change significantly, the addition of diiodooctane suppresses the formation of large domains of PCBM. The AFM and TEM images showed more smooth surfaces and finer mixing with the use of the additive. RSoXS gives evidence of a complex hierarchical morphology, composed of domains of PTB7 crystals, crystalline aggregates, and amorphous mixtures of PTB7 and PCBM domains that give rise to the enhanced device performance.[105] It was shown that PTB7 can form nanostructured aggregates in thin films processed with or without the additive (Figure 9.8). However, the use of the additive suppresses the macroscopic separation of the components, by the formation of a network structure, and the appearance of domains 20–30 nm in size that is key in enhancing the J_{SC}. In a subsequent bilayer and diffusion study, where PCBM was allowed to diffuse into PTB7, the penetration of PCBM into the PTB7 layer was non-Fickian and the miscibility between PTB7 and PCBM was sufficient to dissolve the PTB7 crystals.[106]

9.4   Conclusion

We have reviewed some of the progress in the characterization of the morphology of OPV active layers. While the morphology greatly depends on the materials used, some common denominators emerge. Physical parameters, such as crystalline structure, interaction parameters and miscibility, strongly affect the morphology of the blends. It is evident that controlling the morphology is important for higher efficiency devices. By controlling the thermodynamics, in particular, the interactions between the components, including polymers, fullerenes and solvents, and the interplay of multiple kinetic processes, such as the rate of solvent evaporation, crystallization and phase separation, optimized morphologies for a given system can be achieved.

References

1. C. J. Brabec, N. S. Sariciftci and J. C. Hummelen, *Adv. Funct. Mater.*, 2001, **11**, 15–26.
2. B. C. Thompson and J. M. J. Fréchet, *Angew. Chem., Int. Ed.*, 2008, **47**, 58–77.
3. J. Nelson, *Mater. Today*, 2011, **14**, 462–470.
4. A. C. Mayer, S. R. Scully, B. E. Hardin, M. W. Rowell and M. D. McGehee, *Mater. Today*, 2007, **10**, 28–33.
5. O. Inganäs, F. Zhang and M. R. Andersson, *Acc. Chem. Res.*, 2009, **42**, 1731–1739.
6. Y. Liang and L. Yu, *Acc. Chem. Res.*, 2010, **43**, 1227–1236.
7. G. Dennler, M. C. Scharber, T. Ameri, P. Denk, K. Forberich, C. Waldauf and C. J. Brabec, *Adv. Mater.*, 2008, **20**, 579–583.
8. C. J. Brabec, S. Gowrisanker, J. J. M. Halls, D. Laird, S. Jia and S. P. Williams, *Adv. Mater.*, 2010, **22**, 3839–3856.
9. L. Bian, E. Zhu, J. Tang, W. Tang and F. Zhang, *Prog. Polym. Sci.*, 2012, **37**, 1292–1331.
10. J. You, L. Dou, K. Yoshimura, T. Kato, K. Ohya, T. Moriarty, K. Emery, C.-C. Chen, J. Gao, G. Li and Y. Yang, *Nat. Commun.*, 2013, **4**, 1446.
11. W. Chen, M. P. Nikiforov and S. B. Darling, *Energy Environ. Sci.*, 2012, **5**, 8045–8074.
12. D. M. DeLongchamp, R. J. Kline and A. Herzing, *Energy Environ. Sci.*, 2012, **5**, 5980–5993.
13. F. Liu, Y. Gu, X. Shen, S. Ferdous, H.-W. Wang and T. P. Russell, *Prog. Polym. Sci.*, 2013, **38**, 1990–2052.
14. F. C. Krebs, J. Fyenbo and M. Jørgensen, *J. Mater. Chem.*, 2010, **20**, 8994.
15. R. Søndergaard, M. Hösel, D. Angmo, T. T. Larsen-Olsen and F. C. Krebs, *Mater. Today*, 2012, **15**, 36–49.
16. F. C. Krebs, T. Tromholt and M. Jørgensen, *Nanoscale*, 2010, **2**, 873.
17. J. Nuzi, *C. R. Phys.*, 2002, **3**, 523–542.
18. K. M. Coakley and M. D. McGehee, *Chem. Mater.*, 2004, **16**, 4533–4542.
19. C. J. Brabec, *Sol. Energy Mater. Sol. Cells*, 2004, **83**, 273–292.
20. H. Spanggaard and F. C. Krebs, *Sol. Energy Mater. Sol. Cells*, 2004, **83**, 125–146.
21. Z. He, C. Zhong, X. Huang, W.-Y. Wong, H. Wu, L. Chen, S. Su and Y. Cao, *Adv. Mater.*, 2011, **23**, 4636–4643.
22. S. Yamamoto, A. Orimo, H. Ohkita, H. Benten and S. Ito, *Adv. Energy Mater.*, 2011, **2**, 229–237.
23. K. R. Graham, P. Erwin, D. Nordlund, K. Vandewal, R. Li, G. O. Ngongang Ndjawa, E. T. Hoke, A. Salleo, M. E. Thompson, M. D. McGehee and A. Amassian, *Adv. Mater.*, 2013, **25**, 6076–6082.

24. C. J. Brabec, S. E. Shaheen, C. Winder, N. S. Sariciftci and P. Denk, *Appl. Phys. Lett.*, 2002, **80**, 1288.
25. K. Vandewal, K. Tvingstedt, A. Gadisa, O. Inganäs and J. V. Manca, *Nat. Mater.*, 2009, **8**, 904–909.
26. J. J. M. Halls, C. A. Walsh, N. C. Greenham, E. A. Marseglia, R. H. Friend, S. C. Moratti and A. B. Holmes, *Nature*, 1995, **376**, 498–500.
27. G. Yu, J. Gao, J. Hummelen, F. Wudl and A. Heeger, *Science*, 1995, **270**, 1789–1791.
28. M. T. Dang, L. Hirsch, G. Wantz and J. D. Wuest, *Chem. Rev.*, 2013, **113**, 3734–3765.
29. H. Hoppe and N. S. Sariciftci, *J. Mater. Chem.*, 2006, **16**, 45–61.
30. F. Liu, Y. Gu, J. W. Jung, W. H. Jo and T. P. Russell, *J. Polym. Sci., Part B: Polym. Phys.*, 2012, **50**, 1018–1044.
31. M. A. Ruderer and P. Müller-Buschbaum, *Soft Matter*, 2011, **7**, 5482.
32. S. S. V. Bavel, E. Sourty, G. de With and J. Loos, *Nano Lett.*, 2009, **9**, 507–513.
33. X. Yang, J. Loos, S. C. Veenstra, W. J. H. Verhees, M. M. Wienk, J. M. Kroon, M. A. J. Michels and R. A. J. Janssen, *Nano Lett.*, 2005, **5**, 579–583.
34. A. A. Herzing, L. J. Richter and I. M. Anderson, *J. Phys. Chem. C*, 2010, **114**, 17501–17508.
35. S. D. Oosterhout, M. M. Wienk, S. S. van Bavel, R. Thiedmann, L. J. A. Koster, J. Gilot, J. Loos, V. Schmidt and R. A. J. Janssen, *Nat. Mater.*, 2009, **8**, 818–824.
36. D. R. Kozub, K. Vakhshouri, L. M. Orme, C. Wang, A. Hexemer and E. D. Gomez, *Macromolecules*, 2011, **44**, 5722–5726.
37. B. A. Collins, Z. Li, J. R. Tumbleston, E. Gann, C. R. McNeill and H. Ade, *Adv. Energy Mater.*, 2013, **3**, 65–74.
38. G. Renaud, R. Lazzari and F. Leroy, *Surf. Sci. Rep.*, 2009, **64**, 255–380.
39. D. Chen, A. Nakahara, D. Wei, D. Nordlund and T. P. Russell, *Nano Lett.*, 2011, **11**, 561–567.
40. J. Kiel, A. Eberle and M. Mackay, *Phys. Rev. Lett.*, 2010, **105**, 168701.
41. B. A. Collins, J. E. Cochran, H. Yan, E. Gann, C. Hub, R. Fink, C. Wang, T. Schuettfort, C. R. McNeill, M. L. Chabinyc and H. Ade, *Nat. Mater.*, 2012, **11**, 536–543.
42. F. Liu, C. Wang, J. K. Baral, L. Zhang, J. J. Watkins, A. L. Briseno and T. P. Russell, *J. Am. Chem. Soc.*, 2013, **135**, 19248–19259.
43. J. Jo, S.-I. Na, S.-S. Kim, T.-W. Lee, Y. Chung, S.-J. Kang, D. Vak and D.-Y. Kim, *Adv. Funct. Mater.*, 2009, **19**, 2398–2406.
44. N. D. Treat, M. A. Brady, G. Smith, M. F. Toney, E. J. Kramer, C. J. Hawker and M. L. Chabinyc, *Adv. Energy Mater.*, 2010, **1**, 82–89.
45. D. Chen, F. Liu, C. Wang, A. Nakahara and T. P. Russell, *Nano Lett.*, 2011, **11**, 2071–2078.
46. H.-W. Ro, B. Akgun, B. T. O'Connor, M. Hammond, R. J. Kline, C. R. Snyder, S. K. Satija, A. L. Ayzner, M. F. Toney, C. L. Soles and D. M. DeLongchamp, *Macromolecules*, 2012, **45**, 6587–6599.

47. A. J. Parnell, A. D. F. Dunbar, A. J. Pearson, P. A. Staniec, A. J. C. Dennison, H. Hamamatsu, M. W. A. Skoda, D. G. Lidzey and R. A. L. Jones, *Adv. Mater.*, 2010, **22**, 2444–2447.

48. M. Campoy-Quiles, T. Ferenczi, T. Agostinelli, P. G. Etchegoin, Y. Kim, T. D. Anthopoulos, P. N. Stavrinou, D. D. C. Bradley and J. Nelson, *Nat. Mater.*, 2008, **7**, 158–164.

49. D. M. DeLongchamp, R. J. Kline and A. Herzing, *Energy Environ. Sci.*, 2012, **5**, 5980.

50. D. S. Germack, C. K. Chan, R. J. Kline, D. A. Fischer, D. J. Gundlach, M. F. Toney, L. J. Richter and D. M. DeLongchamp, *Macromolecules*, 2010, **43**, 3828–3836.

51. F. Liu, Z. A. Page, V. V. Duzhko, T. P. Russell and T. Emrick, *Adv. Mater.*, 2013, **25**, 6868–6873.

52. B. A. Collins, E. Gann, L. Guignard, X. He, C. R. McNeill and H. Ade, *J. Phys. Chem. Lett.*, 2010, **1**, 3160–3166.

53. K. Vakhshouri, D. R. Kozub, C. Wang, A. Salleo and E. D. Gomez, *Phys. Rev. Lett.*, 2012, **108**, 026601.

54. J. Rivnay, S. C. B. Mannsfeld, C. E. Miller, A. Salleo and M. F. Toney, *Chem. Rev.*, 2012, **112**, 5488–5519.

55. M. M. Wienk, J. M. Kroon, W. J. H. Verhees, J. Knol, J. C. Hummelen, P. A. van Hal and R. A. J. Janssen, *Angew. Chem., Int. Ed.*, 2003, **42**, 3371–3375.

56. S. E. Shaheen, C. J. Brabec, N. S. Sariciftci, F. Padinger, T. Fromherz and J. C. Hummelen, *Appl. Phys. Lett.*, 2001, **78**, 841–843.

57. H. Hoppe, M. Niggemann, C. Winder, J. Kraut, R. Hiesgen, A. Hinsch, D. Meissner and N. S. Sariciftci, *Adv. Funct. Mater.*, 2004, **14**, 1005–1011.

58. J. K. J. van Duren, X. Yang, J. Loos, C. W. T. Bulle-Lieuwma, A. B. Sieval, J. C. Hummelen and R. A. J. Janssen, *Adv. Funct. Mater.*, 2004, **14**, 425–434.

59. X. Yang, J. K. J. van Duren, R. A. J. Janssen, M. A. J. Michels and J. Loos, *Macromolecules*, 2004, **37**, 2151–2158.

60. M. Brinkmann, *J. Polym. Sci., Part B: Polym. Phys.*, 2011, **49**, 1218–1233.

61. J. A. Lim, F. Liu, S. Ferdous, M. Muthukumar and A. L. Briseno, *Mater. Today*, 2010, **13**, 14–24.

62. Y. Yuan, J. Zhang, J. Sun, J. Hu, T. Zhang and Y. Duan, *Macromolecules*, 2011, **44**, 9341–9350.

63. Y. Kim, S. Cook, S. M. Tuladhar, S. A. Choulis, J. Nelson, J. R. Durrant, D. D. C. Bradley, M. Giles, I. McCulloch, C.-S. Ha and M. Ree, *Nat. Mater.*, 2006, **5**, 197–203.

64. K. Sivula, C. K. Luscombe, B. C. Thompson and J. M. J. Fréchet, *J. Am. Chem. Soc.*, 2006, **128**, 13988–13989.

65. C. Müller, T. A. M. Ferenczi, M. Campoy-Quiles, J. M. Frost, D. D. C. Bradley, P. Smith, N. Stingelin-Stutzmann and J. Nelson, *Adv. Mater.*, 2008, **20**, 3510–3515.

66. S. S. van Bavel, M. Bärenklau, G. de With, H. Hoppe and J. Loos, *Adv. Funct. Mater.*, 2010, **20**, 1458–1463.

67. M. Sanyal, B. Schmidt-Hansberg, M. F. G. Klein, C. Munuera, A. Vorobiev, A. Colsmann, P. Scharfer, U. Lemmer, W. Schabel, H. Dosch and E. Barrena, *Macromolecules*, 2011, **44**, 3795–3800.

68. W. Ma, J. Y. Kim, K. Lee and A. J. Heeger, *Macromol. Rapid Commun.*, 2007, **28**, 1776–1780.

69. P. Schilinsky, U. Asawapirom, U. Scherf, M. Biele and C. J. Brabec, *Chem. Mater.*, 2005, **17**, 2175–2180.

70. R. C. Hiorns, R. de Bettignies, J. Leroy, S. Bailly, M. Firon, C. Sentein, A. Khoukh, H. Preud'homme and C. Dagron-Lartigau, *Adv. Funct. Mater.*, 2006, **16**, 2263–2273.

71. A. Ballantyne, L. Chen, J. Dane, T. Hammant, F. Braun, M. Heeney, W. Duffy, I. McCulloch, D. D. C. Bradley and J. Nelson, *Adv. Funct. Mater.*, 2008, **18**, 2373–2380.

72. M. A. Ruderer, S. Guo, R. Meier, H.-Y. Chiang, V. Körstgens, J. Wiedersich, J. Perlich, S. V. Roth and P. Müller-Buschbaum, *Adv. Funct. Mater.*, 2011, **21**, 3382–3391.

73. G. Li, Y. Yao, H. Yang, V. Shrotriya, G. Yang and Y. Yang, *Adv. Funct. Mater.*, 2007, **17**, 1636–1644.

74. V. D. Mihailetchi, H. Xie, B. de Boer, L. M. Popescu, J. C. Hummelen, P. W. M. Blom and L. J. A. Koster, *Appl. Phys. Lett.*, 2006, **89**, 012107.

75. G. Li, V. Shrotriya, J. Huang, Y. Yao, T. Moriarty, E. Keith and Y. Yang, *Nat. Mater.*, 2005, **4**, 864–868.

76. F.-C. Chen, C.-J. Ko, J.-L. Wu and W.-C. Chen, *Sol. Energy Mater. Sol. Cells*, 2010, **94**, 2426–2430.

77. J. H. Park, J. S. Kim, J. H. Lee, W. H. Lee and K. Cho, *J. Phys. Chem. C*, 2009, **113**, 17579–17584.

78. H. Li, H. Tang, L. Li, W. Xu, X. Zhao and X. Yang, *J. Mater. Chem.*, 2011, **21**, 6563–6568.

79. V. D. Mihailetchi, H. X. Xie, B. de Boer, L. J. A. Koster and P. W. M. Blom, *Adv. Funct. Mater.*, 2006, **16**, 699–708.

80. W. Ma, C. Yang, X. Gong, K. Lee and A. J. Heeger, *Adv. Funct. Mater.*, 2005, **15**, 1617–1622.

81. Y. Zhao, Z. Xie, Y. Qu, Y. Geng and L. Wang, *Appl. Phys. Lett.*, 2007, **90**, 043504.

82. Y. Yao, J. Hou, Z. Xu, G. Li and Y. Yang, *Adv. Funct. Mater.*, 2008, **18**, 1783–1789.

83. A. J. Moulé and K. Meerholz, *Adv. Mater.*, 2008, **20**, 240–245.

84. H.-Y. Chen, H. Yang, G. Yang, S. Sista, R. Zadoyan, G. Li and Y. Yang, *J. Phys. Chem. C*, 2009, **113**, 7946–7953.

85. H. D. Keith and F. J. Padden, *J. Appl. Phys.*, 1963, **34**, 2409–2421.

86. Z. Xu, L.-M. Chen, G. Yang, C.-H. Huang, J. Hou, Y. Wu, G. Li, C.-S. Hsu and Y. Yang, *Adv. Funct. Mater.*, 2009, **19**, 1227–1234.

87. Z. Zhu, D. Waller and R. Gaudiana, *J. Macromol. Sci., Part A*, 2007, **44**, 1249–1253.

88. Z. Zhu, D. Waller, R. Gaudiana, M. Morana, D. Mühlbacher, M. Scharber and C. Brabec, *Macromolecules*, 2007, **40**, 1981–1986.

89. J. Peet, J. Y. Kim, N. E. Coates, W. L. Ma, D. Moses, A. J. Heeger and G. C. Bazan, *Nat. Mater.*, 2007, **6**, 497–500.

90. J. K. Lee, W. L. Ma, C. J. Brabec, J. Yuen, J. S. Moon, J. Y. Kim, K. Lee, G. C. Bazan and A. J. Heeger, *J. Am. Chem. Soc.*, 2008, **130**, 3619–3623.

91. F. Liu, Y. Gu, C. Wang, W. Zhao, D. Chen, A. L. Briseno and T. P. Russell, *Adv. Mater.*, 2012, **24**, 3947–3951.

92. Y. Gu, C. Wang and T. P. Russell, *Adv. Energy Mater.*, 2012, **2**, 683–690.

93. T. Agostinelli, T. A. M. Ferenczi, E. Pires, S. Foster, A. Maurano, C. Müller, A. Ballantyne, M. Hampton, S. Lilliu, M. Campoy-Quiles, H. Azimi, M. Morana, D. D. C. Bradley, J. Durrant, J. E. Macdonald, N. Stingelin and J. Nelson, *J. Polym. Sci., Part B: Polym. Phys.*, 2011, **49**, 717–724.

94. J. T. Rogers, K. Schmidt, M. F. Toney, E. J. Kramer and G. C. Bazan, *Adv. Mater.*, 2011, **23**, 2284–2288.

95. M. Morana, M. Wegscheider, A. Bonanni, N. Kopidakis, S. Shaheen, M. Scharber, Z. Zhu, D. Waller, R. Gaudiana and C. Brabec, *Adv. Funct. Mater.*, 2008, **18**, 1757–1766.

96. D. Di Nuzzo, A. Aguirre, M. Shahid, V. S. Gevaerts, S. C. J. Meskers and R. A. J. Janssen, *Adv. Mater.*, 2010, **22**, 4321–4324.

97. F. Etzold, I. A. Howard, N. Forler, D. M. Cho, M. Meister, H. Mangold, J. Shu, M. R. Hansen, K. Müllen and F. Laquai, *J. Am. Chem. Soc.*, 2012, **134**, 10569–10583.

98. Z. Li and C. R. McNeill, *J. Appl. Phys.*, 2011, **109**, 074513.

99. Y. Sun, G. C. Welch, W. L. Leong, C. J. Takacs, G. C. Bazan and A. J. Heeger, *Nat. Mater.*, 2011, **11**, 44–48.

100. Y. Liang, D. Feng, Y. Wu, S.-T. Tsai, G. Li, C. Ray and L. Yu, *J. Am. Chem. Soc.*, 2009, **131**, 7792–7799.

101. Y. Liang, Y. Wu, D. Feng, S.-T. Tsai, H. J. Son, G. Li and L. Yu, *J. Am. Chem. Soc.*, 2009, **131**, 2008–2009.

102. Y. Liang, Z. Xu, J. Xia, S.-T. Tsai, Y. Wu, G. Li, C. Ray and L. Yu, *Adv. Mater.*, 2010, **22**, E135–E138.

103. H.-Y. Chen, J. Hou, S. Zhang, Y. Liang, G. Yang, Y. Yang, L. Yu, Y. Wu and G. Li, *Nat. Photonics*, 2009, **3**, 649–653.

104. Z. He, C. Zhong, S. Su, M. Xu, H. Wu and Y. Cao, *Nat. Photonics*, 2012, **6**, 593–597.

105. W. Chen, T. Xu, F. He, W. Wang, C. Wang, J. Strzalka, Y. Liu, J. Wen, D. J. Miller, J. Chen, K. Hong, L. Yu and S. B. Darling, *Nano Lett.*, 2011, **11**, 3707–3713.

106. F. Liu, W. Zhao, J. R. Tumbleston, C. Wang, Y. Gu, D. Wang, A. L. Briseno, H. Ade and T. P. Russell, *Adv. Energy Mater.*, 2014, **4**, 1301377.

CHAPTER 10

Charge Generation, Recombination and Transport in Organic Solar Cells

CHENGMEI ZHONG*[a]

[a]University of California, Santa Barbara, CA 93106, USA
*E-mail: sebdsch@physics.ucsb.edu

10.1 Introduction

The organic solar cell (OSC) is a rapidly developing branch of solar cell technologies that has been a major field of research in materials science.[1-3] OSCs have some unique advantages compared to other mature inorganic semiconductor solar cell technologies. First, there is much more variety in the OSC materials design compared to the inorganic counterpart, which enables organic chemists precisely to control the electronic as well as the optical properties of organic semiconductor materials over a very large range by adjusting the molecular structure. For example, by slightly changing the structure of the repeating unit in a conjugated polymer, the bandgap, absorption coefficient and charge mobility can be easily manipulated.[4,5] Second, organic semiconductor materials, especially conjugated polymers, are typically low-cost materials compared to their inorganic counterparts. They generally have very good mechanical processing properties and are soluble in common organic solvents, thus enabling OSCs to be fabricated by large scale

RSC Polymer Chemistry Series No. 17
Polymer Photovoltaics: Materials, Physics, and Device Engineering
Edited by Fei Huang, Hin-Lap Yip, and Yong Cao
© The Royal Society of Chemistry 2016
Published by the Royal Society of Chemistry, www.rsc.org

solution processing techniques such as roll-to-roll processing,[6] which could further enhance their low-cost advantage.

The most important performance parameter for solar cells is the power conversion efficiency (PCE). After many years of development, the PCE of OSC devices has reached the 10% threshold in both single junction devices[7] and multi-junction tandem devices,[8] which is comparable to those of amorphous silicon solar cells.[9] However, in order to push this technology into real large scale applications the PCE must be further improved to make it competitive enough compared to other solar cell technologies, especially after the emergence of perovskite solar cells.[10,11] To further increase the PCE of OSC devices from 10% onwards is a non-trivial task, and it requires deeper understanding of the working principles of the OSC devices. There are three most important aspects in the operation mechanisms that are critical to the performance of OSC devices: charge generation, charge transport, and charge recombination. In this chapter we will review the current understanding of these three particular aspects in the OSC research community.

10.2 The Charge Generation Process in Organic Solar Cells

Like any other solar cell devices, an OSC device also converts photons from sunlight into electron–hole pairs that can be collected for use, however the charge generation mechanism of OSC is very different from those of other types of solar cell. One common source of confusion in the literature is that photo-generated "charges" inside the OSC active layer, before they are collected by the electrodes, are typically not free electrons or holes as in inorganic semiconductors, but rather take the form of positive polarons and negative polarons,[12,13] owing to the more localized energy landscape of organic semiconductors.

The exact picture of the charge generation process in OSCs is still open to debate. Basically, there are two distinct theories on the how the charges are generated in OSCs. The first is the exciton theory. In this theory, a charge pair is not generated directly after the absorption of a photon, but rather a bound electron–hole pair called an exciton is formed first, either in the donor material or in the acceptor material. This exciton must then diffuse to the donor–acceptor interface in order to further dissociate into a mobile charge pair that can later undergo the charge transport process and finally be collected by the electrode. The second theory is the ultrafast charge generation theory. In this theory, a charge pair is directly formed right after photon absorption in a very short amount of time (on the order of 10^2 fs), and exciton formation is not a necessary step in the charge generation process. Although the first theory has been widely accepted as the standard model for describing OSC device physics for some years, in recent years it has been repeatedly challenged by the second theory.

10.2.1   The Exciton Theory of Charge Generation

Organic semiconductor materials typically have dielectric constants in the range of 2–4,[14] which is much lower than those of their inorganic counterparts (>10 [15]), therefore the Coulomb attraction between the opposite charges is much less effectively screened in OSC devices than in their inorganic counterparts. Hence, in the exciton theory the initial excited species created immediately after the absorption of photons are bound electron–hole pairs rather than completely separated electrons and holes, as in the case of inorganic semiconductors. In organic semiconductors, interactions of an excited molecule with neighboring molecules impose reorganization of intermolecular distances and partial polarization of the electronic configuration of the surroundings. This collective response to an excitation is generalized as quasi-particles called excitons.[16] Because intermolecular interactions are usually weak in organic materials, excitons are strongly localized; often the localization is limited to a single molecule or segments of a polymer. Excitons are electrically neutral and bear potential energy that can be released when the molecule/segment returns to the ground state. Electronic states with net spin of 0 or 1 are called singlet excitons or triplet excitons, respectively, and the initial excitons created immediately after photon absorption are singlet excitons. As a bound state, an exciton typically has a binding energy of 0.3–0.5 eV,[17–19] which is much higher than the thermal energy at room temperature. This means that the dissociation of an exciton into free charges (positive polarons and negative polarons) is not a spontaneous process, but requires energy input from the environment.

For OSCs, this energy is provided by the energy offset between the donor material and the acceptor material, specifically the highest occupied molecular orbital (HOMO) as well as the lowest unoccupied molecular orbital (LUMO) energy difference between the donor and the acceptor. This is why, in order for the charge dissociation to happen, an exciton must travel to the donor–acceptor interface. As charge neutral quasi-particles, the excitons can only diffuse to the interface, therefore the rate of charge generation is limited by the diffusion rate of the exciton. Recent studies showed that, even after the exciton has reached the donor–acceptor interface, the charge separation does not occur immediately, but the exciton first transforms into another type of interfacial bound state called the charge transfer state (CT state).[20–23] In this transitional energetic state the electron is partially transferred from the donor molecule to the acceptor molecule, but the electron in the acceptor molecule and the hole in the donor molecule are still loosely bound at the donor–acceptor interface, similar to an exciton. The CT state can then further complete charge separation and finally generate a free charge pair, or it can relax back to ground state in several ways which all result in the loss of charges. The first way is geminate recombination, which happens when the bound charge pair recombines with each other, annihilating both charges and resulting in photon emission; the second is *via* energy transfer to lower triplet energies in either the donor material or the acceptor material.[24]

The whole charge generation process according to the exciton theory is depicted in Figure 10.1.

In general, the exciton theory can well explain why bulk heterojunction (BHJ) structured active layer design can yield much higher performance in OSC devices than bilayer structure design, and why BHJ device performance is very sensitive to donor–acceptor phase separation morphology. As singlet excitons typically have very short lifetime (10^0–10^1 ps),[25,26] the excitons that do not reach the donor–acceptor interface within the lifetime cannot generate charges. Therefore, to achieve efficient charge generation the donor–acceptor phase separation distance cannot be so large as to exceed the mean exciton diffusion distance, while it cannot be too small to ensure connectivity between the same phases. Finding the optimum phase separation morphology of the active layer is a complicated problem; apart from the phase separation distance mentioned above, the purity of the donor phase or the acceptor phase (or the degree of intermixing within a phase)[27,28] may also influence the charge separation efficiency. It has been shown that, by introducing more intermixing (impure phases), the charge generation process can be made more efficient in some BHJ systems,[29] however if the phases in a BHJ is too intimately mixed, resulting in small, isolated domains of either donor or acceptor, the device may suffer from higher geminate recombination and poor charge generation efficiency as a result.[30]

The optimization of BHJ morphology based on the exciton theory has been an intensively studied subject in the field of OSC device physics over the years, however the improvement of device performance in this research direction has proved to be less and less effective. The whole OSC research

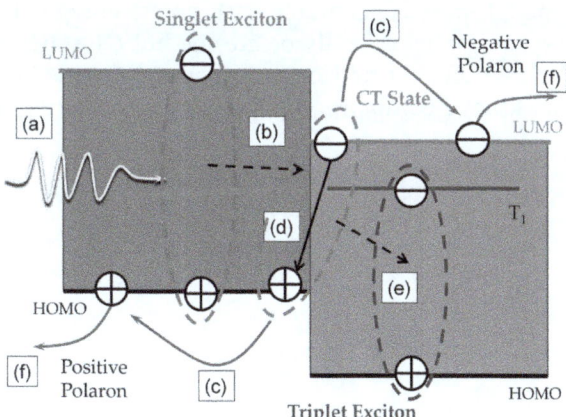

Figure 10.1 Schematics of the charge generation process in OSCs according to the exciton theory. (a) Photon absorption and the formation of a singlet exciton; (b) exciton diffusion to the donor–acceptor interface, forming the CT state; (c) charge separation of the CT state, creating mobile charged species (positive polarons and negative polarons); (d) geminate recombination; (e) CT state energy transfer forming triplet excitons; (f) charge transport.

community is in desperate need of new insights into the device physics of OSCs and novel strategies for device performance improvement. This is one of the main driving forces contributing to the emergence of new theories on the charge generation process. Some try to dig deeper into the CT state and establish a correlation between this intermediate energy state and the charge generation efficiency, while others try to establish a new physical picture of the whole charge generation process.

10.2.2 The CT State, Charge Generation and Geminate Recombination

As discussed in the previous section, it is now widely accepted that the CT state is an important intermediate state before charge separation takes place, and it has become the focus of many studies in recent years. The CT state is actually not a single quasi-steady energy level, but is composed of a series of energy levels similar to the singlet states or triplet states, as depicted in Figure 10.2.[31] The higher the energy level, the greater the spatial delocalization and the shorter the lifetime of the CT state. In general, the CT state will be generated with excess thermal energy due to the energy difference between the exciton and the CT state. The CT state with excess thermal energy is referred to as a hot CT state (CT_n). Two competing charge generation processes can happen after this. The hot CT state can either directly dissociate into the separated charges (CS), or it can thermally relax to the lowest lying CT state $(CT_1$, or CT_0 in some publications), followed by complete charge dissociation.

Given that the CT state is generated with excess energy, it was proposed that this excess energy helps to dissociate the hot CT states directly to the free carriers before they thermalize to CT_1 states.[32] In this hot CT state theory, the primary kinetic competition is between the thermal relaxation and

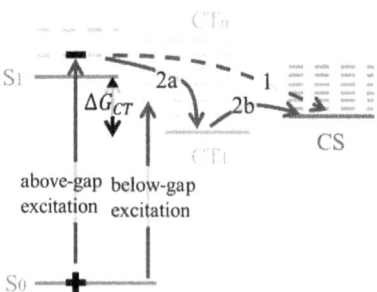

Figure 10.2 Schematics of charge transfer state energy levels. CT_n is the nth level of the hot CT state, CT_1 is the ground level of the CT state, and CS is the energy level of separated charge pairs. The processes in the figure are: (1) direct charge separation from CT_n; (2a) relaxation from CT_n to CT_1; (2b) charge separation of CT_1. Reprinted from ref. 31 with permission from PCCP Owner Societies.

the charge dissociation. Just like singlet or triplet states, higher energy for a CT state means a shorter lifetime; the theoretical lifetime of the hot CT states are typically on the order of several hundred femtoseconds.[33] The charge separation time is supposed to be on the same order of magnitude, at least in order to compete with the relaxation process, which is indeed confirmed by ultrafast spectroscopy measurements (~50 fs[34]). Once relaxed to the ground CT states (CT_1), the charge pairs will primarily recombine because they do not have enough driving force to overcome the Coulomb attraction between them. Bakulin *et al.*[35] provided strong evidence that hot CT states are very important in the charge separation process in OSCs by using a specially designed electro-optical pump-probe experiment. In this experiment the pump laser pulse excites the CT state in the OSC device, lets it relax back to CT_1 for a certain delay time, and then re-excites the CT_1 back to the hot CT state by another infrared "push pulse". It was observed in the experiment that the photocurrent of the OSC devices increased when the push pulse is applied, thus proving that the hot CT states can separate charges more efficiently than relaxed CT states. By adjusting the delay time between the pump and push pulse, the lifetime of CT_n was determined to be within several picoseconds. Jailaubekov *et al.*[36] also provided support for the existence of hot CT states through time-resolved second harmonic generation and time-resolved two-photon photoemission experiments. However, recent internal quantum efficiency (IQE) measurements by Lee *et al.*[37] and van der Hofstad *et al.*,[38] and time-delayed-collection-field experiments by Vandewal *et al.*[39] all suggest that charge carriers are generated exclusively from CT_1, rather than CT_n, for a range of donor and acceptor material combinations, which sparks new controversies on whether CT_n states or CT_1 states are more critical to the charge generation process.

The CT state does not guarantee charge separation, because it will not generate charge pairs if geminate recombination occurs. Therefore geminate recombination is a potential cause of the loss of efficiency in OSC devices and is the focus of many OSC related studies. Experimentally, the geminate recombination dynamics can be traced by probing its products after the recombination, which causes light emission. It should be noted that, before the formation of the CT state, the excitons of either the donor or the acceptor material may also recombine before they reach the interface, resulting in light emission which also counts as geminate recombination, however this is negligible in high efficiency BHJ systems compared to the recombination of CT states. The photoemission of the CT state can be either photoluminescence (PL)[40,41] or electroluminescence (EL),[41,42] and it can be easily separated from the donor or the acceptor exciton PL/EL because of red-shifted spectral peak. Given that an external electric field can help bound charge pairs in the CT state overcome the binding energy and complete charge separation, the geminate recombination of the CT state should be highly electric field dependent, which is consistent with electric field dependent photocurrent generation in many BHJ systems.[43–45] In some cases the energy of the lowest triplet state (T_1) of the donor or the acceptor material is lower than the CT_1

state, and then energy transfer from CT_1 to T_1 may occur;[24,32,46] because triplet decay is usually non-emissive, this kind of energy transfer will result in the quenching of CT state PL.[47] A low lying triplet state will exacerbate geminate recombination and result in extra loss of efficiency in OSC devices, and this explains why in some BHJ systems with ideal HOMO and LUMO energy alignment still results in poor device performance.[48]

Many believe that geminate recombination is an important loss mechanism in OSC devices, however in recent years many high efficiency BHJ systems have been developed with negligible electric field photocurrent dependence and IQE approaching 100% in devices, suggesting that geminate recombination is negligible in these systems.[45,49–51] This raises the question of whether geminate recombination is universal in all BHJ systems. Most recently Seifter *et al.*[52] studied the electric field dependent photocurrent generation dynamics of a state-of-the-art BHJ system through transient photoconductivity and provided strong proof that geminate recombination is negligible in high efficiency BHJ systems.

10.2.3 The Ultrafast Charge Generation Theory

According to the exciton theory, the mobile charges in an OSC device should be generated after the excitons diffuse to the donor–acceptor interface. Therefore the charge generation process is limited by the diffusion time of excitons, which means that charges should be formed starting from 1 to 10 ps[53,54] after initial light excitation. However in recent years there have been many new experimental findings that suggest that charges can be generated in a much shorter timescale,[34,36] a fact that can be explained by the complicated hot CT state theory described in the previous section; however, the explanation is much more straightforward for the ultrafast charge generation theory. Kaake *et al.*[55] discovered unambiguously through transient absorption spectroscopy (TAS, or transient photo-induced absorption spectroscopy) studies that in typical BHJ systems 70% of the total charge carrier population is generated by an ultrafast process that is completed within 100 fs. The timescale of this process is much shorter than the exciton diffusion time, and therefore Kaake *et al.*[55] proposed that there can be ultrafast charge separation in OSCs before the formation of excitons. It should be noted that ultrafast charge transfer from polymer to fullerene was observed many years ago by Brabec *et al.*[56] However, in their study it was not determined whether the products of such charge transfer were bound CT states or free charge carriers. Hwang *et al.*[57] and De *et al.*[58] provided the first evidence of the existence of ultrafast charge carrier generation in polymer:fullerene blends using TAS measurements. The TAS studies by Kaake *et al.*[55,59] developed these findings further into a more systematic theory of the charge generation process in OSCs.

As shown in Figure 10.3,[55] The whole charge generation process according to ultrafast charge generation theory is split into two parts. The first is the ultrafast part described above which happens within 100 fs after light excitation. Within this timescale the excitons are highly delocalized in space because of quantum coherence. The delocalized nature of the excitons means that the distribution of wave functions of electrons and holes can be very far apart from each other, therefore they can get easily separated because of weak Coulomb binding energy, just like in the case of inorganic semiconductors.[55,59] However, the decoherence time for organic materials is generally very short (also within 100 fs at room temperature),[60] the delocalized excitons that did not finish charge separation within the decoherence time will collapse into localized excitons, whose behavior can then be described by the classical exciton theory. When excitons become localized, the second part of the charge generation process begins, which is the same as what the exciton theory describes: excitons diffuse to the donor–acceptor interface, forming a CT state and then the final charge separation. Given that this part of the process is limited by the exciton diffusion, it is a much slower process than the ultrafast process (10^0–10^2 ps). The ultrafast charge generation theory is gaining more and more support from both theoretical and experimental studies. Through the simulations from a new Kinetic Monte Carlo model, Heitzer *et al.*[61] proved that, in BHJ systems, the charge generation process is indeed separated into an ultrafast process (100–200 fs) which is related to the degree of delocalization of the excitons before decoherence, and a slow process limited by exciton dynamics. Nami *et al.*[62] studied BHJ systems with the combination of TAS and electroabsorption spectroscopy[63] and found that in good BHJ systems charge separation can happen within 40 fs, and the separation distance has been determined to be at least 4 nm, which is much larger than the exciton binding distance (2–3 nm). However they argued that this is the result of delocalized electrons inside the fullerene (acceptor) phase, not delocalized excitons which could also be present in the donor phase. Falke *et al.*[64] studied the ultrafast charge generation process in detail with a 15 fs time resolution TAS setup and found evidence of quantum coherence during the process. It was also found in their study that the coherent vibrational motion of the fullerene phase after impulsive optical excitation of the polymer donor phase was the key precursor to the actual charge separation process. After that there have

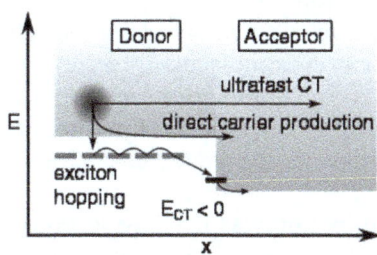

Figure 10.3 Schematics of the physical picture of ultrafast charge generation theory. Reprinted from ref. 54 with permission from the American Chemical Society.

been more publications on ultrafast charge generation theory and its correlation with OSC device performance.[65–69]

One of the chief counterarguments against the ultrafast charge generation theory is that the ultrafast process observed by TAS measurements can also be explained by the complex BHJ film morphology (for example, intermixing between donor and acceptor molecules resulting in impure domains[29,70]) within the concept of classical exciton theory. According to this argument, what appears to be ultrafast charge separation is merely the result of local charge transfer. The dominance of the ultrafast process pathway would then represent a pronounced lack of phase separation between donor and acceptor. To eliminate the confusion around the role of morphology in the ultrafast charge generation process, Kaake *et al.*[71] studied a well-defined bilayer system made with a high performing solution-processed small molecule cast on top of a photopolymerized layer of C_{60}[72] with TAS. It was clearly shown that the majority of carriers are again produced at ~100 fs timescales, ruling out the idea that sub-picosecond charge generation can be understood wholly in terms of localized excitons. This study provided further proof that the excited state is highly delocalized on short timescales in OSC materials. However, there is currently still no direct evidence that mobile charges are generated at <100 fs timescales, and all the evidence so far has been provided indirectly from transient spectroscopy studies, which frequently encounter the difficulty of obtaining clean and indisputable spectral assignment of mobile charges, because the absorption spectra of different excited species (singlet excitons, triplet excitons, polarons in donor, polarons in acceptor, *etc.*) usually have large overlaps with each other. This will remain an important and fundamental problem in the field in the future.

10.3   Charge Recombination in Organic Solar Cells

Recombination is a key source of efficiency loss in OSC devices. There are two types of recombination in OSCs: the first is geminate recombination, which happens before mobile charges are formed, and the second is charge recombination, which is the recombination between mobile charges. Since geminate recombination has been covered in previous chapters, here we focus on the charge recombination.

Charge recombination in OSCs can be classified into three sub-categories: trap-assisted recombination, bimolecular recombination, and auger recombination. Unlike geminate recombination, charge recombination is universal in all types of solar cell and cannot be totally eliminated, in theory. However, in high efficiency OSC devices charge recombination can be greatly reduced to insignificant levels.

Trap-assisted recombination was first proposed by Shockley *et al.*,[73] and it is also known as Shockley–Read–Hall recombination. This recombination model was first used to study the charge recombination in inorganic semiconductors, and later it was introduced into the field of OSC.[74] This type of recombination involves a trapped charge and a mobile charge with the opposite polarity. The trap-assisted recombination rate R_{SRH} can be calculated by eqn (10.1).

$$R_{SRH} = \frac{C_p C_n N_{tr} (np - n_i^2)}{C_n (n + n_1) + C_p (p + p_1)} \tag{10.1}$$

where n and p are mobile negative charge density and mobile positive charge density respectively; n_1 and p_1 are trapped negative charge density and trapped positive charge density respectively; n_i is the intrinsic charge density; N_{tr} is the total charge trap site density; C_n and C_p are the capture probabilities of mobile negative charges and mobile positive charges respectively. In the general case of OSC devices $n \gg n_1$ and $p \gg p_1$ (weak trapping), therefore if the charge traps inside the device are mainly positive charge traps, eqn (10.1) can be simplified to $R_{SRH} = C_n N_{tr} n$. On the other hand, if the charge traps inside the device are mainly negative charge traps, eqn (10.1) can be simplified to $R_{SRH} = C_p N_{tr} p$. Trap-assisted recombination is usually negligible in BHJ solar cells, unless charge traps are deliberately introduced into the active layer.[49,75-77]

Bimolecular recombination is the dominant recombination process in OSC devices. The bimolecular recombination rate R_{bi} can be calculated by eqn (10.2).

$$R_{bi} = \gamma (np - n_i^2) \tag{10.2}$$

where γ is the Langevin coefficient; it is related to the dielectric constant of the material ε, its positive charge mobility μ_p and its negative charge mobility μ_n:

$$\gamma = \frac{q}{\varepsilon} (\mu_p + \mu_n) \tag{10.3}$$

Equations (10.2) and (10.3) describe the process when a pair of opposite charges encounter each other in space and recombine with the help of Coulomb attraction. According to the classical bimolecular recombination model,[78] when the mean free path of charges in a material is smaller than the Coulomb capture radius r_c, bimolecular recombination will take place. The typical mean free path of organic semiconductors is about 1–2 nm, which is much smaller than r_c (10–20 nm at room temperature[79]), therefore bimolecular recombination is universally present in OSC devices.[80,81]

From eqn (10.3) we see that γ is proportional to the charge mobility. A straightforward deduction would be that higher charge mobility would result in worse OSC device performance because of stronger bimolecular recombination, however the actual experimental results show the opposite case. Most OSC systems would benefit from higher charge mobility; this is because higher mobility would lead to faster charge extraction from the active layer to the electrodes, thereby reducing the accumulated charge density inside the active layer. From eqn (10.2) we know that R_{bi} has quadratic dependence on charge density (assuming $n \approx p$, which is usually the case in OSC devices), therefore the reduction in charge density would eventually reduce the R_{bi} in spite of the increase in γ. Detailed OSC device simulation with bimolecular recombination also shows that when the charge mobility is smaller than 10^{-4} cm^2 V^{-1} s^{-1} (assuming balanced transport, $\mu_n \approx \mu_p$), increasing mobility will actually result in lower R_{bi}, while only when mobility is beyond 10^{-2} cm^2 V^{-1} s^{-1} will R_{bi} start to increase with mobility.[82] The other interesting discovery on the bimolecular recombination process in OSC devices is that, although experimentally measured R_{bi} is indeed quadratically dependent on charge density, the value of γ is measured to be several orders of magnitude lower than the value calculated according to eqn (10.2) in many BHJ systems.[83–86] One of the explanations for this unusually small γ value is that the BHJ morphology can help opposite charges travel separately in their respective phases, thus significantly lowering the probability of opposite charges encountering each other, thereby effectively reducing the R_{bi}. This explanation was supported by morphology studies on BHJ devices,[86–88] which showed that the larger the donor–acceptor phase separation, the smaller the R_{bi} in the device. Through device simulations, Deibel *et al.*[89] discovered that the inhomogeneous spatial distribution of charge density inside the BHJ active layer can also result in reduced γ compared to calculated values. This is because the experimentally determined γ value is calculated with eqn (10.2) using spatially averaged charge density values, while in actual devices the charges are accumulated around the respective electrodes, leaving the charge density within the bulk of the active layer much lower than the spatially averaged value. Thus the spatially averaged R_{bi} value would be much lower than the R_{bi} value calculated by putting spatially averaged charge density into eqn (10.2), resulting in the underestimation of γ in the experiments.

Auger recombination[90] is a process that involves the recombination of a positive charge and a negative charge, like bimolecular recombination. However, in this process the excess energy does not convert into heat or radiation as in bimolecular recombination, but is transferred to a second negative charge, exciting it to higher excitation states. The auger recombination rate has a cubic dependence on charge density and theoretically it will only become significant when charge density is several orders of magnitude higher than the typical values during OSC operation, therefore this type of recombination has never been directly observed in OSC devices. Although in some studies cubic dependence on charge density of the recombination rate has been observed, further analysis revealed that this was still actually bimolecular recombination with a charge density dependent γ.[91,92]

10.4 Charge Transport in Organic Solar Cells

The charge transport process in organic semiconductor materials is commonly described by the variable range hopping model.[93–95] In highly ordered crystalline inorganic semiconductor materials, the charge transport can be described by the classical band transport model, where the wave functions of mobile charges are highly delocalized in space, and their transport behavior can be well described by the form of plane waves propagating in the direction of the external electric field. The main obstacle in such a transport process is the phonon scattering effect caused by the thermal vibration of the crystal lattice. As phonon scattering intensity increases with temperature, the charge mobility in inorganic semiconductor materials decreases with temperature. On the other hand, the charge transport behavior in organic semiconductor materials in OSC devices is very different. Organic materials, especially the ones used in OSC active layers, are usually highly disordered amorphous or polycrystalline materials. The charge transport process in organic materials will be disrupted by local changes in polymer chain orientation, molecular stacking, or even the weak interaction between molecules, all of which induce much stronger scattering than phonon scattering. These local effects will introduce localized energy levels within the bandgap of the material, which essentially serve as charge traps. When mobile charges get trapped by these energetic states, they will need external energy input, *e.g.* from heat or an electric field, to detrap and become mobile again. Therefore charge transport in organic materials takes the form of a "hopping" process, which is basically intermittently travel from one charge trap to another, and the mean hopping distance is determined by the spatial distribution of such charge trapping states. As thermal motion is beneficial to detrapping of charges, the charge mobility in organic semiconductor materials actually increases with temperature; the electric field dependence of mobility is also much stronger than in inorganic semiconductors. Currently, the Gaussian disorder model[96] is commonly used in this field to describe the energetic distribution of localized states of charge transport, while Marcus electron transfer theory and Holstein polaron transport theory[97] are used to describe the temperature dependence of charge mobility.

In general there are three distinctive characteristics of charge transport in organic materials that are very different from those in inorganic semiconductor materials. The first is the dispersive transport behavior, or time dependent charge mobility. The localized states in organic materials have a wide distribution in both energy and space; this will result in a wide distribution of charge mobilities *vs.* the charge population. This kind of charge transport behavior is most obvious in time-of-flight mobility measurements,[98,99] where long tails of photocurrent can be observed in organic materials. Dispersive transport can also be studied by ultrafast time dependent charge mobility measurements,[100] where it was shown that the transient mobility value can be several orders of magnitude higher than the average value measured by steady state experiments.[101] The second characteristic is the positive correlation of charge mobility with temperature and the strong electric field dependence which is mentioned above. The third characteristic is the charge density dependent charge mobility. The spatial density of the localized states is limited, therefore when the charge trapping states are gradually filled up with increasing charge density, the charge mobility will increase as a result of less trapping. The charge density dependence of mobility is most obvious in organic field effect transistor devices, because the charge density in different operating conditions can vary by several orders of magnitude.[102]

Our current understanding of the charge transport properties of organic materials is still far from clear. One of the difficulties in studying the charge transport properties of organic materials is to decouple time dependent charge density information from time dependent charge mobility information experimentally. For example, it is difficult for transient photoconductivity measurement, which measures the change in photocurrent over time, to differentiate between the time dependence of charge density caused by charge recombination and the time dependence of charge mobility. Transient spectroscopy measurements qualitatively study the time dependence of charge density alone, therefore it may be possible to decouple the charge density dynamics from charge mobility dynamics by using the two experimental techniques together. However, this would require precise quantitative measurement of charge density by transient spectroscopy, which is a problem that remains to be solved. The other problem in this field that is especially important to OSC research is how the charge trapping states influence the charge transport of organic materials. Street *et al.*[103] studied the shallow charge trapping states in OSC devices through transient photoconductivity measurements and concluded that charge trapping may have a minor influence on OSC device performance. Bailey *et al.*,[104] on the other hand, studied the correlation between positive charge traps and OSC device performance and concluded that such traps can severely hamper the performance of OSC devices. Leong *et al.*[105] further studied the charge trapping behavior of a high efficiency OSC system with a sub-nanosecond transient photoconductivity experimental technique and found that mobile charges gradually get trapped in deeper energy levels. Surprisingly, in this study it was found that, by deliberately introducing shallow traps into the OSC active layer, the charge mobility actually decreased. To address the confusion in this problem a more well-defined material system with known charge trapping energy distribution is needed.

10.5   Conclusion

In this chapter we have reviewed the current understanding in the OSC research community on the three critical processes in OSC device operation: charge generation, charge recombination, and charge transport. The charge generation process in OSCs can be described by either the classical exciton theory or the ultrafast charge generation theory; the former is still widely accepted in the community but the latter has been steadfastly gaining more support in recent years. The charge recombination process in OSC devices is dominantly bimolecular recombination, which is small yet not negligible in high efficiency OSC devices. The charge transport process in OSCs is strongly affected by charge trapping states, however the exact physics is still not totally clear and requires deeper understanding.

References

1. G. Li, R. Zhu and Y. Yang, *Nat. Photonics*, 2012, **6**, 153.
2. S. Günes, H. Neugebauer and N. S. Saricftci, *Chem. Rev.*, 2007, **107**, 1324.
3. J. E. Coughlin, Z. B. Henson, G. C. Welch and G. C. Bazan, *Acc. Chem. Res.*, 2013, **47**, 257.
4. J. Chen and Y. Cao, *Acc. Chem. Res.*, 2009, **42**, 1709.
5. R. Kroon, M. Lenes, J. C. Hummelen, P. W. M. Blom and B. De Boer, *Polym. Rev.*, 2008, **48**, 531.
6. F. C. Krebs, T. Tromholt and M. Jorgensen, *Nanoscale*, 2010, **2**, 873.
7. Y. Liu, J. Zhao, Z. Li, C. Mu, W. Ma, H. Hu, K. Jiang, H. Lin, H. Ade and H. Yan, *Nat. Commun.*, 2014, **5**, 5293.
8. J. B. You, L. T. Dou, K. Yoshimura, T. Kato, K. Ohya, T. Moriarty, K. Emery, C. C. Chen, J. Gao, G. Li and Y. Yang, *Nat. Commun.*, 2013, **4**, 1446.
9. D. E. Carlson and C. R. Wronski, *Appl. Phys. Lett.*, 1976, **28**, 671.
10. M. Liu, M. B. Johnston and H. J. Snaith, *Nature*, 2013, **501**, 395.
11. M. A. Green, A. Ho-Baillie and H. J. Snaith, *Nat. Photonics*, 2014, **8**, 506.
12. A. J. Heeger, N. S. Saricftci and E. B. Namdas, *Semiconducting and Metallic Polymers*, Oxford University Press, 2010.
13. K. Hannewald, V. M. Stojanovic, J. M. T. Schellekens, P. A. Bobbert, G. Kresse and J. Hafner, *Phys. Rev. B*, 2004, **69**, 075211.
14. L. J. A. Koster, S. E. Shaheen and J. C. Hummelen, *Adv. Energy Mater.*, 2012, **2**, 1246.
15. S. M. Sze and K. K. Ng, *Physics of Semiconductor Devices*, John Wiley & Sons, Inc., 2007.
16. M. Pope and C. E. Swenberg, *Electronic Processes in Organic Crystals and Polymers*, Oxford University Press, 1999.
17. M. Scheidler, U. Lemmer, R. Kersting, S. Karg, W. Riess, B. Cleve, R. F. Mahrt, H. Kurz, H. Bässler, E. O. Göbel and P. Thomas, *Phys. Rev. B*, 1996, **54**, 5536.
18. A. Ruini, M. J. Caldas, G. Bussi and E. Molinari, *Phys. Rev. Lett.*, 2002, **88**, 206403.
19. J. L. Bredas, J. E. Norton, J. Cornil and V. Coropceanu, *Acc. Chem. Res.*, 2009, **42**, 1691.
20. C. Deibel, T. Strobel and V. Dyakonov, *Adv. Mater.*, 2010, **22**, 4097.
21. X. Y. Zhu, Q. Yang and M. Muntwiler, *Acc. Chem. Res.*, 2009, **42**, 1779.
22. K. Vandewal, K. Tvingstedt, A. Gadisa, O. Inganas and J. V. Manca, *Nat. Mater.*, 2009, **8**, 904.
23. K. Vandewal, A. Gadisa, W. D. Oosterbaan, S. Bertho, F. Banishoeib, I. Van Severen, L. Lutsen, T. J. Cleij, D. Vanderzande and J. V. Manca, *Adv. Funct. Mater.*, 2008, **18**, 2064.
24. D. Veldman, S. C. J. Meskers and R. A. J. Janssen, *Adv. Funct. Mater.*, 2009, **19**, 1939.
25. P. Peumans, A. Yakimov and S. R. Forrest, *J. Appl. Phys.*, 2003, **93**, 3693.

26. J. Piris, T. E. Dykstra, A. A. Bakulin, P. H. M. van Loosdrecht, W. Knulst, M. T. Trinh, J. M. Schins and L. D. A. Siebbeles, *J. Phys. Chem. C*, 2009, **113**, 14500.

27. A. C. Mayer, M. F. Toney, S. R. Scully, J. Rivnay, C. J. Brabec, M. Scharber, M. Koppe, M. Heeney, I. McCulloch and M. D. McGehee, *Adv. Funct. Mater.*, 2009, **19**, 1173.

28. C. R. McNeill, H. Frohne, J. L. Holdsworth and P. C. Dastoor, *Nano Lett.*, 2004, **4**, 2503.

29. P. Westacott, J. R. Tumbleston, S. Shoaee, S. Fearn, J. H. Bannock, J. B. Gilchrist, S. Heutz, J. deMello, M. Heeney, H. Ade, J. Durrant, D. S. McPhail and N. Stingelin, *Energy Environ. Sci.*, 2013, **6**, 2756.

30. T. M. Clarke and J. R. Durrant, *Chem. Rev.*, 2010, **110**, 6736.

31. F. Gao and O. Inganas, *Phys. Chem. Chem. Phys.*, 2014, **16**, 20291.

32. H. Ohkita, S. Cook, Y. Astuti, W. Duffy, S. Tierney, W. Zhang, M. Heeney, I. McCulloch, J. Nelson, D. D. C. Bradley and J. R. Durrant, *J. Am. Chem. Soc.*, 2008, **130**, 3030.

33. T. Virgili, D. Marinotto, C. Manzoni, G. Cerullo and G. Lanzani, *Phys. Rev. Lett.*, 2005, **94**, 117402.

34. G. Grancini, M. Maiuri, D. Fazzi, A. Petrozza, H. J. Egelhaaf, D. Brida, G. Cerullo and G. Lanzani, *Nat. Mater.*, 2013, **12**, 29.

35. A. A. Bakulin, A. Rao, V. G. Pavelyev, P. H. M. van Loosdrecht, M. S. Pshenichnikov, D. Niedzialek, J. Cornil, D. Beljonne and R. H. Friend, *Science*, 2012, **335**, 1340.

36. A. E. Jailaubekov, A. P. Willard, J. R. Tritsch, W. L. Chan, N. Sai, R. Gearba, L. G. Kaake, K. J. Williams, K. Leung, P. J. Rossky and X. Y. Zhu, *Nat. Mater.*, 2013, **12**, 66.

37. J. Lee, K. Vandewal, S. R. Yost, M. E. Bahlke, L. Goris, M. A. Baldo, J. V. Manca and T. V. Voorhis, *J. Am. Chem. Soc.*, 2010, **132**, 11878.

38. T. G. J. van der Hofstad, D. Di Nuzzo, M. van den Berg, R. A. J. Janssen and S. C. J. Meskers, *Adv. Energy Mater.*, 2012, **2**, 1095.

39. K. Vandewal, S. Albrecht, E. T. Hoke, K. R. Graham, J. Widmer, J. D. Douglas, M. Schubert, W. R. Mateker, J. T. Bloking, G. F. Burkhard, A. Sellinger, J. M. J. Fréchet, A. Amassian, M. K. Riede, M. D. McGehee, D. Neher and A. Salleo, *Nat. Mater.*, 2014, **13**, 63.

40. K. Tvingstedt, K. Vandewal, F. Zhang and O. Inganäs, *J. Phys. Chem. C*, 2010, **114**, 21824.

41. Y. Zhou, K. Tvingstedt, F. Zhang, C. Du, W.-X. Ni, M. R. Andersson and O. Inganäs, *Adv. Funct. Mater.*, 2009, **19**, 3293.

42. K. Tvingstedt, K. Vandewal, A. Gadisa, F. L. Zhang, J. Manca and O. Inganas, *J. Am. Chem. Soc.*, 2009, **131**, 11819.

43. D. Credgington, F. C. Jamieson, B. Walker, T. Q. Nguyen and J. R. Durrant, *Adv. Mater.*, 2012, **24**, 2135.

44. G. F. A. Dibb, F. C. Jamieson, A. Maurano, J. Nelson and J. R. Durrant, *J. Phys. Chem. Lett.*, 2013, **4**, 803.

45. M. Mingebach, S. Walter, V. Dyakonov and C. Deibel, *Appl. Phys. Lett.*, 2012, **100**, 193302.

46. P. C. Y. Chow, S. Albert-Seifried, S. Gelinas and R. H. Friend, *Adv. Mater.*, 2014, **26**, 4851.
47. Y. W. Soon, T. M. Clarke, W. M. Zhang, T. Agostinelli, J. Kirkpatrick, C. Dyer-Smith, I. McCulloch, J. Nelson and J. R. Durrant, *Chem. Sci.*, 2011, **2**, 1111.
48. A. Rao, P. C. Y. Chow, S. Gelinas, C. W. Schlenker, C. Z. Li, H. L. Yip, A. K. Y. Jen, D. S. Ginger and R. H. Friend, *Nature*, 2013, **500**, 435.
49. S. R. Cowan, N. Banerji, W. L. Leong and A. J. Heeger, *Adv. Funct. Mater.*, 2012, **22**, 1116.
50. J. Kniepert, M. Schubert, J. C. Blakesley and D. Neher, *J. Phys. Chem. Lett.*, 2011, **2**, 700.
51. R. A. Street, S. Cowan and A. J. Heeger, *Phys. Rev. B*, 2010, **82**, 121301.
52. J. Seifter, Y. Sun and A. J. Heeger, *Adv. Mater.*, 2014, **26**, 2486.
53. P. E. Shaw, A. Ruseckas and I. D. W. Samuel, *Adv. Mater.*, 2008, **20**, 3516.
54. A. Haugeneder, M. Neges, C. Kallinger, W. Spirkl, U. Lemmer, J. Feldmann, U. Scherf, E. Harth, A. Gugel and K. Mullen, *Phys. Rev. B*, 1999, **59**, 15346.
55. L. G. Kaake, J. J. Jasieniak, R. C. Bakus, G. C. Welch, D. Moses, G. C. Bazan and A. J. Heeger, *J. Am. Chem. Soc.*, 2012, **134**, 19828.
56. C. J. Brabec, G. Zerza, G. Cerullo, S. De Silvestri, S. Luzzati, J. C. Hummelen and S. Sariciftci, *Chem. Phys. Lett.*, 2001, **340**, 232.
57. I. W. Hwang, C. Soci, D. Moses, Z. Zhu, D. Waller, R. Gaudiana, C. J. Brabec and A. J. Heeger, *Adv. Mater.*, 2007, **19**, 2307.
58. S. De, T. Pascher, M. Maiti, K. G. Jespersen, T. Kesti, F. Zhang, O. Inganäs, A. Yartsev and V. Sundström, *J. Am. Chem. Soc.*, 2007, **129**, 8466.
59. L. G. Kaake, D. Moses and A. J. Heeger, *J. Phys. Chem. Lett.*, 2013, **4**, 2264.
60. E. Collini and G. D. Scholes, *Science*, 2009, **323**, 369.
61. H. M. Heitzer, B. M. Savoie, T. J. Marks and M. A. Ratner, *Angew. Chem., Int. Ed.*, 2014, **53**, 7456.
62. S. Gelinas, A. Rao, A. Kumar, S. L. Smith, A. W. Chin, J. Clark, T. S. van der Poll, G. C. Bazan and R. H. Friend, *Science*, 2014, **343**, 512.
63. I. H. Campbell, J. P. Ferraris, T. W. Hagler, M. D. Joswick, I. D. Parker and D. L. Smith, *Polym. Adv. Technol.*, 1997, **8**, 417.
64. S. M. Falke, C. A. Rozzi, D. Brida, M. Maiuri, M. Amato, E. Sommer, A. De Sio, A. Rubio, G. Cerullo, E. Molinari and C. Lienau, *Science*, 2014, **344**, 1001.
65. K. Chen, A. J. Barker, M. E. Reish, K. C. Gordon and J. M. Hodgkiss, *J. Am. Chem. Soc.*, 2013, **135**, 18502.
66. E. R. Bittner and C. Silva, *Nat. Commun.*, 2014, **5**, 3119.
67. G. J. Dutton and S. W. Robey, *J. Phys. Chem. C*, 2013, **117**, 25414.
68. Z. Guo, D. Lee, R. D. Schaller, X. B. Zuo, B. Lee, T. F. Luo, H. F. Gao and L. B. Huang, *J. Am. Chem. Soc.*, 2014, **136**, 10024.
69. R. Tempelaar, F. C. Spano, J. Knoester and T. L. C. Jansen, *J. Phys. Chem. Lett.*, 2014, **5**, 1505.

70. M. Pfannmöller, H. Flügge, G. Benner, I. Wacker, C. Sommer, M. Hansel-mann, S. Schmale, H. Schmidt, F. A. Hamprecht, T. Rabe, W. Kowalsky and R. R. Schröder, *Nano Lett.*, 2011, **11**, 3099.

71. L. G. Kaake, C. Zhong, J. A. Love, I. Nagao, G. C. Bazan, T.-Q. Nguyen, F. Huang, Y. Cao, D. Moses and A. J. Heeger, *J. Phys. Chem. Lett.*, 2014, **5**, 2000.

72. A. M. Rao, P. Zhou, K. A. Wang, G. T. Hager, J. M. Holden, Y. Wang, W. T. Lee, X. X. Bi, P. C. Eklund, D. S. Cornett, M. A. Duncan and I. J. Amster, *Science*, 1993, **259**, 955.

73. W. Shockley and W. T. Read, *Phys. Rev.*, 1952, **87**, 835.

74. M. M. Mandoc, W. Veurman, L. J. A. Koster, B. de Boer and P. W. M. Blom, *Adv. Funct. Mater.*, 2007, **17**, 2167.

75. M. M. Mandoc, F. B. Kooistra, J. C. Hummelen, B. de Boer and P. W. M. Blom, *Appl. Phys. Lett.*, 2007, **91**, 263505.

76. W. L. Leong, G. C. Welch, L. G. Kaake, C. J. Takacs, Y. M. Sun, G. C. Bazan and A. J. Heeger, *Chem. Sci.*, 2012, **3**, 2103.

77. C. H. Duan, W. Z. Cai, B. B. Y. Hsu, C. M. Zhong, K. Zhang, C. C. Liu, Z. C. Hu, F. Huang, G. C. Bazan, A. J. Heeger and Y. Cao, *Energy Environ. Sci.*, 2013, **6**, 3022.

78. P. Langevin, *Ann. Chim. Phys.*, 1903, **28**, 433.

79. C. M. Proctor, M. Kuik and T. Q. Nguyen, *Prog. Polym. Sci.*, 2013, **38**, 1941.

80. U. Albrecht and H. Bassler, *Phys. Status Solidi B*, 1995, **191**, 455.

81. P. W. M. Blom and M. J. M. De Jong, *Philips J. Res.*, 1998, **51**, 479.

82. A. Wagenpfahl, C. Deibel and V. Dyakonov, *IEEE J. Sel. Top. Quantum Electron.*, 2010, **16**, 1759.

83. C. Deibel and V. Dyakonov, *Rep. Prog. Phys.*, 2010, **73**, 096401.

84. A. Pivrikas, N. S. Sariciftci, G. Juska and R. Osterbacka, *Prog. Photovolta-ics*, 2007, **15**, 677.

85. S. Albrecht, W. Schindler, J. Kurpiers, J. Kniepert, J. C. Blakesley, I. Dumsch, S. Allard, K. Fostiropoulos, U. Scherf and D. Neher, *J. Phys. Chem. Lett.*, 2012, **3**, 640.

86. S. Albrecht, S. Janietz, W. Schindler, J. Frisch, J. Kurpiers, J. Kniepert, S. Inal, P. Pingel, K. Fostiropoulos, N. Koch and D. Neher, *J. Am. Chem. Soc.*, 2012, **134**, 14932.

87. J. M. Guo, H. Ohkita, H. Benten and S. Ito, *J. Am. Chem. Soc.*, 2010, **132**, 6154.

88. I. A. Howard, R. Mauer, M. Meister and F. Laquai, *J. Am. Chem. Soc.*, 2010, **132**, 14866.

89. C. Deibel, A. Wagenpfahl and V. Dyakonov, *Phys. Rev. B*, 2009, **80**, 075203.

90. R. A. Street, M. Schoendorf, A. Roy and J. H. Lee, *Phys. Rev. B*, 2010, **81**, 205307.

91. C. Tanase, P. W. M. Blom and D. M. de Leeuw, *Phys. Rev. B*, 2004, **70**, 193202.

92. W. F. Pasveer, J. Cottaar, C. Tanase, R. Coehoorn, P. A. Bobbert, P. W. M. Blom, D. M. de Leeuw and M. A. J. Michels, *Phys. Rev. Lett.*, 2005, **94**, 206601.

93. N. F. Mott, *Philos. Mag.*, 1969, **19**, 835.
94. E. M. Hamilton, *Philos. Mag.*, 1972, **26**, 1043.
95. D. Hertel and H. Bassler, *ChemPhysChem*, 2008, **9**, 666.
96. R. Coehoorn, W. F. Pasveer, P. A. Bobbert and M. A. J. Michels, *Phys. Rev. B*, 2005, **72**, 155206.
97. T. Holstein, *Ann. Physics*, 2000, **281**, 725.
98. T. Tiedje and A. Rose, *Solid State Commun.*, 1981, **37**, 49.
99. A. J. Campbell, D. D. C. Bradley and H. Antoniadis, *Appl. Phys. Lett.*, 2001, **79**, 2133.
100. A. Devizis, A. Serbenta, K. Meerholz, D. Hertel and V. Gulbinas, *Phys. Rev. Lett.*, 2009, **103**, 027404.
101. A. J. Mozer, G. Dennler, N. S. Sariciftci, M. Westerling, A. Pivrikas, R. Osterbacka and G. Juska, *Phys. Rev. B*, 2005, **72**, 035217.
102. C. Tanase, E. J. Meijer, P. W. M. Blom and D. M. de Leeuw, *Phys. Rev. Lett.*, 2003, **91**, 216601.
103. R. A. Street, K. W. Song, J. E. Northrup and S. Cowan, *Phys. Rev. B*, 2011, **83**, 165207.
104. Z. M. Beiley, E. T. Hoke, R. Noriega, J. Dacuna, G. F. Burkhard, J. A. Bartelt, A. Salleo, M. F. Toney and M. D. McGehee, *Adv. Energy Mater.*, 2011, **1**, 954.
105. W. L. Leong, G. Hernandez-Sosa, S. R. Cowan, D. Moses and A. J. Heeger, *Adv. Mater.*, 2012, **24**, 2273.

Multi-Junction Polymer Solar Cells

ALICE FURLAN[a] AND RENÉ A. J. JANSSEN*[a]

[a]Molecular Materials and Nanosystems Eindhoven University of Technology, PO Box 513, 5600 MB Eindhoven, The Netherlands
*E-mail: r.a.j.janssen@tue.nl

11.1 Introduction

11.1.1 Principles of Multi-Junction Polymer Solar Cells

Solar cells based on small organic molecules or polymers have reached power conversion efficiencies (PCEs) in excess of 10%[1-3] and start to approach the 11–12% limits of conversion that have been predicted.[4,5] Although the 11–12% maximum conversion efficiency is not a strict fundamental limit,[6] surpassing this threshold will require innovative steps that likely go beyond existing approaches. One of the main factors limiting the efficiency of organic solar cells is that photons lose a considerable amount of energy in creating free charges.[7,8] The photon energy loss in organic solar cells, defined as the energy difference between the optical bandgap (E_g) and the open-circuit voltage (V_{OC}), $E_{loss} = E_g - qV_{OC}$, is at least 0.6 eV,[5] but even in the most efficient materials is sometimes as high as 0.7 to 0.8 eV. When E_{loss} becomes less than 0.6 eV, the efficiency of free charge generation and power conversion efficiency are usually strongly reduced. The main reason for the high E_{loss} in organic semiconductors compared to inorganic solar materials is their low

RSC Polymer Chemistry Series No. 17
Polymer Photovoltaics: Materials, Physics, and Device Engineering
Edited by Fei Huang, Hin-Lap Yip, and Yong Cao
© The Royal Society of Chemistry 2016
Published by the Royal Society of Chemistry, www.rsc.org

dielectric constant that creates a Coulomb binding energy between the photogenerated holes and electrons.

A viable strategy to improve the efficiency of organic solar cells beyond the current conversion limits is the use of multi-junction configurations. In a tandem cell, which is the simplest multi-junction cell architecture, two absorber layers with different bandgaps are used (Figure 11.1). The maximum power conversion efficiency calculated from detailed balance under AM1.5G solar light for a series constrained device is ~33% for a single junction, ~45% for a double junction, and ~51% for a triple junction.[9] To understand the fundamental advantage of a tandem cell configuration, one must recognize that high energy photons absorbed in a small bandgap semiconductor lose more energy when thermalizing to the bandgap than when they are absorbed in a wide bandgap semiconductor. Conversely, low energy photons that are not absorbed by a wide bandgap semiconductor can be absorbed by the small bandgap material. Hence, a tandem cell can serve to convert a larger part of the solar spectrum but, importantly, it also does this at a higher efficiency. Both thermalization and transmission losses are reduced. It is important to note, however, that the $E_{loss} \geq 0.6$ eV threshold for organic solar cells is not alleviated for the individual subcells. Nevertheless an organic tandem cell can be expected to give an overall reduction of the photon energy loss, compared to a single junction.

In a tandem solar cell the individual two subcells must be interconnected electrically and optically. This requires an intermediate contact that is optically transparent and that has the appropriate work function to contact the adjacent photoactive layers without voltage losses. Two different configurations exist, in which the cells are either connected in series, leading to addition of the voltages, or in parallel, leading to addition in photocurrents. The series connection is much more popular, because no current needs to be extracted *via* the intermediate contact and because it enables utilization of the principal advantages of a tandem configuration outlined above in a simple way. In a series connection, the intermediate contact acts as electron collecting contact for one subcell and as hole collecting contact for the other subcell. Therefore it should have quite different work functions at both sides

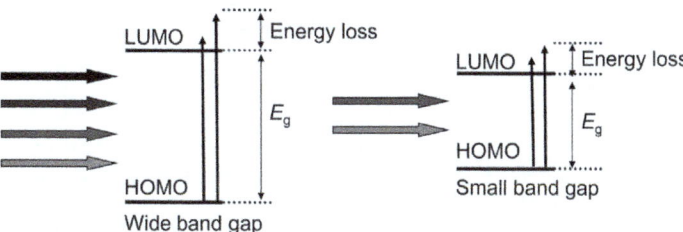

Figure 11.1 Principle of a tandem solar cell. Incoming solar light is spectrally distributed and absorbed in two semiconductor layers with different bandgaps to minimize thermalization energy losses and transmission losses.

to be able to form Ohmic contacts. The intermediate contact layers should also recombine electrons and holes without energy loss to add the voltages of the two subcells. For a series-connected tandem cell, the intermediate contact is often referred to as the recombination layer. Depending on the choice of charge transport layers, a tandem solar cell can be in a regular or an inverted configuration: a schematic of each device layout is shown in Figure 11.2.

11.1.2 Early Developments

The first organic tandem solar cell was published in 1990 by Hiramoto *et al.*,[10] who employed two identical subcells composed of bilayers of a metal-free phthalocyanine and a perylene tetracarboxylic derivative with an ultrathin Au layer to interconnect the two subcells to achieve an almost doubling of the voltage. More than a decade later, in 2002, Forrest *et al.*[11] reported on two, three, and five stacked heterojunction cells consisting of copper phthalocyanine (CuPc) as a donor and perylenetetracarboxylic bis-benzimidazole (PCTBI) as an acceptor. Ultrathin (~5 Å) layers of Ag clusters were placed between the heterojunctions to interconnect the subcells. Similar results were described by Tsutsui *et al.*[12] In 2004, Leo and Pfeifer *et al.*[13,14] introduced a novel interconnection layer consisting of p- and n-doped wide bandgap transport layers that sandwiched an intrinsic (*i*) organic semiconductor

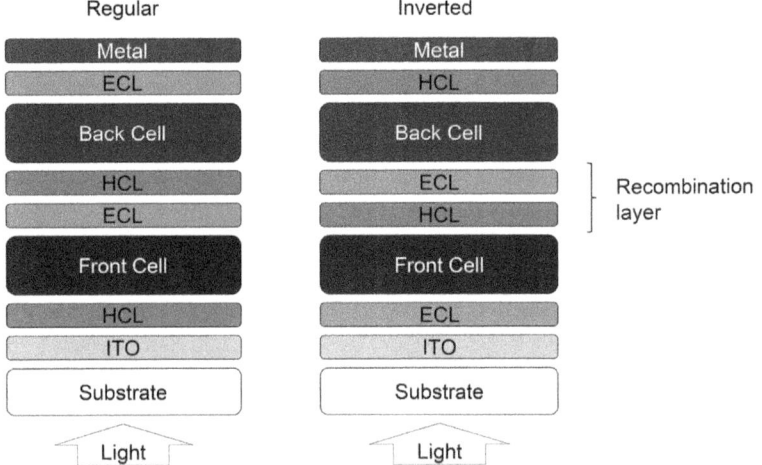

Figure 11.2 Schematic representations of the regular and inverted device layouts for series-connected organic tandem cells. Under operation the indium tin oxide (ITO) electrode is biased positive with respect to the metal electrode in the regular configuration and negative in the inverted configuration. The difference is caused by the positioning of the hole and electron collecting layers (HCL and ECL) with respect to photoactive layers. The intermediate contact formed by HCL and ECL is referred to as the recombination layer. Note that in practice the device layout may vary from these schematics.

consisting of a zinc phthalocyanine:fullerene (ZnPc:C_{60}) blend as photoactive layer. Metal clusters were still used to improve the recombination properties of the doped transport layers. This work forms the basis for the highest certified efficiency of 12%, achieved by Heliatek in 2012 for a triple junction cell that comprises one wide bandgap and two identical small bandgap organic absorbers.[15] Parallel to this development, Forrest *et al.* showed in 2004 that 5.7% efficient tandem cells can be obtained by using two hybrid planar-mixed molecular heterojunction cells in series, also employing doped transport layers.[16] Importantly, these early investigations already considered the absorption of incident light and distribution of the absorbed photons over the two layers using optical modeling.[14,16]

In the first seminal papers small molecule organic semiconductors were used that were deposited by vacuum deposition to create sometimes complex and intricate multilayer stacks. Such multilayer stacks were at that time not achievable *via* solution processing as used for polymer solar cells. As a consequence, early tandem devices comprising polymer layers were hybrid polymer–small molecule devices. Both Sariciftci *et al.*[17] and Lemmer *et al.*[18] published in 2006 a hybrid device consisting of a blend of poly(3-hexylthiophene) (P3HT) and [6,6]-phenyl-C_{61}-butyric acid methyl ester ($PC_{61}BM$) as front subcell and a ZnPc:C_{60} blend or a CuPc/C_{60} bilayer as back subcell. The tandem cells used recombination zones composed of a thin Au interlayer sandwiched between undoped or doped organic semiconductor layers, respectively.[17,18] The first approach to a tandem polymer solar cell was published by Yang *et al.*,[19] who stacked two polymer solar cells processed on individual glass substrates of which one had a semitransparent multilayer electrode. The cells could be connected in series or in parallel. The first monolithic polymer tandem cell, reported by Kawano *et al.*,[20] used a sputtered indium tin oxide (ITO) layer on top of the first polymer–fullerene bulk heterojunction layer, followed by a layer of poly(3,4-ethylenedioxythiophene):poly(styrenesulfonate) (PEDOT:PSS). Because the work function of ITO is too high to collect electrons from fullerenes, a voltage loss was incurred in the front subcell. The first polymer tandem cell that utilized two semiconductors with complementary bandgaps was published in 2006 by Blom *et al.*[21] They described the use of wide bandgap (~1.90 eV) and small bandgap (~1.25 eV) polymer:fullerene bulk heterojunction blends that were solution processed and a recombination contact that consisted of an evaporated LiF/Al/Au stack, covered with PEDOT:PSS.

The next step was the use of solution processed charge transport layers in the recombination contact of polymer solar cells. In 2007, Gilot *et al.* showed that ZnO nanoparticles processed from acetone can be covered with a layer of pH-neutral PEDOT:PSS spincoated from a dispersion in water to fabricate a ZnO/PEDOT:PSS recombination layer that is transparent to light and makes an almost loss-less recombination layer for tandem and triple junction cells.[22] This approach bears similarity to the p- and n-doped wide bandgap transport layers developed for evaporated small molecule tandem cells.[13,14] The ZnO/PEDOT:PSS recombination contact was initially employed in tandem and triple junction solar cells, leading to multilayer devices with up to eight subsequent solution processed layers,[22] but was also used to make six-fold junctions with 17 layers.[23] Also in 2007, Heeger *et al.* published a solution processed polymer tandem solar cell with record efficiency of 6.5% using two different bandgap polymer semiconductors and a solution processed TiO$_x$/PEDOT:PSS intermediate contact.[24] The subsequent developments that have resulted in solution processed multi-junction polymer solar cells that approach 12% efficiency will be outlined in the remainder of this chapter.

11.1.3   Outline

Several excellent reviews[25-34] exist on polymer tandem solar cells and we refer to these publications for a comprehensive overview. In this chapter we focus on the materials requirements for creating efficient polymer solar cells and intermediate contact layers. In the following paragraphs, we will outline the operational principles in more detail. We will then review the most important photoactive materials used in polymer multi-junction cells and outline the material combinations that can used for the recombination layer. We conclude with an overview of recent achievements that have pushed the efficiency to well over 10% and address the progress in processing of large area tandem polymer solar cells.

11.2   Optimization and Characterization of Multi-Junction Polymer Solar Cells

11.2.1   Electrical and Optical Modeling

To reach a high performance, the two subcells of the tandem cell must operate in concert, electrically and optically. To generate and sustain photocurrent in the tandem cell, both subcells must absorb light. In absorbing light, the front subcell will act as an optical filter and reduce the light intensity for the back subcell. Photons that pass both subcells after a first pass can be reflected by the metal back electrode (see Figure 11.2) and have another chance of being absorbed in the second pass. Therefore absorption of light by the back subcell also affects the absorption by the front subcell. A further complication regarding the optics is that the photoactive, charge transport, and interconnection layers have thicknesses that vary from a few to a few hundred nanometers, *i.e.* comparable to the wavelength of light. Because the layers have different refractive indexes (n) and extinction coefficients (k), the multilayer stack acts as a complex optical cavity. In such case detailed optical modeling is necessary to determine the amount of photons absorbed in each layer as first shown by Dennler *et al.* (Figure 11.3).[35] To properly account for the optical interference of all layers the transfer matrix formalism is very useful. With this method it possible to accurately estimate the fraction of absorbed photons in each layer using experimentally determined wavelength-dependent $n(\lambda)$ and $k(\lambda)$ values, that are available *via* ellipsometry or absorption-reflection measurements and Kramers–Kronig transformation.

Next to optical interference there is also electrical interference. In a series-connected tandem solar cell the conservation of charge dictates that the photocurrent through the device must be constant and that the voltages of the subcells add up to give the voltage of the tandem device; this is represented for current density and voltage by the relations:

$$J_{tandem} = J_{front\ subcell} = J_{back\ subcell} \qquad (11.1)$$

$$V_{tandem} = V_{front\ subcell} + V_{back\ subcell} \qquad (11.2)$$

under steady state operation eqn (11.1) and (11.2) are always satisfied. Despite the simple character of these relations their implications are somewhat intricate. Suppose that a tandem cell is at short circuit, *i.e.* $V_{tandem} = 0$. Under these conditions, the two subcells need not be at short circuit, we only know from eqn (11.2) that the bias over the subcells is equal and opposite: $V_{front\ subcell} = -V_{back\ subcell}$. In practice, charge generation in the two subcells is not equal. If one subcell, A, generates more charges than the other subcell, B, and the tandem cell is at short circuit in steady state, the excess charge carriers from A will induce an increased electric field over the limiting subcell B. Effectively this will introduce reverse bias conditions over the limiting cell to increase the photocurrent. This implies that in practice the short-circuit current density (J_{SC}) of the tandem solar cell can be higher than the short-circuit

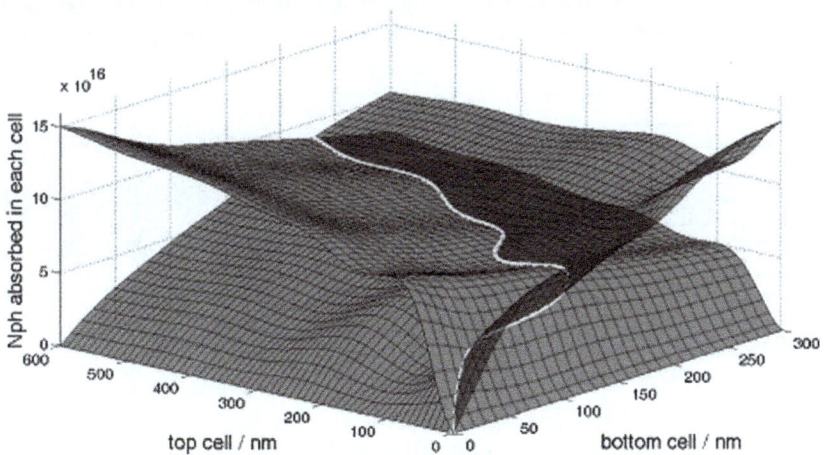

Figure 11.3 Optical modeling of a tandem solar cell. The graph shows the number of photons absorbed in the bottom active layer (blue/paler surface) and in the top active layer (red/darker surface) *vs.* the thickness of the bottom and the top active layers. Reprinted with permission from ref. 35. Copyright 2007, AIP Publishing LLC.

current density of the current-limiting subcell. As most of the excess current generated by the more efficient subcell is lost, there can be an overall loss in efficiency. At open circuit ($J = 0$) the situation is more simple, both subcells are at open circuit and the V_{OC}s of the subcells add up. It must be noted, however, that the V_{OC}s of the subcells are not identical to those when the subcells are under illumination by AM1.5G solar light, because absorption of light by the other subcell reduces the light intensity and changes the spectrum, resulting in slightly lower V_{OC}. If the *J–V* curve of the each of the individual subcells under the effective illumination in the tandem cell is known, the *J–V* curve of the tandem cell can be constructed by adding the voltages at constant current density, employing eqn (11.1) and (11.2). A detailed description of tandem cell device operation has been reported by Blom *et al.*[36] and has been extended for multi-junction solar cells by Forrest *et al.*[37]

To achieve high performing tandem cells it is therefore important to match the two subcells. The first approach, by Dennler *et al.*, considered varying the thickness of the active layers and use optical modeling to find the layer thickness combinations for which the two subcells absorb the same amount of photons.[35] This was later also used by Long *et al.* to describe the absorption of light with metal interlayers.[38] However, such a model only considers the optical absorption of the active materials. In practice, the optimization of the tandem cell must also account for other parameters such as the internal quantum efficiency (IQE) and fill factor (FF) of the subcells. Gilot *et al.* developed a method to incorporate both optical and electrical interactions between the subcells in the optimization (Figure 11.4).[39]

The starting point is to determine the *J–V* characteristics of the two single junction cells as a function of layer thickness and the optical constants (n, k) of the photoactive layers. Using optical modeling of the layer stack of the single junctions, it is then possible to calculate the IQE as the ratio

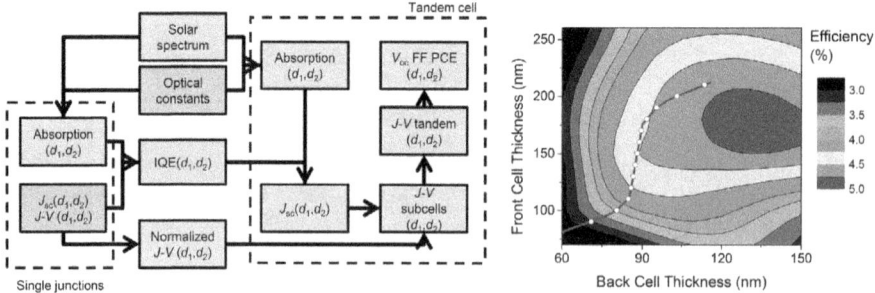

Figure 11.4 (Left) Procedure to optimize a tandem cell based on the experimental optical constants of all layers and the *J–V* characteristics of the single junction cells. (Right) Contour plot of predicted efficiencies of tandem cells calculated by mathematically constructing *J–V* curves. The isoline of matching current-density-generating capacities of the subcells is represented with white circles. The left graph has been reproduced from ref. 39 with permission from Wiley.

between the experimental J_{SC} and the photon flux absorbed by the photoactive layer at each thickness. The IQE is then used, together with optical modeling of the entire layer stack in the tandem cell, to predict the photocurrent generation in each subcell for different layer thicknesses of the photoactive layers. These J_{SC}s, combined with the normalized experimental J–V curves for the single junction cells of the materials at different layer thicknesses, can be combined to predict the J–V curves and solar cell parameters of a series of tandem solar cells with different front and back subcell thicknesses. The calculation is based on the assumption that there is a loss-less Ohmic contact between the subcells and Kirchhoff's law for series-connected circuits and leads to a prediction of the PCE. Figure 11.4 shows that, for the example studied by Gilot *et al.*, just balancing photocurrents would not lead to maximum tandem cell performance. A much more refined model has been developed by Blom *et al.*, who used electrical drift-diffusion modeling of the photoactive layers in combination with optical modeling to predict the active layer combinations that would lead to optimized polymer tandem solar cells performance.[40] Such models were later also used by Jo *et al.*[41]

These approaches optimize the tandem cell by varying the thickness of the active layers. It is also possible to balance the subcells by varying the thickness of charge transport layers. These can act as optical spacers to adjust the electrical optical field in the layer stack such that it maximizes in the photoactive layers. This approach is particularly useful for vacuum processed small molecule tandem cells where the low charge mobility dictates the use of thin photoactive layers of 20–50 nm.[42,43]

These rules and optimization methods assume that the choice of active materials has already been made and will only predict the highest efficiency obtainable within a specific set of photoactive layers. In a study by Dennler *et al.*,[44] design rules for donors in bulk heterojunction tandem solar cells were developed on the basis of Scharber's model for single junctions.[4] The estimates take into account that, for the wide bandgap front subcell, the absorption is reduced because there is no adjacent reflective back contact, and for the small bandgap back subcell some light is filtered out by the top cell. Figure 11.5 shows that optimum tandem cells are expected from stacking of materials of similar E_g, because the optimum is found close to the diagonal. The calculations predict a maximum value of 15% to be achievable for a front subcell with $E_g = 1.6$ eV and a back subcell with $E_g = 1.3$ eV. In a refined study that also accounts for the electrical interaction between the subcells, Blom *et al.* predict a similar realistically achievable efficiency of 14.7% for tandem solar cells.[45] In their case optimal bandgaps are predicted to be $E_g = 1.9$ eV and $E_g = 1.5$ eV.[45]

11.2.2 Characterization of Tandem Cells

As we have seen, the tandem architecture represents a valuable path to increasing the efficiency of polymer tandem solar cells. However, this complex structure, where two subcells are optically and electrically connected, also represents a challenge when it comes to the correct characterization

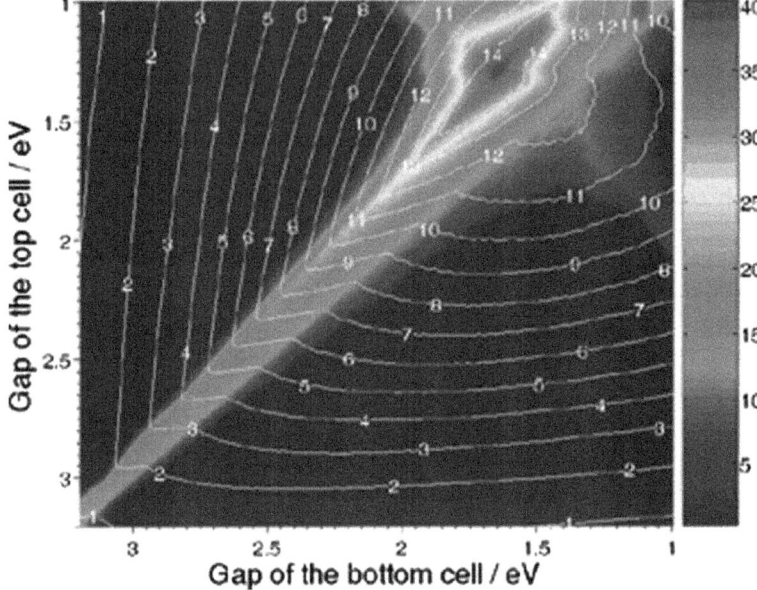

Figure 11.5 Percentage of efficiency increase of a tandem cell over the best single cell (R) for a device comprising a top subcell and a bottom subcell
based on donors each having a LUMO level at −4 eV and each blended
with a fullerene acceptor of LUMO = −4.3 eV. The variables are the
bandgaps of both donors. The lines indicate the efficiency of the tandem devices. Reprinted from ref. 44 with permission from Wiley.

and measurement of the short circuit current of the device. For single junction cells the standard procedure is to measure the *J–V* characteristics under
simulated AM1.5G calculations and correct for the spectral mismatch.[46]
In practice, a good estimate for the short-circuit current, which is most
sensitive to intensity and spectral mismatch, can be obtained by integrating the EQE with the solar AM1.5G spectrum. For single junction organic
solar cells the EQE can be measured with a low intensity probe light and
additional light bias to simulate AM1.5G lighting conditions. While this
method has its own intricacies, especially for organic solar cells,[47] measuring the EQE of a tandem cell is fundamentally hampered by the fact
that, for photocurrent to be measured, both subcells have to be excited
simultaneously. This can be achieved by using (monochromatic) bias light
sources with wavelengths that are primarily absorbed by only one of the
subcells. The additional monochromatic light intensity is adjusted in such
a way that the absorbing subcell is illuminated in excess, making the other
subcell current limiting and assessable for measuring its spectral response.
However as shown by Gilot *et al.*,[48] specific challenges are further caused
by two characteristics of organic solar cells, *i.e.* their sub-linear light intensity dependence of the photocurrent and a field-assisted charge collection.
These properties necessitate that EQE experiments not only are carried

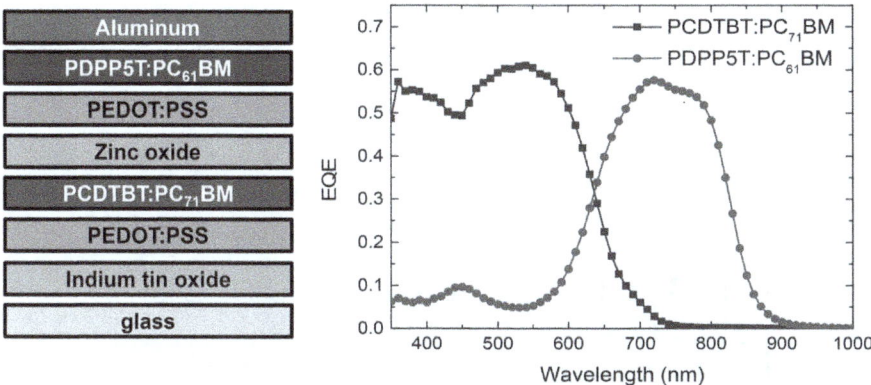

Figure 11.6   Device layout and EQE of a polymer tandem solar cell measured with appropriate optical and electrical bias. The graph clearly shows the complementary spectral contributions of the two subcells. Reproduced from ref. 49 with permission from Wiley.

out under representative illumination conditions but also they require the use of appropriate electrical bias to maintain short-circuit conditions for the addressed subcell. As explained above, the excess illumination will put the subcell that is measured under reverse bias conditions and therefore a compensating forward electrical bias needs to be applied to ensure that the current of that subcell is measured at short circuit. Gilot *et al.* have described a method to determine the magnitudes of the bias illumination and correct bias voltage during EQE measurements of tandem polymer solar cells.[48] An example of EQEs of subcells measured in a tandem cell in this way is shown in Figure 11.6.[49]

Gilot *et al.* further presented two different methods to determine the *J–V* characteristics of the subcells in a two-terminal tandem cell without interfering with the light incoupling of the cell.[50] The first method employs an extra proximity metal electrode that acts as a voltage probe. The second method uses bias-dependent external quantum efficiency measurements of two-terminal tandem solar cells and subsequent integration with the solar spectrum to determine the *J–V* curves of the subcells. The two methods show good mutual agreement.

11.3   Photoactive Layers

Over the last decade there has been enormous progress in tandem polymer solar cells. Advances have been made in terms of understanding, new photoactive materials, new interlayers, and device configurations (regular *vs.* inverted, parallel, semitransparent, *etc.*). In Table 11.1 we provide a list of results obtained up to the end of 2014, following a more or less historical perspective and giving preference to listing the more salient results in terms of PCE. Table 11.1 lists the materials used in the front and back subcell,

Table 11.1 Solution processed tandem polymer solar cells.

Front subcell			Back subcell			Tandem cell		Recombination layer		
Material	E_g^a (eV)	V_{OC} (V)	Material	E_g^a (eV)	V_{OC} (V)	Type^b	PCE (%)	Contact to front subcell	Contact to back subcell	Ref.
PF8TBT:PC$_{61}$BM	1.95	0.90	PTBEHT:PC$_{61}$BM	1.20	0.50	R	0.6	LiF/Al	Au/PEDOT:PSS	21
MDMO-PPV: PC$_{61}$BM	2.10	0.82	MDMO-PPV: PC$_{61}$BM	2.10	0.82	R	1.8	ZnO nanoparticle	n-PEDOT:PSS	22
MDMO-PPV: PC$_{61}$BM	2.10	0.82	P3HT:PC$_{61}$BM	1.90	0.75	R	1.9			
PCPDTBT: PC$_{61}$BM	1.40	0.66	P3HT:PC$_{71}$BM	1.90	0.63	R	6.5	TiO$_x$ sol–gel	PEDOT:PSS(PH500)	24
P3HT:PC$_{61}$BM	1.90	0.60	PTBEHT:PC$_{61}$BM	1.20	0.51	R	0.9	Sm/Au optical spacer	Au/PEDOT:PSS	52
P3HT:PC$_{61}$BM	1.90	0.60	PTBEHT:PC$_{61}$BM	1.20	0.51	R/P	3.0			
PCPDTBT: PC$_{61}$BM	1.40	0.65	P3HT:PC$_{61}$BM	1.90	0.60	R/P	3.1	MoO$_3$/Al	Ag/MoO$_3$	53
P3HT:PC$_{71}$BM	1.90	0.60	PSBTBT:PC$_{71}$BM	1.50	0.67	R	5.8	Al/TiO$_2$ sol–gel	PEDOT:PSS	54
PF10TBT: PC$_{61}$BM	1.95	0.98	PBBTDPP2: PC$_{61}$BM	1.40	0.62	R	4.9	ZnO np	n-PEDOT:PSS	39
P3HT:PC$_{71}$BM	1.90	0.58	PSBTBT:PC$_{71}$BM	1.50	0.62	R/P	4.8	PEDOT/Au	Au/V$_2$O$_5$	55
P3HT:bis PC$_{61}$BM	1.90	0.70	P3HT:PC$_{71}$BM	1.90	0.58	R	5.2	LiF/ITO	MoO$_3$	56
PF10TBT: PC$_{61}$BM	1.95	0.98	PF10TBT:PC$_{61}$BM	1.95	0.98	R	4.1	ZnO np	n-PEDOT:PSS(PH500: DMAE)/Nafion	57
P3HT:PC$_{61}$BM	1.90	0.62	P3HT:PC$_{61}$BM	1.90	0.62	R	3.7	TiO$_2$ np	PEDOT:PSS(PH500)	58
P3HT:PC$_{61}$BM	1.90	0.58	PSBTBT:PC$_{71}$BM	1.50	0.64	I	5.1	MoO$_3$/Al	ZnO-d sol–gel	59
P3HT:PC$_{61}$BM	1.90	0.65	P3HT:PC$_{61}$BM	1.90	0.65	I	2.9	PEDOT:PSS(IPA:BuOH)	ZnO np/C$_{60}$-SAM	60
P3HT:ICBA	1.90	0.83	PSBTBT:PC$_{71}$BM	1.50	0.64	I	6.2	PEDTOT:Au-np	TiO$_2$:Cs	61
P3HT:PC$_{61}$BM	1.90	0.63	P3HT:PC$_{61}$BM	1.90	0.63	I	2.9	MoO$_3$/Ag	Al/Ca	62
P3HT:PC$_{61}$BM	1.90	0.59	P3HT:PC$_{61}$BM	1.90	0.53	R/L	4.1	ZnO	GO:PEDOT:PSS	63

P3HT:ICBA	1.90	0.82	PSBTBT:PC$_{71}$BM	1.50	0.66	R	7.0	Al/TiO$_2$ sol–gel	m-PEDOT:PSS (PH500:NaPSS:DMF)	64
PCDTBT: PC$_{71}$BM	1.88	0.90	PDPP5T:PC$_{61}$BM	1.46	0.58	R	7.0	ZnO np	n-PEDOT:PSS	49
P3HT:ICBA	1.87	0.82	PDPP5T:PC$_{61}$BM	1.46	0.59	I	5.8	PEDOT:PSS(CPP105D:IPA)	ZnO np	65
P3HT:ICBA	1.90	0.85	PBDTT-DPP: PC$_{71}$BM	1.44	0.74	I	8.6	m-PEDOT: PSS(PH500:NaPSS:DMF)	ZnO np	66
P3HT:ICBA	1.90	0.82	PBDTT-C: PC$_{61}$BM	1.56	0.67	I	8.2	PEDOT:PSS(PH1000)	PEIE	67
P3HT:PC$_{61}$BM	1.90	0.60	P3HT:PC$_{61}$BM	1.90	0.60	R	3.6	TiO$_x$ sol–gel/Ca$_2$Nb$_3$O$_{10}$	PEDOT:PSS	68
P3HT:PC$_{61}$BM	1.90	0.63	P3HT:PC$_{61}$BM	1.90	0.63	I	2.1	MoO$_x$/Ag	PEIE	69
SDT-BT:PC$_{71}$BM	1.37	0.60	P3HT:PC$_{71}$BM	1.90	0.60	R	5.2	TiO$_x$ sol–gel	PEDOT:PSS	70
P3HT:PC$_{61}$BM	1.90	0.56	P3HT:PC$_{61}$BM	1.90	0.56	R/L	4.1	GO:SWCNTs	ZnO np	71
P3HT:PC$_{61}$BM	1.90	0.53	P3HT:PC$_{61}$BM	1.90	0.53	I/L	3.5	ZnO np	GO:SWCNTs	71
P3HT:ICBA	1.90	0.84	PBDTT-SeDPP: PC$_{71}$BM	1.38	0.69	I	9.5	PEDOT:PSS	ZnO np	72
P3HT:ICBA	1.90	0.84	PDTP-DFBT: PC$_{61}$BM	1.38	0.70	I	10.6	PEDOT:PSS	PEDOT:PSS	73
PF10TBT: PC$_{61}$BM	1.95	1.04	PDPPTPT: PC$_{61}$BM	1.53	0.78	R	4.6	ZnO np	n-PEDOT:PSS/Nafion	74
PCDTBT: PC$_{71}$BM	1.88	0.87	PMDPP3T: PC$_{61}$BM	1.30	0.61	R	8.9	ZnO np	n-PEDOT:PSS	75
P2:PC$_{71}$BM	1.84	0.88	PDPPTPT: PC$_{71}$BM	1.49	0.78	I	8.4	PEDOT:PSS(CPP105D:IPA)	ZnO-np/CPE(FPQ-Br)	76
PBDTT-FDPP- C$_{12}$:PC$_{61}$BM	1.49	0.76	PBDTT-SeDPP: PC$_{71}$BM	1.38	0.73	I/S	7.4	PFN/TiO$_2$ sol–gel	PEDOT:PSS(PH1000:IPA: DMF)/ PEDOT:PSS	77
PDPP5T-2: PC$_{71}$BM	1.46	0.56	PDPP5T-2: PC$_{71}$BM	1.46	0.56	R	4.9	ZnO np	PH neutral PEDOT:PSS	78
PSEHTT:ICBA	1.82	0.94	PSBTBT:PC$_{71}$BM	1.50	0.68	R	8.4	TiO$_x$ sol gel	GO	79
PBDTFBZS: PC$_{71}$BM	1.81	0.88	PNDTDPP: PC$_{71}$BM	1.36	0.74	I	9.4	PEDOT:PSS	ZnO	80

(continued)

Table 11.1 (continued)

Front subcell			Back subcell			Tandem cell		Recombination layer		Ref.
Material	$E_g{}^a$ (eV)	V_{OC} (V)	Material	$E_g{}^a$ (eV)	V_{OC} (V)	Type^b	PCE (%)	Contact to front subcell	Contact to back subcell	
PTIPSBDT-DFDTQX: PC$_{71}$BM	1.85	0.81	PBPT-8	1.56	0.71	I	7.2	PEDOT:PSS(IPA:surf.)	PEIE	81
PCDTBT: PC$_{61}$BM	1.90	0.85	PCDTBT:PC$_{61}$BM	1.90	0.85	R	3.9	GO-Cs/Al	GO/MoO$_3$	82
P3HT:ICBA PC$_{61}$BM	1.90	0.84	PTB:PC$_{71}$BM	1.60	0.71	I	8.3	PEDOT:PSS	PEI	83
PCDTBT: PC$_{71}$BM	1.90	0.87	PDPP3T:PC$_{71}$BM	1.32	0.67	I	7.3	MoO$_3$	Al doped MoO$_3$	84
P3HT:ICBA PC$_{71}$BM	1.90	0.84	PTB7:PC$_{71}$BM	1.61	0.71	I	7.8	Ag/PEDOT:PSS(surf.)	ZnO np	85
P3HT:ICBA PC$_{71}$BM	1.90	0.83	PTTBDT-FTT: PC$_{71}$BM	1.63	0.77	I	8.4	PEDOT:PSS(surf.)	PEIE	86
PIDT-phanQ: PC$_{61}$BM	1.67	0.85	PIDT-phanQ: PC$_{71}$BM	1.67	0.87	I	7.8	PEDOT:PSS(IPA:surf.)/PH1000	ZnO/C$_{60}$-SAM	87
PCPDTFBT: PC$_{61}$BM	1.44	0.75	PCPDTFBT: PC$_{71}$BM	1.44	0.76	I	7.2			
PIDT-phanQ: PC$_{61}$BM	1.67	0.85	PIDT-phanQ: PC$_{71}$BM	1.67	0.76	I/S	6.7			
PSEHTT:ICBA	1.82	0.91	PSBTBT:PC$_{71}$BM	1.50	0.64	I	8.9	PEDOT:PSS(modified)	PEIE	88
PSEHTT:ICBA	1.82	0.88	PBDT-DPP: PC$_{71}$BM	1.40	0.74	R/S	8.0	ZnO (no details)	PEDOT:PSS	89
PBDTTPD: PC$_{61}$BM	1.71	0.90	PSBTBT:PC$_{71}$BM	1.50	0.64	I	7.5	pH neutral PEDOT:PSS	Li–ZnO	90
PCDTBT: PC$_{71}$BM	1.88	0.89	PDPPTPT: PC$_{61}$BM	1.53	0.80	I/T	6.1	ZnO np/Ag	MoO$_3$	91

Active layer 1	E_g	V_{OC}	Active layer 2	E_g	V_{OC}	Config	PCE	Interlayer	Cathode	Ref
PCDTBT:PC$_{71}$BM	1.88	0.87	PDPP5T-2:PC$_{61}$BM	1.46	0.55	I	7.3	PEDOT:PSS(IPA)/Ag-nanowire	ZnO np	92
P3HT:ICBA	1.90	0.83	PTB7:PC$_{71}$BM	1.61	0.74	I	8.4	PEDOT:PSS(modified)	ZnO np	93
PBDTT-ttTPD:PC$_{71}$BM	1.86	0.84	PTB7:PC$_{71}$BM	1.61	0.74	I	9.4			
PIDTT-DFTQ:PC$_{71}$BM	1.82	0.89	PIDTT-DFTQ:PC$_{71}$BM	1.82	0.92	I/P	8.1	PEDOT:PSS(modified)/Ag	Ag/MoO$_3$	94
PIDTT-DFBT:PC$_{71}$BM	1.78	0.94	PIDTT-DFBT:PC$_{71}$BM	1.78	0.98	I/P	8.7			
P3HT:ICBA	1.90	0.84	PIDTT-DFBT:PC$_{71}$BM	1.78	0.96	I/P	9.2			
P3HT:ICBA	1.90	0.84	PTB7-Th	1.58	0.78	I	9.6	WO$_3$ np/PEDOT:PSS(PH500: DMF:PSS:surf.)	ZnO np	95
PTB7-Th	1.58	0.78	PDTP-DFBT:PC$_{61}$BM	1.40	0.70	I	10.7			
P3HT:ICBA	1.90	0.84	PDTP-DFBT:PC$_{61}$BM	1.40	0.70	I	9.8			
PSEHTT:ICBA	1.82	0.94	PTB7:PC$_{71}$BM	1.61	0.76	I	10.4	pH neutral PEDOT:PSS	Li-ZnO/C$_{60}$-SAM	96
PDCBT:PC$_{71}$BM	1.91	0.80	PBDT-TS1:PC$_{61}$BM	1.55	0.80	I	10.2	MoO$_3$/Ag	ZnO-cp/PFN	97
PTB7-Th:PC$_{71}$BM	1.58	0.79	PTB7-Th:PC$_{71}$BM	1.58	0.79	I	10.8	PEDOT:PSS(PH1000:H$_2$O:surf.)	PEI	98

[a] E_g represents the optical bandgap of the polymer; note that the optical gap of the fullerene (~1.75 eV) may be less.

[b] R = regular configuration, I = inverted configuration, S = semitransparent cell, P = parallel tandem cell, T = top illuminated, L = laminated.

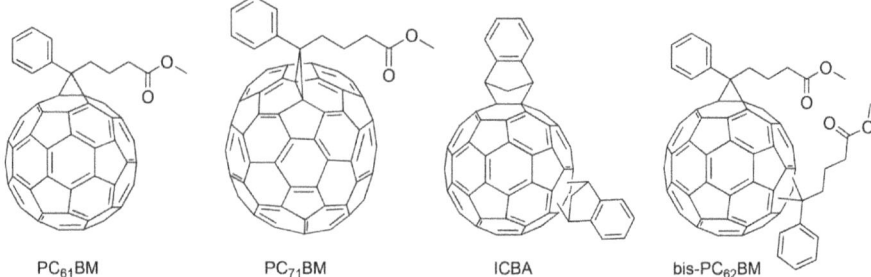

PC$_{61}$BM   PC$_{71}$BM   ICBA   bis-PC$_{62}$BM

Figure 11.7   Structures of fullerene acceptors used in multi-junction polymer solar cells. References to the publications using these acceptors can be found in Table 11.1.

the device configuration (regular or inverted) and the reported PCE of the tandem cell. For each front and back subcell material we also provide the optical bandgap and the V_{OC} of that subcell. The relevance of these two numbers relates to the photon energy loss, $E_{loss} = E_g - qV_{OC}$. The structures of the materials used in the photoactive layers are collected in Figure 11.7 for the fullerene acceptors, in Figure 11.8 for the wide bandgap donors and in Figure 11.9 for the small bandgap donors. The purpose of this section is to give a broad overview of the materials used.

11.3.1   Fullerenes

Four different fullerenes have been used as the acceptor material in polymer tandem solar cells (Figure 11.7); PC$_{61}$BM and PC$_{71}$BM are most commonly used. In terms of energy levels these two are very similar, but owing to the lower overall molecular symmetry the formally forbidden optical transitions of PC$_{61}$BM are partially allowed in PC$_{71}$BM, which increases the absorption coefficient in the 400–700 nm range significantly.[51] Therefore, PC$_{71}$BM can be successfully used to enhance the absorption and photocurrent generation in the wide bandgap subcell. Remarkably, it has also found widespread use in small bandgap subcells. Unless the small bandgap subcell is current limiting, one usually would try *not* to absorb the more energetic photons in that subcell and rather use the less absorbing PC$_{61}$BM. The other two fullerenes, ICBA and bis-PC$_{62}$BM, have a higher LUMO level than PC$_{61}$BM and PC$_{71}$BM and can increase the V_{OC} of the cells, provided that the $E_{loss} \geq 0.6$ eV criterion remains fulfilled. In practice, this works very well for wide bandgap polymer semiconductors with high HOMO levels such as P3HT[56,61] and PSEHTT[79] (see Figure 11.8 for structures).

11.3.2   Wide Bandgap Donors

Table 11.1 shows that, with respect to the wide bandgap subcell, there is still considerable gain possible. To understand this properly, we recall the $E_{loss} \geq 0.6$ eV criterion.[5] The optical bandgaps of PC$_{61}$BM and PC$_{71}$BM

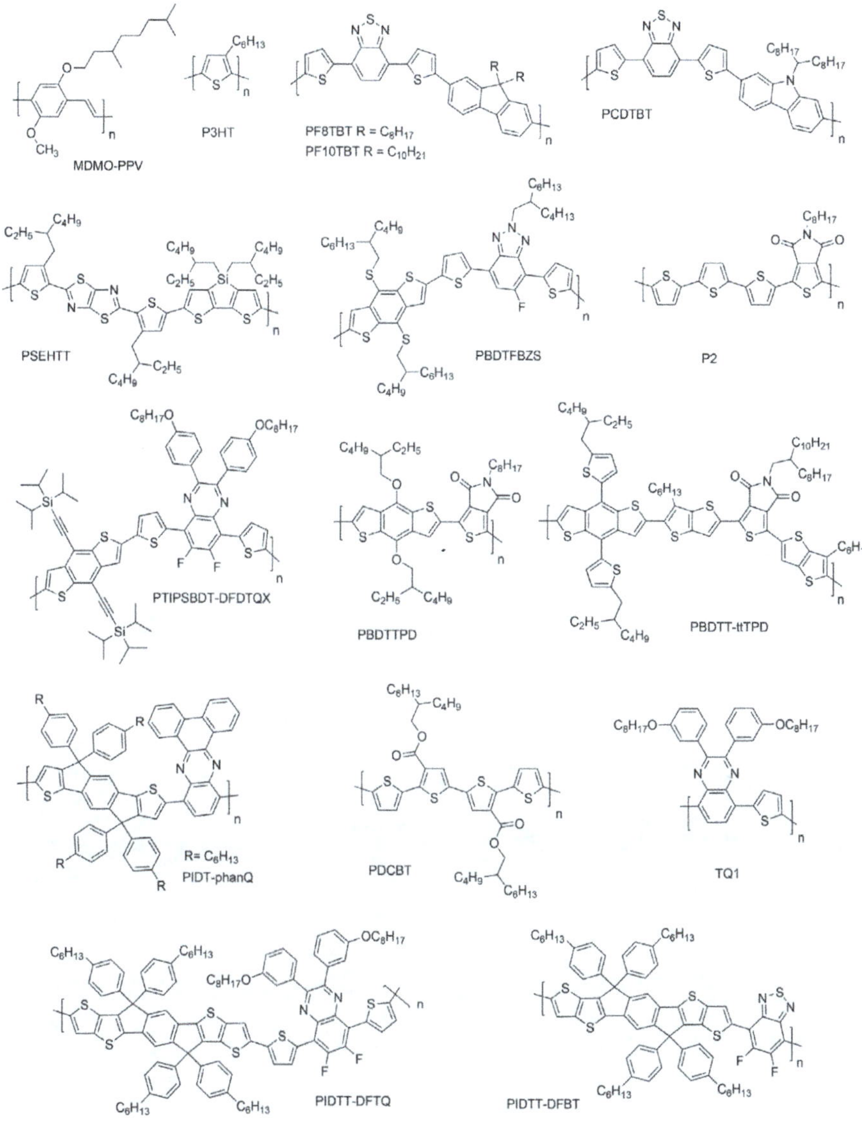

Figure 11.8 Structures of wide bandgap polymers used in multi-junction polymer solar cells. References to the publications using these polymers can be found in Table 11.1.

and that of ICBA are at about 1.7 eV. This implies two design criteria for the wide bandgap subcell. (i) The optical bandgap of the ideal wide bandgap donor should be about 1.7 eV. A higher bandgap would not increase the V_{OC} of the subcell and only reduce the photon absorption. (ii) With a bandgap of 1.7 eV, a V_{OC} of 1.0 V or higher should be attainable. So far, most wide bandgap donor materials have optical bandgaps higher than 1.7 eV, while

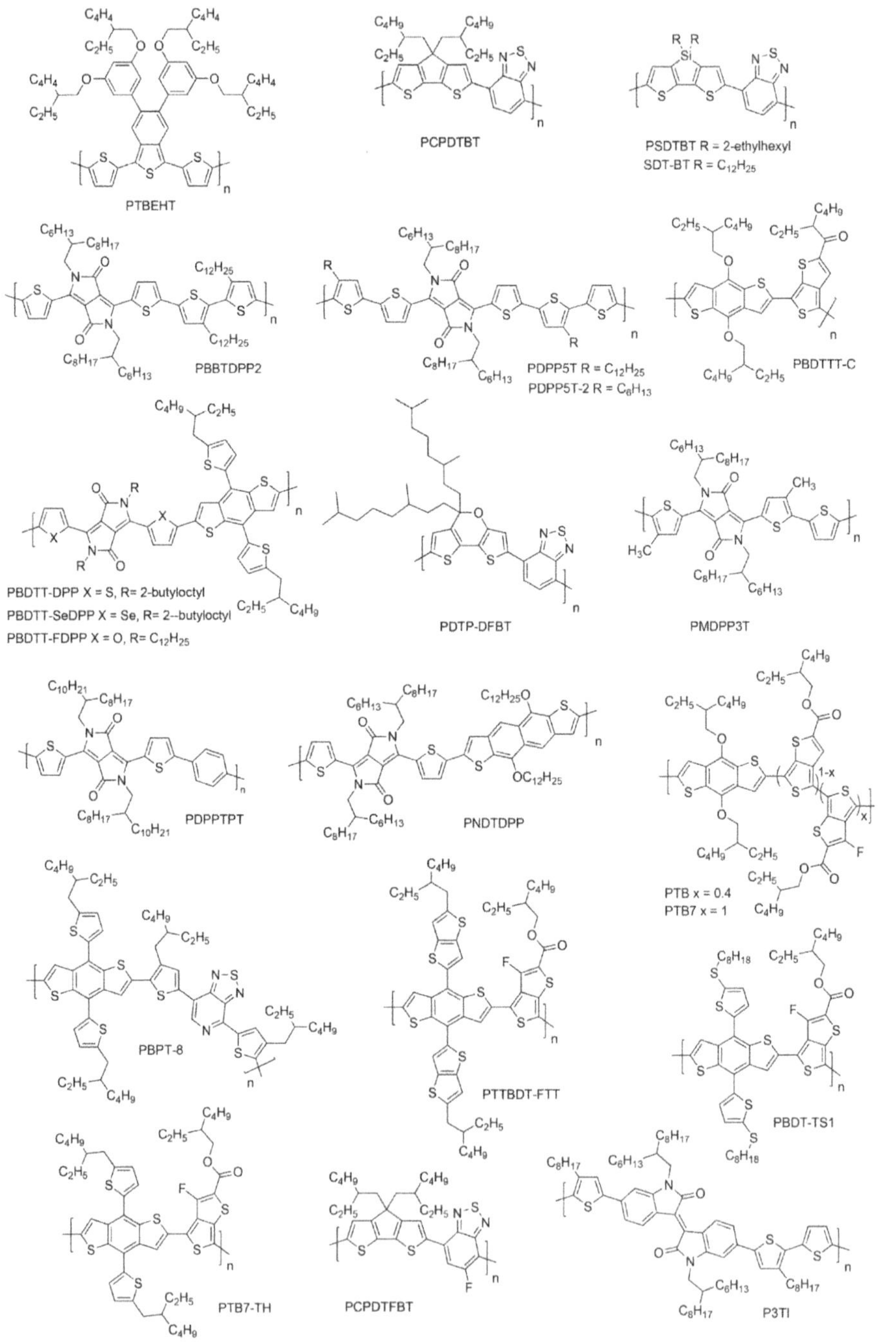

Figure 11.9 Structures of small bandgap polymers used in multi-junction polymer solar cells. References to the publications using these polymers can be found in Table 11.1.

the V_{OC}s are considerably lower than 1.0 V. In fact, taking the average E_{loss} for the wide bandgap subcell of the tandem cells listed gives $E\bar{E}_{loss} = 1.07$ eV. For P3HT:PC_{61}BM, E_{loss} is particularly large (1.3 eV). Also, for P3HT:ICBA, E_{loss} still amounts to 1.06 eV. In terms of energetics, materials such as PIDT-phanQ:PC_{61}BM ($E_{loss} = 0.82$ eV) and PBDTTPD:PC_{61}BM ($E_{loss} = 0.81$ eV) seem definitely more advantageous.

Nevertheless, for creating efficient tandem cells, wide bandgap combinations of polythiophene (P3HT:ICBA) and polycarbazole (PCDTBT:PC_{71}BM) have been good choices in the past. Only more recently have new, specifically designed wide bandgap donors with somewhat deeper HOMO levels, based on thieno-pyrroledione (TPD), quinoxaline (Q), thiazolo[5,4-*d*]thiazole, 4,4,9,9-tetraaryl-4,9-dihydro-*s*-indaceno[1,2-*b*:5,6-*b'*]dithiophene (IDT), and ester-substituted polythiophenes, gained interest (see Figure 11.9 for structures).

11.3.3 Small Bandgap Donors

The variation among the structures used for small bandgap polymers is much larger, but also here some classes can be recognized on the basis of their basic structural similarities. The first class consists of 2,2′-bithiophenes bridged over the 3,3′ positions alternating with benzo[*c*][1,2,5]thiadiazole along the chain. These materials have optical bandgaps ranging from 1.37 to 1.50 eV. The second class relates to the well-known PTB7 polymer and comprises polymers in which a thieno[3,4-*b*]thiophene or 3-fluorothien-o[3,4-*b*]thiophene (T), substituted with a ester solubilizing side chain on the 2 position, is alternating along the polymer chain with a benzo[1,2-*b*:4,5-*b'*] dithiophene (B), substituted on the 4 and 8 positions. These PTB derivatives have optical bandgaps in the range of 1.55 to 1.63 eV, *i.e.* higher than those of the first class. The third class comprises the 2,5-dialkylpyrrolo[3,4-*c*]pyr-role-1,4(2*H*,5*H*)-dione (DPP) based polymers in which the electron-deficient DPP unit alternates with an electron-rich π-conjugated segment. The optical absorption of the DPP polymers can extend well into the near-IR region. In tandem cells DPP polymers with optical bandgaps ranging from 1.30 to 1.50 eV have been used.

The average $\bar{E}_{loss}$ for the small bandgap subcells of the tandem cells listed is 0.87 eV. This clearly demonstrates that V_{OC} and E_g are much better tuned for these small bandgap materials than for the wide bandgap materials where $\bar{E}_{loss} = 1.07$ eV. Particularly promising from an energetic point of view are some of the DPP polymers, where E_{loss} can be as small as ~0.65 eV.

11.4 Recombination Layers

Next to efficient active layer materials, the other critical component of a tandem solar cell is the intermediate contact layer or recombination layer. This layer is responsible for the electrical connection of the subcells and the effective operation of the device. There is no current generation in this area and

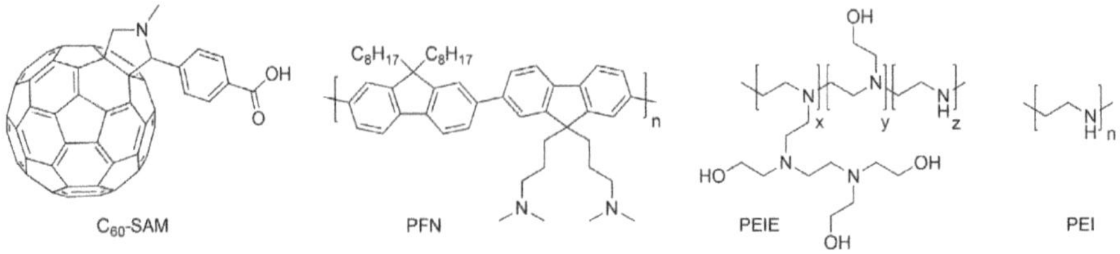

Figure 11.10 Structures of benzoic acid substituted *N*-methylfulleropyrrolidine (C$_{60}$-SAM), poly[(9,9-bis(3'-(*N,N*-dimethylamino)propyl)-2,7-fluorene)-*alt*-2,7-(9,9-dioctylfluorene)] (PFN), ethoxylated polyethylenimine (PEIE), and polyethyleneimine (PEI).

therefore the layers involved need to be transparent to light to avoid absorption losses. Another important requirement for the recombination layer is that it forms Ohmic contacts to the adjacent subcells, one n-type and one p-type. In practice this implies that the recombination layer must have different work functions on either side. Commonly this is achieved by depositing two (or more) layers that individually are suitable to collect electrons or holes. The order of deposition determines the polarity of the device.

As can be seen in Table 11.1 there is a large variety of recombination layers. Typical materials used as n-type layers are solution-processed metal oxides such as ZnO nanoparticles,[22] Li doped ZnO,[90] and sol–gel TiO$_x$.[24] These metal oxides are sometimes functionalized with C$_{60}$ self-assembled monolayers (C$_{60}$-SAMs)[60] or PFN[97] (Figure 11.10) to improve hole blocking.

More recently, polyamines such as PEIE[67] and PEI[83] (Figure 11.10) have been introduced to modify the work function of the adjacent PEDOT:PSS or metal oxide layer to create an effective low work function contact. For the p-type layers that serve to collect holes, PEDOT:PSS is most widely used, together with MoO$_3$,[53] V$_2$O$_5$,[55] WO$_3$,[95] and graphene oxide (GO).[63]

11.4.1 Regular Configuration

In regular polarity devices the electrons that are generated in the front subcell and collected by the n-type layer recombine with the holes that are generated by the back subcell and collected in the p-type layer (see Figure 11.2). The most frequently used material combinations are ZnO/PEDOT:PSS and TiO$_x$/PEDOT:PSS, thanks to their processability from solution. A layer of the two n-type materials ZnO and TiO$_x$ can be obtained in different ways, through a dispersion of nanoparticles or as a sol–gel layer that is converted into the metal oxide after deposition. A sol–gel formulation consists of a metal oxide precursor in solution that undergoes hydrolysis and polycondensation. Full conversion is usually only obtained after deposition with thermal annealing. This is only compatible on top of photoactive layers that can withstand the required annealing temperature (up to 150 °C), such as P3HT:ICBA. Therefore efforts have been directed towards developing metal oxide nanoparticle dispersions that do not need thermal annealing. ZnO nanoparticles[99] that can be processed from acetone or isopropanol (IPA) into closed layers have been used extensively to make multi-junction solar cells.[22] Another option, developed by Hadipour *et al.*, is an extremely thin (3 nm) TiO_x layer that does not require any initial or additional treatment and can be solution-processed into an electron transport layer in air or under a protective atmosphere.[100]

PEDOT:PSS is the most commonly used p-type material, but different formulations are used. To process on top of ZnO, the PEDOT:PSS solution needs to have neutral pH to prevent dissolution of the ZnO. This can been achieved by neutralizing the commercial PEDOT:PSS (PH500) dispersion in water with 2-dimethylaminoethanol (DMAE),[57] or by using a commercial pH-neutral PEDOT:PSS (Orgacon N-1005).[22] The pH-neutral PEDOT:PSS, however, has a lower work function than the commonly used highly acidic PEDOT:PSS dispersion.[57] The work function decreases from 5.05 eV (at pH 1.75) to 4.65 eV (at pH 7).[57] This can cause a voltage loss when combined with photoactive layers that have a deep HOMO level. In practice it difficult to achieve a $V_{OC} > \sim 0.7$ V with pH-neutral PEDOT. The problem can be alleviated, however, by depositing a thin layer of Nafion on the pH-neutral PEDOT:PSS to modify the workfunction.[57,74]

For use on n-type nanocrystalline TiO_2, Yang *et al.* modified a commercial PEDOT:PSS (PH500) formulation with poly(styrenesulfonate) sodium salt (NaPSS) and DMF to enhance conductivity and make it possible to use thicker layers that increase the physical robustness of the intermediate contact,[64] making it compatible with slowly evaporating solvents that are used in the second active layer deposition. More recently, Jang *et al.* replaced PEDOT:PSS by graphene oxide (GO) as a p-type layer on top of TiO_2 to yield tandem cells with 8.4% PCE.[79]

11.4.2 Inverted Configuration

In the inverted configuration, the current flows in the opposite direction when compared to the regular structure, and this means that in the intermediate contact holes will be collected from the front subcell and electrons from the back subcell (see Figure 11.2). The most frequently used material combinations in solution-processed inverted tandem solar cells are PEDOT:PSS/ZnO[60,65,66] and PEDOT:PSS/PEIE[67] (see Table 11.1).

To collect the holes from the front subcell, the p-type material is deposited on the active layer. This is not a trivial step because the polymer active layer surface is hydrophobic and a p-type material spincoated from an aqueous solution or dispersion such as PEDOT:PSS will not form a uniform layer because of de-wetting. Different strategies have been developed to solve this challenge. Kouijzer *et al.* used a mild nitrogen plasma treatment on the P3HT:ICBA active layer and PEDOT:PSS (PH1000) diluted with an alcohol to improve the wetting.[65] A similar approach has been adopted by Kippelen *et al.*, who pre-treated the P3HT:ICBA surface with an oxygen plasma (1 s) to make it more hydrophilic and improved the conductivity of the PH1000 PEDOT:PSS with DMSO.[67] Chen *et al.* have used treatment with 1-hexanol to improve the wettability of a P3HT:ICBA layer for depositing PEDOT:PSS.[101] Another option is to evaporate a thin layer of metal (Ag, Au) to improve the wetting of the PEDOT:PSS.[85] Likewise WO_3 nanoparticles deposited from ethanol have been used as a primer layer for subsequent spincoating of PEDOT:PSS.[95] While each of these strategies gave good PEDOT:PSS layer formation on photoactive layers and excellent performance, the variety of methods that have been developed shows that a universal solution that is valid for all active layer materials remains to be demonstrated.

The n-type layer in the inverted configuration had mostly been ZnO, until Kippelen *et al.* introduced ethoxylated polyethylenimine (PEIE), which is deposited from water and 2-methoxyethanol on a thermally annealed PEDOT:PSS to give a coating of only a few nanometers. This layer induces a modification of the work function of the PEDOT:PSS at the top surface from 4.9 to 3.6 eV, enabling selective electron collection.[67] The effect is thought to originate from a molecular dipole and interface dipole.[102]

11.4.3   Loss-Less Contacts

The p- and n-type materials in the recombination layer must form a loss-less contact to achieve optimal performance, but formation of an Ohmic contact is not straightforward. As explained by Tress and Inganäs, injection barriers caused by a misaligned metal work functions, or extraction barriers resulting from (unintentional) insulating interlayers between metal and active layers, result in a decrease in fill factor. In more extreme cases S-shaped *J–V* characteristics occur.[103] In tandem cells a poor alignment of the energy levels at the junction between the n- and p-type layers in the recombination contact can be another reason for such an S-kink. Gilot *et al.* showed that, for a ZnO/PEDOT:PSS, contact the S-kink can be eliminated by depositing a very thin Ag layer between ZnO and PEDOT:PSS or by UV illumination of ZnO.[22] The effect of UV illumination has been studied in detail and could be attributed to a persistent photodoping (Figure 11.11).[104,105] UV illumination of ZnO creates free electrons and holes. The holes recombine with O–2 defects, and molecular O_2 is desorbed from the nanoparticle surface. As a result trapped holes and free electrons remain. The free electrons are easily detected with electron spin resonance (ESR); by the filling of the conduction band they cause a blue shift of the ZnO bandgap *via* the Burstein–Moss effect and an increase of the conductivity (Figure 11.11). The effect of UV illumination can be reversed by exposing the ZnO to oxygen from the air. The effect of UV illumination does enhance the number of free charge carriers in one of the layers and allows creation of an Ohmic contact. This mechanism is likely also operating for TiO_2 where a similar effect of illumination on the *J–V* characteristics has been observed.[54,58,68,70]

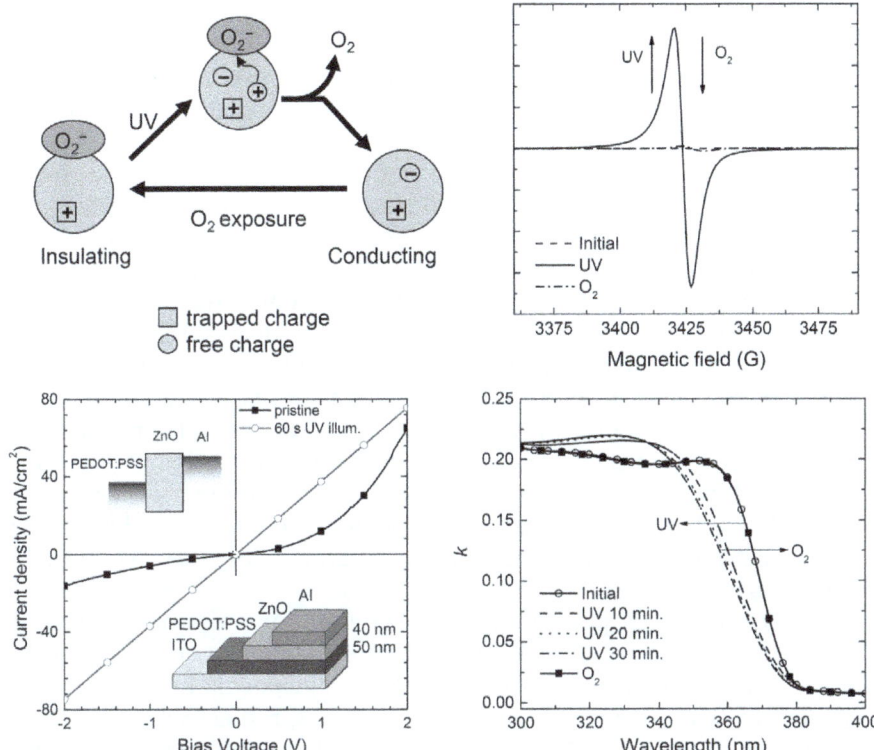

Figure 11.11 Mechanism for persistent photodoping of ZnO particles by UV light. ESR of pristine ZnO nanoparticles in vacuum before and after UV illumination and subsequently after exposure to air. Change in resistance to Ohmic behavior for a ZnO/PEDOT:PSS junction upon UV exposure. Burstein–Moss effect (blue shift) occurs upon UV illumination by emptying the valence band and filling the conduction band. Adapted with permission from ref. 104. Copyright 2007, AIP Publishing LLC. Adapted with permission from ref. 105. Copyright (2010) American Chemical Society.

11.5 Advancing the Efficiency of Solution Processed Multi-Junction Cells

11.5.1 Polymer Tandem Cells

Following the demonstration of solution processed tandem polymer solar cells in 2007,[22,24] the performance of these devices has developed significantly. In this section, we highlight some of the advances made since then.

In 2009, Yang *et al.* published a 5.8% efficient tandem cell consisting of P3HT:PC$_{71}$BM and PSBTBT:PC$_{71}$BM.[54] The materials used were similar to those used by Heeger's group, but Yang's team reversed the stack design and placed the wide bandgap semiconductor in the front subcell. Yang *et al.* also showed that the use of a highly conductive PEDOT in the intermediate contact can result in a significant overestimation of the PCE because the recombination layer can collect charges from the subcells on either side over a larger area than the nominal area defined by the overlap of the top and bottom electrodes. This implied that previously reported PCEs were possibly overestimated.

Despite the favorable efficiencies, the tandem cells published by the groups of Heeger and Yang did not yet benefit from the principal advantage of a tandem configuration.[24,54] In these tandem cells the V_{OC} of the wide band-gap P3HT:PC$_{71}$BM subcell (~0.63 V) is less than that of the small bandgap PCPDTBT:PC$_{71}$BM or PSBTBT:PC$_{71}$BM subcell (~0.67 V). Hence, thermalization losses are actually increased compared to a single junction of the small bandgap cell. The increased efficiency with respect to a single junction originates from an increased spectral coverage and higher EQE. The almost equal voltages of the subcell materials available at that time prompted the development of parallel connected tandem cells by the groups of Blom,[52] Xie,[53] and Yang,[55] resulting in 4.8% PCE. In a parallel tandem configuration a common middle electrode is used, and rather than adding the voltages of the subcells the photocurrents are added. The design criterion is then to have subcells with matched V_{OC}s.

In 2009, Gilot et al. published a first series connected tandem polymer solar cell that really benefitted from reduced thermalization losses for the higher energy photons employing wide (1.95 eV) and small (1.40 eV) bandgap subcells that provided V_{OC}s of 0.98 and 0.62 V, respectively.[39] Gilot also showed that the PCE of 4.9% for the tandem cell matched the expected performance based on electrical and optical modeling, indicating that the ZnO/PEDOT:PSS recombination contact did not represent a significant loss.

One of the difficulties in fabricating polymer tandem solar cells is the presence of the thin and fragile interconnecting layer. As a consequence, the back must usually be processed from fast-evaporating, low-boiling point solvents such as chloroform to prevent damage to the front subcell. This limits the possibility of using high efficiency blends and their optimal processing conditions, which often include slow drying. To solve this problem, Yang et al. developed a modified formulation of PEDOT:PSS (m-PEDOT) which could be deposited in greater thicknesses without creating Ohmic losses between the subcells. The new interconnecting layer was physically more robust and allowed employment of an efficient small bandgap PSBTBT:PC$_{71}$BM layer, processed from chlorobenzene. Combining this blend with P3HT:ICBA in the front subcell led to a record efficiency of 7.0% for a regular polarity tandem configuration.[64] A similarly high efficiency of 7% was achieved shortly afterwards by Gevaerts et al., who used efficient wide (PCDTBT:PC$_{71}$BM) and small (PDPP5T:PC$_{61}$BM) bandgap polymer–fullerene subcells and a ZnO/PEDOT:PSS recombination layer.[49]

In 2010, Yang et al. published the first inverted configuration tandem polymer solar cell using P3HT:PC$_{61}$BM and PSBTBT:PC$_{71}$BM subcells.[59] As a recombination layer, Yang et al. used thermally evaporated MoO$_3$ and Al, covered with a sol–gel ZnO layer to reach a PCE of 5.1%. Almost at the same time, Jen *et al.* published a solution-processed inverted tandem solar cell consisting of two P3HT:$PC_{61}BM$ subcells.[60] The recombination layer consisted of PEDOT:PSS, deposited using isopropanol (IPA) and butanol (BuOH) in the aqueous dispersion to improve the wetting, and a ZnO layer that was decorated with a C_{60}-SAM (see Figure 11.10) to improve the selectivity of the contact. In 2011, Yang *et al.* showed the beneficial use of plasmonic effects in an inverted P3HT:ICBA/PSDTBT:$PC_{71}BM$ tandem polymer solar cell configuration by blending Au nanoparticles into the interconnecting layer to reach PCE = 6.2%.[61] This cell was also the first tandem polymer solar cell that used P3HT with ICBA rather than with $PC_{61}BM$ or $PC_{71}BM$ to increase the V_{OC} of the front subcell. In 2012, Kouijzer *et al.* published an inverted tandem cell based on wide bandgap P3HT:ICBA and small bandgap PDPP5T:$PC_{61}BM$ subcells with a solution-processed PEDOT:PSS/ZnO interconnection layer with PCE = 5.8%.

Following these promising developments, the inverted tandem cell design attracted attention and gained importance. Two significant developments were published in the course of 2012. The first was the development of a new efficient small bandgap (~1.44 eV) DPP polymer, PBDTT-DPP (see Figure 11.9), and its use as back subcell in an inverted tandem cell together with a P3HT:ICBA small bandgap cell to increase the PCE of polymer tandem solar cells to 8.6%.[66] The second was the introduction of ethoxylated polyethylenimine (PEIE) in an efficient inverted tandem cell, eliminating the need to use a metal oxide (Figure 11.12). PEIE lowers the work function of the underlying layer. In the intermediate contact the work function of the PEDOT:PSS is reduced from 4.9 eV to 3.6 eV, allowing an Ohmic contact for electron collection.[67] The result was a 8.2% tandem solar cell using a P3HT:ICBA front subcell, a small bandgap PBDTTT-C:$PC_{61}BM$ back subcell and a novel recombination contact material.

The 8.6% PCE achieved by Yang *et al.* showed that designing materials tailored to operate in tandem cells can lead to considerable improvements.

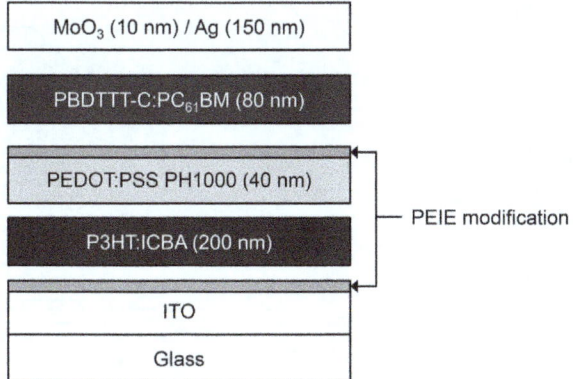

Figure 11.12 Device structure of the polymer tandem solar cells with PEDOT:PSS (PH1000) modified by PEIE as the recombination layer. Reproduced from ref. 67.

Along this line, Yang's group developed a selenium DPP type polymer with a ~1.38 eV bandgap that increased the PCE in an inverted tandem cell to 9.5%.[72] Next, in a collaboration with Sumitomo, they pushed the PCE beyond the 10% limit with a 10.6% cell.[73] This outstanding result was obtained by designing a new small bandgap polymer, PDTP-DFBT (see Figure 11.9), that absorbs light up to 900 nm and gave PCE = 7.9% in single junction solar cells. The authors also compared two different tandem solar cells where the small bandgap polymer in the back subcell is combined with $PC_{61}BM$ or with $PC_{71}BM$, affording PCEs of 10.6% and 10.2%, respectively. The difference is due to the absorption of high energy photons by the back subcell, in which the $PC_{71}BM$ "steals" photons from the front subcell, resulting in a lower EQE contribution from the latter. This demonstrates the importance of subcell matching for highly efficient tandem solar cells.

Further, in 2013, a new small bandgap (1.30 eV) DPP polymer, PMDPP3T (see Figure 11.9), was designed by Li *et al.*, starting from the known PDPP3T and introducing two methyl groups on the thiophene to tune the energy levels carefully for improved charge generation.[75] The resulting single junction solar cells reached 7.0% efficiency with absorption in the near infrared up to 960 nm. The use of this material in a regular tandem cell yielded devices with PCE up to 8.9%. The improvement of the photocurrent of the low bandgap cell, by pushing the bandgap to lower values, meant that the current generation in the two subcells was not matched and the wide bandgap material ($PCDTBT:C_{71}BM$) limited the overall efficiency. This limitation was lifted by using an additional small bandgap cell on top of the existing structure. This 1 + 2 triple junction structure has the advantage of using the excess charge generation by the small bandgap polymer to create an extra voltage, added in the series connection (Figure 11.13). With optimized layer thicknesses, the 1 + 2 triple junction reached a PCE of 9.6%.[75]

Several groups have reported new and specifically designed materials for tandem cells. In this respect it is not only of interest to design small bandgap

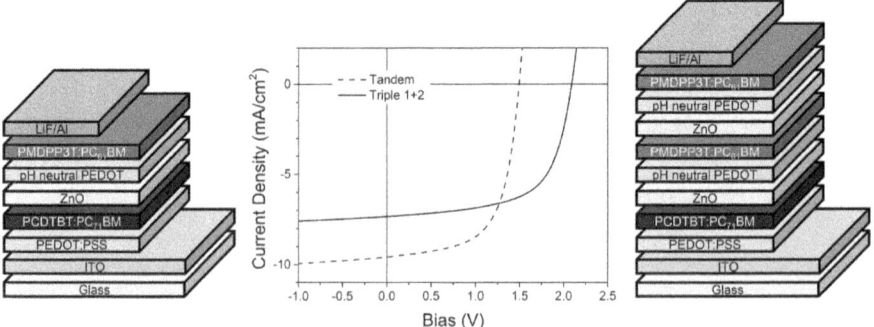

Figure 11.13 Device configuration of the tandem cell and the corresponding 1 + 2 triple junction cell that reached PCEs of 8.9% and 8.6%, respectively. Adapted with permission from ref. 75. Copyright (2013) American Chemical Society.

polymers for the back subcell, but, as was shown by Li *et al.*,[75] the wide bandgap polymers used in the front subcell also need improvement. Along this line, Hwang *et al.* introduced a new wide bandgap polymer (PTIPDBDT-DFDTQX, see Figure 11.8) based on (triisopropylsilyl)ethynyl (TIPS) substituted benzodithiophene (BDT) and a fluorinated quinoxaline. The new polymer had a bandgap of about 1.85 eV and gave PCEs up to 6.1% in single junction solar cells.[81] Combining it with a small bandgap polymer and using the novel PEDOT:PSS/ PEIE recombination layer developed by Kippelen *et al.*,[67] they obtained a PCE of 7.2% in an inverted tandem solar cell.[81] In a related approach, Peng *et al.* developed a new polymer consisting of an electron rich dialkylthiol-substituted BDT and an electron deficient monofluorinated benzotriazole (FBZ).[80] The resulting PBDTFBZS polymer (see Figure 11.8) has an optical bandgap of around 1.8 eV and a deep HOMO level that resulted in high V_{OC}s, up to 0.9 V, providing a PCE of 7.7% in single junction cells. This represents a significant improvement compared to existing wide bandgap material combinations such as P3HT:ICBA, and in combination with a complementary DPP-based low bandgap material a tandem solar cell of 9.4% was obtained.[80]

Also in 2013, new intermediate contact layers were implemented in tandem solar cells. Heeger *et al.* showed the relevance of interfacial engineering by introducing a conjugated polyelectrolyte (CPE) on top of a PEDOT:PSS/ ZnO recombination layer to decrease the work function and thereby improve the electron collection.[76] This effect has been observed in both single junction and tandem solar cells, leading to a PCE improvement from 7.2% for the tandem cell without CPE to 8.6% with CPE. Interesting progress was also reported by Jang *et al.*, who introduced a new recombination layer consisting of sol–gel TiO_x (40 nm) and graphene oxide (GO) (10 nm).[79]

In 2014, the research on polymer tandem solar cells gained considerable momentum and several groups have reported very high efficiencies. Tandem cells with PCEs ranging from 3.9% to 8.9% were obtained in regular and inverted configurations using interconnection layers comprising GO/ GO-Cs/Al/MoO_3,[82] PEDOT:PSS/PEI,[83] MoO_3/Al-doped MoO_3,[84] PEDOT:PSS/ ZnO,[85,89,93] PEDOT:PSS/PEIE,[86,88] PEDOT:PSS/ZnO/C_{60}-SAM,[87] PEDOT:PSS/ Li-ZnO,[90] ZnO/Ag/MoO_3,[91] PEDOT:PSS/Ag nanowire/ZnO[92] and a variety of wide and small bandgap materials. The large variety in recombination layers and their processing details, which provide such high PCEs in tandem cells, demonstrates that there are many options for creating an efficient interconnection layer but at the same hints at the challenge that a robust and universally applicable technique has not yet been established.

By the end of 2014, a number of polymer tandem cells reached PCEs very close to or even exceeding 10%. Among these are a series of tandem cells developed by Yang *et al.* as reference cells for a record high triple junction cell of 11.6% that will be discussed in detail in Section 11.5.3.[95] The reference tandem cells use wide (1.90 eV), medium (1.58 eV), and small (1.40 eV) bandgap photoactive layers in each of the three possible mixed configurations to give 9.6%, 10.7%, and 9.8% inverted tandem cells (see Table 11.1).[95] The novelty is the use of WO_3 nanoparticles as an interlayer between the photoactive layer and the PEDOT:PSS/ZnO recombination contact. Jang *et al.* published a 10.4% tandem cell that used a recombination contact of PEDOT:PSS and Li-doped ZnO covered with C_{60}-SAM. Hou *et al.* used a novel wide bandgap polythiophene (PCDBT, Figure 11.8) with a deep HOMO level and small bandgap BDT-TT derivative to obtain PCE = 10.2%.[97] Finally, in a very interesting publication, Lee *et al.* showed a 10.8% efficient homo-tandem cell in which both photoactive layers contain a PTB7-Th:PC$_{71}$BM bulk heterojunction and a recombination layer of PEDOT:PSS/PEI.[98] They used an intriguing one-step coating procedure in which the bulk heterojunction and the PEI of the recombination layer were simultaneously deposited onto the ITO bottom contact and the intermediate PEDOT:PSS layer. The self-organized recombination layer formed *via* a spontaneous vertical phase separation as a result of the different surface energies.[98] The principle is shown in Figure 11.14.

11.5.2 Small Molecule Tandem Cells

In addition to solution-processed polymer solar cells, solution-processed small molecule cells have made important developments in the last few years. In fact the best solution-processed small molecule cells have reached similar performance levels to their macromolecular counterparts. This has also resulted in attempts to develop solution-processed small molecule organic solar cells. Initially, hybrid approaches were chosen in which one part of the cell is deposited from solution and the other *via* evaporation. Forrest *et al.* describe a solution-processed and solvent-vapor-annealed small bandgap squaraine dye in a bilayer front subcell with evaporated C_{70} and a wide bandgap graded subphthalocyanine (SubPc):C_{70} back subcell, to reach a tandem cell with 6.6% efficiency.[106] In a subsequent paper, Forrest *et al.* use a solution-processed functionalized squaraine dye in both subcells, in combination with evaporated C_{60} and charge transport layers, to reach a 6.2% efficient tandem cell.[107] A related approach has been described by Palomares *et al.*, who described a 2.2% efficient tandem cell comprising a solution-processed front subcell consisting of a bulk heterojunction of DPP(TBFu)$_2$:PC$_{71}$BM and an evaporated bilayer back subcell that employs a squaraine donor and C_{60} acceptor.[108]

A fully solution processed and highly efficient small molecule tandem cell has been reported by Yang *et al.* They used a π-conjugated oligothiophene derivative (SMPV1, Figure 11.15). Single junction solar cells based on SMPV1:PC$_{71}$BM reach PCE = 8.0%. A homo-tandem solar cell based on SMPV1 was constructed with a recombination layer that comprised a bilayer of a conjugated polyelectrolyte, demonstrating PCE = 10.1%.[109]

11.5.3 Polymer Multi-Junction Cells

A further increase in the efficiency of polymer solar cells can be achieved by extending the multi-junction configuration from two to three or more photoactive layers. Multi-junction cells with more than two absorber layers are readily obtained using thermal evaporation of photoactive and

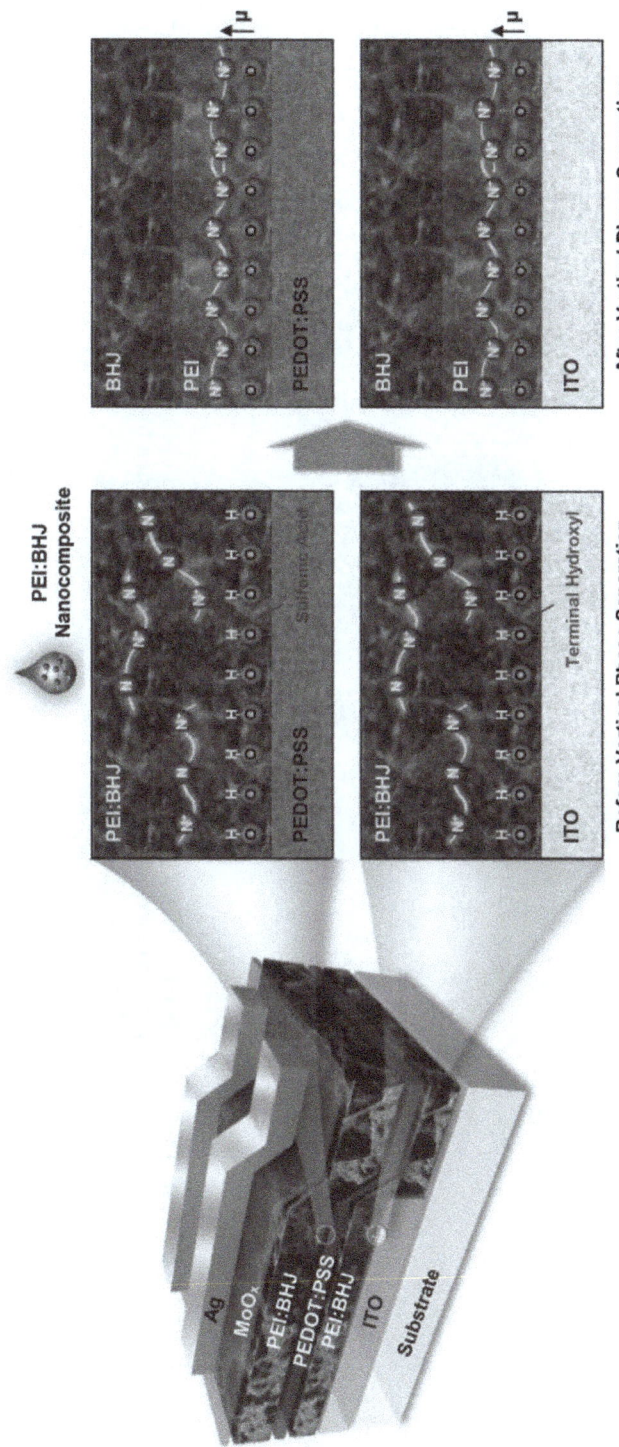

Figure 11.14 Tandem structure created by self organization of the PEI:bulk heterojunction on PEDOT:PSS and ITO surfaces. Reproduced from ref. 98 with permission from Wiley.

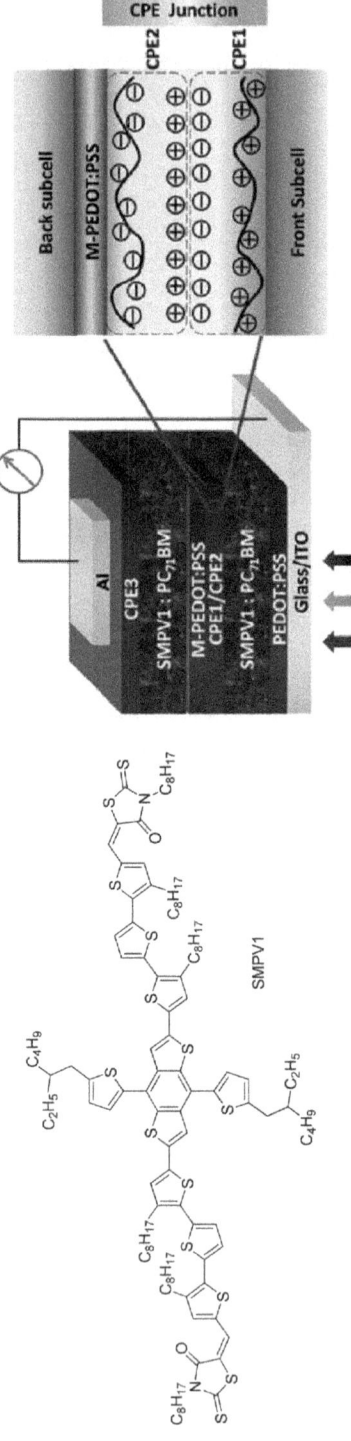

Figure 11.15 Chemical structure, tandem cell device structure and proposed principle of the self-assembled bilayer recombination junction built from two conjugated polyelectrolytes (CPEs). Reprinted by permission from Macmillan Publishers Ltd: Scientific Reports, ref. 109, copyright (2013).

interconnection layers. In 2002, Forrest *et al.* had already shown that three- and five-fold bilayer stacks of CuPc and PCTBI could reach PCEs of 2.3% and 1.0%, respectively.[16] Adachi *et al.* reported on a series of multi-junction cells with chloroaluminium phthalocyanine (ClAlPc) as donor and C_{60} as acceptor, resulting in V_{OC}s of 3.50 V for five-fold and 5.89 V for 10-fold junctions.[110] Using evaporated small molecules in planar-mixed heterojunction photoactive layers, Forrest *et al.* demonstrated in 2014 tandem and triple junction cells with 10.0% and 11.1% efficiency, respectively.[111] The configuration of the triple junction consists of front and back green-absorbing cells and a complementary NIR-absorbing middle cell.[111] Also, the highest efficiencies, up to 12%, reported for evaporated cells comprise triple junctions in 1 + 2 configurations, as explained in Section 11.1.2.[15]

Polymer multi-junction cells have also achieved noticeable attention. In Table 11.2 we provide an overview of the multi-junction cells that had been published up to the end of 2014. In 2007, Gilot *et al.* showed that it is possible fully to solution process three photoactive layers and two recombination layers based on ZnO/PEDO:PSS and obtain a working triple junction solar cell with V_{OC} = 2.19 V.[22] By the same procedure, six-fold junction cells with

Table 11.2   Solution processed multi-junction polymer solar cells.

Cell configuration[a]	E_g^b (eV)	V_{OC} (V)	Type[c]	PCE (%)	Recombination layer front/back	Ref.
MDMO-PPV:PC$_{61}$BM (1,2)	2.10	2.19	3-fold R	2.1	ZnO/n-PEDOT:PSS	22
P3HT:PC$_{61}$BM (3)	1.90					
MDMO-PPV:PC$_{61}$BM (1–5)	2.10	3.58	6-fold R	2.1	ZnO/n-PEDOT:PSS	23
P3HT:PC$_{61}$BM (6)	1.90					
P3HT:PC$_{61}$BM (1–3)	1.90	1.73	3-fold R	2.0	Al/MoO$_3$	112
P3HT:PC$_{61}$BM (1–3)	1.90	1.66	3-fold I	2.3	PEDOT:PSS-(IPA:BuOH)/ZnO np/C$_{60}$-SAM	60
PF10TBT:PC$_{61}$BM (1)	1.95	2.33	3-fold R	5.3	ZnO/n-PEDOT:PSS/	74
PDPPTBT:PC$_{61}$BM (2,3)	1.53				Nafion	
PCDTBT:PC$_{71}$BM (1)	1.88	2.09	3-fold R	9.6	ZnO/n-PEDOT:PSS	75
PMDPP3T:PC$_{61}$BM (2,3)	1.20					
PDPP5T-2:PC$_{71}$BM (1–3)	1.46	1.63	3-fold R	3.9	ZnO/n-PEDOT:PSS	78
PDPP5T-2:PC$_{71}$BM (1–4)	1.46	2.03	4-fold R	3.0	ZnO/n-PEDOT:PSS	78
P3HT:ICBA (1)	1.90	0.84	3-fold I	11.6	PEDOT:PSS/ZnO	95
PTB7-Th:PC$_{71}$BM (2)	1.58	0.78			WO$_3$/PEDOT:PSS/	
PDTP-DFBT:PC$_{71}$BM (3)	1.40	0.70			ZnO	
PSEHTT:ICBA (1)	1.82	0.94	3-fold I	11.8	pH neutral	96
PTB7:PC$_{71}$BM (2)	1.60	0.76			PEDOT:PSS/	
PMDPP3T:PC$_{71}$BM (3)	1.30	0.60			Li-ZnO/C$_{60}$-SAM	

[a]The numbers between parentheses refer to the subcell number in the stack, with (1) being the front subcell.
[b]E_g represents the optical bandgap of the polymer; note that of the fullerene may be less.
[c]R = regular configuration, I = inverted configuration.

V_{OC} = 3.58 V were also obtained.[23] In these cells, all subcells contained an MDMO-PPV:PC$_{61}$BM blend except for the last subcell, which was based on a P3HT:PC$_{61}$BM bulk heterojunction. Both three- and six-fold junction cells gave PCEs of about 2.1%. In 2009, Kwong *et al.* published a series-connected triple junction cell consisting of three P3HT:PC$_{61}$BM layers fabricated using transparent vacuum processed Al(1 nm)/MoO$_3$(30 nm) recombination layers, giving V_{OC} = 1.73 V and PCE = 2.0%.[112] The first inverted triple junction cell was reported by Jen *et al.* in 2010.[60] They also used three P3HT:PC$_{61}$BM layers but reversed the polarity of the device and introduced a solution-processed PEDOT:PSS/ZnO/C$_{60}$-SAM recombination layer. The V_{OC} = 1.66 V and PCE = 2.3% were in the same range as for the regular cells reported by Kwong *et al.*[112] Although each of these results clearly demonstrated the viability of solution-processed multi-junction cells, their efficiencies were not equivalent to the developments in the field, and solution-processed multi-junctions were considered as a curiosity rather than a clear path to reaching high efficiencies.

In 2013, Esiner *et al.* developed a more efficient triple-junction solar cell. By using wide bandgap (1.95 eV) and small bandgap (1.53 eV) polymers with deep HOMO levels, Esiner *et al.* achieved a triple junction polymer tandem cell in a 1 + 2 configuration with a relatively high V_{OC} of 2.33 V.[74] To reach the high V_{OC} it was necessary to treat the ZnO/n-PEDOT layer with an additional Nafion layer to increase the work function.[57,74] The PCE of 5.3% attained was still moderate, but the high V_{OC} enabled the use of this cell for direct photoelectrochemical water splitting with two Pt electrodes operating close to the maximum power point of the triple-junction, giving a solar-to-hydrogen energy conversion efficiency of about 3.1%. Later in 2013, Li *et al.* demonstrated the first efficient triple junction cell with PCE = 9.6%, using a 1 + 2 configuration, as shown in Figure 11.13.[75] Triple and quadruple homo multi-junction cells were reported by Li *et al.*[78] In these cells the same photoactive layer, based on a small bandgap PDPP5T-2:PC$_{71}$BM blend, was incorporated in each subcell and a solution-processed ZnO/pH neutral PEDOT:PSS interconnection layer was used.

In 2014, Yang *et al.* demonstrated the first efficient triple-junction polymer solar cells that incorporated three subcells with different bandgaps (Figure 11.16).[95] The materials used are P3HT:ICBA (1.90 eV), PTB7:PC$_{71}$BM (1.58 eV) and PDTP-DFBT:PC$_{71}$BM (1.40 eV), which, when combined, absorb light from 300 to 900 nm and show a PCE of 11.6%. The cell was optimized using optical modeling. Importantly, this triple-junction solar cell outperformed any of the single or tandem junctions that can be made with the same materials. This triple cell not only represents three different bandgap subcells, but these also show the required V_{OC}s, such that higher energy photons are collected that lose less energy in thermalization. The same approach for a triple-junction cell was shown by Jang *et al.* using PSEHTT:ICBA (1.90 eV), PTB7:PC$_{71}$BM (1.58 eV) and PMDPP3T:PC$_{71}$BM (1.30 eV).[96] With these subcells the spectral coverage was further extended to ~930 nm and Jang *et al.* reached an efficiency of 11.8% which—at the end of 2014—represented the highest reported efficiency for a solution-processed organic solar cell.

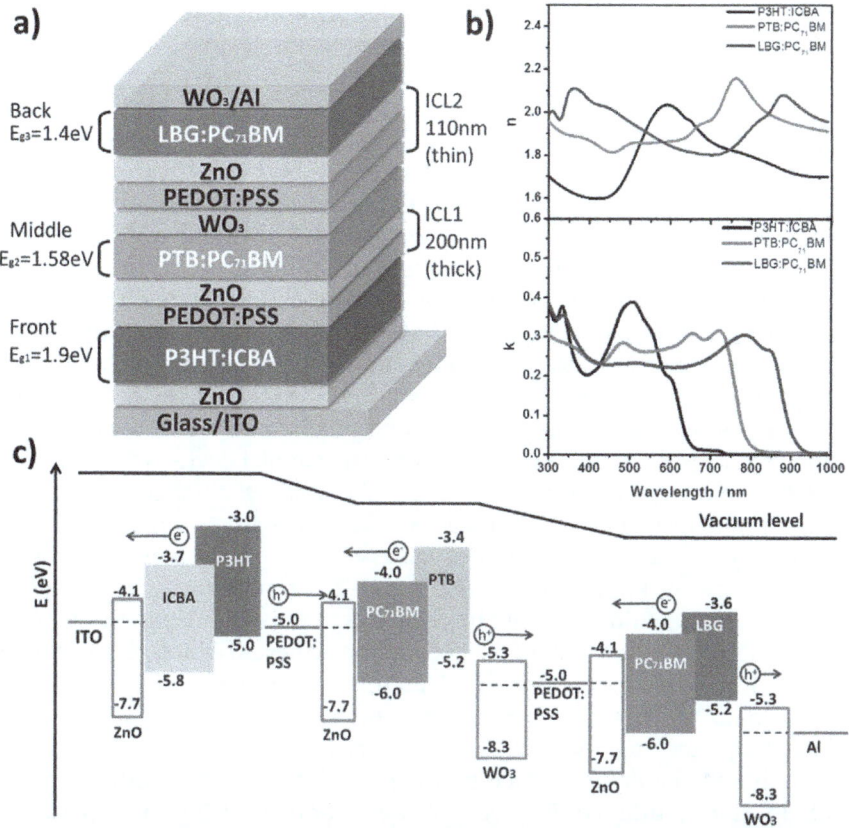

Figure 11.16 (a) Device configuration of triple-junction solar cell. (b) Optical constants of the photoactive layers. (c) Energy levels of the materials used. Reproduced from ref. 95 with permission from Wiley.

These advances in triple-junction polymer solar cells show the potential for this type of device. Considering that in these triple junctions the photon energy loss (E_{loss}) in each subcell is at least 0.7 eV, and sometimes even above 1.0 eV (Table 11.2), it can be expected that by developing new materials in which E_{loss} is more tailored to 0.6 eV it will be possible to increase the efficiency of triple-junction solar cells considerably. In our view efficiencies in the 15–20% range are within reach.

11.6 Special Device Configurations

The improvements described above have been obtained *via* the design and development of new semiconducting polymers, with wide, medium, and small bandgaps, and by the improvement of the interconnecting layers in regular or inverted series-connected multi-junction cells. In addition to the series-connected subcells in a monolithic stack configuration, several alternative device architectures are possible.

The first alternative to series-connected cells is the parallel three-terminal cell configuration. Parallel configuration polymer tandem solar cells were first described by Blom et al., who demonstrated the use of a tailored optical spacer that allows independent optimization of both the electronic and the optical properties of the tandem cell.[52] Other examples of three-terminal tandem cells have been demonstrated by Wang et al.[53] and Yang et al.,[55] who used a common middle Au anode to extract the current. The merits of parallel- vs. series-connected polymer tandem solar cells have been studied theoretically and experimentally by Pacios et al.[113] Modeling revealed maximum PCEs of ~15% for series and ~13% for parallel stack configurations with optimized bandgaps for the front and back subcells.[113] Pacios et al. argued that, owing to resistive losses at the interlayer and the amount of current that the device has to manage, the cell width must be lower for the parallel device, which causes a reduction in the geometrical fill factor of solar modules.[113] High-performance parallel tandem solar cells comprising a semi-transparent front subcell and a microcavity assisted top-illuminated back subcell were demonstrated by Jen et al.[94] The device architecture takes advantage of microcavity light-trapping effects by using an ultra-thin Ag film as an intermediate transparent electrode. As a result, a very high PCE of 9.2% could be achieved, which represents the best result reported for organic parallel tandem solar cell.

Both parallel- and series-connected multi-junction cells can be constructed by stacking individual semi-transparent solar cells.[19,114–116] This approach was first demonstrated by Yang et al.[19] More recently, Inganäs et al. used an amphiphilic conjugated polymer modified ITO coated glass substrate as the Ohmic electron-collecting cathode and PEDOT:PSS (PH1000) as the hole-collecting anode to make semitransparent cells.[114] Using a wide bandgap (TQ1:PC$_{71}$BM) front cell and a small bandgap (P3TI:PC$_{71}$BM) back subcell in combination with reflective mirror at the back, PCEs of 5.3% and 5.6% were obtained in parallel- and series-connected configurations, respectively (the structures are shown in Figures 11.8 and 11.9).[114] They further predicted that by stacking five separately prepared semi-transparent cells on top of each other it would be possible to obtain a higher photocurrent than in an optimized standard solar cell. In a similar approach Aernouts et al. show a four-terminal configuration of two single junction modules with complementary absorbing active layers.[115] Optical simulations of the new device structure, which has an experimental PCE of 6.5%, show that its energy output can be higher that the complementary standard tandem cell by more than 32% over a year considering central European solar irradiance conditions because of the independence on balancing the currents or voltages of tandem subcells.[115] In a related, intriguing cell configuration, Inganäs et al. use independent subcells in a V-shaped geometry such that photons that are not absorbed are reflected towards the other subcell (Figure 11.17a).[117] The cells can have complementary absorption spectra on both sides. The V-shaped configuration offers benefits in terms of enhanced absorption via multiple reflections.

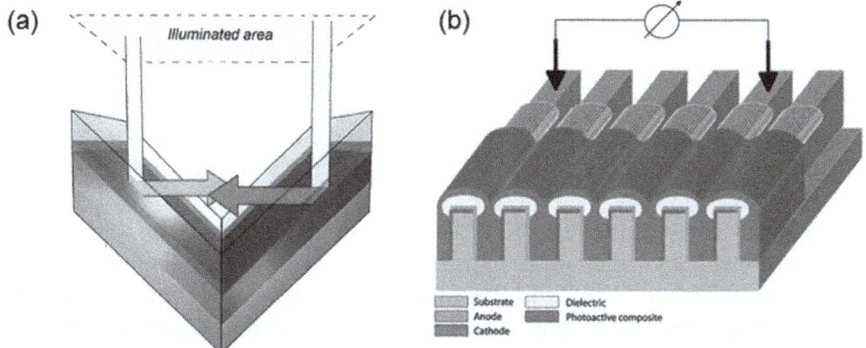

Figure 11.17 Schematic layouts of: (a) V-shaped reflective tandem cell where single cells are reflecting the nonabsorbed light on the adjacent cell, and (b) series-connected photovoltaic nanomodule. Reprinted with permission from ref. 117, copyright 2007, AIP Publishing LLC and from ref. 118 with permission from Wiley.

Niggemann *et al.* presented a novel architecture for a photovoltaic nano-module with a series interconnection of organic photovoltaic cells on the scale of several hundred nanometers, resulting in 1390 interconnected elementary cells per millimeter (Figure 11.17b).[118] Under solar illumination, a 7.9 mm long photovoltaic nanomodule generates a V_{OC} of 880 V and PCE of 0.008%, based on a P3HT:PC$_{61}$BM active layer. In a related approach, Malliaras *et al.* interconnected 300 solar cells in series, with a period of 50 mm, and achieved a V_{OC} of 90 V and PCE of 0.3%.[119] More recently, Lee *et al.* developed a miniature high-voltage source from polymer solar cells with charge-transporting molybdenum oxide (MoO$_x$) integrated in a serial architecture through sacrificial layer assisted patterning, achieving V_{OC} of 24 V and PCE of 0.5% for 50 interconnected cells.[120]

11.7 Processing Issues for Multi-Junction Polymer Solar Cells

11.7.1 Laboratory Scale Devices

Although multi-junction polymer solar cells are a valid strategy for reaching high PCEs, the increased number of layers that have to be processed on top of each other increases the complexity of the fabrication process. The first important factor is solvent orthogonality. Each of the layers has to be processed from a solution that will not damage or wash away any of the previously deposited layers. For processing on top of a bulk heterojunction polymer:fullerene, only a few solvents can be used in practice: water, acetone, and alcohols. Typically, the photoactive layers are hydrophobic and this represents an obstacle to the deposition of materials from aqueous solutions because of poor wettability. Therefore, extra measures have to be taken such as modifying the formulation of the solution or treating the surface of the photoactive layer. Another issue is related to the choice of the layers that will be stacked in a multi-junction cell. Materials that need post-processing such as thermal or solvent annealing can only be processed on top of layers that can withstand the same treatment. A typical example is the use of modified PEDOT:PSS in inverted solar cells. This has been shown to work well on top of P3HT:ICBA because both materials need an annealing step at 150 °C,[66] but using it on a temperature sensitive active layer would lead to the degradation of the latter during the annealing process. As mentioned in Section 11.4, the same considerations also limit the use of sol–gel processed metal oxides for intermediate contacts on active layer blends that degrade upon annealing.

As discussed in Section 11.4.3, the use of metal oxides such as TiO_x and ZnO as electron transport layers in polymer solar cells is often accompanied by the need for UV light to create an Ohmic contact.[22,54,58,68,70] The UV light leads to persistent photodoping of the metal oxides and increases the charge carrier density, enabling creation of Ohmic contacts (see Figure 11.11). Of course, the dependence on UV light is a drawback in practice and therefore recombination layers that do not need UV doping, such as those based on a PEDOT:PSS interface with PEIE[67] or PEI,[98] are important.

11.7.2   Large Area and Printed Multi-Junction Cells

Until recently, studies on mass-produced organic photovoltaic devices lagged far behind those on lab-scale devices. This prompted Brabec *et al.* to investigate six different commercially available and some proprietary conjugated polymers and systematically study their potential in organic tandem solar cells. PCEs of 3.4–6.6% were obtained for tandem devices processed under ambient conditions using doctor-blading (knife coating).[121] Subsequently, Brabec *et al.* revealed the feasibility of achieving high efficiencies for tandem polymer solar cells with an inverted architecture using doctor-blading (knife coating) at temperatures ≤80 °C either on glass or polyethylene terephthalate (PET) substrates.[122] Using a proprietary wide bandgap polymer with $PC_{61}BM$ in the front subcell and a small bandgap PDPP5T-2:$PC_{61}BM$ as the back subcell, they achieved PCEs of 7.7% on glass and 5.4% on PET.[122] The key achievement in these contributions lies in the fact that knife coating shows compatibility with roll-to-roll printing.[121,122]

The first tandem cell modules were published by Brabec *et al.*[123] Solar cells based on ternary blends of P3HT, PSDTBT, and $PC_{61}BM$ were used in a homo tandem configuration to give a PCE of 4.7% with V_{OC} = 1.06 V. The inverted device used a recombination layer consisting of ZnO and PEDOT:PSS (Heraeus HIL3.3). By using laser patterning, involving three laser-patterned lines, three-cell and 10-cell modules with fully additive V_{OC}s of 3.10 V and 10.33 V were obtained. The PCEs dropped somewhat to 4.2% and 3.3% respectively.[123] In a next step, Krebs *et al.* demonstrated that it is possible to fabricate flexible polymer tandem solar cell modules having an active area of above 50 cm^2 completely by sequential wet processing on a roll-to-roll line under ambient conditions.[124] The final module comprises 14 stacked layers deposited with different techniques such as flexographic printing, rotary screen printing, and slot–die coating to accommodate the specific requirements of the different inks (Figure 11.18). For 25 m of in-line processed foil, comprising 500 modules and ~4000 tandem cells, virtually all modules were found to be functional.[124] The cell comprises wide bandgap ($MH301:PC_{61}BM$) and small bandgap ($MH306:PC_{61}BM$) photoactive layers. The PCE for an eight-cell module is about 1.8%.

This effort demonstrates the feasibility of upscaling these complex devices, but the Krebs *et al.* stress that more resources should be guided toward the development of inks specifically formulated to facilitate the roll-to-roll processing.[124] These inks should be formulated by taking into account, on top of electrical and optical performance, viscosity, shelf life, thermal expansion coefficients, high elasticity and surface adhesion of all the layers. Furthermore, efficient sequential processing requires that each layer be dried in an oven before the next one is deposited and this happens commonly around 80 to 100 °C, which limits the choice of materials to use.

In a comprehensive study, Brabec and Krebs *et al.* have addressed a cost analysis for single junction and tandem polymer solar cells based on

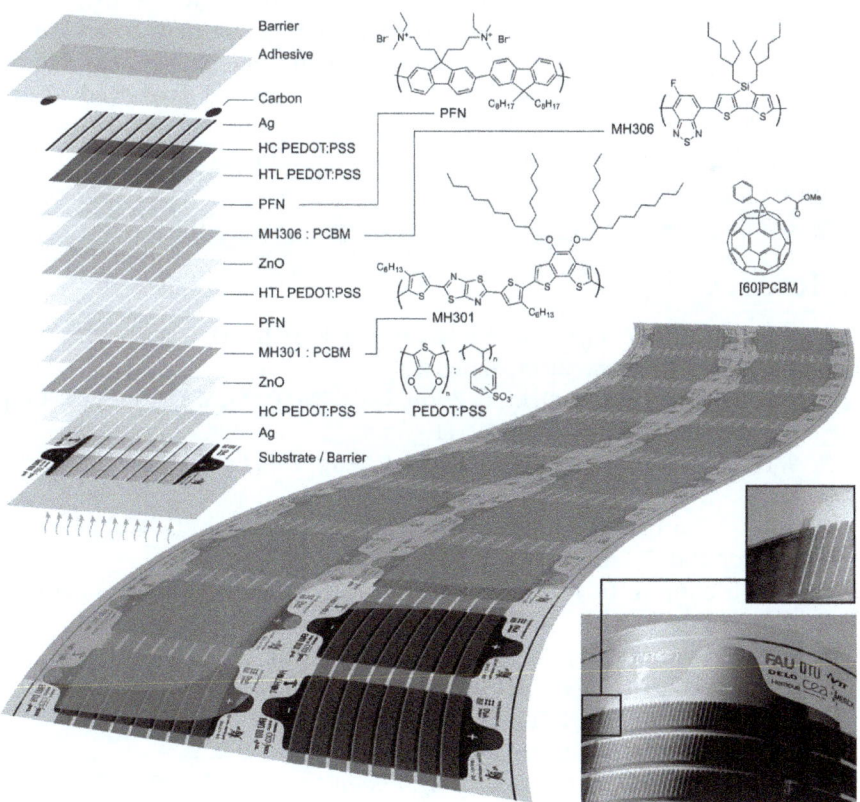

Figure 11.18 Complete 14-layer printed tandem stack developed by Krebs *et al.* Reproduced from ref. 124.

manufacturing data in large roll-to-roll coating pilot experiments.[125] They compare different models designated as current status, upscaling, and industrial scenarios. Within the current status scenario they further consider low and high cost models that differ with regard to the generation of absorber materials used. In the current status (kW regime), the module costs are estimated to be 34.56 € m^{-2} in the low cost model and 61.71 € m^{-2} in the high cost model. They project, however, that upon upscaling (MW regime) and reaching the industrial production regime (GW regime), these costs can be reduced considerably to 12.81 € m^{-2} and 6.79 € m^{-2}, respectively.[125]

11.8   Conclusions

Multi-junction polymer cells have evolved in the past 10 years from a scientific curiosity into a potential new technology for solar energy conversion. With PCEs approaching 12%, triple junction polymer solar cells outperform the best single junction and tandem polymer cells, albeit by a small difference. Future efficiencies in the range of 15–20% can be expected. In addition, first steps towards large area devices made *via* printing have been successful. These important accomplishments show the viability of the multi-junction approach, but formidable challenges remain to creating a commercially attractive technology. More efficient polymer semiconductors tailored to wide, medium, and small optical bandgaps, improved recombination layers, novel device designs, better encapsulation and barrier layers, and advances in roll-to-roll processing technologies will be required to achieve this goal.

References

1. Y. Liu, J. Zhao, Z. Li, C. Mu, W. Ma, H. Hu, K. Jiang, H. Lin, H. Ade and H. Yan, *Nat. Commun.*, 2014, **5**, 5293.
2. J.-D. Chen, C. Cui, Y.-Q. Li, L. Zhou, Q.-D. Ou, C. Li, Y. Li. and J.-X. Tang, *Adv. Mater.*, 2015, **27**, 1035–1041.
3. S.-H. Liao, H.-J. Jhuo, P.-N. Yeh, Y.-S. Cheng, Y.-L. Li, Y.-H. Lee, S. Sharma and S.-A. Chen, *Sci. Rep.*, 2014, **4**, 6813.
4. M. C. Scharber, D. Mühlbacher, M. Koppe, P. Denk, C. Waldauf, A. J. Heeger and C. J. Brabec, *Adv. Mater.*, 2006, **18**, 789.
5. D. Veldman, S. C. J. Meskers and R. A. J. Janssen, *Adv. Funct. Mater.*, 2009, **19**, 1939.
6. R. A. J. Janssen and J. Nelson, *Adv. Mater.*, 2013, **25**, 1847.
7. P. K. Nayak, G. Garcia-Belmonte, A. Kahn, J. Bisquert and D. Cahen, *Energy Environ. Sci.*, 2012, **5**, 6022.
8. P. K. Nayak and D. Cahen, *Adv. Mater.*, 2014, **26**, 1622.
9. A. S. Brown and M. A. Green, *Prog. Photovoltaics*, 2002, **10**, 299.
10. M. Hiramoto, M. Suezaki and M. Yokoyama, *Chem. Lett.*, 1990, **19**, 327.
11. A. Yakimov and S. R. Forrest, *Appl. Phys. Lett.*, 2002, **80**, 1667.

12. K. Triyana, T. Yasuda, K. Fujita and T. Tsutsui, *Jpn. J. Appl. Phys.*, 2004, **43**, 2352.

13. J. Drechsel, B. Männig, F. Kozlowski, D. Gebeyehu, A. Werner, M. Koch, K. Leo and M. Pfeiffer, *Thin Solid Films*, 2004, **451–452**, 515.

14. B. Maennig, J. Drechsel, D. Gebeyehu, P. Simon, F. Kozlowski, A. Werner, F. Li, S. Grundmann, S. Sonntag, M. Koch, K. Leo, M. Pfeiffer, H. Hoppe, D. Meissner, N. S. Sariciftci, I. Riedel, V. Dyakonov and J. Parisi, *Appl. Phys. A*, 2004, **79**, 1.

15. Heliatek, http://www.heliatek.com (accessed March 2014).

16. J. Xue, S. Uchida, B. P. Rand and S. R. Forrest, *Appl. Phys. Lett.*, 2004, **85**, 5757.

17. G. Dennler, H.-J. Prall, R. Koeppe, M. Egginger, R. Autengruber and N. S. Sariciftci, *Appl. Phys. Lett.*, 2006, **89**, 073502.

18. A. Colsmann, J. Junge, C. Kayser and U. Lemmer, *Appl. Phys. Lett.*, 2006, **89**, 203506.

19. V. Shrotriya, E. H.-E. Wu, G. Li, Y. Yao and Y. Yang, *Appl. Phys. Lett.*, 2006, **88**, 064104.

20. K. Kawano, N. Ito, T. Nishimori and J. Sakai, *Appl. Phys. Lett.*, 2006, **88**, 073514.

21. A. Hadipour, B. de Boer, J. Wildeman, F. B. Kooistra, J. C. Hummelen, M. G. R. Turbiez, M. M. Wienk, R. A. J. Janssen and P. W. M. Blom, *Adv. Funct. Mater.*, 2006, **16**, 1897.

22. J. Gilot, M. M. Wienk and R. A. J. Janssen, *Appl. Phys. Lett.*, 2007, **90**, 143512.

23. J. Gilot, Polymer tandem solar cells, PhD thesis, Eindhoven University of Technology, 2010, ISBN: 978-90-386-2279-8.

24. J. Y. Kim, K. Lee, N. E. Coates, D. Moses, T.-Q. Nguyen, M. Dante and A. J. Heeger, *Science*, 2007, **317**, 222.

25. A. Hadipour, B. de Boer and P. W. M. Blom, *Adv. Funct. Mater.*, 2008, **18**, 169.

26. T. Ameri, G. Dennler, C. Lungenschmied and C. J. Brabec, *Energy Environ. Sci.*, 2009, **2**, 347.

27. M. K. Siddiki, J. Li, D. Galipeau and Q. Qiao, *Energy Environ. Sci.*, 2010, **3**, 867.

28. S. Sista, Z. Hong, L.-M. Chen and Y. Yang, *Energy Environ. Sci.*, 2011, **4**, 1606.

29. Y. Yuan, J. Huang and G. Li, *Green*, 2011, **1**, 65–80.

30. J. You, L. Dou, Z. Hong, G. Li and Y. Yang, *Prog. Polym. Sci.*, 2013, **38**, 1909.

31. T. Ameri, N. Li and C. J. Brabec, *Energy Environ. Sci.*, 2013, **6**, 2390.

32. O. Adebanjo, P. P. Maharjan, P. Adhikary, M. Wang, S. Yang and Q. Qiao, *Energy Environ. Sci.*, 2013, **6**, 3150.

33. O. Adebanjo, B. Vaagensmith and Q. Qiao, *J. Mater. Chem. A*, 2014, **2**, 10331.

34. J. Gilot and R. A. J. Janssen, Tandem and multi-junction organic solar cells, in *Organic Solar Cells: Fundamentals, Devices, and Upscaling*, ed. B. P. Rand and H. Richter, Pan Stanford Publishing Pte. Ltd, 2014.

35. G. Dennler, K. Forberich, T. Ameri, C. Waldauf, P. Denk, C. J. Brabec, K. Hingerl and A. J. Heeger, *J. Appl. Phys.*, 2007, **102**, 123109.
36. A. Hadipour, B. de Boer and P. W. M. Blom, *Org. Electron.*, 2008, **9**, 617.
37. B. E. Lassiter, C. K. Renshaw and S. R. Forrest, *J. Appl. Phys.*, 2013, **113**, 214505.
38. Y. Long, L. Shen, S. Ruan, W. Yu, Y. Wang, Q. Zeng and J. Luo, *Appl. Phys. Lett.*, 2012, **100**, 103304.
39. J. Gilot, M. M. Wienk and R. A. J. Janssen, *Adv. Mater.*, 2010, **22**, E67.
40. D. J. D. Moet, P. de Bruyn, J. D. Kotlarski and P. W. M. Blom, *Org. Electron.*, 2010, **11**, 1821.
41. Y. M. Nam, J. Huh and W. H. Jo, *Sol. Energy Mater. Sol. Cells*, 2011, **95**, 1095.
42. R. Schueppel, R. Timmreck, N. Allinger, T. Mueller, M. Furno, C. Uhrich, K. Leo and M. Riede, *J. Appl. Phys.*, 2010, **107**, 044503.
43. M. Riede, C. Uhrich, J. Widmer, R. Timmreck, D. Wynands, G. Schwartz, W.-M. Gnehr, D. Hildebrandt, A. Weiss, J. Hwang, S. Sudharka, P. Erk, M. Pfeiffer and K. Leo, *Adv. Funct. Mater.*, 2011, **21**, 3019.
44. G. Dennler, M. C. Scharber, T. Ameri, P. Denk, K. Forberich, C. Waldauf and C. J. Brabec, *Adv. Mater.*, 2008, **20**, 579.
45. J. D. Kotlarski and P. W. M. Blom, *Appl. Phys. Lett.*, 2011, **8**, 053301.
46. J. M. Kroon, M. M. Wienk, W. J. H. Verhees and J. C. Hummelen, *Thin Solid Films*, 2002, **403**, 223.
47. D. J. Wehenkel, K. H. Hendriks, M. M. Wienk and R. A. J. Janssen, *Org. Electron.*, 2012, **13**, 3284.
48. J. Gilot, M. M. Wienk and R. A. J. Janssen, *Adv. Funct. Mater.*, 2010, **20**, 3904.
49. V. S. Gevaerts, A. Furlan, M. M. Wienk, M. Turbiez and R. A. J. Janssen, *Adv. Mater.*, 2012, **24**, 2130.
50. J. Gilot, M. M. Wienk and R. A. J. Janssen, *Org. Electron.*, 2011, **12**, 660.
51. M. M. Wienk, J. M. Kroon, W. J. H. Verhees, J. Knol, J. C. Hummelen, P. A. van Hal and R. A. J. Janssen, *Angew. Chem., Int. Ed.*, 2003, **42**, 3371.
52. A. Hadipour, B. de Boer and P. W. M. Blom, *J. Appl. Phys.*, 2007, **102**, 074506.
53. X. Guo, F. Liu, W. Yue, Z. Xie, Y. Geng and L. Wang, *Org. Electron.*, 2009, **10**, 1174.
54. S. Sista, M.-H. Park, Z. Hong, Y. Wu, J. Hou, W. L. Kwan, G. Li and Y. Yang, *Adv. Mater.*, 2010, **22**, 380.
55. S. Sista, Z. Hong, M.-H. Park, Z. Xu and Y. Yang, *Adv. Mater.*, 2010, **22**, E77.
56. J. Sakai, K. Kawano, T. Yamanari, T. Taima, Y. Yoshida, A. Fujii and M. Ozaki, *Sol. Energy Mater. Sol. Cells*, 2010, **94**, 376.
57. D. J. D. Moet, P. de Bruyn and P. W. M. Blom, *Appl. Phys. Lett.*, 2010, **96**, 153504.
58. W.-S. Chung, H. Lee, W. Lee, M. J. Ko, N.-G. Park, B.-K. Ju and K. Kim, *Org. Electron.*, 2010, **11**, 521.
59. C.-H. Chou, W. L. Kwan, Z. Hong, L.-M. Chen and Y. Yang, *Adv. Mater.*, 2011, **23**, 1282.

60. S. K. Hau, H.-L. Yip, K.-S. Chen, J. Zou and A. K.-Y. Jen, *Appl. Phys. Lett.*, 2010, **97**, 253307.
61. J. Yang, J. You, C.-C. Chen, W.-C. Hsu, H. Tan, X. W. Zhang, Z. Hong and Y. Yang, *ACS Nano*, 2011, **5**, 6210.
62. D. W. Zhao, L. Ke, Y. Li, S. T. Tan, A. K. K. Kyaw, H. V. Demir, X. W. Sun, D. L. Carroll, G. Q. Lo and D. L. Kwong, *Sol. Energy Mater. Sol. Cells*, 2011, **95**, 921.
63. V. C. Tung, J. Kim, L. J. Cote and J. Huang, *J. Am. Chem. Soc.*, 2011, **133**, 9262.
64. J. Yang, R. Zhu, Z. Hong, Y. He, A. Kumar, Y. Li and Y. Yang, *Adv. Mater.*, 2011, **23**, 3465.
65. S. Kouijzer, S. Esiner, C. H. Frijters, M. Turbiez, M. M. Wienk and R. A. J. Janssen, *Adv. Energy Mater.*, 2012, **2**, 945.
66. L. Dou, J. You, J. Yang, C.-C. Chen, Y. He, S. Murase, T. Moriarty, K. Emery, G. Li and Y. Yang, *Nat. Photon.*, 2012, **6**, 180.
67. Y. Zhou, C. Fuentes-Hernandez, J. W. Shim, T. M. Khan and B. Kippelen, *Energy Environ. Sci.*, 2012, **5**, 9827.
68. L. Chang, M. A. Holmes, M. Waller, F. E. Osterloh and A. J. Moulé, *J. Mater. Chem.*, 2012, **22**, 20443.
69. J. W. Shim, Y. Zhou, C. Fuentes-Hernandez, A. Dindar, Z. Guan, H. Cheun, A. Kahn and B. Kippelen, *Sol. Energy Mater. Sol. Cells*, 2012, **107**, 51.
70. J. Kong, J. Lee, G. Kim, H. Kang, Y. Choi and K. Lee, *Phys. Chem. Chem. Phys.*, 2012, **14**, 10547.
71. V. C. Tung, J. Kim and J. Huang, *Adv. Energy Mater.*, 2012, **2**, 299.
72. L. Dou, W.-H. Chang, J. Gao, C.-C. Chen, J. You and Y. Yang, *Adv. Mater.*, 2013, **25**, 825.
73. J. You, L. Dou, K. Yoshimura, T. Kato, K. Ohya, T. Moriarty, K. Emery, C.-C. Chen, J. Gao, G. Li and Y. Yang, *Nat. Commun.*, 2013, **4**, 1446.
74. S. Esiner, H. van Eersel, M. M. Wienk and R. A. J. Janssen, *Adv. Mater.*, 2013, **25**, 2932.
75. W. Li, A. Furlan, K. H. Hendriks, M. M. Wienk and R. A. J. Janssen, *J. Am. Chem. Soc.*, 2013, **135**, 5529.
76. J. Jo, J.-R. Pouliot, D. Wynands, S. D. Collins, J. Y. Kim, T. L. Nguyen, H. Y. Woo, Y. Sun, M. Leclerc and A. J. Heeger, *Adv. Mater.*, 2013, **25**, 4783.
77. C.-C. Chen, L. Dou, J. Gao, W.-H. Chang, G. Li and Y. Yang, *Energy Environ. Sci.*, 2013, **6**, 2714.
78. N. Li, D. Baran, K. Forberich, M. Turbiez, T. Ameri, F. C. Krebs and C. J. Brabec, *Adv. Energy Mater.*, 2013, **3**, 1597.
79. A. R. B. M. Yusoff, W. J. da Silva, H. P. Kim and J. Jang, *Nanoscale*, 2013, **5**, 11051.
80. K. Li, Z. Li, K. Feng, X. Xu, L. Wang and Q. Peng, *J. Am. Chem. Soc.*, 2013, **135**, 13549.
81. J.-H. Kim, C. E. Song, H. U. Kim, A. C. Grimsdale, S.-J. Moon, W. S. Shin, S. K. Choi and D.-H. Hwang, *Chem. Mater.*, 2013, **25**, 2722.
82. Y. Chen, W.-C. Lin, J. Liu and L. Dai, *Nano Lett.*, 2014, **14**, 1467.
83. J. Lee, H. Kang, J. Kong and K. Lee, *Adv. Energy Mater.*, 2014, **4**, 1301226.

84. J. Liu, S. Shao, G. Fang, J. Wang, B. Meng, Z. Xie and L. Wang, *Sol. Energy Mater. Sol. Cells*, 2014, **120**, 744.

85. P.-N. Yeh, T.-H. Jen, Y.-S. Cheng and S.-A. Chen, *Sol. Energy Mater. Sol. Cells*, 2014, **120**, 728.

86. J.-H. Kim, C. E. Song, B. S. Kim, I.-N. Kang, W. S. Shin and D.-H. Hwang, *Chem. Mater.*, 2014, **26**, 1234.

87. C.-Y. Chang, L. Zuo, H.-L. Yip, C.-Z. Li, Y. Li, C.-S. Hsu, Y.-J. Cheng, H. Chen and A. K.-Y. Jen, *Adv. Energy Mater.*, 2014, **4**, 1301645.

88. A. R. B. M. Yusoff, S. J. Lee, H. P. Kim, F. K. Shneider, W. J. da Silva and J. Jang, *Adv. Funct. Mater.*, 2014, **24**, 2240.

89. A. R. B. M. Yusoff, S. J. Lee, F. K. Shneider, W. J. da Silva and J. Jang, *Adv. Energy Mater.*, 2014, **4**, 1301989.

90. A. R. B. M. Yusoff, S. J. Lee, J. Kim, F. K. Shneider, W. J. da Silva and J. Jang, *ACS Appl. Mater. Interfaces*, 2014, **6**, 13079.

91. D. Gupta, M. M. Wienk and R. A. J. Janssen, *ACS Appl. Mater. Interfaces*, 2014, **6**, 13937.

92. N. Li, T. Stubhan, J. Krantz, F. Machui, M. Turbiez, T. Ameri and C. J. Brabec, *J. Mater. Chem. A*, 2014, **2**, 14896.

93. J.-H. Kim, J. B. Park, F. Xu, D. Kim, J. Kwak, A. C. Grimsdale and D.-H. Hwang, *Energy Environ. Sci.*, 2014, **7**, 4118.

94. L. Zuo, C.-C. Chueh, Y.-X. Xu, K.-S. Chen, Y. Zang, C.-Z. Li, H. Chen and A. K.-Y. Jen, *Adv. Mater.*, 2014, **26**, 6778.

95. C.-C. Chen, W.-H. Chang, K. Yoshimura, K. Ohya, J. You, J. Gao, Z. Hong and Y. Yang, *Adv. Mater.*, 2014, **26**, 5670.

96. A. R. B. M. Yusoff, D. Kim, H. P. Kim, F. K. Shneider, W. Jose da Silva and J. Jang, *Energy Environ. Sci.*, 2015, **8**, 303.

97. Z. Zheng, S. Zhang, M. Zhang, K. Zhao, L. Ye, Y. Chen, B. Yang and J. Hou, *Adv. Mater.*, 2015, **27**, 1189–1194.

98. H. Kang, S. Kee, K. Yu, J. Lee, G. Kim, J. Kim, J.-R. Kim, J. Kong and K. Lee, *Adv. Mater.*, 2015, **27**, 1408–1413.

99. W. J. E. Beek, M. M. Wienk, M. Kemerink, X. Yang and R. A. J. Janssen, *J. Phys. Chem. B*, 2005, **109**, 9505.

100. A. Hadipour, R. Müller and P. Heremans, *Org. Electron.*, 2013, **14**, 2379.

101. P.-N. Yeh, S.-H. Liao, Y.-L. Li, H.-R. Syue and S.-A. Chen, *Sol. Energy Mater. Sol. Cells*, 2014, **128**, 240.

102. Y. H. Zhou, C. Fuentes-Hernandez, J. Shim, J. Meyer, A. J. Giordano, H. Li, P. Winget, T. Papadopoulos, H. Cheun, J. Kim, M. Fenoll, A. Dindar, W. Haske, E. Najafabadi, T. M. Khan, H. Sojoudi, S. Barlow, S. Graham, J.-L. Brédas, S. R. Marder, A. Kahn and B. Kippelen, *Science*, 2012, **336**, 327.

103. W. Tress and O. Inganäs, *Sol. Energy Mater. Sol. Cells*, 2013, **117**, 599.

104. F. Verbakel, S. C. J. Meskers and R. A. J. Janssen, *J. Appl. Phys.*, 2007, **102**, 083701.

105. G. Lakhwani, R. Roijmans, A. J. Kronemeijer, J. Gilot, R. A. J. Janssen and S. C. J. Meskers, *J. Phys. Chem. C*, 2010, **114**, 14804.

106. B. E. Lassiter, J. D. Zimmerman, A. Panda, X. Xiao and S. R. Forrest, *Appl. Phys. Lett.*, 2012, **101**, 063303.

107. B. E. Lassiter, J. D. Zimmerman and S. R. Forrest, *Appl. Phys. Lett.*, 2013, **103**, 123305.

108. W. Cambarau, A. Viterisi, J. W. Ryan and E. Palomares, *Chem. Commun.*, 2014, **50**, 5349.

109. Y. Liu, C.-C. Chen, Z. Hong, J. Gao, Y. Yang, H. Zhou, L. Dou, G. Li and Y. Yang, *Sci. Rep.*, 2013, **3**, 3356.

110. Y. Zou, Z. Deng, W. J. Potscavage, M. Hirade, Y. Zheng and C. Adachi, *Appl. Phys. Lett.*, 2012, **100**, 243302.

111. X. Che, X. Xiao, J. D. Zimmerman, D. Fan and S. R. Forrest, *Adv. Energy Mater.*, 2014, **4**, 1400568.

112. D. W. Zhao, X. W. Sun, C. Y. Jiang, A. K. K. Kyaw, G. Q. Lo and D. L. Kwong, *IEEE Electron Device Lett.*, 2009, **30**, 490.

113. I. Etxebarria, A. Furlan, J. Ajuria, F. W. Fecher, M. Voigt, C. J. Brabec, M. M. Wienk, L. Slooff, S. Veenstra, J. Gilot and R. Pacios, *Sol. Energy Mater. Sol. Cells*, 2014, **130**, 495.

114. Z. Tang, Z. George, Z. Ma, J. Bergqvist, K. Tvingstedt, K. Vandewal, E. Wang, L. M. Andersson, M. R. Andersson, F. Zhang and O. Inganäs, *Adv. Energy Mater.*, 2012, **2**, 1467.

115. R. Gehlhaar, D. Cheyns, L. van Willigenburg, A. Hadipour, J. Gilot, R. Radbeh and T. Aernouts, *Proc. SPIE*, 2013, **8830**, 88300I.

116. W.-T. Lin, Y.-T. Lin, C.-H. Chou, F.-C. Chen and C.-S. Hsu, *Sol. Energy Mater. Sol. Cells*, 2014, **120**, 724.

117. K. Tvingstedt, V. Andersson, F. Zhang and O. Inganäs, *Appl. Phys. Lett.*, 2007, **91**, 123514.

118. M. Niggemann, W. Graf and A. Gombert, *Adv. Mater.*, 2008, **20**, 4055.

119. Y.-F. Lim, J.-K. Lee, A. A. Zakhidov, J. A. DeFranco, H. H. Fong, P. G. Taylor, C. K. Ober and G. G. Malliaras, *J. Mater. Chem.*, 2009, **19**, 539.

120. S.-M. Cho, C.-M. Keum, H.-L. Park, M.-H. Kim, J.-H. Bae and S.-D. Lee, *Jpn. J. Appl. Phys.*, 2014, **53**, 042301.

121. N. Li, D. Baran, K. Forberich, F. Machui, T. Ameri, M. Turbiez, M. Carrasco-Orozco, M. Drees, A. Facchetti, F. C. Krebs and C. J. Brabec, *Energy Environ. Sci.*, 2013, **6**, 3407.

122. N. Li, D. Baran, G. D. Spyropoulos, H. Zhang, S. Berny, M. Turbiez, T. Ameri, F. C. Krebs and C. J. Brabec, *Adv. Energy Mater.*, 2014, **4**, 1400084.

123. N. Li, P. Kubis, K. Forberich, T. Ameri, F. C. Krebs and C. J. Brabec, *Sol. Energy Mater. Sol. Cells*, 2014, **120**, 701.

124. T. R. Andersen, H. F. Dam, M. Hösel, M. Helgesen, J. E. Carlé, T. T. Larsen-Olsen, S. A. Gevorgyan, J. W. Andreasen, J. Adams, N. Li, F. Machui, G. D. Spyropoulos, T. Ameri, N. Lemaître, M. Legros, A. Scheel, D. Gaiser, K. Kreul, S. Berny, O. R. Lozman, S. Nordman, M. Välimäki, M. Vilkman, R. R. Søndergaard, M. Jørgensen, C. J. Brabec and F. C. Krebs, *Energy Environ. Sci.*, 2014, **7**, 2925.

125. F. Machui, M. Hösel, N. Li, G. D. Spyropoulos, T. Ameri, R. R. Søndergaard, M. Jørgensen, A. Scheel, D. Gaiser, K. Kreul, D. Lenssen, M. Legros, N. Lemaitre, M. Vilkman, M. Välimäki, S. Nordman, C. J. Brabec and F. C. Krebs, *Energy Environ. Sci.*, 2014, **7**, 2792.

CHAPTER 12

Semi-Transparent Polymer Solar Cells for Power Generating Window Applications

HIN-LAP YIP*[a] AND ALEX K.-Y. JEN[b]

[a]State Key Laboratory of Luminescent Materials and Devices, South China University of Technology, Guangzhou, China; [b]Department of Materials Science and Engineering, University of Washington, Seattle, USA
*E-mail: msangusyip@scut.edu.cn

12.1   Introduction

Although the uncertainty of fossil fuel energy sources drives rigorously the development of the current photovoltaic (PV) technologies based on silicon and inorganic thin films, their high cost of production and installation can be a "limiting factor" preventing them being economically competitive with other energy sources. As a result, new PV technologies based on organic or organic/inorganic hybrid materials are emerging as an inexpensive alternative for renewable energy.[1,2] The promise of polymer-based organic photovoltaics (OPVs) is an ultra-low-cost technology that could be produced by printing the organic materials onto a large, thin, lightweight, flexible, plastic substrate using fast and cheap manufacturing processes that are scalable.[3] The implications are that solar cell technology could soon start to be integrated into structures rather than being installed on top of them. Roof tiles could be made to generate electricity themselves, as could the casings of portable electronic

RSC Polymer Chemistry Series No. 17
Polymer Photovoltaics: Materials, Physics, and Device Engineering
Edited by Fei Huang, Hin-Lap Yip, and Yong Cao
© The Royal Society of Chemistry 2016
Published by the Royal Society of Chemistry, www.rsc.org

devices such as mobile phones and the coatings on vehicles. Another important feature of polymer solar cells is that they can be made with decent optical transparency and tunable colors, making it a unique technology for potential use as power-generating semitransparent films for building windows, foldable curtains, automobile windshields, and other architectural and fashion elements (Figure 12.1).[4]

Currently there are two types of semitransparent photovoltaic (ST-PV) products on the market and both are based on silicon PV technology. The most commonly adopted ST-PV modules are based on either monocrystalline-Si or polycrystalline-Si cells mounted on glass, with the distance between the cells depending on the desired transparency level and the criteria for electricity production. The space between the cells transmits diffuse daylight and therefore provides both shading and natural lighting while producing electricity. As the ST-PV modules are built with high efficiency crystalline Si (c-Si) cells, the overall performances are relatively good for this type of ST-PV. However, the major disadvantage is its limitation for esthetic applications because the cells used are not intrinsically semitransparent and the color of the ST-PV cannot be tuned to achieve a pleasing appearance. Therefore this type of ST-PV module is usually installed on the ceilings of buildings instead of the side walls as power generating solar windows because the latter usually require a more appealing appearance. The second type of ST-PV is based on amorphous Si (a-Si) thin-films.[5] Thin-film materials can be made directly into modules by sputtering the cell material and transparent conductive oxide (TCO) electrodes onto a substrate of glass. The thickness of the amorphous Si can be easily controlled and therefore can be made very thin to achieve semitransparent films. Compared to the crystalline Si based ST-PVs, the appearance of a-Si based ST-PVs is optically more homogeneous and visually more comfortable. Therefore it is more suitable for integration

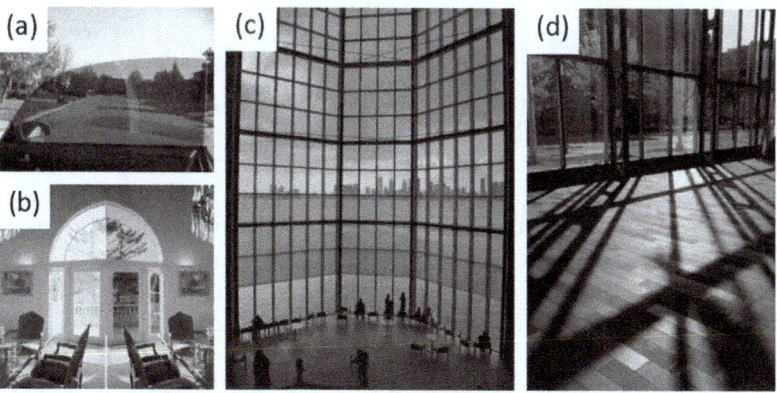

Figure 12.1 Potential applications of semitransparent (ST)-OPVs as power generating windows for (a) automobiles, (b) residential buildings, (c) landmark buildings, and (d) architectural integration with esthetic purposes.

as solar windows in buildings. However, the performance, including both transparency and efficiency, of this type of ST-PV is much lower than that of c-Si based ST-PVs. Another limitation of a-Si ST-PVs is their poor color tuning ability; typically the ST-PV appears black in color, whereas the a-Si film is brown–orange in color when the thickness of the film decreases.

In contrast to silicon, the absorption properties of organic semiconductors can be tuned by modifying their chemical structures,[6] and ultrathin films can be easily prepared to achieve high transparency.[7] Therefore, semitransparent organic photovoltaic cells (ST-OPVs) have great potential to be used as power generating windows for buildings. The best OPVs reported to date are composed of a layer of polymer donor and fullerene acceptor bulkheterojunction (BHJ) film sandwiched between a transparent electrode, such as indium tin oxide (ITO), and a metal electrode. Interlayers are always inserted between the BHJ and the electrodes to improve the charge selectivity and also to determine the polarity of the solar cells. In such devices, the incoming light enters from the ITO transparent electrode and is reflected back by the metal mirror electrode, therefore doubling the optical path within the light harvesting layer. In the case of ST-OPVs, the mirror electrode is replaced by another transparent electrode, which allows the incoming light to be transmitted through the cell in order to obtain the semitransparent appearance. However, in such cases the light absorption of the BHJ film is reduced because only small portion of light is reflected by the back electrode, and therefore the photocurrent and power conversion efficiency (PCE) of the ST-OPV are reduced.

As a result, in contrast to opaque cells, the performance characterization of ST-OPVs depends not only on PCE but also their average visible light transmittance (AVT). Unfortunately, these two are competing parameters so the grand challenge for improving the overall performance of ST-OPVs is to develop new materials and device engineering approaches that allow simultaneous achievement of both high PCE and high AVT. In addition, when ST-OPVs are used as power generating windows for buildingintegrated photovoltaics, their color rendering properties also need to be considered, to ensure visual comfort. The choice of transparent electrode pairs and interlayers is also important, and the overall goal is to reduce absorption loss at these layers in order to achieve an optimal balance between light harvesting and light transmission. The development of low bandgap polymers that can selectively harvest the infrared light and allow the transmission of most of the visible light is another important strategy that can improve the PCE but without sacrificing the AVT. Given that relatively complicated light management is required for ST-OPVs, optical modeling becomes a powerful tool to provide guidelines on designing the device architecture in order to achieve the desired optical and electrical properties simultaneously for tailored power generating window applications. In this chapter, we will provide an overview on the recent development of ST-OPV technology and also discuss the important factors that determine their performance.

12.2 Optical Assessment

Semitransparent photovoltaic windows are considered to be a potential solution for integrating PV into both existing and new buildings to balance energy generation with daylight requirements. In order to fulfil the requirement of the architectural design of the exterior, as well as color perception in the indoor environment, the ST-PV system should be available in a variety of designs and colors. Color neutrality, color rendering and visible light transmittance are common parameters that are used to determine the color properties of current architectural glass and glazing to ensure the visual comfort of occupants, and therefore these parameters are also critical when designing ST-PV windows. For most applications, ST-PV with limited chromatic alternations of the light source is preferred because it allows viewing of outside scenes with natural colors. However, in some applications where esthetic and decorative purposes are required, ST-PV windows with tunable colors are preferred.

12.2.1 Color Rendering Properties

Color perception and visual comfort in a room are influenced by the spectral power distribution of the visible range (370–740 nm) of daylight transmitted into the room. The illumination quality of the daylight is affected by the spectral transmission properties of the window and is characterized by parameters such as color render index (CRI), correlated color temperature (CCT) and the (x,y) coordinates in the CIE chromaticity diagram. The CRI is a quantitative measurement of the ability of a light source to reveal the colors of an object in comparison with a reference or natural light source, and it is calculated by comparing the color rendering of the test source to that of a "perfect" source, which is a blackbody radiator, using the *Test Color Method*.[8] In the case of ST-OPVs, the CRI measures the capability of the light transmitted by the ST-OPV to produce the true colors of the objects viewed. The CRI can range from 0 to 100, where higher numbers represent a better color rendering capacity while smaller numbers represent transmitted light that is less capable of rendering colors properly. The CRI of the radiation from a black body, such as the sun or an incandescent lamp, is defined as 100.

In addition to CRI, another important parameter for assessing the optical property of a light source is the color temperature, which corresponds to the temperature of the blackbody radiator. The trace of the blackbody emission as the function of temperature in a particular color space is known as the Planckian locus or the blackbody locus (Figure 12.2a). When the color coordinate of a light source is considered in the CIE 1931 color space chromaticity diagram, which is specifically designed to represent the colors perceptible to the human eye, the correlated color temperature (CCT) of the light source is defined as the color temperature of the nearest point on the Planckian locus. The CCT of an incandescent lamp is about 2700 K, which generates a yellow–orange color, and that of the sunlight above the atmosphere is about

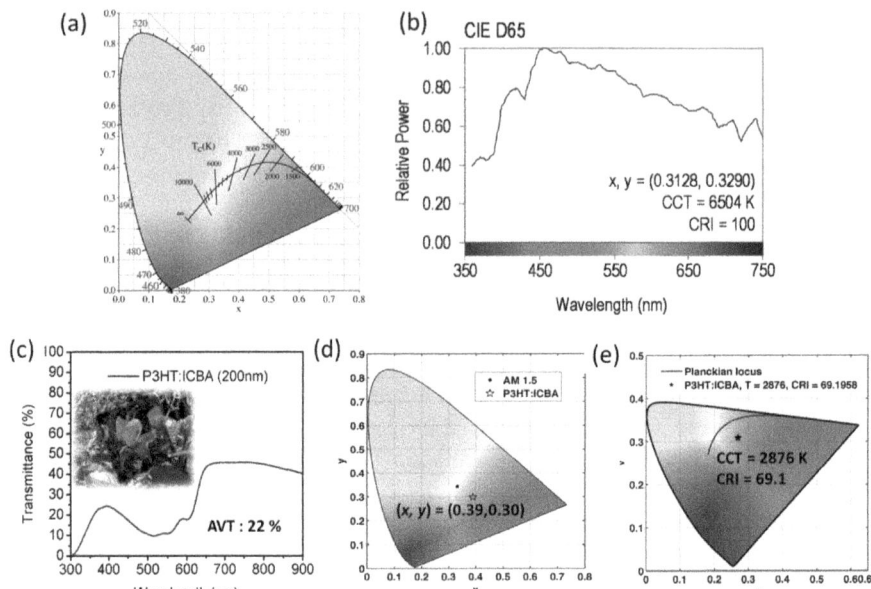

Figure 12.2   (a) The CIE 1931 x,y chromaticity space, also showing the chromaticities of blackbody light sources of various temperatures (Planckian locus), and lines of constant correlated color temperature. (b) Spectral power distribution of CIE Standard Illuminant D65. (c) Transmittance curve of a ST-OPV device based on P3HT:ICBA BHJ; the inset shows a photo of the ST-OPV. (d) The corresponding color coordinates of the ST-OPV shown in the CIE 1931 chromaticity diagram. (e) The corresponding color coordinates of the ST-OPV shown in the CIE 1960 (u,v) chromaticity diagram. The CCT was obtained by finding the closest point to the Planckian locus and the CRI was obtained by the Test Color Method.

6500 K and appears to be bluish white. The CIE Standard Illuminant D65 (Figure 12.2b) is a commonly used standard illuminant that tries to portray standard illumination conditions in the open air in different parts of the world. It represents average daylight and has a CCT of approximately 6500 K (exactly 6504 K). The CIE 1931 color space chromaticity coordinates of D65 are $x = 0.3128, y = 0.3290$, which is close to a "white-point" in the CIE diagram.

    The optical parameters, including AVT, CRI, CCT and CIE (x,y) coordinates, for a ST-OPV cell based on the commonly used P3HT:fullerene bulk heterojunction active layer are shown in Figure 12.2(c–e) as an example to illustrate how those optical properties affect the appearance and performance of an ST-OPV. The PCEs of the ST-OPV cell and the corresponding opaque cell are ~3% and ~5%, respectively. Given that the P3HT film is purple in color with a strong and localized absorption in the visible range of 500–600 nm, a corresponding dip was found in the transmission spectrum of the P3HT-based ST-OPV with an AVT (370–740 nm) of 22% (Figure 12.2c). A photograph of the

ST-OPV cell placed in front of a pink flower is shown in the inset of Figure 12.2c. Its obvious color neutrality was not obtained in such a ST-OPV cell and the colors of the flower and background deviate from those of the actual objects. The CIE (x,y) coordinates are (0.39,0.30), which deviate from the white point and fall into the purple region of the chromaticity diagram. The result is in good agreement with the purple appearance of the ST-OPV cell. A relatively poor CRI of 69.1 was obtained with a CCT of 2876 K, which are significantly different from the standard D65 daylight illuminant. Therefore, the overall optical property for P3HT-based ST-OPV is not satisfied when color-critical applications are required, and new materials and advanced device engineering are required to make ST-OPVs suitable for real-life solar window applications.

12.2.2   Optical Simulations

The application of ST-OPVs as power generating windows for buildings typically requires customized properties with a balance between the optical transparency and electricity generation. This involves sophisticated design of the ST-OPV device structure and the material system needs to be chosen wisely in order to fulfil the application criteria. The establishment of an optical modeling method that can predict precisely the relationship between light transmission and absorption properties in the ST-OPV is very important because it provides a means to guide the design of ST-OPVs with tailored properties. The most commonly used optical model for OPV is the transfer matrix method (TMM), which is a mathematical formalism that can be applied to estimate the electric field distribution within a multilayer OPV device with the assumptions of planar interfaces and total isotropy for all layers.[9] This calculation only requires knowledge of the optical properties of the materials specified by the complex refractive index $[n(\lambda) = \eta(\lambda) + i\kappa(\lambda)]$. The imaginary part, κ, is known as the extinction coefficient and is responsible for absorption in the medium; the real part, η, determines the wavelength and speed of light in the material. The optical constants for different components of the device can be acquired by variable angle spectroscopic ellipsometry. Because charge generation is affected by the spatial distribution of the optical electric field within the device, a careful selection of film thickness and the choice of different layers, including the active layer, interlayers and transparent electrodes, will affect the reflection, absorption and transmission properties of the ST-OPV.[10]

An example of the modeled electric field intensity profiles of an opaque cell and the corresponding ST-OPV cell are given in Figure 12.3(a).[11] The ST-OPV presented is composed of a PBDTTT-C-T:PC$_{71}$BM BHJ film sandwiched between an ITO and an ultrathin Ag film (12 nm) transparent electrode, while for the opaque cell a thick Ag film was used as the mirror electrode. The simulations indicate that in both cases the distributions of the electric field inside the solar cells are highly wavelength dependent and inhomogeneous

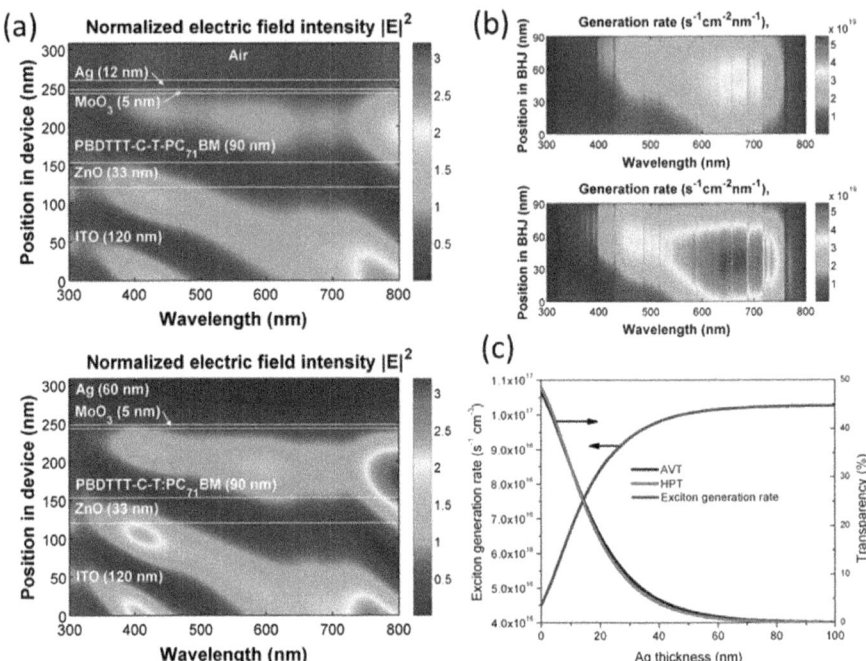

Figure 12.3 (a) Simulations for the electric field intensity profile $|E|^2$ in ST-OPV devices with ultrathin Ag transparent electrode (top figure) and thick Ag mirror electrode (bottom figure). (b) Simulations of the exciton generation rate in ST-OPV devices with the Ag transparent electrode (top figure) and thick Ag mirror electrode (bottom figure). (c) Dependence of the exciton generation rate on the thickness of the Ag electrode from optical modeling (blue curve, bottom left). The dependence of AVT (black curve, top left) and human perception of transmittance (HPT, red curve, top left) of devices simulated from the optical model at various Ag thicknesses are shown.

as a result of the interference effect. In the case of the opaque cell with a mirror Ag electrode, higher electrical field intensity was observed within the BHJ layer as more light was reflected back at the interface between the active layer and the mirror electrode. In contrast, the ultrathin transparent Ag film allows a portion of light to transmit through it, which reduces the degree of light reflection at the electrode interface, and therefore lower electric field intensity was formed in the BHJ layer. The exciton generation rate at a particular wavelength and the position in the active layer can also be calculated from the product of the modulus squared of the electric field intensity, refractive index and the absorption coefficient of the active layer at the specific wavelength and position. Finally, a profile of the exciton generation rate within the whole active layer in both the opaque cell and ST-OPV cell can be generated, and they are shown in Figure 12.3(b). As expected, the overall charge generation rate in the opaque device is higher than that of the ST-OPV cell, implying that a larger photocurrent and PCE can be obtained in the opaque cell. However, for power generating window applications, the transmission of the ST-OPV also needs to be considered to evaluate its overall performance. Based on the TMM model, the transmission curve of the ST-OPV can also be simulated and the AVT of the ST-OPV cell can therefore be calculated by averaging the transmittance of the visible wavelengths. A plot of the relationship between the simulated charge generation rate and the AVT, depending on the thickness of the top Ag electrode is provided in Figure 12.3(c), showing that they are inversely proportional to each other. As the PCE is directly related to the charge generation rate of the solar cell, we can therefore predict the relationship between the highest achievable PCE and AVT. The success of optical simulations in portraying the optical behaviors of ST-OPVs makes it possible to build devices with predictable performance and transparency and provides very useful guidelines for designing future ST-OPV systems.

12.3 Transparent Electrodes for ST-OPV

The three major components for constructing a ST-OPV are the transparent electrodes (TE), interlayers and the polymer:fullerene BHJ film as the light harvesting layer. A typical ST-OPV device structure is TE/interlayer/BHJ/interlayer/TE. The interlayers play an important role in defining the interfacial properties and they serve multiple functions that include tuning of the energy level alignment at the electrode–active layer interfaces, improving charge selectivity at the interfaces and also defining the polarity of the electrodes. By switching the position of the hole and electron selective interlayers in the device, one can build the ST-OPV with a normal structure or an inverted structure. Detailed discussions about the interlayer effects in OPV devices can be found in other reviews.[12] The choice of the pairs of transparent electrodes significantly affects the performance of ST-OPVs. Several types of transparent electrode have been studied for ST-OPVs and the selection principles are based on the optical transparency, electrical conductivity and the processing property of the electrode.

12.3.1 Transparent Conductive Oxides

Transparent conductive oxides such as indium-tin oxide (ITO), fluorine-doped tin oxide (FTO) and Al-doped ZnO (AZO)[13,14] are the most commonly used transparent electrodes for optoelectronic devices. Among them, ITO has been extensively applied for both opaque cells and ST-OPV cells owing to its low sheet resistance (10–15 Ω sq^{-1}) and high optical transmittance (~90%). ITO can be coated on either glass substrate or plastic substrate and can be used as the bottom transparent electrode in conjunction with other types of transparent electrode as the top electrode. Indeed, in one of the earliest demonstrations of the concept of the ST-OPV device, ITO was used as both the bottom and top electrodes, facilitated by a lamination method to stick together two individually

prepared stacks of glass/ITO/Cs$_2$CO$_3$/P3HT:PCBM and PEDOT:PSS/ITO/
glass.[15] The advantage of this structure is that the glass substrates at both sides
of the device provide an effective way to encapsulate the cell and prevent moisture and oxygen from penetrating into the active layer; these are identified as
some of the major degradation factors for OPVs. However, despite the relatively good transparency and electrical conductivity of ITO, its relatively high
cost and poor mechanical properties on plastic substrate make it an inappro-
priate choice as a transparent electrode for cheap and flexible OPVs. Therefore,
ultimately, new transparent electrodes with the desired optical, electrical and
processing properties are required to replace ITO in ST-OPVs.

12.3.2   Conducting Polymers

Conducting polymers such as PEDOT:PSS are other widely explored transparent electrodes for organic optoelectronic devices. They are of relatively
low cost, highly flexible and also compatible with large scale production
based on printing methods.[16,17] The conductivity of PEDOT:PSS can be tuned
by varying the doping content and the microstructure of the PEDOT:PSS
film. In some optimized cases, electrical conductivities up to 4000 S cm^{-1}
with transmittance over 80% can be achieved.[18] Conducting polymers can
be used in conjunction with graphene or carbon nanotubes to improve the
optical and electrical properties of the hybrid transparent electrodes further.[19,20] The most frequently adopted ST-OPV structure is glass/ITO/ZnO/
BHJ/PEDOT:PSS, which is composed of a highly conductive PEDOT:PSS
(Clevios™ PH500 or PH1000, Heraeus) film as the top transparent electrode
and ITO as the bottom electrode.[21-25] The PEDOT:PSS film can be prepared
by either spincoating, spray coating or roll-to-roll coating from its solution to
form the top electrode.[26] To demonstrate the potential of constructing ITO-
free, fully solution-processed devices, highly conductive PEDOT:PSS films
were applied as both the transparent electrodes in an ST-OPV.[27-29] It was
found that, although the PEDOT:PSS film provided decent transparency and
a good processing window as a solution processed transparent electrode, its
limited electrical conductivity significantly affected the performance of the
ST-OPV, with a major drawback in the ohmic loss due to the high resistivity
of the electrode. The effect becomes even more serious when the size of the
cell increases. Therefore, PEDOT:PSS film by itself is not an ideal transparent
electrode and cannot be used for large area solar panels; an additional charge
collecting electrode such as a printed metal grid is required to improve the
charge collection efficiency.[30,31]

12.3.3   Ultrathin Metal Films

Ultrathin metal film (UTMF) is another widely studied candidate for the
transparent electrode in ST-OPV. Metal films are highly conductive and duc-
tile; when the thickness of metal film reduces to under ~20 nm, it starts

to become optically transparent and can be applied as a transparent electrode. The transmittance typically increases with decreasing film thickness but thinner films also exhibit higher resistivity due to electron scattering from the surface and grain boundaries of the films. To enhance the optical transmittance of the UTMF, a dielectric layer with high refractive index (n) can be introduced below and above the UTMF layer to form a dielectric/metal/dielectric (DMD) multilayer structure. The multilayer structure can be designed to exhibit multiple reflections and interferences to achieve optimal transmittance of the UTMF. Typically, transparent metal oxides and other large bandgap semiconductors with high refractive index are chosen as the dielectric layers. DMD structures such as $WO_3/Ag/WO_3$,[32,33] $ZnS/Ag/WO_3$,[34] $MoO_3/Ag/WO_3$,[35] $V_2O_5/Ag/V_2O_5$,[36] and $ZTO/Ag/ZTO$[37] have been explored as the top transparent electrodes for ST-OPVs. These DMD structures can be used as either the transparent anode or the cathode, depending on the choice of the dielectric layer (hole or electron selective layer) sandwiched between the BHJ and the UTMF. Typical hole selective layers include MoO_3, WO_3 and V_2O_5, while TiO_2 or ZnO can be applied as the electron selective layer in the DMD transparent electrode.[38] The optical and electrical properties of the DMD transparent electrode can also be tuned facilely by adjusting the thickness of the D and M layers. Optimized DMD transparent electrodes can simultaneously achieve high conductivity, high transmittance and good flexibility. The OPV devices that incorporate the DMD transparent electrode show good potential to outperform those with the ITO electrode.[39] However, although the DMD structures have many advantages, their major deficiency is their high production cost because DMD transparent electrodes are fabricated under the vacuum deposition process. The ideal preparation method for the new transparent electrode should be compatible with solution-based roll-to-roll coating processes.

12.3.4 Metal Nanowires

Recently, new transparent electrodes based on Ag nanowire films have been extensively studied for use in ST-OPVs.[40–47] Metal nanowires with tunable diameters and lengths can be synthesized and deposited from solution phase on to a substrate to form an electrical conductive nanowire network. The transmittance of a nanowire film decreases with the nanowire diameter (D) and the fraction of surface covered by the nanowires; in contrast, the electrical conductivity increases with the D and covered area.[48] Therefore, optimizing the optoelectronic properties of the nanowire films requires minimizing the sheet resistance at a given D and covered area. Some of the best Ag nanowire films show sheet resistance down to $10 \ \Omega \ sq^{-1}$ with a transmittance of 90%, which is comparable to or even better than ITO.[49] The Ag nanowires are also fully compatible with solution-based roll-to-roll coating processes. The combined advantages of high conductivity, high transparency, good mechanical flexibility and good processing properties make the Ag nanowires the most promising transparent electrode for flexible OPVs.

Figure 12.4(a) illustrates the fabrication process of ST-OPV employing Ag nanowires as both the bottom and top transparent electrodes. The bottom Ag nanowires film was prepared by blade coating followed by laser abrasion to define the area of the electrode. The charge selective layer and active layer were then deposited subsequently and the ST-OPV device completed with coating another layer of Ag nanowires film as the top transparent electrode. Figure 12.4(b) shows a photograph of the completed ST-OPV device. This particular ST-OPV showed a PCE of ~3% with a high transmittance of over 50% at a wavelength of 550 nm (Figure 12.4c),[47] while the PCE of the corresponding opaque cell using Ag nanowires and thick Ag film as bottom and top electrodes, respectively, was ~4%. The successful demonstration of the use of Ag nanowires as transparent electrodes to fully replace ITO in ST-OPVs paves

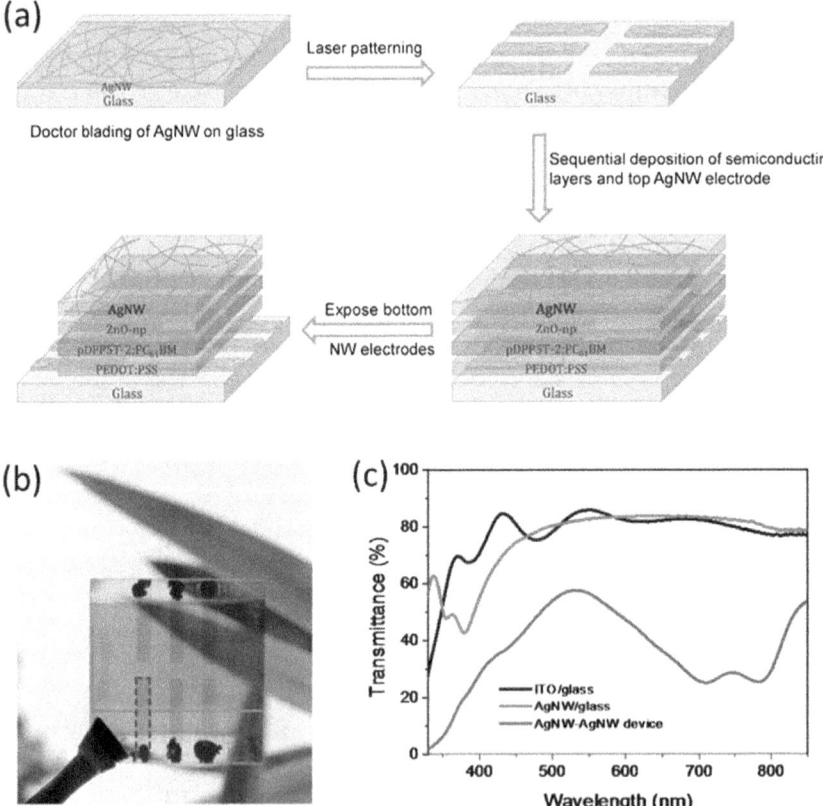

Figure 12.4   (a) Schematic illustration of the full solution processing of the ST-OPV using Ag nanowires as both the bottom and top transparent electrodes. (b) A digital photograph of the as-fabricated ST-OPV device. (c) Transmittance spectra of Ag nanowire-coated glass and commercial ITO-coated glass, and the ST-OPV with Ag nanowires as both top and bottom transparent electrodes. Reprinted with permission from ref. 47. Copyright 2009 American Chemical Society.

the way for the commercialization of ST-OPV products. A future challenge will be further lowering the cost of the metal nanowire-based transparent electrode by replacing Ag nanowire with another cheaper metal, such as Cu nanowires.[50]

12.4   Low Bandgap Polymers

The major dilemma in designing an efficient ST-OPV is that the visible transmittance and the photocurrent are competing factors. One potential strategy to circumvent this problem is to incorporate PV materials that absorb mainly NIR light; this enables the majority of the visible light to pass through the devices while maintaining a good light harvesting property. Inorganic semiconductors cannot fulfil this requirement because they always exhibit broad band absorption. However, organic semiconductors show great potential for fulfilling this strict requirement because the absorption spectrum can be tuned to localize at the NIR region by novel chemical design. The key design concept for low bandgap polymers is to introduce an electron donating block (D) and electron accepting block (A) on the backbone of the polymer, forming alternating D–A type polymers. The bandgap of the polymer is determined by the intramolecular charge transfer characteristics between the D and A; the stronger the D and A, the smaller the bandgap. In addition, the absorption spectrum of D–A polymers is typically localized at a particular range of wavelengths instead of being spread over a wide spectrum, making it possible to design new materials that only absorb NIR light. A few examples of the absorption spectra of polymer semiconductors with different bandgaps are shown in Figure 12.5, which reveals that the bandgap and absorption spectrum of the polymers can be readily tuned to suit ST-OPV applications.

A wide range of novel low bandgap polymers with bandgaps ranging from 1.4 to 1.8 eV have been employed as the donor material in the light harvesting layer for ST-OPVs. The chemical structures of these polymers are shown in Figure 12.6 and the corresponding ST-OPV device structures and performance data are summarized in Table 12.1. Beiley *et al.* employed a high performance polymer, PBDTTPD, to construct ST-OPV, and a composite film based on ZnO nanoparticles and Ag nanowires was used as the top transparent electrode.[51] PBDTTPD is a relatively large bandgap polymer with absorption peak and edge at 610 nm and 700 nm, respectively.[52] Champion opaque cells showed a high PCE of 8.5% with a high V_{OC} of 0.95 V, J_{SC} of 12.9 mA cm^{-2} and FF of 0.69, whereas the PCE of the ST-OPV dropped to 5% with a major loss in J_{SC} and FF to 8.8 mA cm^{-2} and 0.60, respectively. Although the absorption of the polymer was concentrated in the visible range, a decent AVT (400–700 nm) of 37% was obtained in the corresponding ST-OPV. A photograph of the device is shown in Figure 12.7(a). Both the PCE and the AVT of the ST-OPV are significantly better than those based on commercially available P3HT donor polymer, suggesting that the development of new materials is fundamental to improving the optoelectronic properties of ST-OPVs.

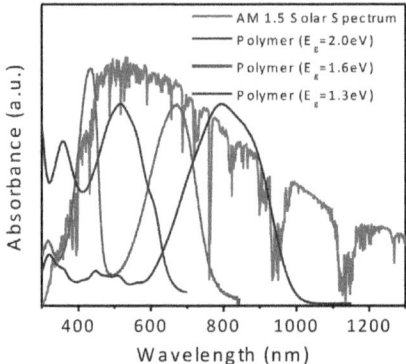

Figure 12.5   Solar radiation spectrum under AM1.5 conditions, and the absorption spectra of three conjugated polymers with different bandgaps.

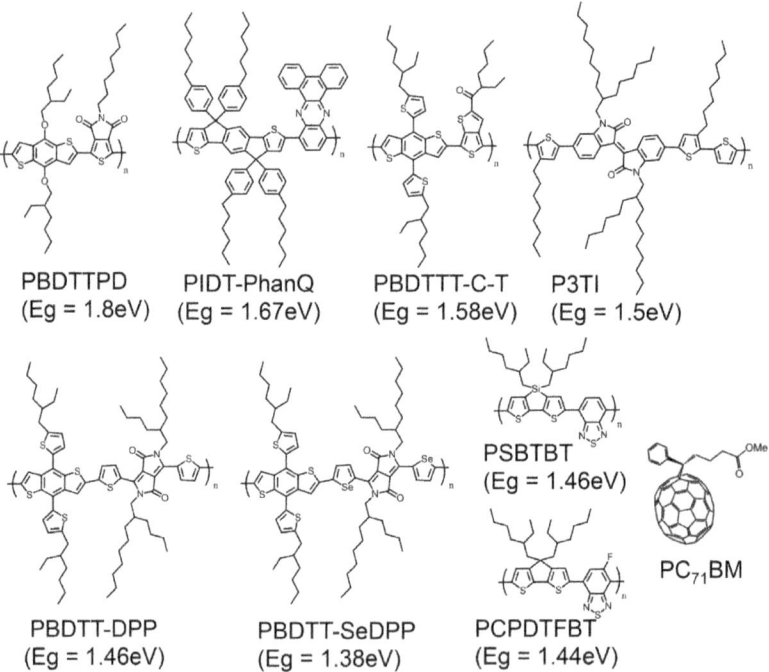

Figure 12.6   The chemical structures of the conjugated polymer donors and fullerene acceptor that have been employed as a light harvesting layer in ST-OPV.

Other high performance low bandgap polymers, including PIDT-phanQ and PBDTTT-C-T, have also been explored as donor materials in ST-OPVs.[7,11] The optoelectronic properties, including PCE, AVT and the color rendering properties of these devices, have been fully characterized. The absorption of both polymers extended to the NIR region, with the absorption edge reaching

Table 12.1   A summary of different ST-OPV device structures with corresponding performances and optical properties.

ST-OPV device structure	Opaque cell PCE	ST-OPV PCE	Transmittance	CRI@CCT	CIE (x,y)	Ref.
ITO/MoO$_3$/PBDTTPD:PC$_{71}$BM/ZnO/AgNW	8.5%	5.0%	AVT = 37%	—	—	51
ITO/PEDOT/PIDT-phanQ:CP$_{71}$BM/C$_{70}$-bis/UTMF(Ag-20 nm)	6.6%	4.2%	AVT = 32%	98@4711 K	(0.359,0.355)	7
ITO/ZnO/PBDTTT-C-T:PC$_{71}$BM/MoO$_3$/UTMF(Ag-12 nm)	7.6%	5.6%	AVT = 28%	97@6156 K	(0.319,0.335)	11
ITO/PFPA-a/P3Ti:PC$_{71}$BM/PEDOT:PSS(PH 1000)	6.2%	2.2%	~50%@550 nm	—	—	53
ITO/PEDOT/PBDTT:DPP:PC$_{71}$BM/TiO$_2$/AgNW	6.5%	4.0%	~66%@550 nm	—	—	45
ITO/PEDOT/PBDTT:SeDPP:PC$_{71}$BM/TiO$_2$/AgNW	7.2%	4.5%	AVT = 58%	—	—	54
ITO/PEDOT/PSB'TBT:PC$_{71}$BM/LiCoO$_2$/Al(3 nm)/AZO	5.5%	2.8%	~25%@550 nm	83@5307 K	(0.338,0.377)	14
ITO/ZnO/PCPDTFBT:PC$_{71}$BM/PEDOT:PSS/UTMF(Ag-10 nm)	6.6%	5.0%	AVT = 47%	99@5365 K	(0.330,0.350)	55

Figure 12.7   Photographs of ST-OPV devices reported in: (a) ref. 51, reproduced from ref. 51 with permission from Wiley; (b) ref. 7; (c) ref. 45, reproduced from ref. 45 with permission from the American Chemical Society, and (d) ref. 61.

750 nm and 780 nm for PIDT-phanQ and PBDTTT-C-T, respectively. In both cases, UTMF was used as the top transparent electrode and the thicknesses of the UTMF were tuned to evaluate the relationship between PCE and transmittance. The results revealed that the PCE and AVT linearly oppose each other, which clearly shows the tradeoff between the transparency and photon collection.[11] An optical model was also established which can precisely predict the transmittance and performance relationship and also the color rendering properties of the ST-OPVs; this provides a very powerful means to design the appropriate device structures to achieve desired optoelectronic properties. The optimized ST-OPV devices show high PCE, over 5.5%, with AVT close to 30%; the transmitted light also demonstrated extraordinary transparency and color rendering capacities with very high CRIs of >97 (a photograph of the corresponding ST-OPV is shown in Figure 12.7b), which is suitable for color-sensitive solar window applications.

Given that a major part of the absorption spectrum of the polymers discussed above is still located in the visible region, it is difficult to achieve ST-OPV with both high AVT and PCE simultaneously. Therefore, polymers with even lower bandgaps and further red-shifted absorption spectra are required. A series of polymers with bandgaps between 1.4 and 1.5 eV have been studied; these polymers show strong absorption localized in the range of 650–850 nm and weak absorption in the visible range, making them suitable for ST-OPV. In the case of P3TI, the backbone of the polymer is composed of a strong acceptor unit, isoindigo, and the thiophene donor units resulted in a bandgap of 1.5 eV.[53] The corresponding ST-OPV showed a relatively low PCE of 2.2%, which was probably due to the unoptimized device structure. However, a high transmittance of 50% at a wavelength of 550 nm was achieved, due to the low absorbance of the polymer in the visible region. PBDTT-DPP and PBDTT-SeDPP represent another class of high performance low bandgap polymers with a backbone composed of a weak benzodithiophene (BDT) donor and a strong diketopyrrolopyrrole (DPP) acceptor.[45,54] Substitution of the thiophene unit in the case of PBDTT-DPP with selenophene in the case of PBDTT-SeDPP further decreases the bandgap from 1.46 eV to 1.38 eV with absorption extended from 850 nm to 900 nm. In both cases, high PCE and transmittance were achieved. A very encouraging result was obtained for the PBDTT-SeDPP based ST-OPV, with PCE of 4.5% and a recorded high AVT of 58%. A photograph of a representative ST-OPV device is shown in Figure 12.7(c).

In addition to employing a strong accepter unit for low bandgap polymers in the above-mentioned materials, introduction of a strong donor unit, such as cyclopentadithiophene (CPD), in the D–A polymer structure is another way to produce low bandgap polymers. One classic example is PSBTBT, of which the backbone is composed of Si-substituted CPD donor and benzothiadiazole (BT) acceptor, forming a low bandgap polymer with bandgap of 1.46 eV. Initial results showed that ST-OPV based on PSBTBT had good color rendering properties, with CRI of 83. However the device performance values, including the transmittance and PCE, are both quite low.[14] To improve the performance of ST-OPVs further, a modified polymer, PCPDTFBT, with CPD donor and fluorine-substituted BT acceptor was introduced. The opaque device based on PCPDTFBT showed an improved performance with PCE of 6.6%.[55] By optimizing the ST-OPV device structure, a PCE of 5.0% with a high AVT of 47% was demonstrated. In addition, an extremely good CRI of 99 at a CCT of 5365 K and CIE (x,y) coordinates of $(0.33,0.35)$ were achieved, which are among the best of all the reported ST-OPV devices. These results suggest that the utilization of high performance low bandgap materials in ST-OPVs is a very promising approach to optimizing both efficiency and transparency. The future challenge will be further pushing the bandgap of these polymers to below 1.4 eV to harvest a larger portion of the NIR light.

12.5   Semitransparent Tandem Solar Cells

One of the limiting factors for the PCE of OPVs is insufficient absorption of solar radiation, due to the narrow absorption bandwidth of PV polymers, and limited film thickness, due to the low charge carrier mobility of semiconducting polymers. Single junction OPV devices also suffer thermalization losses of hot charge carriers generated by high energy photons, which limits the V_{OC} of the devices. One strategy used to minimize this loss, and also to improve the light harvesting bandwidth, is to construct tandem OPV devices by stacking individual cells with complementary absorption characteristics.[56] The typical configuration of a tandem cell consists of a front cell with a large bandgap polymer and a rear cell with a low bandgap polymer connected by a charge recombination layer. Recent studies have suggested that in some cases applying sub-cells composed of the same type of polymer can also enhance the OPV performance under optimized conditions.[57] Based on the tandem cell approach, OPVs with a very high PCE of over 11% have been demonstrated.[58,59]

The tandem cell concept can also be adopted to improve ST-OPV devices, and the performance data are summarized in Table 12.2.[53,60,61] Tang *et al.* demonstrated the tandem cell idea by stacking two individual ST-OPV cells composed of different light harvesting layers, as illustrated in Figure 12.8(a).[53] The cells can be connected externally by using either serial or

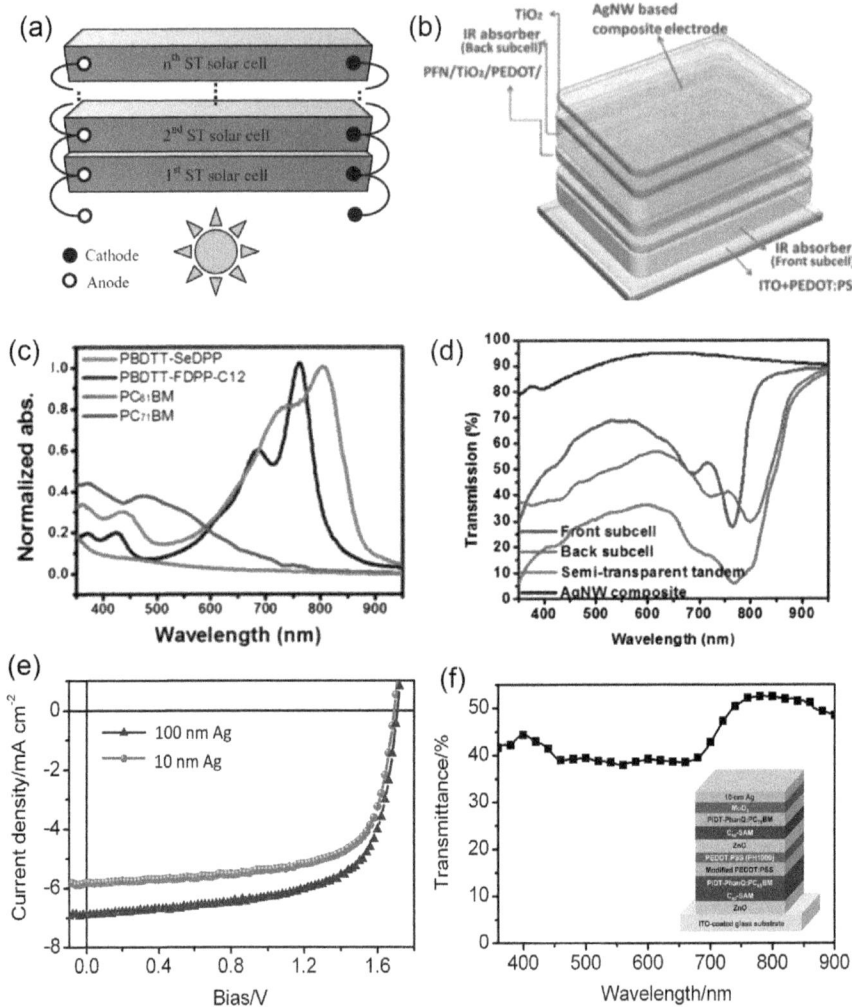

Figure 12.8 (a) Schematic device architecture of a tandem ST-OPV with sub-cells connected in parallel; it can also be connected in series by alternatively connecting the cathode and anode of the sub-cells. Reproduced from ref. 53 with permission from Wiley. (b) Tandem ST-OPV device structure, (c) absorption spectra of donor and acceptor materials, and (d) the transmission spectra of the tandem ST-OPV device from ref. 60. Reproduced from ref. 60 with permission from The Royal Society of Chemistry. (e) J–V curves of opaque and semitransparent tandem devices under standard AM1.5G illumination at 100 mW cm^{-2} and (f) tandem ST-OPV device structure and its transmission spectrum from ref. 61.

Table 12.2   A summary of different tandem ST-OPV devices and their corresponding performances and optical properties.

Tandem ST-OPV device structure		V_{OC} (V)	J_{SC} (mA cm^{-2})	FF	PCE (%)	AVT (%)	CRI@CCT	CIE (x,y)	Ref.
Front subcell	Rare subcell								
TQ1: PC$_{71}$BM	P3TI: PC$_{71}$BM	1.46	4.5	0.67	4.4	25	—	—	53
PBDTT-FDPP-C$_{12}$: PC$_{61}$BM	PBDTT-SeDPP: PC$_{61}$BM	1.46	7.2	0.61	6.4	43	—	—	60
PBDTT-FDPP-C$_{12}$: PC$_{61}$BM	PBDTT-SeDPP: PC$_{71}$BM	1.47	8.4	0.59	7.3	30	—	—	60
PIDT-phanQ: PC$_{61}$BM	PIDT-phanQ: PC$_{71}$BM	1.70	5.8	0.67	6.7	40	97@5622 K	(0.330, 0.339)	61

parallel configurations. For the front cell, a BHJ layer based on a large bandgap polymer, TQ1, and PC$_{71}$BM was used to harvest visible light while a rare sub-cell based on P3TI:PC$_{71}$BM was used to harvest NIR light. The tandem ST-OPV connected in series showed a PCE of 4.4%, which is improved when compared to the individual ST-OPV based on P3TI:PC$_{71}$BM, which showed a PCE of only 2.2%, as mentioned in a previous section. However, the AVT also decreased to ~25% owing to the absorption of visible light from the front cell, and there were also additional losses of transmittance from the glass substrate, because multiple substrates were used in this type of configuration.

Another more commonly used tandem structure is based on sub-cells connected through an interconnection layer. Chen *et al.* demonstrated tandem ST-OPVs using a multilayer configuration of PFN/TiO$_2$/PEDOT to electrically connect the sub-cells and Ag nanowires as the top transparent electrode.[60] A schematic drawing for the device structure is shown in Figure 12.8(b). In both the sub-cells, low bandgap polymers including PBDTT-FDPP-C$_{12}$ and PBDTT-SeDPP were employed. The major difference in chemical structure between these two polymers is the flank group between the BDT and DPP units, which is furan in the case of PBDTT-FDPP-C$_{12}$ and selenophene for PBDTT-SeDPP. The absorption spectra of the two polymers, PC$_{61}$BM and PC$_{71}$BM, are shown in Figure 12.8(c). Owing to the weaker electron donating power of furan, the absorption of PBDTT-FDPP-C$_{12}$ was blue-shifted by ~50 nm compared to PBDTT-SeDPP but the major absorption still localized in the NIR range of 650–800 nm. A tandem ST-OPV composed of a PBDTT-FDPP-C$_{12}$:PC$_{61}$BM front cell and PBDTT-SeDPP:PC$_{61}$BM rear cell showed a high PCE of 6.4% with an AVT of 43%. These performances can be tuned by replacing the PC$_{61}$BM in the rear cell with PC$_{71}$BM. Given that PC$_{71}$BM has better absorption properties in the visible range than PC$_{61}$BM, it helps to increase the photocurrent, and this led to a very high PCE of 7.3% for the corresponding ST-OPV. However, owing to the absorption loss in visible light, the AVT of the device decreased to 30% and the transmittance spectrum is shown in Figure 12.8(d) for reference.

Considering that $PC_{61}BM$ and $PC_{71}BM$ possess very different absorption characteristics in the visible region, they can be combined with a single type of low bandgap polymer to achieve complementary absorption in different subcells of the tandem cell. This idea was demonstrated by using PIDT-phanQ as the donor polymer to fabricate tandem ST-OPVs.[61] The ST-OPV composed of a PIDT-phanQ:$PC_{61}BM$ (70 nm) front cell and a PIDT-phanQ:$PC_{71}BM$ (85 nm) rear cell connected by a PEDOT:PSS/PH1000/ZnO interconnection layer and a 10 nm thick Ag film as the top transparent electrode. The *J–V* characteristics of the tandem ST-OPV and a corresponding opaque tandem device with a thick Ag electrode are shown in Figure 12.8(e). A high V_{OC} and FF of 1.7 V and 0.67 V, respectively, were achieved in both cases, and only a slight decrease in J_{SC} from 6.9 to 5.8 mA cm^{-2} and PCE from 7.8 to 6.7% were observed in the tandem ST-OPV device. The transmittance curve of the tandem ST-OPV is shown in Figure 12.8(f), and a relatively flat transmission between 400 and 700 nm was achieved, leading to an AVT of ~40%, which is very high considering a high PCE was obtained. The color rendering property of this device is also very good, showing CRI of 97 at a CCT of 5622 K and CIE (x,y) coordinates of (0.330,0.339), which is very close to the "white point" and produces a color neutral ST-OPV as shown in Figure 12.7(d). These promising performance results from ST-OPVs clearly suggest that organic-based PV technology offers a huge potential for power generating window applications.

12.6   Photonic Crystal-Enhanced ST-OPV

The introduction of low bandgap polymers has proven to be an efficient approach to improving the performance of ST-OPVs by harvesting NIR photons. However, because only a relatively thin harvesting layer can be used in ST-OPV devices, owing to the low carrier mobilities of semiconducting polymers, only part of the NIR light spectrum is absorbed. Therefore, it is important to develop new methods to re-harvest the NIR photons that are lost, to further improve the performance of ST-OPVs. Coupling the ST-OPV with a photonic crystal reflector that can reflect light with selected wavelengths back to the cell was demonstrated as an effective approach to enhance the light harvesting property of ST-OPVs.[62-67] The one-dimensional (1-D) photonic crystal reflector, or distributed Bragg reflector, is a mirror structure which consists of an alternating sequence of layers of two different optical materials with high refractive index (n_H) and low refractive index (n_L), respectively. In order to achieve constructive interference of all the reflected light at the interfaces of the two different materials, the optical thicknesses (d) are typically chosen to be quarter-wavelength long, that is, $n_H d_H = n_L d_L = \lambda/4$, where λ is the operating wavelength. The reflectivity achieved is determined by the number of layer pairs and by the contrast in refractive index between the layer materials. The reflection bandwidth is mainly determined by the refractive index contrast. By careful design of the multilayers, it is even possible to achieve non-periodic photonic crystal structures that can simultaneously reflect light at different wavelengths, such as in the UV and NIR regions.[66,67]

This advanced optical management concept was nicely demonstrated by Betancur *et al.*[67] They introduced an optimized non-periodic 1-D photonic crystal composed of alternating LiF/MoO$_3$ layers with the appropriate thicknesses selectively to enhance the UV and NIR reflection for ST-OPVs. The device structures for the typical and photonic crystal-enhanced ST-OPV (pc-ST-OPV) are shown in Figure 12.9(a). A high performance low bandgap polymer, PTB7, with absorption up to 750 nm was used for the light harvesting layer and a 10 nm Ag thin-film was used as the top transparent electrode. The opaque device showed a PCE of 7.3%, while the PCE of the ST-OPV device dropped to 3.6% with an AVT of 32%, mainly due to the reduction in J_{SC} from 14.0 to 8.5 mA cm^{-2}. The pc-ST-OPV recovered part of the J_{SC}, to 10.5 mA cm^{-2}, leading

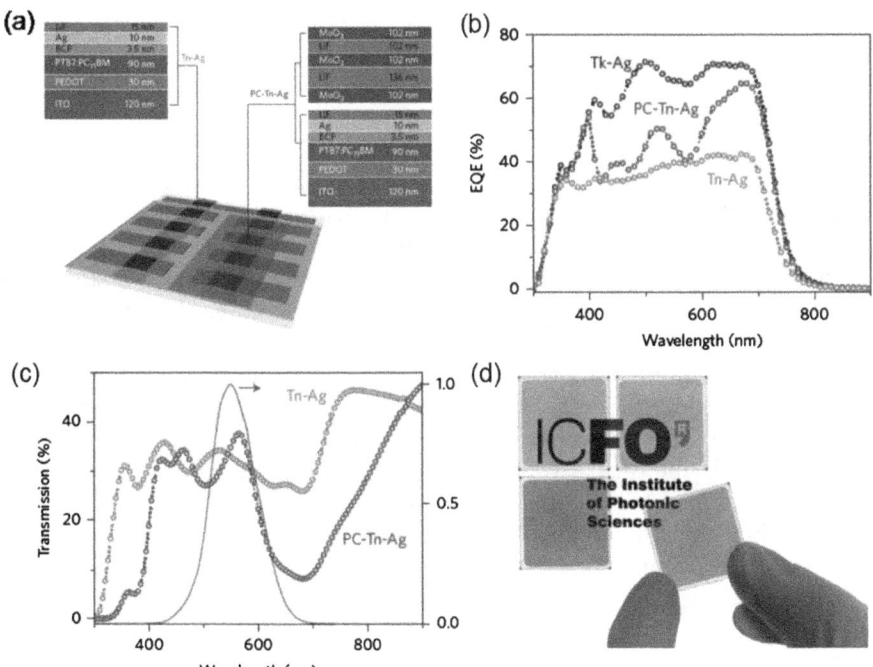

Figure 12.9   (a) Schematic view of the typical ST-OPV device structure and the photonic crystal-enhanced ST-OPV device structure. (b) Experimentally measured EQEs for the opaque, ST-OPV and pc-ST-OPV devices. (c) Experimentally measured transmission spectra of the ST-OPV and pc-ST-OPV devices. (d) Photographs of four pc-ST-OPV devices fabricated with different relative layer thickness to tune the color of the device. Reproduced from ref. 67 with permission from *Nat. Photon.*

to an improved PCE of 5.2% while maintaining the AVT of ~29%. The optical enhancement effect of the photonic crystal was evaluated by comparing the EQE spectrum and transmittance curves of the three different devices, which are shown in Figure 12.9(b) and (c), respectively. The EQE spectra of the ST-OPV device shows a plateau across the visible range with average EQE of ~40%, which is significantly lower than that of the opaque device, which shows EQE up to 70%. For the pc-ST-OPV, the EQE below 400 nm and above 650 nm are recovered, which is in good agreement with the specific design of the photonic crystal reflector that aimed to enhance UV and NIR reflectivity. The transmission curve of the pc-ST-OPV shows that the transmittance in the UV and NIR regions drops significantly, which also echoes the EQE results, suggesting that the photonic crystal effectively reflects light of the targeted wavelengths while leaving the transmittance in the visible range relatively unaffected. To further demonstrate the versatility of using photonic crystals for optical management in ST-OPVs, the appearance of ST-OPV devices may be tuned by adjusting the photon crystal structures, which alters the shape of the transmission curves, and therefore devices with different colors can be achieved (shown in Figure 12.9d). The capability for facially tuning the color of the ST-OPVs makes it suitable for solar window applications with esthetic considerations.

12.7 Conclusions

Integration of semitransparent photovoltaic products as power generating windows in buildings is a new market that will continue to grow because of the increasing demand for sustainable energy. Unlike conventional PV modules, whose performance mainly depends on the PCE, PV systems for solar window applications are evaluated by their colors, transparency levels and efficiencies. Current ST-PV technology based on Si may not fulfil the requirement for many window applications because their optical properties are intrinsically difficult to tune. This offers a huge opportunity for polymer-based PV because the optical absorption of polymers can be easily tuned by their chemical structures. It is even possible to design polymers with absorption localized only in the infrared region, making them capable of achieving high efficiency without sacrificing the transparency in the visible range. The unique optical properties offered by ST-OPV have already triggered increasing research efforts focusing on this new field. Over the past few years, tremendous progress has been made in improving the performance of ST-OPVs through the development of new materials and device architectures. However, many challenges remain to be solved before ST-OPVs can be considered for real-life applications. These challenges include the development of new generation polymers with absorption extend to 1000 nm to improve further the light harvesting properties of ST-OPVs; development of low cost, solution processible transparent electrodes with the desired optical and electrical properties; development of simple solution preparation methods to produce high quality photonic crystals to enhance the optical properties of

ST-OPVs; and, finally, development of an efficient encapsulation strategy and evaluation of the lifetime of ST-OPVs. By overcoming these hurdles, we can expect the commercialization of ST-OPVs in near future.

References

1. C. J. Brabec, N. S. Sariciftci and J. C. Hummelen, *Adv. Funct. Mater.*, 2001, **11**, 15.
2. B. C. Thompson and J. M. J. Frechet, *Angew. Chem., Int. Ed.*, 2008, **47**, 58.
3. F. C. Krebs, *Sol. Energy Mater. Sol. Cells*, 2009, **93**, 394.
4. F. Guo, T. Ameri, K. Forberich and C. J. Brabec, *Polym. Int.*, 2013, **62**, 1408.
5. J.-H. Yoon, J. Song and S.-J. Lee, *Sol. Energy*, 2011, **85**, 723.
6. Y.-J. Cheng, S.-H. Yang and C.-S. Hsu, *Chem. Rev.*, 2009, **109**, 5868.
7. C.-C. Chueh, S.-C. Chien, H.-L. Yip, J. F. Salinas, C.-Z. Li, K.-S. Chen, F.-C. Chen, W.-C. Chen and A. K. Y. Jen, *Adv. Energy Mater.*, 2013, **3**, 417.
8. N. Lynn, L. Mohanty and S. Wittkopf, *Build. Environ.*, 2012, **54**, 148.
9. L. A. A. Pettersson, L. S. Roman and O. Inganas, *J. Appl. Phys.*, 1999, **86**, 487.
10. T. Ameri, G. Dennler, C. Waldauf, H. Azimi, A. Seemann, K. Forberich, J. Hauch, M. Scharber, K. Hingerl and C. J. Brabec, *Adv. Funct. Mater.*, 2010, **20**, 1592.
11. K.-S. Chen, J.-F. Salinas, H.-L. Yip, L. Huo, J. Hou and A. K.-Y. Jen, *Energy Environ. Sci.*, 2012, **5**, 9551.
12. H.-L. Yip and A. K.-Y. Jen, *Energy Environ. Sci.*, 2012, **5**, 5994.
13. A. Bauer, T. Wahl, J. Hanisch and E. Ahlswede, *Appl. Phys. Lett.*, 2012, **100**, 073307.
14. A. Colsmann, A. Puetz, A. Bauer, J. Hanisch, E. Ahlswede and U. Lemmer, *Adv. Energy Mater.*, 2011, **1**, 599.
15. J. Huang, G. Li and Y. Yang, *Adv. Mater.*, 2008, **20**, 415.
16. T. T. Larsen-Olsen, R. R. Sondergaard, K. Norrman, M. Jorgensen and F. C. Krebs, *Energy Environ. Sci.*, 2012, **5**, 9467.
17. J.-W. Kang, Y.-J. Kang, S. Jung, D. S. You, M. Song, C. S. Kim, D.-G. Kim, J.-K. Kim and S. H. Kim, *Org. Electron.*, 2012, **13**, 2940.
18. N. Kim, S. Kee, S. H. Lee, B. H. Lee, Y. H. Kahng, Y. R. Jo, B. J. Kim and K. Lee, *Adv. Mater.*, 2014, **26**, 2268.
19. R. Po, C. Carbonera, A. Bernardi, F. Tinti and N. Camaioni, *Sol. Energy Mater. Sol. Cells*, 2012, **100**, 97.
20. D. S. Hecht, L. Hu and G. Irvin, *Adv. Mater.*, 2011, **23**, 1482.
21. Q. Dong, Y. Zhou, J. Pei, Z. Liu, Y. Li, S. Yao, J. Zhang and W. Tian, *Org. Electron.*, 2010, **11**, 1327.
22. F. Nickel, A. Puetz, M. Reinhard, H. Do, C. Kayser, A. Colsmann and U. Lemmer, *Org. Electron.*, 2010, **11**, 535.
23. H. P. Kim, H. J. Lee, A. R. B. M. Yusoff and J. Jang, *Sol. Energy Mater. Sol. Cells*, 2013, **108**, 38.
24. J. E. Lewis, E. Lafalce, P. Toglia and X. Jiang, *Sol. Energy Mater. Sol. Cells*, 2011, **95**, 2816.

25. Y. Zhou, F. Li, S. Barrau, W. Tian, O. Inganäs and F. Zhang, *Sol. Energy Mater. Sol. Cells*, 2009, **93**, 497.
26. F. C. Krebs, S. A. Gevorgyan and J. Alstrup, *J. Mater. Chem.*, 2009, **19**, 5442.
27. S. K. Hau, H.-L. Yip, J. Zou and A. K. Y. Jen, *Org. Electron.*, 2009, **10**, 1401.
28. Y. Zhou, H. Cheun, S. Choi, C. Fuentes-Hernandez and B. Kippelen, *Org. Electron.*, 2011, **12**, 827.
29. J. Czolk, A. Puetz, D. Kutsarov, M. Reinhard, U. Lemmer and A. Colsmann, *Adv. Energy Mater.*, 2013, **3**, 386.
30. J. Zou, H.-L. Yip, S. K. Hau and A. K.-Y. Jen, *Appl. Phys. Lett.*, 2010, **96**, 203301.
31. M. Manceau, D. Angmo, M. Jorgensen and F. C. Krebs, *Org. Electron.*, 2011, **12**, 566.
32. C. Tao, G. Xie, C. Liu, X. Zhang, W. Dong, F. Meng, X. Kong, L. Shen, S. Ruan and W. Chen, *Appl. Phys. Lett.*, 2009, **95**, 053303.
33. C. Tao, G. Xie, F. Meng, S. Ruan and W. Chen, *J. Phys. Chem. C*, 2011, **115**, 12611.
34. D. Han, H. Kim, S. Lee, M. Seo and S. Yoo, *Opt. Express*, 2010, **18**, A513.
35. F. Li, S. Ruan, Y. Xu, F. Meng, J. Wang, W. Chen and L. Shen, *Sol. Energy Mater. Sol. Cells*, 2011, **95**, 877.
36. L. Shen, Y. Xu, F. Meng, F. Li, S. Ruan and W. Chen, *Org. Electron.*, 2011, **12**, 1223.
37. T. Winkler, H. Schmidt, H. Flügge, F. Nikolayzik, I. Baumann, S. Schmale, T. Weimann, P. Hinze, H.-H. Johannes, T. Rabe, S. Hamwi, T. Riedl and W. Kowalsky, *Org. Electron.*, 2011, **12**, 1612.
38. J. Zou, C. Z. Li, C. Y. Chang, H. L. Yip and A. K. Y. Jen, *Adv. Mater.*, 2014, **26**, 3618.
39. K.-S. Chen, H.-L. Yip, J.-F. Salinas, Y.-X. Xu, C.-C. Chueh and A. K. Y. Jen, *Adv. Mater.*, 2014, **26**, 3349.
40. J.-Y. Lee, S. T. Connor, Y. Cui and P. Peumans, *Nano Lett.*, 2010, **10**, 1276.
41. J. Krantz, T. Stubhan, M. Richter, S. Spallek, I. Litzov, G. J. Matt, E. Spiecker and C. J. Brabec, *Adv. Funct. Mater.*, 2013, **23**, 1711.
42. M. Reinhard, R. Eckstein, A. Slobodskyy, U. Lemmer and A. Colsmann, *Org. Electron.*, 2013, **14**, 273.
43. Y.-J. Kang, D.-G. Kim, J.-K. Kim, W.-Y. Jin and J.-W. Kang, *Org. Electron.*, 2014, **15**, 2173.
44. J. H. Yim, S.-Y. Joe, C. Pang, K. M. Lee, H. Jeong, J.-Y. Park, Y. H. Ahn, J. C. de Mello and S. Lee, *ACS Nano*, 2014, **8**, 2857.
45. C.-C. Chen, L. Dou, R. Zhu, C.-H. Chung, T.-B. Song, Y. B. Zheng, S. Hawks, G. Li, P. S. Weiss and Y. Yang, *ACS Nano*, 2012, **6**, 7185.
46. F. Guo, X. Zhu, K. Forberich, J. Krantz, T. Stubhan, M. Salinas, M. Halik, S. Spallek, B. Butz, E. Spiecker, T. Ameri, N. Li, P. Kubis, D. M. Guldi, G. J. Matt and C. J. Brabec, *Adv. Energy Mater.*, 2013, **3**, 1062.
47. F. Guo, P. Kubis, T. Stubhan, N. Li, D. Baran, T. Przybilla, E. Spiecker, K. Forberich and C. J. Brabec, *ACS Appl. Mat. Interfaces*, 2014, **6**, 18251.
48. S. Ye, A. R. Rathmell, Z. Chen, I. E. Stewart and B. J. Wiley, *Adv. Mater.*, 2014, **26**, 6670.

49. M. Song, D. S. You, K. Lim, S. Park, S. Jung, C. S. Kim, D.-H. Kim, D.-G. Kim, J.-K. Kim, J. Park, Y.-C. Kang, J. Heo, S.-H. Jin, J. H. Park and J.-W. Kang, *Adv. Funct. Mater.*, 2013, **23**, 4177.

50. H. Wu, L. Hu, M. W. Rowell, D. Kong, J. J. Cha, J. R. McDonough, J. Zhu, Y. Yang, M. D. McGehee and Y. Cui, *Nano Lett.*, 2010, **10**, 4242.

51. Z. M. Beiley, M. G. Christoforo, P. Gratia, A. R. Bowring, P. Eberspacher, G. Y. Margulis, C. Cabanetos, P. M. Beaujuge, A. Salleo and M. D. McGehee, *Adv. Mater.*, 2013, **25**, 7020.

52. Y. Zhang, S. K. Hau, H.-L. Yip, Y. Sun, O. Acton and A. K.-Y. Jen, *Chem. Mater.*, 2010, **22**, 2696.

53. Z. Tang, Z. George, Z. Ma, J. Bergqvist, K. Tvingstedt, K. Vandewal, E. Wang, L. M. Andersson, M. R. Andersson, F. Zhang and O. Inganäs, *Adv. Energy Mater.*, 2012, **2**, 1467.

54. L. Dou, W.-H. Chang, J. Gao, C.-C. Chen, J. You and Y. Yang, *Adv. Mater.*, 2013, **25**, 825.

55. C.-Y. Chang, L. Zuo, H.-L. Yip, Y. Li, C.-Z. Li, C.-S. Hsu, Y.-J. Cheng, H. Chen and A. K. Y. Jen, *Adv. Funct. Mater.*, 2013, **23**, 5084.

56. T. Ameri, G. Dennler, C. Lungenschmied and C. J. Brabec, *Energy Environ. Sci.*, 2009, **2**, 347.

57. J. You, C.-C. Chen, Z. Hong, K. Yoshimura, K. Ohya, R. Xu, S. Ye, J. Gao, G. Li and Y. Yang, *Adv. Mater.*, 2013, **25**, 3973.

58. C.-C. Chen, W.-H. Chang, K. Yoshimura, K. Ohya, J. You, J. Gao, Z. Hong and Y. Yang, *Adv. Mater.*, 2014, **26**, 5670.

59. A. R. B. M. Yusoff, D. Kim, H. P. Kim, F. K. Shneider, W. J. da Silva and J. Jang, *Energy Environ. Sci.*, 2015, **8**, 303.

60. C.-C. Chen, L. Dou, J. Gao, W.-H. Chang, G. Li and Y. Yang, *Energy Environ. Sci.*, 2013, **6**, 2714.

61. C.-Y. Chang, L. Zuo, H.-L. Yip, C.-Z. Li, Y. Li, C.-S. Hsu, Y.-J. Cheng, H. Chen and A. K. Y. Jen, *Adv. Energy Mater.*, 2014, **4**, 1301645.

62. W. Yu, L. Shen, Y. Long, W. Guo, F. Meng, S. Ruan, X. Jia, H. Ma and W. Chen, *Appl. Phys. Lett.*, 2012, **101**, 153307.

63. W. Yu, L. Shen, P. Shen, Y. Long, H. Sun, W. Chen and S. Ruan, *ACS Appl. Mat. Interfaces*, 2014, **6**, 599.

64. W. Yu, L. Shen, P. Shen, F. Meng, Y. Long, Y. Wang, T. Lv, S. Ruan and G. Chen, *Sol. Energy Mater. Sol. Cells*, 2013, **117**, 198.

65. Y. Galagan, M. G. Debije and P. W. M. Blom, *Appl. Phys. Lett.*, 2011, **98**, 043302.

66. F. Pastorelli, P. Romero-Gomez, R. Betancur, A. Martinez-Otero, P. Mantilla-Perez, N. Bonod and J. Martorell, *Adv. Energy Mater.*, 2015, **5**, 1400614.

67. R. Betancur, P. Romero-Gomez, A. Martinez-Otero, X. Elias, M. Maymo and J. Martorell, *Nat. Photon.*, 2013, **7**, 995.

CHAPTER 13

Solution Processed Organic Photovoltaics (OPVs)

HONGSEOK YOUN[a] AND L. JAY GUO[*b]

[a]Department of Mechanical Engineering, Hanbat National University, Deajeon, 305–719, Korea; [b]Department of Electrical Engineering and Computer Science University of Michigan, Ann Arbor, MI 48109, USA
*E-mail: guo@umich.edu

13.1 Introduction

Nowadays, sustainable and environmentally friendly manufacturing processes are gaining more attention in the solar cell industries. Among various types of photovoltaics, organic and polymer photovoltaics have been regarded as one of the third generation power sources, and they offer the advantages of low cost, flexibility and light weight. These fascinating advantages are not only useful in the use of mobile electronics as a new power source but also for scalable devices such as building integrated photovoltaics (BIPV), as shown in Figure 13.1. However, the low power conversion efficiency (PCE) of the OPVs has been one of the serious limitations. Amazingly, in the last few years, the power conversion efficiency (PCE) of the organic photovoltaics has been improved dramatically, to more than 10%.

In terms of cost and manufacturing technologies for scalable devices, because the functional materials of OPVs are likely to dissolve in common organic solvents, they can be easily printed or coated onto the flexible polymer substrates such as polyethylene terephthalate (PET) and polyethylene

RSC Polymer Chemistry Series No. 17
Polymer Photovoltaics: Materials, Physics, and Device Engineering
Edited by Fei Huang, Hin-Lap Yip, and Yong Cao
© The Royal Society of Chemistry 2016
Published by the Royal Society of Chemistry, www.rsc.org

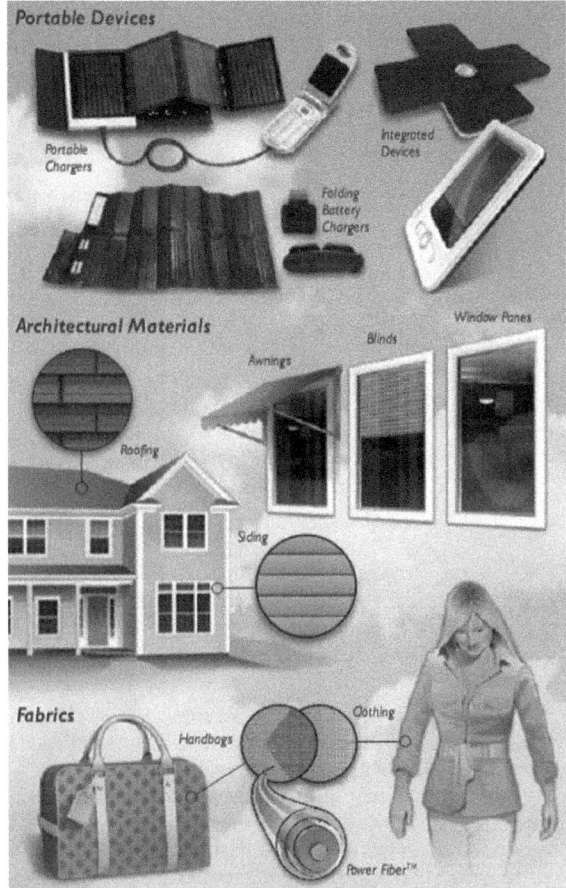

Figure 13.1 Various applications of organic photovoltaics. The images were captured at http://www.konarka.com, copyright © Konarka Inc.

naphthalate (PEN). Therefore, in principle, large area OPVs can be fabricated by simple printing technologies such as roll-to-roll, inkjet, screen and spray coating systems.[1–4] This chapter introduces various printing technologies and deals with solution processible functional materials. Finally, we will discuss device and fabrication issues in scalable OPVs.

13.2 Material Cost Issues in OPVs

As mentioned above, OPVs are one of the promising types of photovoltaic in terms of the low-cost advantage. However, the materials costs of OPVs occupy around 80% of the overall module cost per m^2, as shown in Figure 13.2. More seriously, as shown in Table 13.1, the estimated actual total module cost to produce 1 watt exceeds 8 euros (the total cost was 5.35 € to produce 660 mW). Therefore, the material wastage and losses in the process should

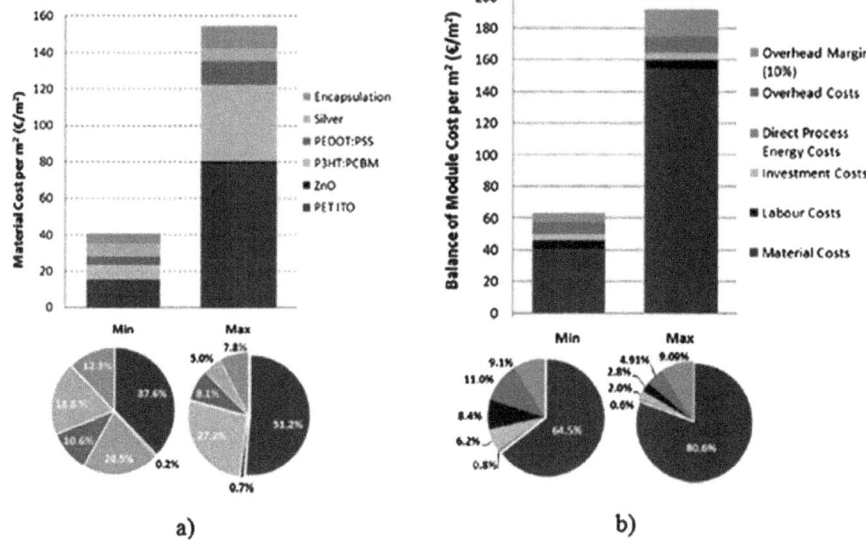

Figure 13.2   (a) Estimation of the material costs of polymeric solar cells; (b) estimated costs of polymeric solar cells. Reproduced from ref. 5 with the permission of RSC Publications, © 2014.

Table 13.1   Optimized cost structure in terms of materials usage and processing time for the manufacturing of polymer solar cell modules. The calculation is the actual cost for the manufacture of one 16 × 13 mm module with an active area of 360 cm² and includes associated materials losses. Power outputs for these modules are up to 660 mW (AM1.5G, 1000 W m⁻²). Reproduced from ref. 3 with the permission of RSC Publications, © 2014.

Material	Material cost/€	Processing cost/€	Total/€
Barrier	0.4575	0.03173	0.4892
Pressure sensitive adhesive	0.1918	0.03173	0.2236
PET-ITO	2.6077	0.21111	2.8188
ZnO	0.0582	0.16667	0.2249
P3HT-PCBM	0.4492	0.16667	0.6159
PEDOT:PSS(EL-P 5010)	0.2311	0.16667	0.3978
Silver (PV410)	0.4120	0.16667	0.5787
Total	4.4078	0.9412	5.3491

be minimized and we need to select carefully the appropriate fabrication technologies to enable sustainable and green manufacturing processes for highly efficient devices.

The so-called inverted structure is favorable for the application of practical printing technologies and to ensure long-term durability. A typical inverted device structure consists of indium tin oxide (ITO) as a cathode and zinc oxide (ZnO) as an electron transport layer, P3HT:PCBM as a photoactive layer, PEDOT:PSS as a hole transport layer and a silver anode, which are all printable.

In the inverted structure, the ZnO, P3HT:PCBM and PEDOT:PSS are initially prepared in the liquid phase. These functional layers can be prepared by various coating technologies. However, the non-liquid layers such as ITO and the reflective electrode are commonly deposited in vacuum chambers, which is time consuming and results in material wastage due to a low material utilization rate during the evaporation. In particular, the material cost of the ITO can amount to 50% of the total material cost because the indium is a rare element on the earth and the low materials cost is very sensitive to the market. Therefore, alternative electrodes for use instead of ITO are urgently required for not only OPVs but also organic displays. In this regard, our recent development of an ultrathin Al-doped Ag film may provide an excellent candidate for a transparent conductor that can be prepared easily by sputtering or evaporation.[6]

13.3 Fabrication Technologies Toward Low-Cost and Scalable OPVs

The functional layers having a liquid phase can be fabricated by slot–die, inkjet, screen and blade coating methods. However the functional layers not only have different coating characteristics but also have different film thicknesses, and therefore it is necessary to select the appropriate fabrication method for each functional layer.

13.3.1 Slot–Die Coating Process

Originally, the slot–die coating system was developed to achieve uniform coating films using various types of photoresist in the large-scale displays industry. A slot–die system utilizes a slot nozzle which extrudes solution from an ink reservoir by a static pumping system, as shown in Figure 13.3. To realize uniform film thickness over the whole wet film area, a constant pumping rate is a key parameter. The film thickness can be controlled by the pumping rate and speed of movement of the coating system. Finally, the thickness can be determined by the mass flow of the solution onto the substrate. However, the thickness of the electron extraction layers should be controlled within a few nanometers in an OPV. To guarantee thinner films (*e.g.* <10 nm), specialized syringe pumping has been frequently used to minimize the volume of the solution.[7] In addition, although the slot–die system is useful for achieving a continuous and uniform film, it is limited to making only striped patterns. This is the reason why a typical module consists of striped cells, as shown in Figure 13.3. In slot–die coating, a continuous and stable meniscus between the nozzle and the substrate is important. As can be seen in Figure 13.4, air pockets in the meniscus will create unwanted discontinuous coating lines. Therefore, to create a robust coating bead, it is necessary to control the instability and impurities of the coating solutions.

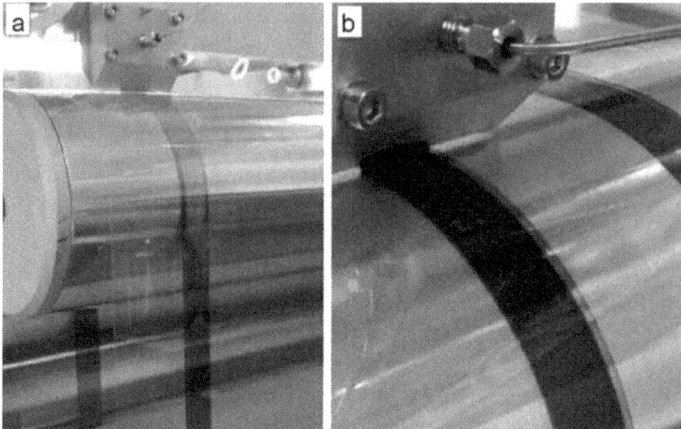

Figure 13.3   Photographs of a roll-to-roll (R2R) coating experiment in progress. (a) Coating of the front bulk heterojunction (BHJ) material (the drying process of the film is visible). (b) Coating of the back BHJ material. Reproduced from ref. 8 with the permission of Elsevier, © 2014.

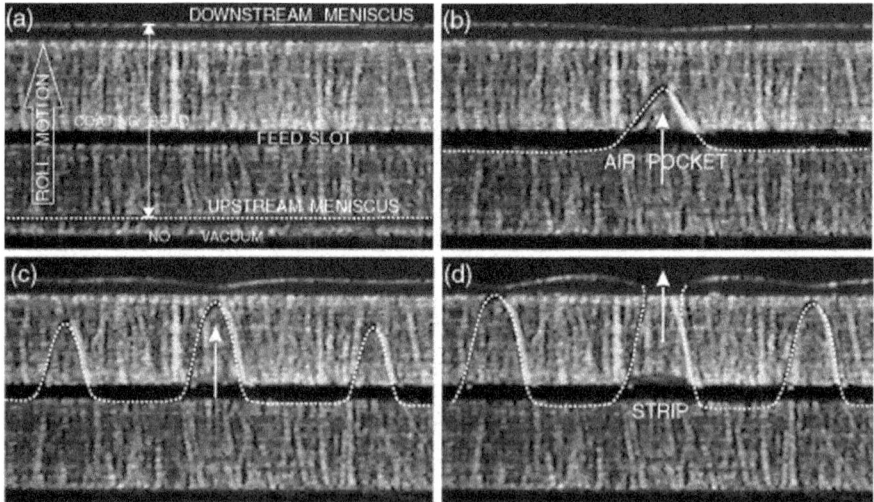

Figure 13.4   View through the glass roll of the coating bead as the flow rate is reduced. The roll speed was fixed capillary number (Ca) = 0.12 and no vacuum was applied to the upstream free surface. The cylinder was moving from bottom to top in each photograph. As the film thickness falls, the upstream free surface moves towards the feed slot and becomes three-dimensional. As the flow rate is decreased even further, the "V" pattern grows until the coating bead breaks. Reproduced from ref. 9 with the permission of Elsevier, © 2014.

In one report, a ZnO nanoparticle layer, P3HT:PCBM and a PEDOT:PSS layer were slot–die coated sequentially onto ITO-PET, and the performance of the device fabricated by the slot–die coating system exhibited around 2% PCE (the module area was 96 cm^2).[3]

13.3.2   Inkjet Printing Process

As discussed above, slot–die coating is unable to create arbitrary patterns. Inkjet printing is appropriate for creating such patterns. Therefore, inkjet printing technologies are frequently used for display applications such as color filters and organic light-emitting diode (OLED) displays. In the OPVs, however, relatively large patterns are required in the light-absorption area. Therefore, a smooth film and printing speed should be considered for scalable OPVs.

To fabricate OPVs by this printing system, the ink formulation needs to be modified to control the droplet volume and meniscus of the solution. In addition, the inkjet printing process should allow good film morphology without the coffee ring effect, as shown in Figure 13.5.

There have been some trials involving fabrication of OPVs by the inkjet printing process. Usually, P3HT:PCBM is prepared by the inkjet printing process. The performance of a device fabricated by the inkjet printing process was around 1.25% at a size of 18 cm^2.[11]

13.3.3   Traditional Roll-to-Roll Printing Process

There are three types of printing method commonly used in traditional roll-to-roll printing: gravure, flexography, and offset printing, as summarized in Table 13.2. Gravure printing is widely applied to fabricate OPVs because it is able to create a continuous film *via* small printing cell patterns.

The engraved roll surface uses many small cells to transfer the inks to the substrate. The inks are loaded into the cells from the ink reservoir and then metered by a doctor blade. They are transferred onto the substrate and finally merged to create a continuous wet film on the substrate.

In the general coating process, the ink viscosity and the surface energy of the substrate are prominent factors responsible for creating a uniform wet film. If the viscosity is too high or the surface energy of the substrate is small, the cell patterns will remain. When the substrate is treated by O$_2$ plasma, the surface energy of the substrate can be modified. Thus, the coating performance can be greatly improved with respect to the conditions of the solution and substrate.

During the coating process, unwanted instabilities can occur and are mainly related to the viscosity and surface tension of the solution, as well as the coating speed. The dimensionless capillary number is useful to characterize the various instabilities such as ribbing, cascading, mist, *etc.*

$$C_a = \frac{\mu V}{\sigma} \tag{13.1}$$

where μ is the viscosity of fluid, V the flow rate (roll speed), and σ the surface tension. These coating instabilities can be regarded as coating failures

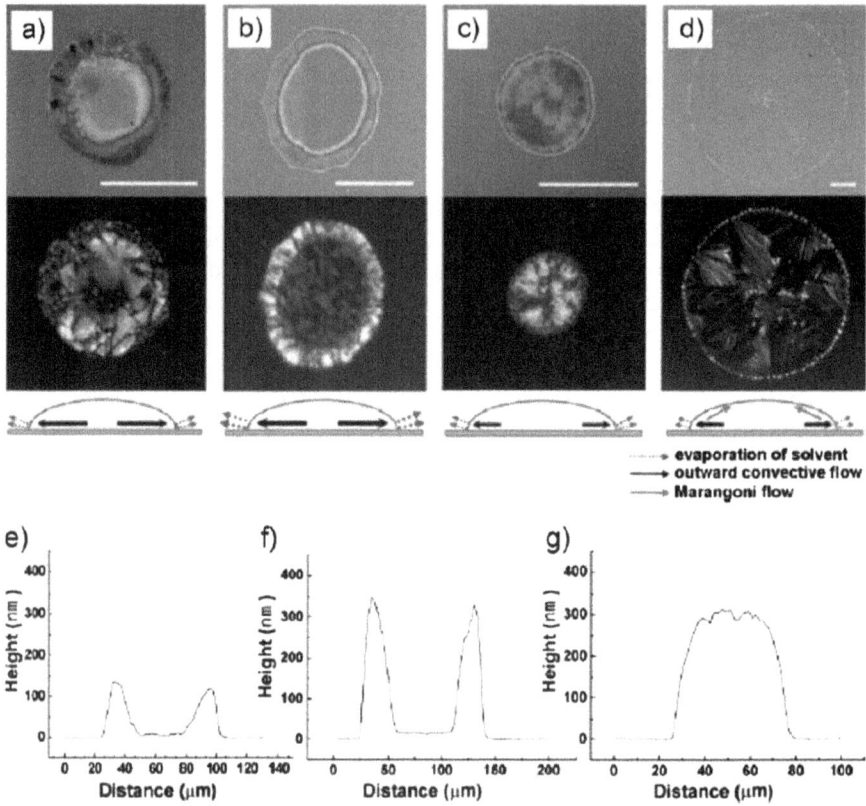

Figure 13.5   Optical microscopy and polarized images of inkjet-printed drop-
lets of TIPS pentacene, using chlorobenzene as the major solvent
mixed with 25% of the minor solvents (a) chlorobenzene, (b) hexane,
(c) *o*-dichlorobenzene, and (d) dodecane. (Scale bar represents 50
mm.) The profilometery images of single dots using chlorobenzene
mixed with 25% minor solvents, (e) chlorobenzene, (f) hexane, (g)
o-dichlorobenzene, are also shown. Reproduced from ref. 10 with the
permission of Wiley-VCH Verlag GmbH & Co. KGaA, © 2014.

Table 13.2   Comparison of roll-to-roll printing processes.

	Gravure	Flexography	Offset
Resolution (µm)	20–75	75	30–50
Thickness (nm)	<50	<50	<1000
Viscosity (mPa s)	50–200	50–500	20k–100k
Coating speed (m^2 s^{-1})	60	10	20
Polymer ink	o	o	×

or defects which will be harmful to the device. To regulate the coating instabilities, the capillary number should be properly chosen with respect to the
operating speed, according to Figure 13.6.

To address the issue of instability in the printing process, one can modify the ink formulation and the surface of the substrate. If irregular ripples
or streaks are observed along the coating direction (called "ribbing instability"), typically the problem can be solved by reducing the capillary number,
see for example point 1 or 3 in Figure 13.6 for small initial Ca numbers. This
can be done by reducing the ink viscosity or increasing surface tension; see
eqn (13.1). In the case of a large initial capillary number, a higher coating
speed is the best way to improve coating quality (*e.g.* point 2 in Figure 13.6).
Thus, one needs to decide which parameters to adjust with respect to the
initial capillary number. As another example, if many ripples are observed in
the direction perpendicular to the coating direction (referred to as "cascading coating instability"), one can efficiently improve the coating quality by
reducing the coating speed.

In one report in the literature, the performance of a device fabricated by
the gravure printing system exhibited around 0.6% PCE (the device area was
4.5 mm^2).[13] Typically, P3HT:PCBM and PEDOT:PSS layers are slot–die coated
sequentially onto the ITO-PET.

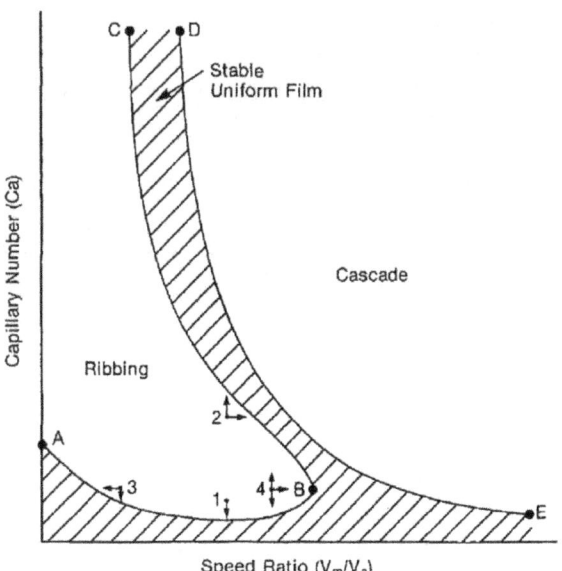

Figure 13.6   Typical stability diagram for a metered film in reverse roll coating.
Reproduced from ref. 12 with the permission of Wiley-VCH Verlag
GmbH & Co. KGaA, © 2014.

13.4　Materials for Functional Layers

13.4.1　Flexible Substrates

Many plastic substrates are available for use in flexible OPVs. As summarized in Table 13.3, high optical transmittance, heat resistance and solvent resistance are requirements for the flexible and scalable OPVs. Therefore, PET and PEN are most frequently used as the substrates in OPVs.

As mentioned previously, the surface conditions of the substrate are important when solutions such as PEDOT:PSS or other inks are coated. Basically, it is not easy to coat aqueous based PEDOT:PSS onto the hydrophobic polymer substrate. Therefore, various techniques can be applied to modify the surface of the substrate, *e.g.* plasma treatment and treatment with various chemicals such as primers or coating promoters. The surface energy can be divided into the disperse energy and polar energy. The sum of the two terms is the total surface energy. For example, the surface energy of the substrate can be enhanced by O_2 plasma. The surface energy is increased with respect to the treatment time with O_2 plasma, as shown in Table 13.4.

13.4.2　Silver Back Electrode

Silver pastes are commonly used in photovoltaics applications as top electrodes. The typical resistivity of the commercial highly conductive Ag paste is around $4.5{\times}10^6$ Ω cm^{-1} (see http://www2.dupont.com). In the case of OPVs, silver paste is also used as an anode. Because this silver paste has a relatively large viscosity, it is commonly screen printed. However, the silver paste consists of many micron-sized silver flakes which are easily able to penetrate to the underlying polymer layer; this can cause an electrical shorting problem

Table 13.3　Comparison of polymer materials for flexible substrates.[a]

	PET	PEN	PC	PAR	PES	PI
Optical clarity	o	o	◎	o	o	×
Upper operating temperature	△	o	△	◎	o	◎
Dimensional stability	o	o	△	△	△	o
Surface roughness	×	×	o	o	o	o
Solvent resistance	o	o	×	×	×	o
Moisture absorption	o	o	△	△	×	×
Young's modulus	o	o	△	△	△	△

[a]◎: Excellent, o: good, △: fair, ×: poor.

Table 13.4　The surface energy of ITO substrates after O_2 plasma treatment.

Exposure time (min)	0	30	60	120
Disperse (mN m^{-1})	24.71	34.85	36.09	36.03
Polar (mN m^{-1})	0.92	11.25	16.96	16.78
Total (mN m^{-1})	25.63	46.1	53.05	52.81

when the device is driven. To address this issue, a thick charge transport layer (800 nm–2 μm) is required, such as highly conductive PEDOT:PSS. However, this thick PEDOT:PSS layer is undesirable because of its higher resistance and reduced light transmission. Therefore, researchers have reported recently that a highly conductive and highly reflective top electrode showed better efficiency without a thick protective layer if produced *via* the transferring method.[14]

Recently, we have been able to achieve a highly conductive, solution processed Ag back electrode by using organometallic or nanoparticle based conductive silver inks, with resistivity around 3×10^6 Ω cm, which is comparable to the resistivity of bulk silver.[15] We found that the conductivity of the Al layer strongly depends on the annealing temperature because organic residues in the silver film can affect the conductivity. To achieve high conductivity of the silver electrode, a temperature of 180 °C is required for removal of the organic residues and growth of the silver grains. Direct printing of electrodes using organometallic inks frequently causes contamination due to penetration of the inks into the underlying organic layers during the printing and annealing processes. In addition, as mentioned above, the metal inks usually require high sintering temperatures to remove the residual organics that may not be compatible with the plastic substrate. To address these limitations, we developed a multi-layer roll transferring (MRT) approach, where a highly conductive and solution processed Ag electrode is prepared separately from the rest of the organic layers; following this, the device is completed by transferring the Ag electrode to the organic layers.[15] The MRT processed metal electrode has an excellent resistivity (3.4 μΩ cm) comparable to that of a thermally evaporated silver film. The performances of the devices fabricated by the MRT process were comparable to those of metal evaporated devices.[15]

13.4.3 Active Layer and Coating Issues

To characterize the coating performances of the various solutions, we developed roll-to-roll and roll-to-plate coating systems. The solution can be extruded *via* a slit nozzle added to the roll-to-roll system which functions similarly to the nozzle in the slot–die system. The film thickness can be controlled by the coating speed and slit gap. PEDOT:PSS, P3HT:PCBM, PBDTTT:PCBM, ZnO nanoparticle and interface dipole solutions have been tested by using this coating system.

The photoactive materials are usually dissolved into common organic solvents, as summarized in Table 13.5. In particular, dichlorobenzene and chlorobenzene are mostly used as the solvent. Because these solvents have relatively high boiling temperatures, they can provide enough time to create bulk heterojunction crystalline P3HT:PCBM.[16] However, the long evaporation time can cause unwanted flows, such as convection flows or Marangoni flows, during the drying period. These are due to the differences in the convection rate and surface tension over the whole wet coating area. That is why printing industries commonly utilize fast volatile solvents, not only to increase manufacturing speed but also to improve the printing quality.

Table 13.5   Properties of common solvents used in OPVs.

	Boiling temperature (°C)	Vapor pressure (mmHg at 20 °C)	Surface tension (mN m^{-1})	Density (g cm^{-3})
Toluene	110.6	22	28.4	0.87
Chlorobenzene	131	11.8	33.6	1.11
Dichloroben-zene	180	1.2	26.84	1.3
NMP (*N*-methyl-2-pyrrolidone)	202	0.5	40.79	1.03
Isopropanol	82.3	44	23	0.786
Water	100	17.5	72.8	1

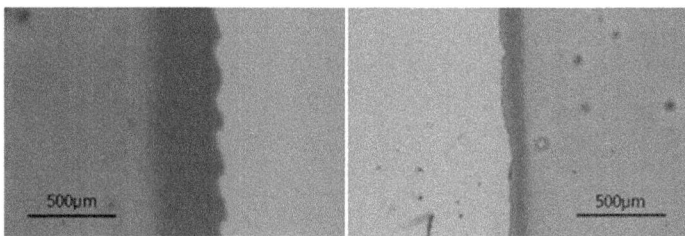

Figure 13.7   Microscope pictures of the printed pattern edge area of the conducting polymer PEDOT:PSS, showing a naturally dried edge (left) and an edge dried using hot wind drying (right).

Because the flow issues during drying directly affect the coating quality and uniformity, basic coating tests are required using various solvents for the functional materials in OPVs.

The surface tension usually results in a thicker film at the edge of the material under natural drying conditions, which presents similar challenges in inkjet printing applications, such as the "coffee ring" effect. However, by facilitating the evaporation of the solution using a heat blower, such as a heat gun, or hot plate, the flow issue during the drying time can be improved, as shown in Figure 13.7.

Basically, the film thickness can be determined by the viscosity of the solution, among other factors. Given that the viscosity of the solution is proportional to the concentration,[13] when we change the solution concentration to control the thickness, the film thickness can be roughly estimated by the following relations:

$$h = 1.34 \left(\frac{\mu V_{\text{blade}}}{\sigma} \right)^{2/3} R \qquad (13.2)$$

where V_{blade} represents the coating speed, μ and σ are the viscosity and surface tension of the solution, and R is the curvature radius of the down-stream meniscus. For example, assume that the viscosity is linearly proportional to the concentration of the solution and all other parameters are fixed. The thickness of the P3HT:PCBM layer is 300 nm using 1 wt% solution. To achieve 150 nm

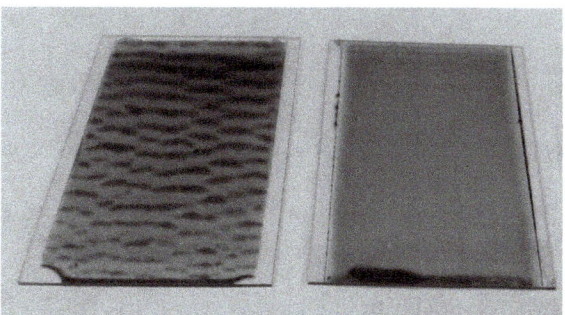

Figure 13.8 Coated film images made by the blade–slit coating method at different coating speeds. PBDTTT film which was coated at a speed of 2 cm s^{-1} shows irregular patterns (left). The film quality was greatly improved by reducing the coating speed (0.8 cm s^{-1}).

thickness of the P3HT:PCBM layer, the viscosity should be reduced to 36% of the initial viscosity. Thus, the concentration of the solution needs to be reduced to around 0.36 wt%.

In practical experiments, however, it is hard to control the viscosity precisely because the viscosity is not always linearly proportional to the solution concentration. However, we can also control the thickness by adjusting the coating speed. This is similar to spincoating, where adjustment of the coating speed is the easiest way to control thickness.

P3HT:PC$_{61}$BM is widely reported to have been fabricated by various printing technologies because it is relatively robust and has good printing properties. However, the efficiency has been limited to around 1–2% PCE (in the case of all printed OPVs). Recently, various low bandgap polymers (*e.g.* PBDTTT) have been developed, and the devices exhibit 7–10% PCE. To improve the morphology of PBDTTT and PC$_{71}$BM, small molecule DIO is commonly added. Moreover, even though a longer evaporation time (solvent annealing) improves crystallinity in the case of P3HT:PCBM, PBDTTT:PC$_{71}$BM does not require additional evaporation time after film casting. However, the typical thicknesses of these low bandgap polymers are around 100 nm, which is much thinner than P3HT:PC$_{61}$BM. Thus, the realization of a thin and uniform film is much more challenging. For example, the coated film shows irregular ripples (cascading) along the transverse direction of the coating, as mentioned above. We can estimate this coating instability by using the appropriate capillary number. To reduce cascading, a favorable and effective method is to control the coating speed, as shown in Figure 13.8.

13.4.4 Interfacial Layer (PEO, PEIE)

In conventional OLEDs or OPVs, the electron injection layers are commonly fabricated by vacuum processes such as thermal evaporation under high vacuum conditions, owing to the vulnerability of these layers to oxygen and moisture in the air. Devices not employing these materials usually show poor device efficiencies. Electron injection layers such as LiF,[17] CsF,[18] NaF,[19] and Cs$_2$CO$_3$[20] are only a few nanometers thick. Moreover, because the ultrathin electron injection/extraction layers are very sensitive to the surface conditions, such as the roughness of the underlying layers, one cannot directly apply these reactive materials in the solution process. Therefore, these ultrathin electron injection layers are not appropriate for a solution processed device. There have been many efforts to replace these reactive materials by stable and liquid phase materials such as water soluble polymers having amino groups or organic surfactants. Polyethylene oxide (PEO) is a polymer that is well-known as an organic surfactant that efficiently creates an interface dipole because the molecule has lone pair electrons in the ethylene oxide monomer. Polyethylenimine (PEIE), known as a surface promoter, has amino groups in the chains, and the amino groups can create interfacial dipoles with contact layers.[21] These interface dipoles can reduce the work function of the metal or the energy levels by shifting the vacuum energies; this is very useful in reducing the electron injection/extraction barriers.[22] Moreover, because these materials are relatively stable in the air, they can be fabricated by practical solution processing in an air environment.

In the inverted structure, ultrathin (<10 nm) interface dipole layers are employed to reduce energy barriers between the photoactive layer and the electron transport layer, or to reduce the work function of the cathode. The solution of the interfacial materials does not have sufficient viscosity, owing to the extremely low concentration of the solid, and the solvents are very different from those used for conventional photoactive materials. Therefore, to realize the ultrathin film of the interfacial layer, specialized coating methods are required. Only a few papers have reported that these ultrathin layers were successfully coated; for example, Youn *et al.* reported a roll-to-roll cohesive coating method, as shown in Figure 13.9.[23] Because this cohesive coating method only utilizes the cohesive force of the solution, the amount of solution ejected from the nozzle can be efficiently minimized. Applying such a method has successfully demonstrated devices at a relatively large scale (7 × 8 cm).[23]

13.4.5   Hole Transport Layer (HTL)/Electron Transport Layer (ETL)

PEDOT:PSS is a well-known conducting polymer; it has a good hole transporting behavior in OPVs and OLEDs. The conductivity of the PEODT:PSS can be decided by the ratio of the PEDOT and PSS. Generally, a higher portion of PEDOT gives higher conductivity, as summarized in Table 13.6. There are several commercialized products of PEDOT:PSS as shown in Figure 13.10. The typical viscosity of PEDOT:PSS (Clevious P) is around 30–150 mPa s. In the case of less solid concentrations of the polymer, the viscosities are lower, around 20 mPa s. To achieve a uniform PEDOT:PSS film, the PET substrate is pre-cleaned and treated with O_2 plasma.

PEDOT:PSS can be used as a transparent and flexible electrode instead of ITO. Because ITO is a brittle metal oxide, it is vulnerable under bending stress. By adding DMSO, for example, to the solution, followed by post-treatment using methanol, the conductivity can be enhanced dramatically

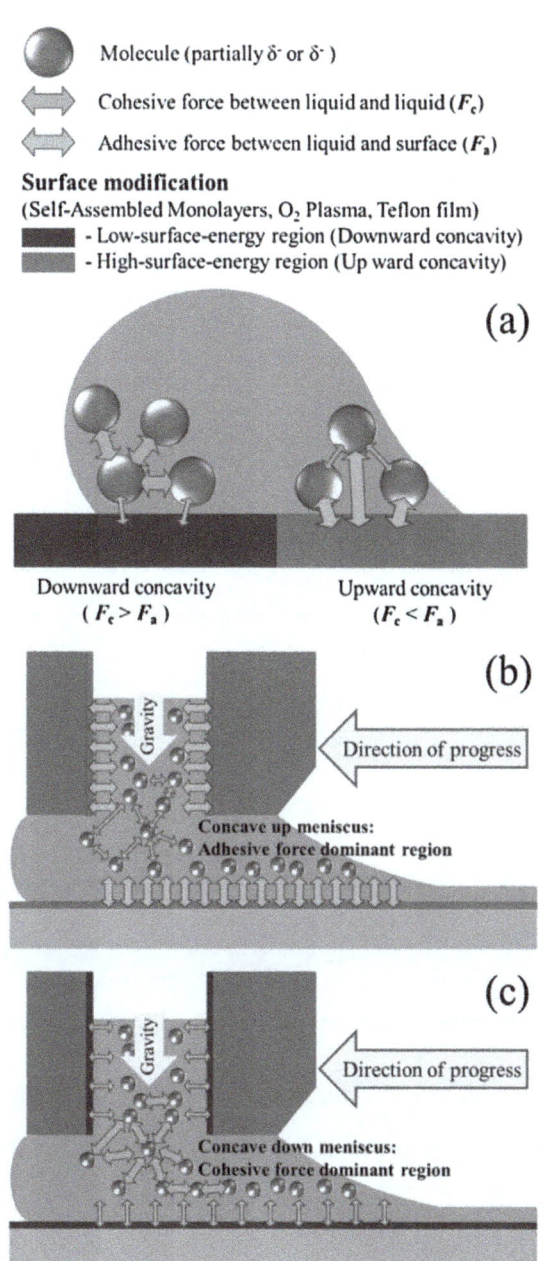

Figure 13.9 Illustrations of the cohesive coating mechanism and details of the coating flows. (a) Adhesive (F_a) and cohesive (F_c) force diagram between hydrophobic and hydrophilic surfaces. Contact angles depend on the relative forces between solvent molecules and surfaces. In the case of downward concavity, the cohesive force is higher than the adhesive force, whereas upward concavity means that the adhesive force is higher than the cohesive force. (b) The adhesive force dominant condition. Adhesive forces are enhanced by surface energy modifications, such as O_2 plasma treatment. Therefore, the flow rate is reduced by adhesive forces. (c) Cohesive force dominant condition. Adhesive forces are relatively reduced by self-assembled monolayers (SAMs). Thus, the flow rate is increased. Reproduced from ref. 23 with the permission of Wiley-VCH Verlag GmbH & Co. KGaA, © 2014.

Table 13.6   PEDOT:PSS specification. All information and data were based on http://www.heraeus-clevios.com.

Property or characteristic	Clevios™ PH1000	Clevios™ P VP AI4083	Clevios™ P VP CH8000
Solid content	1.1–1.3%	1.3–1.7%	2.4–3.0%
Resistivity	1.0×10^{-2} Ω cm^{-1} (doped with DMSO)	500–5000 Ω cm	5.0×10^{4} Ω cm^{-1} -3.0×10^{5} Ω cm^{-1}
Viscosity	15–50 mPa s	5–12 mPa s	2–20 mPa s
Form	Liquid	Liquid	Liquid
Odor	Odorless	Odorless	Odorless
Color	Dark blue	Dark blue	Dark blue
PEDOT:PSS ratio	1:2.5 (by weight)	1:6 (by weight)	1:20 (by weight)
pH (at 20 °C)	1.5–2.5	1.2–2.2	1.0–2.0
Density (at 20 °C)	1 g cm^{-3}	1 g cm^{-3}	1 g cm^{-3}
Boiling point	Approx. 100 °C	Approx. 100 °C	Approx. 100 °C
Work function	5.0 eV	5.2 eV	5.2 eV

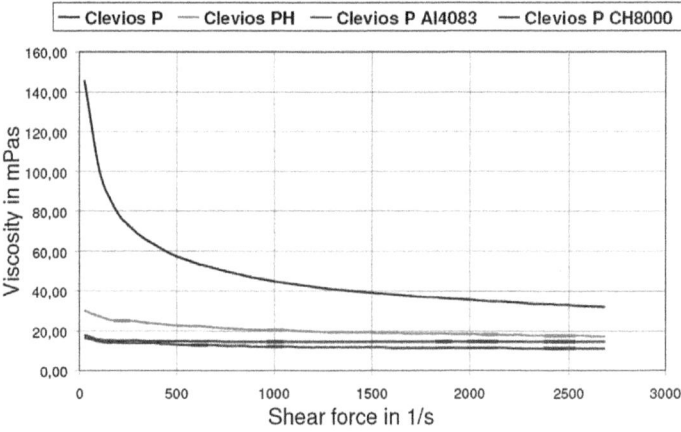

Figure 13.10   Viscosity *vs.* shear rate for various types of PEDOT:PSS.

(>1000 S cm^{-1}). To coat water-borne PEDOT:PSS solution on ITO-coated PET, the substrate is pre-cleaned by deionized water, acetone and isopropyl alcohol sequentially, and subsequently the substrate is usually treated by O$_2$ plasma. In addition, without surface modifications, one can coat PEDOT:PSS by adding a surfactant to promote the wetting. The well-known wetting agent Dynol 604 (Airproducts) can be added to improve the wetting of the PEDOT:PSS solution. The wetting angle is much reduced by the use of additives with PEDOT:PSS, as shown in Figure 13.11.

A ZnO nanoparticle layer is widely used as an electron transport layer. The ZnO nanoparticle has excellent electron transport behavior because it has favorable electron mobility and the conduction band level is nicely matched with the LUMO level (lowest unoccupied molecular orbital) of the PCBM. It

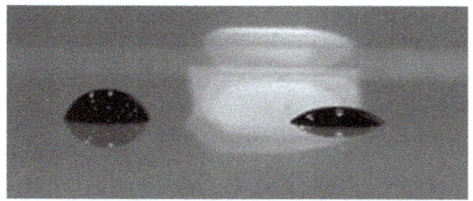

Figure 13.11 Images of PEDOT:PSS wetting. PEDOT:PSS with surfactant (Dynol 604) shows better wetting on the PET substrate without using the O_2 plasma treatment.

Figure 13.12 Images of ZnO nanoparticle synthesis. The transparent precursor solution (left) becomes translucent (middle) and turbid (right) during the synthesis of ZnO nanoparticles.

also has a good hole blocking behavior in devices owing to its deep valence band level (7.5 eV). The ZnO nanoparticles are usually synthesized from the reaction of zinc acetate and potassium hydroxide and precipitated by dehydration on a hotplate, as shown in Figure 13.12. The nanoparticles (with an average size of 5 nm) are gathered by centrifugation and then redistributed in 1-butanol. The typical thicknesses of ZnO nanoparticle layers fabricated by the blade coating method are around 20–50 nm on the ITO cathode. Unlike the synthesized ZnO sol–gel, the crystallinity of the ZnO nanoparticle is already established after the synthesis, therefore it does not require an annealing process for crystallization.

13.5   Issues in Scalable OPVs

13.5.1   Effect of Device Size

When the device size increases, the device performance is affected as a result of manufacturing or device issues. First, with increasing device size, the possibility of failure resulting from defects will increase. Thus, scalable OPVs require robust and defect-free processing. These defects, such as pin-holes and particulates, cause unwanted charge shunt paths. Low bandgap polymer

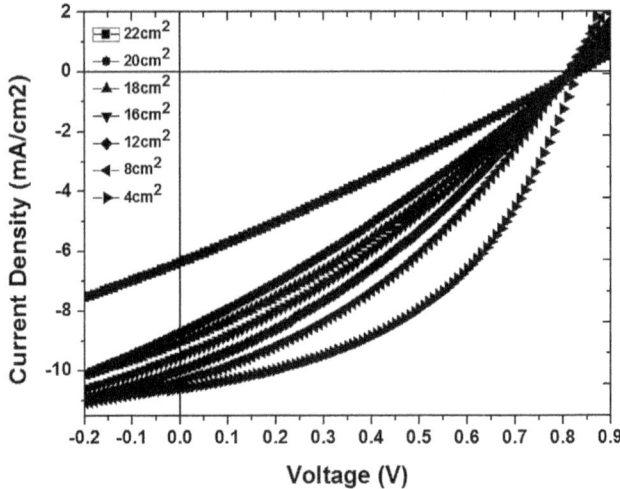

Figure 13.13   The *J–V* device performance of PIDT-PhanQ:PC$_{71}$BM cells with to varying device size (4–22 cm^2).

Table 13.7   The device performance of PIDT-PhanQ:PC$_{71}$BM cells of different sizes.

Device size (cm^2)	J_{SC} (mA cm^{-2})	V_{OC} (V)	FF (%)	PCE (%)	R_{series} (Ω cm^2)	R_{shunt} (Ω cm^2)
22	6.34	0.84	26.67	1.42	112.62	160.14
20	8.72	0.84	27.77	2.03	76.99	130.88
6	9.04	0.83	29.96	2.24	66.72	152.87
12	9.52	0.83	30.63	2.42	59.64	155.71
8	9.98	0.83	32.49	2.69	51.03	177.69
4	10.35	0.82	35.90	3.04	40.00	216.25

PIDT-PhanQ devices (ITO/ZnO/PIDT-PhanQ:PC$_{71}$BM/MoO3/Ag) have been fabricated to evaluate the device size effect. When the device size increases, the shunt resistance significantly reduces in devices having an area greater than 4 cm^2, as shown in Figure 13.13 and Table 13.7. Apart from the defect issues, the resistance of the transparent electrode causes power loss due to the large series resistance of the device.

We can evaluate these two resistances from the *J–V* characteristics of the devices. Basically, the series resistance can be obtained from the curve-fitting of the open circuit condition and the shunt resistance can be acquired from the curve-fitting of the short circuit condition. Because the thin organic functional layers are weak and soft, they are sensitive to the processing conditions and to dusts and contamination in the atmosphere. Therefore, first and foremost, to address these issues in scalable devices, all the constituent layers should be perfect over the whole area after device fabrication to maintain the high shunt resistance of the device. Defects are detrimental to large area devices because they can essentially sink the photo-generated currents and result in poor device performance (reduced fill factor). Second, to reduce the series resistance for the transparent electrode, one could use alternative highly conductive electrodes instead of the ITO. Recently, many researchers have studied alternative transparent conductive electrodes such as metal mesh,[25] metal nanowires, carbon nanotubes (CNTs) and graphenes, or combinations of them.[26]

13.5.2 Isolation of Defects

If a device has defects, the device will not operate properly. A severe case will produce an electrical shorting problem, or the defects may cause shunt loss so that the PV cell cannot attain a good fill factor, due to the small shunt resistance. These issues will be critical and will impact the manufacturing cost. If a tool can be developed to define localized defects and further curing methods are available, this could be a practical solution in the OPV field to allow future large scale manufacturing.

There are special tools for detecting defects that use non-contact imaging techniques, such as the light beam induced current (LBIC) mapping method, dark lock-in thermography imaging, electroluminescence (EL) imaging, and photoluminescence (PL) imaging.[24] These can be used to obtain information on the devices, such as localized regions of leakage from particles and pinholes, as shown in Figure 13.14.

To isolate the defects, pulsed laser has been widely used in inorganic Si-based solar cells. Because the OPV layers are much softer than those of Si-based devices, a simple razor blade may be used to isolate the defects in the laboratory if the defect can be detected by the naked eye (Figures 13.15 and 13.16). The device performance is greatly improved by isolation of the defect, as shown in Figure 13.16. In particular, not only will the shunt resistance be increased, but the series resistance will also be improved by isolating the defects from the active region of the device (Table 13.8).

Figure 13.14 LBIC image acquired in roll-to-roll processing showing a defect (the resolution is 100 micron, the size of the imaged module is 305 × 250 mm, and the data acquisition time is ~1 s). Reproduced from ref. 24 with the permission of Wiley-VCH Verlag GmbH & Co. KGaA, © 2014.

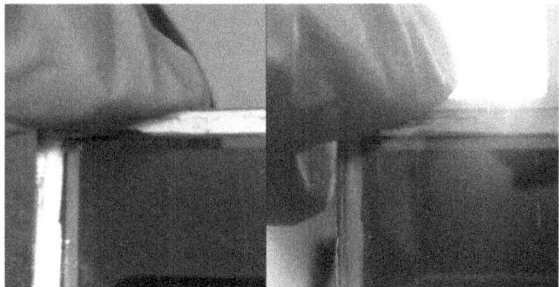

Figure 13.15   Particulate on a PIDT-PhanQ polymer layer isolated by a razor blade.

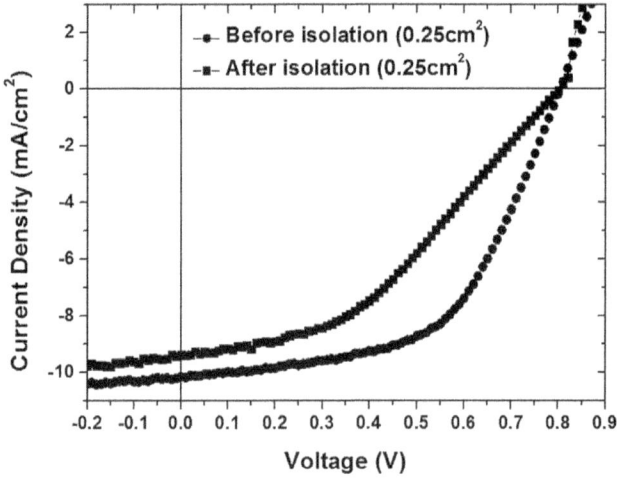

Figure 13.16   The *J–V* device performances of PIDT-PhanQ:PC$_{71}$BM cells following isolation of defects using a razor blade.

Table 13.8   The device module performance of PIDT-PhanQ:PC$_{71}$BM.

Device	J_{SC} (mA cm^{-2})	V_{OC} (V)	FF (%)	PCE (%)	R_{series} (Ω cm^2)	R_{shunt} (Ω cm^2)
Without defect isolation	9.37	0.81	39.79	3.02	24.04	443.24
Defect isolated	10.15	0.81	55.38	4.55	22.98	664.19

13.6   Conclusion

The OPVs are considered to be one of the promising third generation power sources owing to their great potential as low-cost and scalable devices. The power conversion efficiency has been greatly improved in recent years to exceed 10%. However, OPVs are still facing manufacturing and stability issues. Much more work needs to be done before the OPVs can become practical mobile and building integrated photovoltaic cells. Many researchers are focusing on the development of simple and versatile manufacturing methods to realize low-cost and scalable OPVs. In addition, cost-effective and robust encapsulating technologies are needed to ensure the long-term stability of OPVs. With more efforts devoted to addressing these issues, we hope to see practical OPV power sources in the future.

References

1. F. C Krebs, S. A. Gevorgyan and J. Alstrup, A roll-to-roll process to flexible polymer solar cells: model studies, manufacture and operational stability studies, *J. Mater. Chem.*, 2009, **19**, 5442–5451.
2. H. J. Park, M.-G. Kang, S. H. Ahn and L. J. Guo, A Facile Route to Polymer Solar Cells with Optimum Morphology Readily Applicable to a Roll-to-Roll Process without Sacrificing High Device Performances, *Adv. Mater.*, 2010, **22**, E247–E253.
3. F. C. Krebs, T. Tromholt and M. Jørgensen, Upscaling of polymer solar cell fabrication using full roll-to-roll processing, *Nanoscale*, 2010, **2**, 873–886.
4. Y. Galagan, J.-E. J. M. Rubingh, R. Andriessen, C.-C. Fan, P. W. M. Blom, S. C. Veenstra and J. M. Kroon, ITO-free flexible organic solar cells with printed current collecting grids, *Sol. Energy Mater. Sol. Cells*, 2011, **95**, 1339–1343.
5. S. Beaupré and M. Leclerc, PCDTBT: en route for low cost plastic solar cells, *J. Mater. Chem. A*, 2013, **1**, 11097–11105.
6. C. Zhang, D. Zhao, D. Gu, H. Kim, T. Ling, Y.-K. Wu and L. J. Guo, An Ultrathin, Smooth, and Low-Loss Al-Doped Ag Film and Its Application as a Transparent Electrode in Organic Photovoltaics, *Adv. Mater.*, 2014, **26**, 5696–5701.
7. J. Alstrup, M. Jørgensen, A. J. Medford and F. C. Krebs, Ultra Fast and Parsimonious Materials Screening for Polymer Solar Cells Using Differentially Pumped Slot-Die Coating, *ACS Appl. Mater. Interfaces*, 2010, **2**, 2819–2827.
8. T. T. Larsen-Olsen, T. R. Andersen, B. Andreasen, A. P. L. Böttiger, E. Bundgaard, K. Norrman, J. W. Andreasen, M. Jørgensen and F. C. Krebs, Roll-to-roll processed polymer tandem solar cells partially processed from water, *Sol. Energy Mater. Sol. Cells*, 2012, **97**, 43–49.
9. O. J. Romero, W. J. Suszynski, L. E. Scriven and M. S. Carvalho, Low-flow limit in slot coating of dilute solutions of high molecular weight polymer, *J. Non-Newtonian Fluid Mech.*, 2004, **118**, 137–156.
10. M. Singh, H. M. Haverinen, P. Dhagat and G. E. Jabbour, Inkjet Printing—Process and Its Applications, *Adv. Mater.*, 2010, **22**, 673–685.
11. J. Jung, D. Kim, J. Lim, C. Lee and S. C. Yoon, Highly Efficient Inkjet-Printed Organic Photovoltaic Cells, *Jpn. J. Appl. Phys.*, 2010, **49**, 05EB03.
12. D. J. Coyle, C. W. Macosko and L. E. Scriven, The fluid dynamics of reverse roll coating, *AIChE J.*, 1990, **36**, 161–174.

13. M. M. Voigt, R. C. I. Mackenzie, S. P. King, C. P. Yau, P. Atienzar, J. Dane, P. E. Keivanidis, I. Zadrazil, D. D. C. Bradley and J. Nelson, Gravure printing inverted organic solar cells: The influence of ink properties on film quality and device performance, *Sol. Energy Mater. Sol. Cells*, 2012, **105**, 77–85.

14. D. Angmo, J. Sweelssen, R. Andriessen, Y. Galagan and F. C. Krebs, Inkjet Printing of Back Electrodes for Inverted Polymer Solar Cells, *Adv. Energy Mater.*, 2013, **3**, 1230–1237.

15. H. Youn, T. Lee and L. J. Guo, Multi-film roll transferring (MRT) process using highly conductive and solution-processed silver solution for fully solution-processed polymer solar cells, *Energy Environ. Sci.*, 2014, **7**, 2764–2770.

16. G. Li, V. Shrotriya, J. Huang, Y. Yao, T. Moriarty, K. Emery and Y. Yang, High-efficiency solution processable polymer photovoltaic cells by self-organization of polymer blends, *Nat. Mater.*, 2005, **4**, 864–868.

17. C.-I. Wu, G.-R. Lee and T.-W. Pi, Energy structures and chemical reactions at the Al/LiF/Alq$_3$ interfaces studied by synchrotron-radiation photoemission spectroscopy, *Appl. Phys. Lett.*, 2005, **87**, 212108.

18. G. Li and J. Shinar, Combinatorial fabrication and studies of bright white organic light-emitting devices based on emission from rubrene-doped 4,4′-bis(2,2′-diphenylvinyl)-1,1′-biphenyl, *Appl. Phys. Lett.*, 2003, **83**, 5359.

19. J. Lee, Y. Park, D. Y. Kim, H. Y. Chu, H. Lee and L.-M. Do, High efficiency organic light-emitting devices with Al/NaF cathode, *Appl. Phys. Lett.*, 2003, **82**, 173.

20. J. Huang, Z. Xu and Y. Yang, Low-Work-Function Surface Formed by Solution-Processed and Thermally Deposited Nanoscale Layers of Cesium Carbonate, *Adv. Funct. Mater.*, 2007, **17**, 1966–1973.

21. Y. Zhou, C. F. Hernandez, J. Shim, J. Meyer, A. J. Giordano, H. Li, P. Winget, T. Papadopoulos, H. Cheun, J. Kim, M. Fenoll, A. Dindar, W. Haske, E. Najafabadi, T. M. Khan, H. Sojoudi, S. Barlow, S. Graham, J.-L. Brédas, S. R. Marder, A. Kahn and B. Kippelen, A Universal Method to Produce Low-Work Function Electrodes for Organic Electronics, *Science*, 2012, **336**, 327–332.

22. H. Youn and M. Yang, Solution processed polymer light-emitting diodes utilizing a ZnO/organic ionic interlayer with Al cathode, *Appl. Phys. Lett.*, 2010, **97**, 243302.

23. S. Shin, M. Yang, L. Jay Guo and H. Youn, Roll-to-Roll Cohesive, Coated, Flexible, High-Efficiency Polymer Light-Emitting Diodes Utilizing ITO-Free Polymer Anodes, *Small*, 2013, **9**, 4036–4044.

24. R. Søndergaard, M. Hösel, D. Angmo, T. T. L.-Olsen and F. C. Krebs, Roll-to-roll fabrication of polymer solar cells, *Mater. Today*, 2012, **15**, 36–49.

25. M.-G. Kang, M.-S. Kim, J. Kim and L. J. Guo, Organic solar cells using nanoimprinted transparent metal electrode, *Adv. Mater.*, 2008, **20**, 4408–4413.

26. J. G. Ok, M. K. Kwak, C. M. Huard and L. J. Guo, Photo Roll Lithography for continuous and scalable patterning, with application in transparent conductor fabrication, *Adv. Mater.*, 2013, **25**, 6554–6561.

Subject Index